AF809056

The Pattern of Progress

How Radically
Cheaper, Easier, Faster, Better
Technologies and Ideas
Rapidly Change the World

Bradd Libby, Ph.D.

The Pattern of Progress

Copyright © 2025 Bradd Libby & RethinkX, Inc.

All rights reserved. No part of this book may be reproduced, stored in a retrieval system, or transmitted by any means—electronic, mechanical, photocopying, recording, or otherwise—without written permission from the author or publisher, except for brief quotations in reviews or articles.

No part of this book may be used or reproduced in any manner for the purpose of training artificial intelligence technologies or systems. In accordance with Article 4(3) of the Digital Single Market Directive 2019/790, RethinkX expressly reserve this work from the text and data mining exception.

ISBN: 978-1-0670464-5-3

Published by RethinkX

61 Grosvenor Street

London W1k 3JE

United Kingdom

This is a work of non-fiction. The events, descriptions, and ideas contained herein are presented as accurately as possible, and the views and opinions expressed are those of the author. Any errors are unintentional.

First Edition: 2025

Original cover illustration: www.vecteezy.com/members/sm_wizard

RethinkX, Inc
For information about special discounts for bulk purchases, author appearances, or permission requests, please contact:
www.rethinkx.com/contact • www.rethinkx.com

◆ ◆ ◆

"Til Johan, jeg håper du liker denne boken like mye som jeg likte lignende bøker når jeg var på din alder"

Bradd Libby

◆ ◆ ◆

Foreword by Tony Seba

The Pattern of Progress is an extraordinary journey through the seemingly chaotic history of technology, but with a crucial difference: it doesn't just explore what happened, but how and why. This book is the culmination of years of research.

In our books *Rethinking Humanity: Five Foundational Sector Disruptions, the Lifecycle of Civilizations, and the Coming Age of Freedom (2020)* and *Stellar: A World Beyond Limits, and How to Get There (2025)*, James Arbib and I showed that what really drives progress for humanity is the creation of knowledge. Knowledge that is applied in two fundamental ways - for practical use in the form of technology and for how we organize society.

Throughout history, new technologies have repeatedly changed how we obtain energy, how we move goods and people, how we communicate, how we produce food, medicines, and materials, and more. This pattern holds across centuries, continents, and cultures. This pattern holds at all levels, from molecules to products to sectors to society. And this pattern will continue into the future.

Recognizing this pattern has been my quest for two decades. If you have seen me present before on the subject of technology disruptions, you perhaps have heard me start a presentation by saying: "I want to take you into the future (of transportation, energy, etc) but before I do that I want to take you into the past." There is a simple reason why I start that way - in order to see where the world is going, we need to know where it came from and understand the dynamics of disruption. We only have examples from the past to study, from which we might gain insight into how disruptions happen.

I have also said publicly before that I had looked at every technology I could think of and found that whenever a new technology gets good enough, and cheap enough, and significantly outperforms an incumbent along an important key metric, that technology will be rapidly adopted by the marketplace.

Every technology. Every single time. And Bradd's work has helped me prove it.

My quest led me to co-found RethinkX with James Arbib in 2017 and start what I think of as the science of change, disruption and transformation. Bradd joined us a year later and led a project to formally collect as much historic data about technology disruption. He was uniquely positioned to do this. I've never met anyone who could connect the ripple effect of disruptions on other disruptions in space and time like Bradd could.

Bradd and the RethinkX research team have meticulously compiled, examined and made sense of more than 1500 examples of new technologies and business models. They gathered data about how quickly they got cheaper and better, which innovations and scientific discoveries converged to enable these new technologies, how quickly the new technology was adopted by the market, how quickly the old incumbent technology was displaced, and the social, economic, and environmental implications of those disruptions. These technologies stretch from the invention of cooking to the rise of artificial intelligence.

The Pattern of Progress is Bradd's distillation of that research effort. This book contains about 150 of those examples, starting as far back as humans walking, controlling fire, and developing stone tools. Like a technology mapmaker, he shows how seemingly disparate innovations – the steam engine, the cotton gin, the birth control pill, the shipping container, even the humble ballpoint pen – not only follow a fundamental underlying pattern, but are also woven together to form the very fabric of our society and global economy. The book educates, entertains and enlightens the reader with example after example of technological progress in Energy, Transportation, Food and Agriculture, Information and Communications, Materials, Labor as well as healthcare, entertainment, and sports.

Recognizing this pattern helps you understand that we are on the cusp of transformations so fundamental that they will redefine how we organize our societies, how we run our economies, how we interact with the planet, and even what it means to be human. The coming disruptions in energy, transportation, food, labor, materials, and healthcare aren't just incremental improvements to the system of extraction we have iterated on for thousands of years. They are the keys to unlocking a Stellar World – one built not upon scarcity, but superabundance. But the journey won't be easy. Recognizing the pattern of how technological progress occurs, and why it proceeds swiftly and causes disruptions, will therefore be crucial if we are to successfully navigate the difficult terrain ahead.

The Pattern of Progress is an invitation to understand the past so that we can better imagine – and build – the most amazing future for humanity.

- Tony Seba

Table of Contents

Table of Contents

The Pattern of Progress

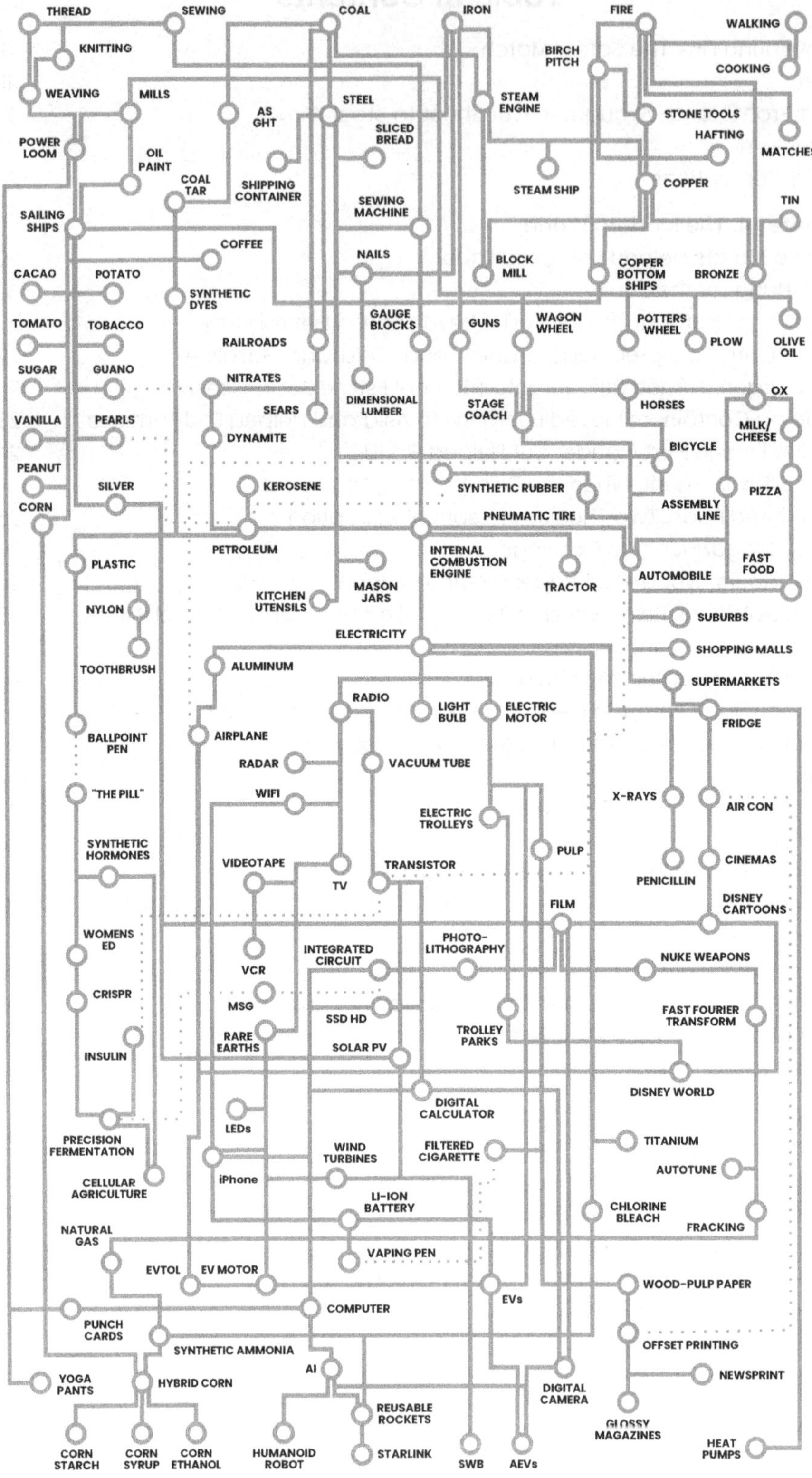

Chapter 1:

The Dynamics of Disruption

◆ ◆ ◆

"The farther back you can look, the farther forward you are likely to see."

Winston Churchill

◆ ◆ ◆

Talking to Aliens about Yogurt: Is Rapid Change Universal?

> *The better we understand the processes that drive technological, economic, and social change, the better prepared we will be to meet the future*

If aliens came down to Earth and we figured out some way to communicate with them, what could we talk to them about? Basic mathematics could be high on the list. If these aliens could build an interstellar spaceship, then they probably had ideas like zero and counting. Which means that maybe they invented multiplication and fractions and perhaps the geometry of circles, squares, and triangles and the relations between them, like the number *pi*, just like we did thousands of years ago.

They would probably understand most of the chemical elements that we know. Hydrogen would be a good place to start that conversation since, as far as we can tell, it is by far the most abundant element in the Universe. If they do not know about hydrogen, it is difficult to see how a conversation about chemistry could continue forward. And if they came from a planet made of solids, liquids, and gases, like our own, then they might know about carbon, nitrogen, oxygen, silicon, and the things made from those elements. Diamonds are pure carbon, so if there is carbon on their world, they might know about diamonds. Carbon and oxygen make carbon monoxide and carbon dioxide. Oxygen and hydrogen make water. Nitrogen and hydrogen make ammonia. Do they know of those things? What else? Gold, maybe. Because it does not interact with oxygen, gold can be found in its pure form in nature. But aluminum readily binds with oxygen, so even we did not figure out how to isolate aluminum in large amounts until about one hundred fifty years ago. I think it would be interesting to know if they have figured out how to get aluminum.

At some point we should talk to them about the elements that we have never found in nature, but which we have made artificially, including ones most people have never heard of – like tennessine and oganesson. Have they made those too? Have they made ones even heavier? What about those chemistry equations you might have learned in school, like the ideal gas law, PV = nRT. Remember that one? Do you know what it means? Do they? We have known about prime numbers for thousands of years, but most of these chemical discoveries are only from the last few hundred.

Once we tested their knowledge of things that we discovered thousands of years ago and the ones that we found in the past couple of centuries, maybe we should ask about those that we have figured out just in the past century or few decades. What's the mass of an electron? Do they have any thoughts about Einstein's theory of relativity? Are photons waves or particles? $E=mc^2$ would be a good equation to run past them. It is both profound and simple, as many profound things are. Do they know that silicon, a semiconductor, can be made to work as an electric switch? If they don't know that one, wow, they are going to thank us for sharing that. That alone would probably make the whole trip worth it.

But what would they already know about our history, our societies, and the Earth's environment? Could they know anything even before they met us? Did organisms wherever they came from develop through evolution by natural selection? Is evolution a mechanism in the Universe as fundamental as the ideal gas law? And if they had thought of ideas like 'mutation and selection' and 'survival of the fittest', then did they also invent ideas like Capitalism and Social Darwinism? Have they ever had slavery? Have they gotten rid of it? (When you think about, have we?) Do they have equal rights? (Again, do we?) Did some of them ever fight in a war to force others of them to abide by ideas they did not share? Have they had social awakenings? Political upheavals? Disruptive innovations? Technology revolutions? Actually, that's probably more interesting than if they have found aluminum.

On our world, new ideas and new technologies have been the motor of progress, yet we understand too little about how new inventions or concepts develop, or how a new idea replaces an old one. Why did the first landing of humans on the Moon happen in 1969 and not 1959 or 1979? I think the fact that we discovered that silicon is a semiconductor played a significant role in that. I think the sudden popularity of surfing after World War II played a small role too. (I'll explain later.) Did the end of the Atlantic trade in enslaved persons happen because some people suddenly had a great spiritual and moral awakening in the 1850s, or mostly because we found a way to power machines by steam so that those machines could do much more work than human muscles? Have famines become rare in the past couple of decades because we suddenly became better stewards of the environment and radically better at agriculture, or simply because of improvements in international shipping? Innovation, Matt Ridley wrote in *How Innovation Works,* "is the reason most people today live lives of prosperity and wisdom compared with their ancestors, the overwhelming cause of the great enrichment of the past few centuries, the simple explanation of why the incidence of extreme poverty is in global freefall for the first time in history: from 50 per cent of the world population to 9 per cent in my lifetime." Do aliens have a word for 'extreme poverty'?

Technology shapes society, but at the same time, society shapes technology. As Peter Norton wrote in his 2008 book *Fighting Traffic,* "Different people see evolving technologies differently. They fight for their vision of the technology's future. And if one vision prevails, those who fought for it portray the technology's history as an evolution by technological selection — a survival of the fittest technology."

The rapid and transformative adoption of new technologies, new ideas, new behaviors, and new business models has followed a repeatable pattern for at least hundreds of years, maybe tens of thousands or millions. It might be a general feature of collective decision-making in many sentient creatures, not just us *Homo sapiens*. And it is likely to be a key dynamic, if not *the* key dynamic, that governs whether in the next decades we will produce unparalleled progress and shared prosperity, solving some of our greatest environmental and social challenges if we harness its power well, or whether we will slide into an extended period of stagnation and decline if we do not.

Disruption is all around you. You might have slept last night on cotton sheets or brushed your teeth this morning with a nylon-bristle toothbrush, used a smartphone or tablet, rode in a car or on an electric bicycle, ate something with sugar in it, and turned on a light. We are seeing now dramatic changes in both the capabilities and the economics of incredibly powerful technologies that have the potential to profoundly transform how we obtain, store, and use energy, how we move ourselves and our goods around, and how we communicate with each other. These massive changes have already begun. In fact, many of the key transformations have been underway for several decades, in areas like computer processing power, data storage, communications, digital photography, lithium-ion batteries, artificial intelligence, robotics, gene sequencing, precision fermentation, wind power, and solar photovoltaics.

Technologies change, but human needs do not. We need oxygen, water, sleep, and food. For at least hundreds of thousands of years we have kept ourselves warm with fire and used it to cook our food. We have sought or built shelters. We have communicated with others, even leaving messages for those who will be born long after we have passed away. We have formed complex social structures that allow us to live close to each other and have been far better off for doing so. We have moved ourselves and our goods. We have made clothing. We have made art. We have worshipped higher powers.

All these needs are fundamental. We always have had, and will always need to have, answers to a few basic questions:

- How do we get **food** to eat?
- What are the **key materials** we prize and know how to use?
- How do we get the **power, energy, and labor** to do work and keep ourselves comfortable?
- How do we **travel** both to transport ourselves and the goods we use?
- How do we **communicate and store information**?

Scientists estimate that our species is at least 200,000 years old. We have engaged in agriculture, in comparison, for only about ten thousand years. Put another way, for at least 95% of our history, we have fed ourselves without intentionally growing plants or raising animals. We now get most of our food calories from just a handful of animal and plant species, only about five of each — cows, pigs, chickens, wheat, corn, rice, the potato, and a few other things — all of which were domesticated thousands of years ago. But we are just now unlocking the potential of using microorganisms to produce virtually any molecule we want, including but not limited to those that nature has discovered, and not only edible molecules but also fibers, dyes, fragrances, and functional paints, in what can only rightly be described as a 'Second Domestication' of plants and animals.

For thousands of years people relied on muscle power, our own muscles and those of animals, to propel ourselves, to move our goods, and to do our work. Most of our energy, in the form of light, heat, and electricity, has come from burning things, like dung, wood, coal, oil, and natural gas. But now the cost of harnessing energy without burning anything is cheaper, for the first time in history, than the cost of burning fuels. For the first time, we are on the edge of a "clean disruption" in energy, to borrow the title of a best-selling book by entrepreneur Tony Seba.

In fact, now we are seeing revolutions in our answers to *all* the fundamental questions listed above. Transportation, energy, food, materials, information, communications, healthcare, and repetitive (or otherwise difficult, unpleasant, or dangerous) labor are undergoing revolutions based on the convergence of a few key technologies. And all these revolutions are happening simultaneously and interacting with each other.

Solar photovoltaic (PV) panels, which were once so expensive that it only made sense to use them on space satellites, fell in price more than 80% from 2010 to 2020. They are now the *cheapest* source of new electric power in most places in the world. Likewise, the price of wind power fell 46% in that decade and the price of lithium-ion batteries a staggering 87% since 2010. A 2020 report from Tony Seba and Dr. Adam Dorr suggests that these technologies might fall by similar percentages again over the course of the 2020s.

It is possible that we will soon see the cost of producing electric power from rooftop panels and storing that power in a household battery system become cheaper than even the *transmission cost* of utility-generated electric power. When this happens, it will not really matter which source of power a utility company chooses to generate its electricity, since it will not be able to compete with a rooftop solar system even if the generation cost was *free*. Seba calls this price, the price at which generating electricity on-site on demand is cheaper than the transport cost of centrally generated power, 'Generation On Demand' parity, or GOD parity. And if locally generated power is cheaper than utility-transmitted power, then even God will not be able to stop consumers from installing as many solar panels and lithium-ion battery backup systems as possible, as quickly as possible. (OK, maybe a homeowner's association could stop them, but God Himself could not.) Unencumbered, we will switch from one technology to the other just as quickly as we cast aside film cameras for digital cameras, or flip phones for smartphones.

Cars and trucks powered by electricity are rapidly growing in popularity. In 2024, more than one in five passenger car sales worldwide was electric. Tesla, which few people had even heard of ten years prior, was the world's largest car company by stock market capitalization. Vehicles that drive themselves as a taxi service are already on the road. If the power for them comes from clean, cheap sources, like solar PV panels and wind turbines, the effects on geopolitics and the environment would be staggering.

Batteries and electric motors are enabling a revolution in smaller vehicles too, for example, electric bicycles, electric scooters, and China's best-selling Hongguang MINI EV car, which costs about $5000, making vehicle ownership available to more people than ever before. Making most cars drive themselves could eliminate the need for vehicle ownership just as the mobile phone has made ownership of landline phones, desktop calculators, and stand-alone cameras, obsolete. Why own a (depreciating) vehicle if you could rent one by the ride as needed, through your phone, at a lower cost? And if a vehicle can come to you and take you to a supermarket without any intervention on your part, then why have supermarkets at all? Why not order groceries and have a self-driving delivery wagon simply roll them to you? Every retail space would suddenly be useful only as a warehouse.

Autonomous driving unlocks dramatically more efficient use of vehicles as certainly as unlocking aluminum enables soft drink cans and airplanes. Instead of being parked about 95% of the time, a much-smaller number of vehicles (and a number of much-smaller vehicles) could be in use most of the time and rarely parked. We would get a lot more 'bang for our buck' for steel, glass, and rubber.

This would reshape how we use land in cities by freeing up portions of urban areas that are devoted to roads and parking and allowing people to travel more easily, including those who cannot drive now due to age or infirmity, with much less impact on the environment and peoples' wallets. When I talk about dramatically more efficient use of resources, I'm not talking about small percentages, but large multiples. (Hopefully aliens understand these concepts, too.)

At the same time that we are seeing a revolution in road transportation, we are also beginning to see a disruption of the production of food and materials, like leather, meat, and milk, that amounts to far more than just another 'agricultural revolution'. For example, when it comes to obtaining milk, we currently rely on a couple of species first domesticated thousands of years ago, mostly cattle, and also a few goats and camels. But there is ice cream and yogurt on the shelves right now that is made with cow's milk proteins that were not taken from a cow. The genetic material responsible for making the milk protein was put into a microorganism so that we can 'brew' milk proteins by precision fermentation with basically the same technique that we have been fermenting alcohol for millennia. Chicken meat has already been sold in restaurants and in dog snacks that was grown by 'cellular agriculture' in a bioreactor, made from cells from the end of a feather that fell off a living bird, not taken from a dead one.

Just as 'clean transportation' by a smaller number of electrically powered vehicles might reduce the need for urban parking, refined oil products, steel, human labor, and other inputs, 'modern food' made from precision fermentation and cellular agriculture could free up rural land for other uses, all while producing the same, and even better, food products, medicines, and materials, with a fraction of the amount of energy, water, fertilizer, repetitive human labor, and other inputs. With the end of industrial agriculture, we might finally see the end of human slavery, too.

Companies like SpaceX and Rocket Labs are making reusable rockets that are dramatically lowering the price of lifting material into orbit around the Earth, or deep into the interplanetary space beyond. SpaceX was founded in 2002 and according to Sam Korus, a technology analyst at investment firm ARK Invest, in less than twenty years the company lifted into orbit as many satellites as all other organizations on Earth combined. A centerpiece of SpaceX's efforts is Starlink, a constellation of thousands of small communications satellites that will ensure every place on Earth always has a high-speed data connection, even when flying in an airplane or out on the open ocean. High-quality digital conferencing between any two points on Earth, at any time, could cut the need for travel as much as autonomous taxis and reusable rockets are helping to enable cheap, clean transportation of people and goods.

To see where these 'clean disruptions' in energy, transportation, food, materials, and information might take us in a decade or two, it is a good idea for us to understand how profound technological, economic, political, and social changes have happened in the past. These are powerful technologies I am talking about, and the implications they could have on human well-being, health, social structure, and the environment are almost too difficult to comprehend.

A container of milk inoculated with the right microorganisms will quickly turn into yogurt. The mechanism that governs this is perhaps the most-powerful technology ever known: the self-reinforcing feedback loop. One cell turns into two and two into four. It does not take too many doublings for the system to fill the container, or to cause it to overflow.

The self-reinforcing feedback loop is the mechanism behind the rise of the radically cheaper, easier, faster, better technologies and ideas that have rapidly changed the world over and over again. It is the driving force behind human population growth and our economic growth, behind radically more powerful computer chips and cheaper lithium-ion batteries, the course of pandemics, nuclear reactions, behind growing chicken meat in steel vats, and much else, including, as I just said, the production of yogurt.

So if aliens chose today to come down to Earth, it might be wise to skip right over the chit-chat about the ideal gas law and get straight to the subject that is most likely to guide our future, and the one we should start with in this book: Tell us everything you know about self-reinforcing feedback loops, and don't hold back, because the more we learn about them, the more we want to learn about them.

◆ ◆ ◆

"History doesn't crawl; it leaps."
Nassim Nicholas Taleb

◆ ◆ ◆

Virtuous Cycles: Drivers of Growth

Changes over the coming decades will be driven, as fast and dramatic changes are, by very powerful 'self-reinforcing' feedback loops. For example, for decades computers have gotten more powerful with each successive generation. Now, they can hold conversations and make images from just a few words of prompting. But computers themselves are used to design the circuits and write the software for the next generation of computing technology. So, as computers get more powerful it becomes easier to make computers that are even more powerful. This is a self-reinforcing process that has helped the computer revolution play out faster and faster over time. In just the past couple of years, companies like OpenAI have made computer programs that are able to write computer programs, dramatically cutting the time, effort, and cost of developing software. The feedback loop keeps spinning faster.

The observation, first made in 1965 by late Intel co-founder Gordon Moore, that computer circuits hold twice as many transistors every eighteen months has remained true for decades. Processors are dramatically more powerful today than they were not long ago. Presumably in the future they will be even more powerful still. And it is not just the raw computing power per chip that has increased over time, but as Moore also predicted the cost has declined precipitously as well, astronomically improving our data processing ability.

A wide variety of other technologies have seen similar improvements in performance (or again, performance relative to cost), which have held in a pattern so predictable that academics have dubbed the relationship the 'experience curve', or 'Wright's Law', after Theodore Paul Wright, an aeronautical engineer who observed in the 1930s "that every doubling of production of US aircraft brought down prices by 13%. Since then, a similar effect has been found in nearly every technology area, going back to the Ford Model T." (Naam, 2019)

For technologies that conform to Wright's Law, each successive doubling in the cumulative number of units that have ever been produced causes the marginal cost of production to decrease by a certain percentage. Put more simply, the more you make of something, the cheaper it gets to make even more of it. A self-reinforcing feedback loop. And the cost comes down by a certain percentage (usually about 15-20%) every time the total number of units you have ever made doubles.

The chart overleaf shows the cost of a Ford Model T compared to the cumulative number produced up to that time for the years 1909 (the first full year of production) through the last year of production, 1927, when the fifteen millionth unit rolled off the assembly line. While the earliest Model T's cost nearly $25,000 (in constant 2020 dollars), within less than fifteen years, the price had come down by a factor of about ten. It is no wonder that the Model T got more popular as more and more of them were produced.

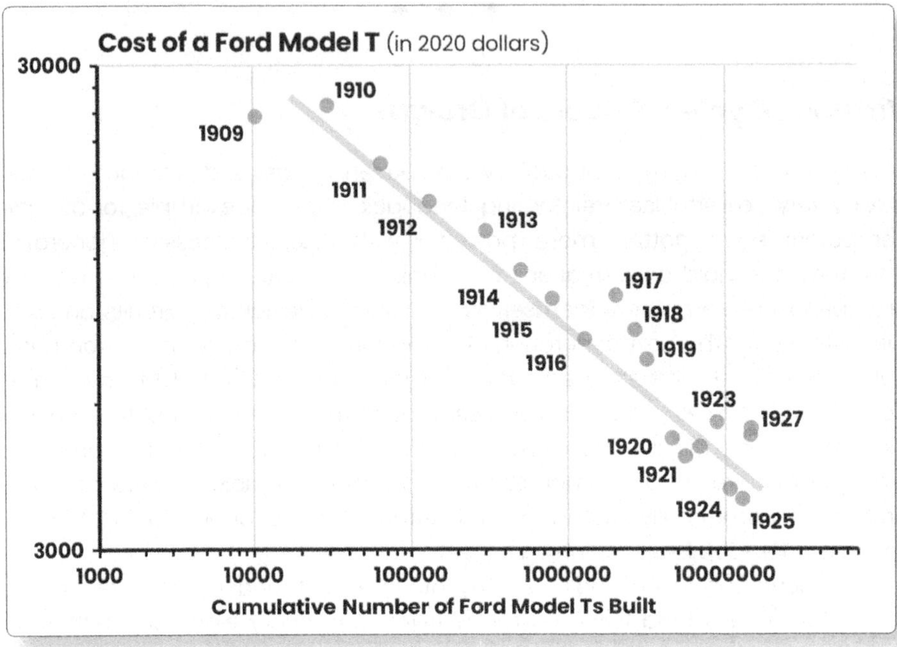

Figure 1 – Cost of a Ford Model T vs. the Cumulative Number of Ford Model Ts produced from 1909 to 1927

If you are not a 'numbers person', this chart can be a little strange to read. First, the horizontal axis is not 'time', it is the 'cumulative number of units' ever produced. The vertical axis is 'cost of producing one more unit' (that is, the 'marginal cost of production'). Each point is placed from finding, for each year, the cumulative number of units ever produced up to that year and the cost of producing one more unit in that year. This might seem like some sort of trick, but it's not. It's simply an answer to the question, "In a given year, knowing how much cumulative experience we have at making these things, how much will it cost to make one more?"

A second thing that is a little strange about this chart is that the axes are on "logarithmic" scales, which means that every 10X increase is equally spaced. So, the tick marks are '1000', '10000', '100000', etc., rather than '1', '2', '3' and so on.

On this scale, the general drop in price of the Model T is apparent over its entire lifetime, with some exceptional years when the US was involved in World War I and toward the end of the life of the model, as Ford was reconfiguring his factories to begin making his new vehicle, the Model A. Henry Ford's focus on bringing down the cost of production paid off. By 1914, just five years after his moving assembly line started, his Model T was selling more than all other automakers combined.

This simple, almost-straight-line relationship between cumulative production and marginal cost for most of the years is even more remarkable given that many dramatic events affected the US during this time. As mentioned, there was a world war, but also the Spanish Flu influenza pandemic, the Panic of 1910-1911 (an economic depression), presidents from the both the Republican (Taft and Harding) and Democratic (Wilson) parties, the women's suffrage movement, and the start of Prohibition, as well as numerous other radical technological changes, like in electricity, radio, and airplanes, that placed increasing demands on some of the same materials used to make cars. And yet the price of Ford's cars marched downward almost every year.

Even more remarkable is that this almost-straight-line relationship holds true for many technical products today. Wright's Law was developed using data from the early airplane industry, but it still applies to the latest fighter jets nearly a century later. The graph below shows the amount of labor required (in thousands of hours) to produce an F-35 fighter jet compared to the cumulative number of F-35s ever produced, according to the US General Accounting Office. As with the Model T, which had several variants including a touring car, a roadster, and a runabout, the F-35 has several major variants, built for the US military and for various NATO members, some of whom have demanded special configurations.

A wide variety of other products have seen the same dynamics. Aluminum, which used to be more expensive than gold, is now essentially disposable. Both computer transistors and gigabytes of data storage are now a tiny fraction of their costs from a couple of decades ago. The same is true of solar panels and lithium-ion batteries. And in 2012, Dutch doctor Mark Post presented a $325,000 hamburger patty that he grew from beef cells without killing the animal. Today, similar products are much less expensive than they were ten years ago. In early 2025, the London-based company Meatly began selling "chick bites" dog treats made, in part, from cultivated chicken cells. With just a few more doublings in cumulative production, these new technologies I have been talking about could be substantially cheaper than they are now, enabling and requiring us to rethink large parts of how we get energy, materials and food, move around, and communicate.

The mechanics of why this relationship hold for many different technologies as different as a Model T is from an F-35 fighter jet are not completely understood. Buying raw materials in bulk means getting them cheaper per unit than buying smaller quantities. Making a larger number of parts means investing in more-sophisticated tools or automated processes, or even that you might design and build whole new fabrication lines. And as you learn how to build these things you just get faster at building them.

As well, one of the best-known self-reinforcing feedback loops is the 'word of mouth' process. Customers who are happy with a purchase might recommend the product or service to their friends, who might try the product for themselves and, in turn, recommend it to their friends as well. As long as each purchaser 'recruits' at least one other purchaser, the number of purchasers grows and grows. A similar process happens, for example, when neighbors have solar photovoltaic panels on

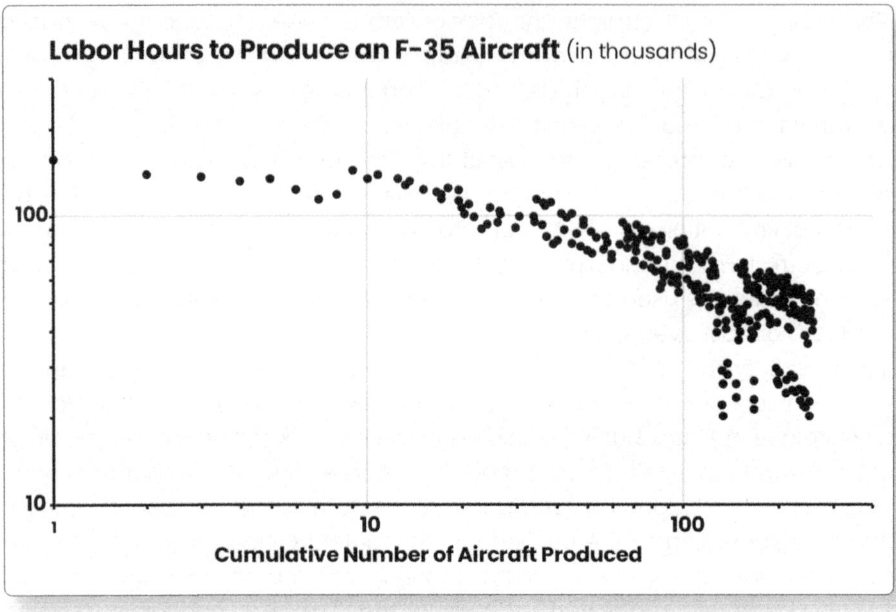

Figure 2 – Labor Hours Required to Produce an F-35 Aircraft compared to the Number of F-35 Aircraft Ever Produced. *(USGAO, June 2018)*

their roof or an electric car in their driveway. Observing a neighbor who is happy with their purchase might make us more likely to make a similar purchase. In a 2016 interview with the French company now called TotalEnergies, futurist Ramez Naam said: "Every year that solar and wind and batteries and electric vehicles get cheaper, it makes political action easier too. It's a positive feedback loop".

Exponentially increasing performance (as with computers), improving economics (due to Wright's Law), and self-reinforcing social processes (as with the observability of neighbors' solar panels or cars) work together to make technological disruptions play out quickly.

Unfortunately for incumbent producers, these powerful feedback loops work the other way too. The loss of market share, the reduction in economies of scale, and the unravelling of network effects can make incumbent technologies radically less appealing over a short period of time in a self-reinforcing vicious cycle.

◆ ◆ ◆

"Look back over the past, with its changing empires that rose and fell, and you can foresee the future too."

Marcus Aurelius

◆ ◆ ◆

Vicious Cycles: The 'Death Spiral'

Self-reinforcing loops are not always virtuous or beneficial. Vicious cycles and so-called 'death spirals' are also demonstrations of self-reinforcing dynamics. The same network effects and economies of scale that make a new technology more appealing and lower in cost as it gets more popular unravel to make it less appealing and higher in cost as the technology later gets less popular.

Figure 3 – Self-reinforcing feedback loops in the switch from one technology to another.
Source: RethinkX

The virtuous feedback loops that drive the adoption of the new technology, and the vicious feedback loops that bring the old technology down, are numerous. As a new industry rises, more demand for the new product drives greater economies of scale, lower production costs, greater public acceptance, higher revenues, higher margins, more investment, and greater government support, among other things.

Likewise, as an old industry declines, financial capital gets harder to access, making it more expensive to borrow money. Higher finance costs hurt the profitability of the industry and of every individual factory, making it more likely that factories shut down, further hurting the investment prospects of the industry.

These self-reinforcing dynamics conspire to bring change about "faster than expected". Global technology think tank RethinkX (of which I am a member) wrote that "feedback loops amplify one another, accelerating the adoption of new technology in a virtuous cycle while accelerating the abandonment of old technology in a vicious cycle. The result of these systems dynamics is that disruption tends to unfold with surprising speed".

As film-based photography became less popular, being a film-based photographer became more difficult. Electrical grids face a similar dynamic – they suffer when customers simply use less electricity. The principle is very simple: the cost of operating the entire grid in a given month, divided by the amount of electricity that passes through that grid in the month, determines the cost of the grid per unit of electricity delivered. As home solar panels and battery systems become more popular, less electricity is delivered over the regional grid, and therefore the cost per-unit of delivering electric power goes up, making it even more tempting for consumers to install additional panels and storage.

Not everyone has the same requirements for an electrical system, or for photographic equipment for that matter. Professional photographers usually have the most extreme needs – the best-quality photos, the physically smallest or lightest camera, the largest number of photos that can be taken – and are willing to pay to get the equipment that meets their needs. Other people take photos rarely or never. But most people are amateur photographers and want equipment that meets their basic needs at a price they are willing to pay.

Likewise, with electricity consumption, lighting, car travel, telephone usage, Wi-Fi data usage, steel, pizza, meat consumption, computer processing, or virtually any other aspect of life, there is a 'bell curve' of consumption, with some people or organizations demanding large amounts of product (or the highest-quality products) and others demanding far less.

In his highly influential 1962 book *Diffusion of Innovations*, Everett M. Rogers presented his theory about how new innovations are adopted in the marketplace. He suggested that a small percent of a population are 'Innovators', the people

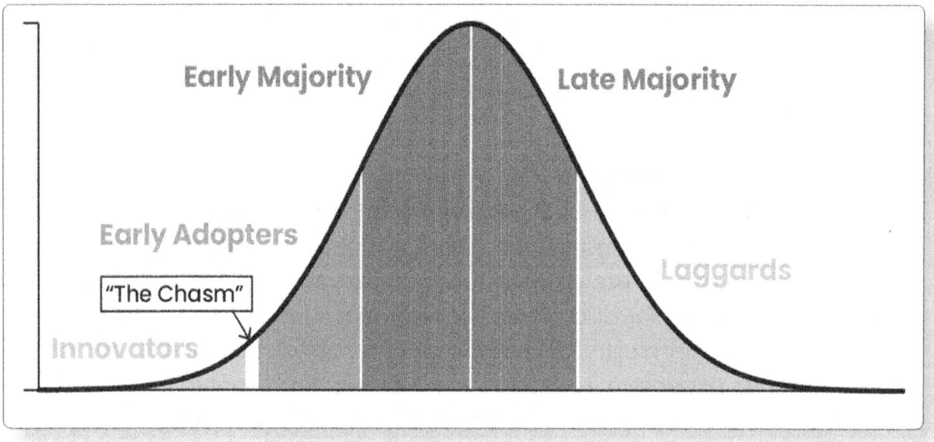

Figure 4 – The Bell Curve of Adopters

with the highest propensity to adopt the product. About 15% fall into the tranche he calls 'Early Adopters' and about one-third fall into each of the 'Early Majority' and 'Late Majority' categories. The remainder of people (or organizations) are 'Laggards', those least inclined to adopt.

Fundamentally, Rogers argues, this bell curve is the reason why we see an S-shaped curve of market adoption for most new technologies that become widespread. A small portion of the population adopts the innovation. With the easiest customers acquired, the innovation reaches a 'chasm', the point at which it will either appeal to a much wider customer base or fade into obscurity. (Geoffrey Moore's best-selling 1991 book *Crossing the Chasm* deals with this specific issue.) If successful, the innovation then spreads through the bulk of the population, slowing as the pool of potential customers becomes saturated.

We have seen this 'pattern of progress' repeatedly, in many different fields, stretching over millennia. A new technology takes over a market over a short time, in an S-shaped profile, often expanding the market greatly in the process. And the

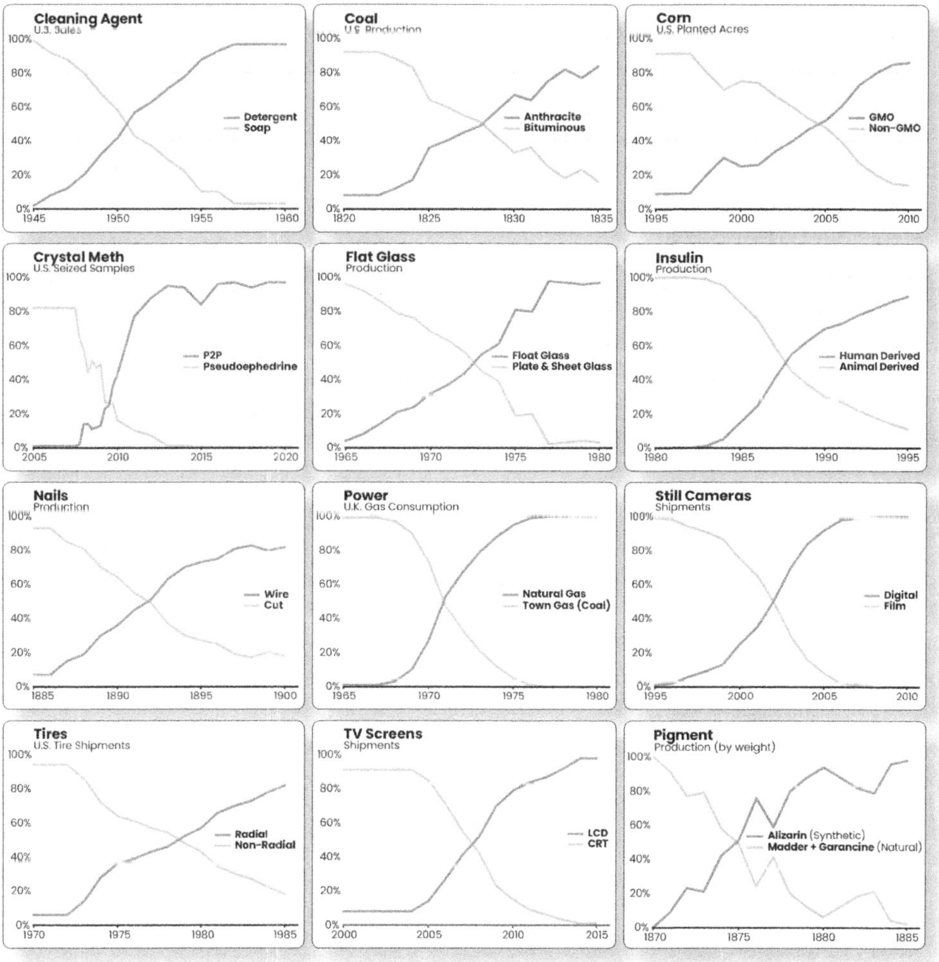

Figure 5 – Some Historical Examples of Disruption. *Source: RethinkX*

old technology declines, also in an S-shaped profile. The two curves of the new innovation's adoption and the old innovation's decline form an X-shaped curve.

We will revisit all these examples in this book – fabric dyes, insulin, video cameras, corn, nails, coal, car tires – and many others. The better we understand the pattern of progress for new technologies and ideas, the better we might be at shaping their evolution and development.

Since 2018, researchers (including myself) at the think tank RethinkX have been constructing a dataset representing hundreds of examples of technology-driven disruptions: collecting and compiling basic information for each about when and where the underlying scientific discoveries (if any) were made, which other technologies or scientific discoveries converged to enable the new technology, who invented the technology, when the first commercial product was introduced to the market, when the first successful product began to get a foothold in the market, when it saturated the market, which scientific discoveries and other technologies were enabled by it, what key social, economic or cultural consequences or outcomes resulted from it, and the business model(s) that each innovation employed or enabled.

The dataset is not restricted to purely historical examples, but also includes technologies that are currently undergoing rapid changes in performance and cost and that are being rapidly adopted right now: solar photovoltaic (PV) panels, wind turbines, electric vehicles of various kinds, graphics processing units (GPUs), lithium-ion batteries, genetic technologies, and many others, as well as their incumbent alternatives.

Sources of these data include published academic books, popular press books, Ph.D. dissertations, patent filings, peer-reviewed journal articles, national statistical abstracts, and publicly available sources.

This dataset forms the basis of the content presented in this book.

◆ ◆ ◆

"Here's the question that I've asked myself for fifteen years: Why do smart people in smart organizations consistently fail to lead or anticipate technology disruptions? ... Can we anticipate, can we forecast, more or less, disruptions to come?"

— Tony Seba

◆ ◆ ◆

The Dynamics of Disruption

Disruption has been a hot business buzzword since the publication of late Harvard Business School professor Clayton Christensen's best-selling book *The Innovator's Dilemma: When New Technologies Cause Great Firms to Fail* in 1997. Christensen expanded his body of work over the next decades into a series of books and a research institute that bears his name, the Christensen Institute. In his seminal work, he looked at a variety of fast-changing industries, from computer hard drives to hydraulic construction machinery and steel 'mini-mills'.

And he noticed a pattern: new technologies often perform worse than the existing market-dominating product, but they serve at least one niche customer segment better. Because the new product does not serve the bulk of the market well, it is initially mocked by incumbents as being a 'toy'. However, as the new product is improved over time, it moves upmarket in capabilities, eventually appealing to the bulk of the population and taking over the entire industry. As a result, the market-leading company loses its dominant position to the new upstart run by outsiders.

Silicon Valley entrepreneur and author Tony Seba has shown that Christensen's explanation is only one of a variety of possible ways a market can get overturned by a disruption. Specifically, Christensen's observation that a technology that initially underperforms the market-leading solution and only appeals to customers with a niche need, but then move upmarket, is what Seba calls a 'Disruption from Below'.

In studying disruptions we have seen a similar pattern again and again that we show in stylized form in Figure 6 below. A new innovation is introduced, but dismissed by incumbents as their market is still growing. Often by only taking a few percent market share, the disruptor causes sales of the incumbent to peak. And then, rapidly, often over just the course of a decade or so, the incumbent's sales collapse as the disruptor takes off. This is not a one-for-one replacement. Ultimately, the size of the new market can be many times the size of the old one.

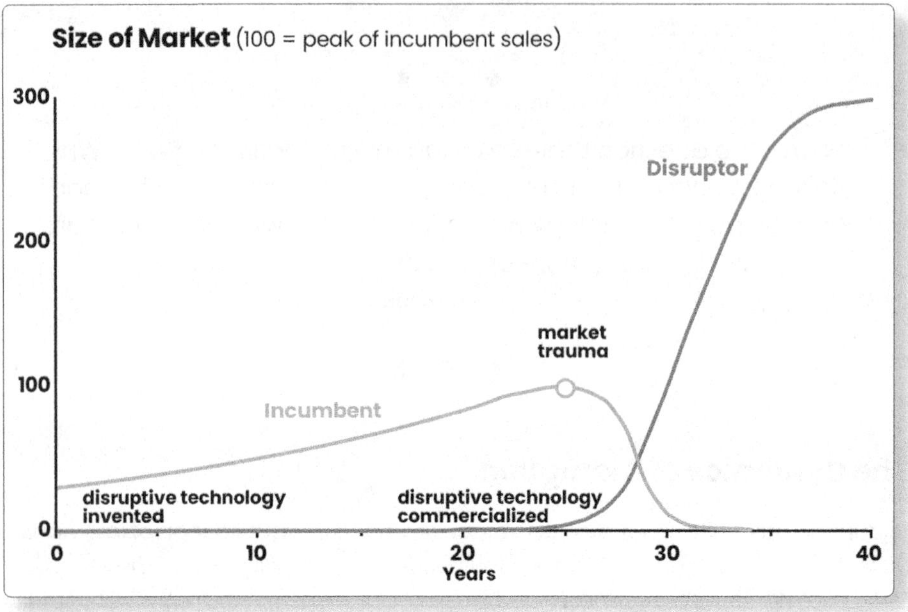

Figure 6 – The Pattern of Progress. Typical adoption dynamics of an incumbent and a disruptor.

There are different mechanisms by which these disruptions occur:

Disruption from Below

As Seba describes it, a 'Disruption from Below' is when "a new product is initially inferior to mainstream products, but improves its performance while decreasing costs at a faster rate than incumbent products. This product may initially serve the needs of niche markets, and as it improves its utility, it expands into new markets. Eventually it overtakes and disrupts incumbent products and markets."

All of this happens not because incumbents failed to pay attention to their customers nor because of a lack of knowledge by incumbents. In fact, in many cases the incumbent that got disrupted created the technology that disrupted them. Kodak was not just aware of digital cameras, they *invented* them. But Christensen claims that Kodak ignored the potential of digital photography because they, as a company, were too focused on serving existing film customers and were not paying enough attention to the early adopters of what would become the new normal.

Often, when incumbents realize that the new upstart is stealing their market share, they respond by retreating upmarket to the higher-margin products. As 'mini-mills' disrupted steel making in the 1970s, large integrated mills saw their lower-end, less-profitable market segments like rebar ('reinforcement bar', the rods that are put into concrete to strengthen it) taken away. To maintain their business, they simply focused on the higher-quality, higher-cost, and higher-profit products, like sheet metal for automobile bodies until, over time, mini-mills began producing those products too.

Disruption from Above

But not all new technologies are like this. Some products, like the smartphone, are initially *better performing* and *more expensive* than the incumbent offerings. Seba calls this situation a 'Disruption from Above'.

"A new product is superior but more expensive than competing products in the mainstream market. In time, however, the cost of the product is lowered until it becomes less expensive than incumbent products", Seba writes. "The smartphone is a recent example. When the Apple iPhone came out at about $600 in 2007, experts said that it was not disruptive. Who would want to buy a $600 phone when they could buy a $100 Nokia cell phone? What they did not understand is that a smartphone is not just a phone. An iPhone is a platform that allows us to do hundreds of things, including finding a date, getting driving directions, doing online banking, and, yes, making phone calls. The smartphone is not and never was a one-to-one substitute for the conventional cell phone. The electric vehicle (EV) is another example of a disruption from above. The EV is a superior product in a number of ways, not just an electric version of an ICE [internal combustion engine] car."

The iPhone too was mocked by incumbents. In an interview in April 2007, shortly after Apple announced the impending release of their phone, Microsoft CEO Steve Ballmer remarked, "there's no chance that the iPhone is going to get any significant market share. No chance." Ballmer cited not only the high price, but also the lack of a physical keyboard, which would make Apple's phone less appealing to business customers. Yet another case of a CEO paying too-close attention to his existing customer base.

Unlike a 'Disruption from Below', when a better performing but more expensive disruptive product enters the scene, incumbents do not have the option to retreat upmarket – they simply watch their customer base quickly evaporate.

Note that Ballmer's criticism of the iPhone focused on its high price. When a 'Disruption from Below' is mocked as a "toy", it is usually because of the low quality or low capabilities of the product. But when a 'Disruption from Above' is mocked, it is often for being "a toy for the wealthy", or for people 'with more money than brains', or similar.

Big Bang Disruption

A third kind of disruption, Seba notes, is when "a new product is *better, faster,* and *cheaper* than mainstream products on the day it is launched." (italics added) "Examples include Google Maps with driving directions API, which disrupted the then growing GPS market served by companies like Tom Tom and Garmin."

Seba says that 'Big Bang' disruptions are "a model often seen in the healthcare industry where new drugs or treatments are approved and outperform existing products. It is a model we expect to see for Transport as a Service – robotaxis once they are approved for widespread use on the roads." When autonomous electric cars can pick up passengers without a human driver, as they are already starting to do, the cost per-mile of moving people or goods would be significantly less than owning a car oneself. Essentially overnight, it will be in peoples' financial best interest to simply stop driving and to use 'Transport as a Service' (TaaS) instead.

Architectural Disruption

In the final case Seba identifies, "a new product radically changes the way products and services are produced, managed, delivered, and sold." For example, compared to centralized electricity production by utility companies, "solar energy and batteries flip the architecture of electricity: they enable millions of customers to generate, store, manage, and trade electricity. When the cost curves of solar and batteries (plus sensors, power electronics, software, and new business models) converge, the central generation model is disrupted. At that point, the architecture of energy flips from central generation to distributed generation."

In these four forms of disruption, historians identify four key signposts new technologies pass in the process of being adopted: ideation, invention, innovation, and diffusion. 'Ideation' is when a new product or technology is first described, often in science-fiction literature or other artwork. The first functional device is the step of 'invention'. Commercial 'innovation' happens when the first product is launched. The 'diffusion' step is when the product is widely adopted in the marketplace.

From the perspective of the incumbent industry, the dynamics of the collapsing business also form an S-shaped decline that unfolds rapidly. As demand for the old product falls in an S-shaped profile, the existing customer base dries up, leaving only the customers who derive the greatest benefit from the incumbent product and the laggards who are unwilling or unable to switch to the new product. Declining economies of scale in a diminished market mean that the unit price the incumbent can (and must) charge tends to go up, somewhat counteracting the declining unit sales. This tends to cause industry-wide revenues to decline less dramatically, in percentage terms, than the unit sales. Since the incumbent needs to invest little in research & development of upgraded offerings, or even in the maintenance of existing production facilities, the industry often experiences a period of heightened profitability, a sort of 'last hurrah' before the incumbent industry's ultimate end.

A New Defense Landscape Every Decade

In April 2001, late US Secretary of Defense Donald Rumsfeld wrote a memo called 'Thoughts for the 2001 Quadrennial Defense Review'. Declassified in 2009, it shows how quickly global geopolitics change by comparing the start of each decade of the 1900s. The memo reads, in part:

"If you had been a security policy-maker in the world's greatest power in 1900, you would have been a Brit, looking warily at your age-old enemy, France."

> *"By 1910, you would be allied with France and your enemy would be Germany…"*
>
> *"By 1930… the Great Depression was underway, and the defense planning standard said 'no war for ten years.' Nine years later World War II had begun."*
>
> *"By 1950… the Atomic Age had dawned… By 1970… we were anointing the Shah as our protégé in the Gulf region."*
>
> *"By 1980, the Soviets were in Afghanistan, Iran was in the throes of revolution" while "by 1990, the Soviet Union was within a year of dissolution… and almost no one had heard of the internet. Ten years later, Warsaw was the capital of a NATO nation".*
>
> The point was, at any point for the last century or longer, technology, economics and geopolitics could be completely rewritten in only a decade or so.

Principles of Disruption: Signs of Impending Radical Change

> *When you learn to recognize the characteristics that many technology-driven disruptions share, you will be better at spotting disruptions before and as they happen*

In our efforts to catalog and compile examples of technologies and ideas that rapidly changed the world, we have identified traits that many of them share. Together, they form what we call the 'pattern of progress'. Here are some of these key characteristics:

A new idea is conceived and described, sometimes decades or even centuries before a single functional version is ever produced.

The aerospace company SpaceX is revolutionizing space travel by making rockets re-usable, which brings down the cost of each flight dramatically. In a 2015 interview on *The Late Show with Stephen Colbert,* the CEO of SpaceX said: "If you throw away the rockets every time, it's crazy expensive to go to space. But if you can re-fly the rockets, it could be comparable to air flight in its costs."

The sight of the lower stages of their rockets falling back down to Earth, orienting themselves upright, and then landing safely on the ground or on a barge in the open ocean with their engines roaring to life in the last few seconds before touchdown is unforgettable, even to people who have only seen them through a computer screen.

But the concept was not invented by SpaceX – it came from science fiction. Landing a rocket on a barge looks remarkably like a scene from a 1959 Russian science fiction movie, "The Sky Calls". SpaceX's barges have been given whimsical names like 'Of Course I Still Love You' and 'Just Read the Instructions', names which themselves are taken from the science fiction novels of Iain M. Banks.

Figure 7 – Still frame from *The Sky Calls*, a 1959 Soviet science fiction film, (left) and a landing of a SpaceX rocket stage on a barge in 2016 (right)

Science fiction has long been the means to float new technological ideas and their possible social implications.

In January 1946, cartoon detective Dick Tracy showed off his two-way wrist radio for the first time. By 1964, he had upgraded to a two-way wrist TV. Both devices, though just drawings, made many young people realize that communications tools that were then large and expensive might one day be small, mobile, personal, affordable, and ubiquitous. But maybe the idea was not that original: in January 1926, electrical engineer Nikola Tesla said in an interview with *Collier's* magazine, "When wireless [radio] is perfectly applied the whole earth will be converted into a huge brain, which in fact it is, all things being particles of a real and rhythmic whole. We shall be able to communicate with one another instantly, irrespective of distance. Not only this, but through television and telephony we shall see and hear one another as perfectly as though we were face to face, despite intervening distances of thousands of miles; and the instruments through which we shall be able to do this will be amazingly simple compared with our present telephone. A man will be able to carry one in his vest pocket."

The biggest thing Nikola Tesla got wrong was that, in general, men no longer wear vests.

◆ ◆ ◆

"A lot of the economy is inefficiency in the processes we use."
— **Tony Seba**

◆ ◆ ◆

Figure 8 – Postcard issued in 1930 by Echte Wagner, a German margarine company, showing women of the future talking on video-screen phones.

The notebooks of Leonardo da Vinci, made in the late 1400s, include drawings of a helicopter, an airplane, a parachute, and a military tank, some of them more practical in design than the others. It is with the word or picture that we first bring new ideas into the world. Even if an idea is only described in a rudimentary way, if the idea is compelling enough, it gets people's attention for what it is and why it might be useful. It might then be decades, or even longer, for the idea to ever become a functional first prototype, but the initial concept starts with a spoken word, in writing, or in a sketch.

A technological invention, a new social movement, or even a new business model provides an improvement in performance or a reduction in cost (or both) over an existing, incumbent solution

Overcoming inefficiency and inability, ultimately, are the sources of disruption. An incandescent light bulb only emits a few percent of the energy it consumes in the form of light. The rest becomes heat. Compact fluorescent bulbs are more efficient than incandescents, and LEDs (light-emitting diodes) are more efficient still. Each of these technologies have displaced the ones that came before them because they are more efficient.

Radical innovations enable us to do something better, cheaper, easier, or faster. They enable us to do something we were not able to do previously at all (like fly through the air or walk on the Moon). Or, in some cases, they remove steps, parts, or other products entirely.

Cattle or chickens only convert a few percent of their feed into meat, milk, or eggs. Growing meat in a bioreactor, as has already been done and sold to the public, is much more efficient in energy and resources. Genetically engineered micro-organisms can produce the same proteins found in cow's milk without needing to spend the money or do the work to care for the cow. Ice cream made from cow-free cow's milk is already on the shelves. Just as the digital camera took away the need to buy film and to wait and pay to develop it in order to see an image, the micro-organism will take away the cow to get milk protein.

The technology gets good enough and cheap enough to yield an offering that is much better, in some key way, than the old one

Tony Seba says "10X has always caused a disruption. I'll repeat: a 10X difference in cost for the same product or service has always caused a disruption - every single time". Another way to think of a '10X' disruption is a new technology or way of doing things that uses only a fraction of a key input as the best existing method. And when a technology can deliver better performance for a given price, the desire to adopt it and abandon the old technology can be overwhelming.

The disruption is unlocked by a convergence of technologies or ideas

To invent the automobile, you need at least a motor, a steel chassis, and durable tires. If you want to move these heavy pieces around an enclosed factory, electric motors would help, since they do not emit any toxic gases indoors. What is the earliest time in human history when a person could have assembled these things?

By 1870 less than 5% of all iron products were steel (as opposed to cast iron or wrought iron), but by 1880 it was over 20% and by 1890 over 50%. (Ayers, 1989) Gottlieb Daimler made his first four-stroke gasoline engine small enough to fit on a chassis in 1883, about the same time that Karl Benz made the engine for his first car. John D. Rockefeller formed Standard Oil Company in 1865 and by 1879, just fourteen years later, controlled 90% of America's refining capacity. If you wanted to make an electric car, the first commercially successful direct current (DC) motors were not available until the early 1870s, and alternating current (AC) motors until the late 1880s, but even then, lead-acid batteries were very heavy for the charge they held, and so were impractical for automobiles.

All these things seem to point to somewhere around the mid-1880s for the first time an automobile was possible, which is exactly when Benz patented his 'Motorwagen'. Even then, he used steel-spoked wheels and solid rubber tires, as John Dunlop's pneumatic tire was not invented until 1888. (When Henry Ford made his first car, the Quadricycle, in 1896, he used pneumatic tires, since they gave a much smoother ride on rough ground than a solid rubber tire.)

In a 2020 presentation to the North Carolina Department of Transportation, Seba asked, "Why did the smartphone happen in 2007? ... why didn't the iPhone come out 2005 or 2009? Now both the iPhone and the Android came out the same year 2007. Not '05 or '09. What was the magic of 2007? 2007 is when all the technologies that made the smartphone possible converged, so all the techno-computing power, broadband, data storage, touch screen, all these technologies that made a $600 smartphone possible happened in 2007."

The new technology changes the basis of competition

Clayton Christensen in his best-selling book *The Innovator's Dilemma*, defined the 'basis of competition' of a product as "the criteria by which customers choose one product over another". In the early 1980s, the dominant 5.25-inch hard disk drive was very limited in its storage capacity, although the capacity was improving every year. Smaller, 3.5-inch drives had even less storage capacity than 5.25-inch drives and they were also more expensive on a per-megabyte basis. Nevertheless, 3.5-inch drives quickly conquered the personal computer market. "In 1985," Christensen wrote, "less than 1 percent of the desktop market had switched to the 3.5-inch format... By 1989," he continued, "only four years after the 3.5-inch product had appeared as a faint blip on the radar screen of the market, it accounted for 60 percent of drive sales."

Why? According to Christensen, "the smaller 3.5-inch drive allowed computer manufacturers to reduce the size, or desktop footprint, of their machines." The smaller disk drives were *good enough* and *cheap enough* to enable computers that were physically smaller, taking up less space on a desktop, even though they were more expensive than 5.25-inch drives and had less capacity.

In desktop computers, but also in portable devices like laptops, digital cameras, music players and tablets, we have seen this phenomenon again and again – new generations of data storage devices have enabled new products by excelling in some characteristic where older generations of devices did not: storage capacity, physical size, reliability, shock resistance, and price, as Christensen mentioned, but also access time, data transfer speed, power consumption (which affects battery life), and weight. For each new generation of data storage devices, the primary criterion by which customers chose one product over another changed.

The new technology changes the key performance metrics

Because a disruptive innovation changes the basis of competition, it also changes the key metrics by which the performance of the technology is measured. When James Watt greatly improved the steam engine, he needed to describe how much work his invention could do compared to animal labor, so he invented the term 'horsepower' to quantify how much water his pumps could lift, how far, and in what time. Likewise, the output of electric motors that were invented more than a century later could have been measured in horsepower, but since these motors could also be used for non-mechanical uses like lighting or heating, Carl Wilhelm Siemens, developer of the open-hearth steel furnace, proposed the measure of electric power to be the Watt, named after James Watt. In turn, as light-emitting diode (LED) lighting displaced incandescent bulbs, manufacturers have turned to 'lumens' and 'color temperature' as key metrics of the performance of lighting instead of wattage. From horsepower to watts to lumens, each new generation of technology changes the key metrics by which the technology is judged.

Industry insiders often develop the key technologies that enable the disruption, but an industry outsider begins the disruption

It was Steve Sasson, an employee of Kodak, the world's largest photography company at the time, who invented the first hand-held, self-contained digital camera in 1975. He said it was 'the greatest mistake of his life' to refer to his camera as "filmless" in his presentation to Kodak executives, as they realized that eliminating film would also get rid of the largest source of the company's profit.

Understandably then, Kodak did not bring the first consumer digital camera to the market. That distinction went in 1990, fifteen years later, to Dycam, a California start-up company founded only two years prior. Their Model 1 cost $995, took photos that were just 376 by 240 pixels, and could hold only thirty-two of those images. But that camera, also called the Logitech Fotoman, revolutionized photography forever.

Repeatedly we see that insiders invent powerful technologies, but outsiders take advantage of them. "Many of the inventions that Europeans used to colonize the world (the compass, gunpowder, paper money, the printing press) were of Chinese origin," Manuel De Landa wrote in *A Thousand Years of Nonlinear History*, "while Europe's accounting techniques and instruments of credit (which are often cited as examples of her unique "rationality") came from Islam." But it is difficult to doubt that Europeans took greater advantage of these ideas than the people who invented them.

There is often an 'incubation period' between successive steps in a disruption

A prototype is produced when the invention has the technological ability to be functional and the first commercial product is released when the components of the invention get cheap enough for there to be a market for the product. But, in her 2002 book *Technological Revolutions and Financial Capital*, economist Carlota Perez wrote, "radical innovations can occur at any time, though their gestation period can be long. Due to the relative autonomy or the production of science and technology, there are always potential innovations in various fields waiting in the wings. At any point in time, therefore, the spaces of the scientifically conceivable and the technologically possible are much wider than those of the economically viable or the socially acceptable."

Karl Benz developed a two-stroke gasoline engine in 1873 and the world's first production automobile, his Patent-Motorwagen, went into production in 1886, thirteen years later. On the other side of the Atlantic, Henry Ford built his first car, the Quadricycle, in June 1896 and released his first mass-produced car, the Model T, in October 1908, a little over twelve years later. There can be an extended time from when a radical innovation occurs to when it begins to impact the market.

The new idea causes 'market trauma' when it reaches just a few percent of the market (usually 3% to 10%), causing revenue, profit, and investment problems for incumbents

Incumbents cannot afford to pay attention to every new competing idea that comes along. However, once a product reaches some market share, satisfies the first tranche of customers, and then leaps to a wider audience, the innovation begins to earn the attention (and ire) of the incumbents. Often this threshold is breached when the new idea constitutes just a small portion of the market. This the story of many incumbents: even when they are slightly past their peak, they are still convinced that no one in the world is powerful enough to defeat them.

A 'take-off' point is reached, after which comes a period of rapid adoption

Regardless of how long it takes for an idea to become a prototype, for that prototype to become to a commercial product, or for that product to begin to take over a market, once that commercial product proves, in principle, it can serve the vast majority of a market, if not the entire market, the time from when that 'take-off point' is reached to when the product dominates the market is often very short, typically a decade or two.

Becquerel discovered radiation in the 1890s, and it was not until the 1940s (a half-century later) that the first atomic weapon was detonated, but in the fifteen years starting around 1950, the US nuclear arsenal went from essentially nothing to its peak size.

The adoption of a disruption often follows the same basic pattern: a period of slow growth, an S-shaped period of rapid growth, and then a plateau (perhaps followed by the replacement by a new technology)

Technologies live the gangster's life: most of them don't die of old age. They are introduced to the scene as the new hot thing, rise quickly to come out on top for a brief but glorious time, and then one day turn around to find themselves staring down the barrel of the new disruptor. We have seen a similar pattern repeatedly – a period of exponential growth turns into an S-shape of maturation, there is a time of peak popularity, and then a brutal decline. Perhaps you have a box of old compact discs somewhere, a drawer full of computer cables that do nothing, or an out-of-date mobile phone. They are not sitting in a drawer because you thought they were worthless at the time you put them there, now are they?

Disruptions often begin in the Business-to-Business (B2B) market

Because a new technology is often high cost relative to performance, the earliest adopters are those who have the greatest need for it, who will use it to its greatest extent, or who will get the most benefit from it. Those are often businesses rather than individual consumers. For example, spreadsheets and word processors were developed for business users in the 1980s but are now on many personal computers. Magnetic videotape was originally developed for the television industry in the 1950s but found widespread adoption in home entertainment systems in the 1980s. Today, companies like Perfect Day (which makes milk proteins), Geltor (collagen), Colorifix (dyes), and Bolt Threads (leather) are making products that you might one day find in your home, but their products are sold as ingredients to other businesses, not to you directly.

An 'incumbent death spiral' results in increasing costs, and decreasing competitiveness, for the old industry

As a disruption plays out, the new technology does not just get better, the old one gets worse. The same dynamics behind network effects and economies of scale that help to drive down the cost and to drive up the performance of the new technology, cause the old one's value proposition to unravel.

We have seen examples like this with tetraethyl lead, whale oil, whale meat, and Blockbuster video stores. And we expect to see it with many more in the coming years, including milk, beef, and electricity made from burning fuel.

The disruptive innovation greatly expands a market (that is, the new thing is not a 'one-for-one' replacement for the old) and it expands the 'possibility space' of things that can be done economically

Steam ships did not just replace sailing ships, they greatly expanded the amount of cargo carried annually as well as trade routes and the ports that were served. A century later, jet engines rewrote the map of places served by propeller airplanes.

Digital cameras mean that far more still images and videos are made annually now than at any time in history. Artificial dyes and artificial fibers mean that many people have far more clothing than we have ever had before. As Dr. Adam Dorr said in his TEDx talk in Glasgow, Scotland, in October 2021, "disruptions do more than just swap a new technology in for an old one. Electricity wasn't just cheaper whale oil. Automobiles weren't just faster horses. Farming wasn't just more fruitful hunting and gathering. These better tools transformed energy, transportation, and food and civilization along with them."

Often, an *intermediate technology* is developed to forestall the impending disruption

It is not just the disruptors who realize the potential of, and try to take advantage of, a new technology. In fact, the tendency for an incumbent technology to experience "a wave of improvements when a new competing technology appears" (in the words of Nicola De Liso of the University of Salento, Italy, and co-workers) is so common as to have earned the name the 'sailing-ship effect', after the rush to develop faster sailing ships in the years after steam ships were introduced.

Sailing ships are far from the only technology that has seen this effect. When the first consumer digital cameras were available that could compete commercially with film cameras, the film industry began adding digital features to their cameras, like a digital preview window or offering a compact disc (CD) of images with film processing. 'Hybrid' cars used electric motors and a (relatively small) battery pack to move a vehicle on short trips and an internal-combustion engine for longer trips. Some electric typewriters incorporated digital features like built-in spelling dictionaries that were enabled by the kinds of processing chips in home computers. And 'turboprop' airplane engines use a turbine to drive a propeller. All these are examples of attempts to take advantage of new technologies – digital cameras, lithium-ion batteries, computer chips, and turbines engines – to help mature technologies compete with new disruptors.

De Liso and co-workers argue as well that the 'sailing ship effect' is rational from the incumbent's perspective, as it allows them to squeeze a few more years of profitability out of a dying technology.

Sudden 'step' changes move us from one 'punctuated equilibrium' to another

Longevity offers little protection against disruption. Neanderthals were the dominant hominin species in Europe for several hundred thousand years, far longer than we have occupied that land. But over a relatively brief period about forty thousand years ago, we displaced them completely, perhaps being a significant factor in what drove them to extinction. Having gotten to Europe first and having lived there the longest did nothing to protect them. Likewise, our hunter-gatherer lifestyle served us well until suddenly, with the invention of agriculture about ten thousand years ago, we quickly switched to a new way of life.

During the 1800s, according to the website *Our World in Data*, the chances a woman in Finland (one of the few countries for which detailed records exist from that time) would die due to pregnancy averaged about 1-in-120 per conception. But now Finland's maternal death rate is two hundred times lower per pregnancy and women also tend to have fewer children. Maternal mortality went from being a significant cause of death to largely a non-concern in many countries in about a generation.

Existing technologies, ideas, and conditions might dominate for centuries, even millennia, until suddenly they do not. We used the horse as our primary means of transport for thousands of years until they automobile made them obsolete for that purpose over little more than a decade in the early 1900s. Zoroastrianism was the predominant religion of Persia until the arrival of Islam. The transatlantic trade in enslaved people from Africa to the New World ran for about three hundred and fifty years, conducted primarily by four nations: Britain, the Netherlands, Spain, and Portugal. And then, over the course of about fifteen years from the early 1850s through the middle of the 1860s, the trade simply stopped, never to resume.

The disruptive innovation often changes the key skills valued by an industry and by the companies in it

Because radically cheaper, easier, faster, or better ways to do something are often based on innovative technologies, new engineering or science, and different business models, the disruptive industry often requires and values skills that the old one did not.

Digital cameras are based on sensors whose production requires facilities and skills similar to the semiconductor industry, while the production of photographic film relies more on chemical engineering. Similarly, being a digital photographer often necessitates knowledge of software-based tools that are radically different from the knowledge of film development and manipulation.

Fully battery-electric vehicles (BEVs) require far more knowledge of electrical engineering and computer programming than cars based on internal combustion engines. Likewise, direct sales to customers and direct service of vehicles requires BEV companies to have certain kinds of employees in-house that incumbent automobile manufacturers outsource to their dealer network.

Changing the way an industry functions changes the key skills valued by that industry.

In a disrupted world, advantages can become disadvantages, disadvantages can become advantages, and assets can become liabilities

Companies work hard to build some sort of 'moat', a term popularized by extraordinary investor Warren Buffet to describe a sustainable competitive advantage. But disruptive technologies and disruptive business models can cause old moats to to dry up quickly and become a grave. Buffet has historically favored investing in small-town newspapers, those whose market is not large enough to support many competitors and whose printing facilities are expensive to operate. But as websites like *Craigslist* and Google siphoned away newspapers' classified advertisement business, operating expensive printing facilities to serve a local market has become a much less attractive proposition.

In their annual filing for the year 2000, video rental company Blockbuster estimated that "about 70% of the U.S. population lives within a ten-minute drive of one of our stores." However, getting DVDs delivered by mail from Netflix meant that the entire concept of *driving* somewhere to borrow a video was on the way out. Over the period of just a few years in the early 2000s, Blockbuster's large number of locations went from an irreproducible business advantage to an enormous burden, hastening their bankruptcy filing in 2011. In a future with ubiquitous autonomous delivery vehicles, restaurants, shopping centers, and even supermarkets also might find that having physical locations are burdensome.

'Moats' are not always physical assets, sometimes they are simply the mental association that customers attribute to a brand (what marketers call 'positioning'). Burger King prides itself on not using fillers in their patties, claiming instead that their hamburgers are "100% Beef". That has made a marketing a challenge for their 'Impossible Whopper', a beef-free burger made by California-based Impossible Foods that is surprisingly similar in flavor and texture to Burger King's meat-based patties. In 2019 Impossible Foods' vegetarian burger patty, a product that was only introduced in 2016, had become "the most popular late-night delivery snack in America", according to GrubHub, a popular food delivery service.

As autonomous, electric-powered delivery vehicles (some of which look like little more than insulated carrying cases on wheels) become more popular, Burger King might find itself fighting disruptions on multiple fronts – trying to undo their image as a source of only "100% Beef" burgers in customers' minds while simultaneously finding the distinctive décor of their physical locations increasingly irrelevant.

Disruptions unleash ripple effects, some of which are unintended

A 2017 article on the Vox news website said that in the ten years after the launch of the iPhone in 2007, chewing gum sales in the US dropped about 15%. They reason that gum is often an impulse purchase made when waiting in line, but now while waiting "we're so consumed with our phones that we're not reaching for a pack of gum". (Molla, 2017)

This might seem like a trivial example, but it might not be. Dramatic and unanticipated consequences are the norm when introducing new ideas and technologies. The conversion of the Roman population to Christianity saw the decline of the scroll and the rise of the 'codex', the earliest form of book. The reasons for this are not clear but some suspect that it is easier to preach with a book in one hand than when trying to hold a scroll. The automobile was not simply a faster horse; it enabled the drive-in theater, the suburb, the supermarket, the strip mall, and the first canned dog food, Ken-L Ration, made from the meat of suddenly-obsolete horses, among many other things. And the decline in syphilis due to the widespread availability of penicillin after World War II, combined with the availability of oral contraceptives, likely contributed to the Sexual Revolution of the 1960s and 1970s.

None of these effects were intended. None were even anticipated. In fact, chemist Russell Marker, who was the first to synthesize progesterone insists, according to Bernard Asbell in his book *The Pill: A Biography of the Drug that Changed the World*, that he was not even trying to make an oral contraceptive at all. Asbell writes "prevention of pregnancy as a goal never crossed his mind". Marker was simply trying to make progesterone for scientific and therapeutic use. But by doing that, he not only revolutionized birth control, but also women's educations and careers and much of our modern society as a result.

Disruptions enable and reinforce each other in an interdependent network

Lithium-ion batteries first found use in consumer electronics like cameras and laptops, but are now used to power cars. If in the 2010s you wanted to know the future of the car industry, you needed to be paying attention to the consumer electronic business, as cars were increasingly becoming 'smartphones on wheels'. Likewise, light-emitting diodes (LEDs) were first used as tiny indicator lights for electronic devices but now are used for household illumination and even super-bright outdoor flood lights.

Both technologies – lithium-ion batteries and LED lights – make the other one more valuable. Combined with cheap and powerful solar panels, these three technologies could help unlock new revolutions in power usage.

Understanding the future of any one technology, and its implications for wider society, means understanding all the key technologies and their interconnections.

Let us now examine some of the key technologies that are shaping our future answers to the questions related to materials, food, energy, transportation, communications.

◆ ◆ ◆

"I'd put my money on the sun and solar energy. What a source of power! I hope we don't have to wait until oil and coal run out before we tackle that."

— Thomas Edison, 1931

◆ ◆ ◆

Key Technologies: Solar Power, Batteries, AI, Fermentation, and More

Physicists tell us that the universe is made of energy, matter, and information. In all three of those areas, new and clearly disruptive technologies are emerging that are reshaping how we obtain and store energy, how we obtain food and materials and what we can do with them, and how we store, process, and use information.

Solar power and wind power are not just cheaper than they ever have been before, they are now cheaper in most places than any other power source, and are getting cheaper still. Combined with the fact that lithium-ion batteries are getting cheaper and better-performing as well, we are on the cusp of having abundant clean energy at zero marginal cost for much of the time.

Precision fermentation, the ability to produce any specific molecule in essentially the same way that we brew alcohol using yeast, and cellular agriculture (growing large amounts of meat and fat cells by taking a small sample of tissue from a single organism) promise to end the use of animals and some plants to get the products we want. We could still get protein, fats, and carbs, meat, milk, eggs, fibers, leather, and more, just without using animals, or much land or resources.

And cheap computational power, sensors, and data storage are enabling computers to do increasingly complex tasks that were formerly only done by people, including moving goods around warehouses, automating retail check-out systems, and making artwork. When Artificial Intelligence (AI) can drive taxis and trucks, or even delivery robots and self-piloting drone aircraft, we might see cheaper goods delivery, the virtual end of car ownership, a drastic drop in car crashes, and a complete reshuffling of real estate values.

Solar Photovoltaics (PV), Wind Power, and Lithium-ion Batteries Could Make Clean Energy Cheap, Clean, and Superabundant

The price of solar photovoltaic (PV) modules has declined so dramatically in the past forty years that it is difficult to even make a graph that is easy to understand. We took numbers from the website *Our World in Data* which consisted of the cost of a solar PV module (in constant 2022 US dollars per Watt) and the cumulative amount of solar panels installed worldwide, in megawatts (or, millions of Watts). Perhaps the same kind of 'learning curve' graphs we made for the Ford Model T and for the F-35 fighter jet would be helpful here.

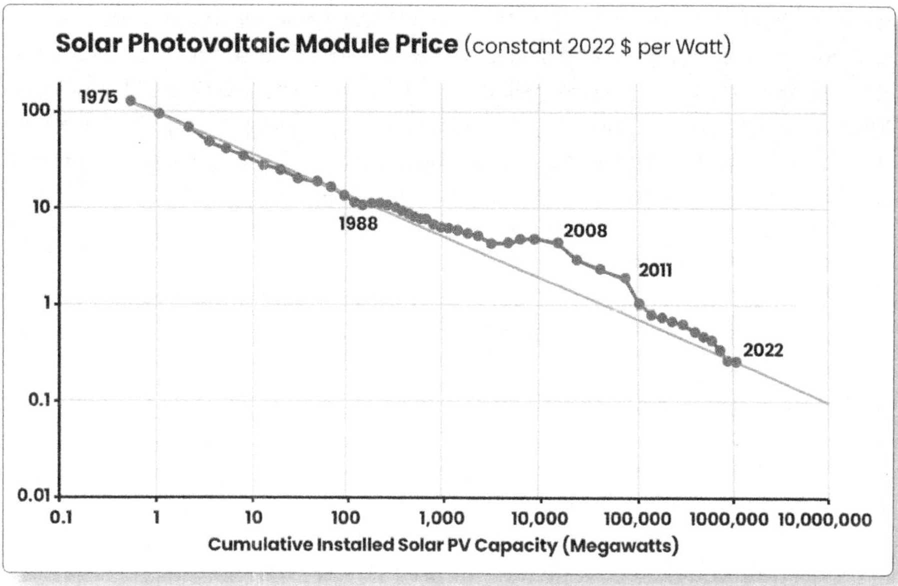

Figure 9 – Price of solar photovoltaic modules. *Source: Our World in Data*

As of 1975, the cost per watt of panels was over $125. One year later, when just over one megawatt of solar PV panels had ever been installed worldwide, the price had fallen below $100 per watt for the first time. By 1980, just four years later, the price per watt had come down to under $35 and by 1988, it was under $11, a decrease of more than 90% in just thirteen years. By 2022, the price of solar panels was about $0.25 per watt, about 1% of the price in 1982, a 100X decline in the span of four decades. And there is no sign of slowing down. The International Energy Agency, headquartered in Paris, estimates the cost will drop another 40% from 2021 to 2030.

The first commercial solar cell was produced by Bell Labs in 1954. It took more than sixty years between then and when solar made up just 2% of US electric production in 2017. By 2021, just four years later, solar power accounted for nearly 4% of US electricity, a doubling, and the Energy Information Administration, a US government agency devoted to collecting statistics about American energy production, estimated in February 2025 that it will reach 8% in 2026. It would take only three more doublings – to 16% to 32% and to 64% – for solar PV to constitute most of the electricity in the US. One more doubling after that would mean that the amount of solar power produced in the US could exceed the current total level of electricity production. After that is all gravy.

How long might it take for this to happen? In the 2010s, the total installed PV capacity worldwide grew at about 40% per year. In the 2020s so far, it has averaged about 20% growth per year. At the higher growth rate, the solar capacity installed doubles about every two years, or about six years total to get three doublings. Even at the lower growth rate, production of solar power in the US could equal all of current US electricity consumption, in principle, within about fifteen years.

The force underlying this remarkable growth, of course, is the self-reinforcing feedback loop. As the cost of solar power has gone down, the rate of new installations has gone up. This is basic economics: the cheaper something gets, the more of it that people will buy. But for new technologies, the converse relationship is also true. The more of something that people buy, the cheaper it gets, due to Wright's experience curve. Author Tim Harford wrote in an article for the BBC that "the learning curve creates a feedback loop that makes it harder to predict technological change. Popular products become cheap and cheaper products become popular."

On a watt-for-watt basis, new solar power installations are already cheaper than new coal-fired and gas-fired power plants. Soon, they are likely to be cheaper than even providing the *fuel* for existing coal and gas plants. After that, cheaper than the even the cost of *transmitting* power from a coal-fired plant or gas-fired plant. At that point, it does not matter which fuel is used at the remote plant as the cost of generating power locally will always be cheaper. Tony Seba calls this "GOD parity", or 'Generation on Demand parity" – even if fuel was free, even God could not make electricity cheaper remotely than PV could locally. The whole electric system would be pushed from its century-plus-old business model based on centralized production from fuel-powered plants to a new one of distributed production based on solar panels.

The sun does not always shine, but that is not so bad because other sources of renewable power like wind power are also experiencing sharp declines in their cost, which is fortunate as the wind often blows in the mornings and evenings when solar power is not the strongest.

The decline in the cost of wind power has not been as dramatic as for solar power but is still impressive. In 2017, one watt of solar panels cost less than 10% of the price in 1995. The amount of money that could buy one panel then could buy

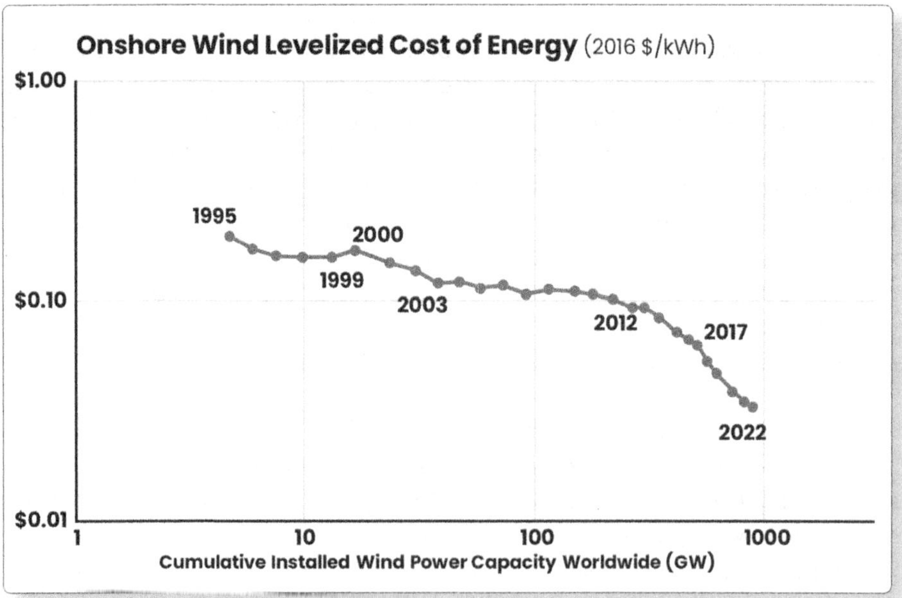

Figure 10 – Cost of wind power. *Source: Our World in Data*

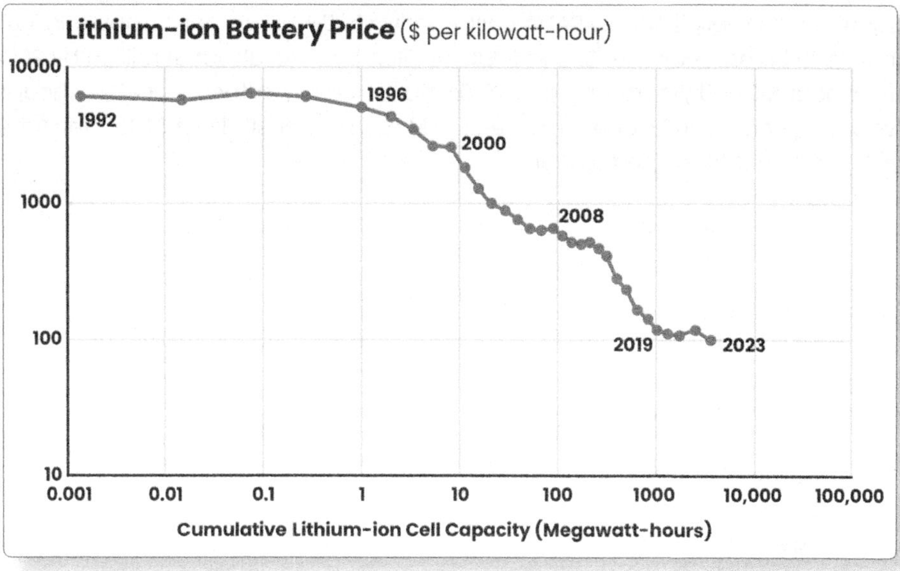

Figure 11 – Lithium-ion battery price per kilowatt-hour. *Source: Our World in Data*

more than a dozen twelve years later. In comparison, in 2017 one kilowatt-hour of power from onshore wind turbines was less than 25% of the 1995 price. Essentially, wind turbines are having a 'buy one, get three free' sale.

Solar panels and wind turbines become much more valuable when there is a cheap, easy way to store their power. Fortunately, the price of lithium-ion batteries has also come down dramatically in the past forty years, even as their performance had increased. Lithium-based batteries were proposed by British chemist M. Stanley Whittingham while working for Exxon. The rechargeable lithium-ion battery was patented in 1977. And the first practical lithium-ion battery, invented by John B. Goodenough and Akira Yoshino, was brought to the market by Sony in 1991, used first in cameras and laptops. Battery costs have fallen more than 97% since then. (For their work, Whittingham, Goodenough, and Yoshino won the Nobel Prize in Chemistry in 2019.) And batteries did not just get cheaper – the power density increased, number of cycles through which they could be recharged increased, and the charging time decreased. By 2003, lithium-ion batteries were cheap enough and good enough to power cars. By 2017, they were being deployed to store power for regional electrical grids.

These three technologies – solar photovoltaic panels, wind turbines, and lithium-ion batteries – combined have the power to radically transform the world's energy system. Together with cheap computer processing and cheap sensors, they could change much else besides.

Artificial Intelligence through Cheap Sensors, Computer Processing, and Data Storage Could End Driving and Much of Human Labor

In terms of cost reduction, few technologies compare to what the computer processor and data storage have done in the past half-century. According to *Our World in Data*, a website that collects a wide variety of information, computer processor memory (that is, RAM) cost nearly $5000 billion per terabyte in 1970, under $20 billion before 1980, under $100 million by 1990, about $1 million by 2000, and about $10,000 per terabyte in 2010. That is literally a factor of 100X cheaper every *ten years*. In 2023, one terabyte of RAM memory cost about $1,000, ten times cheaper than thirteen years prior.

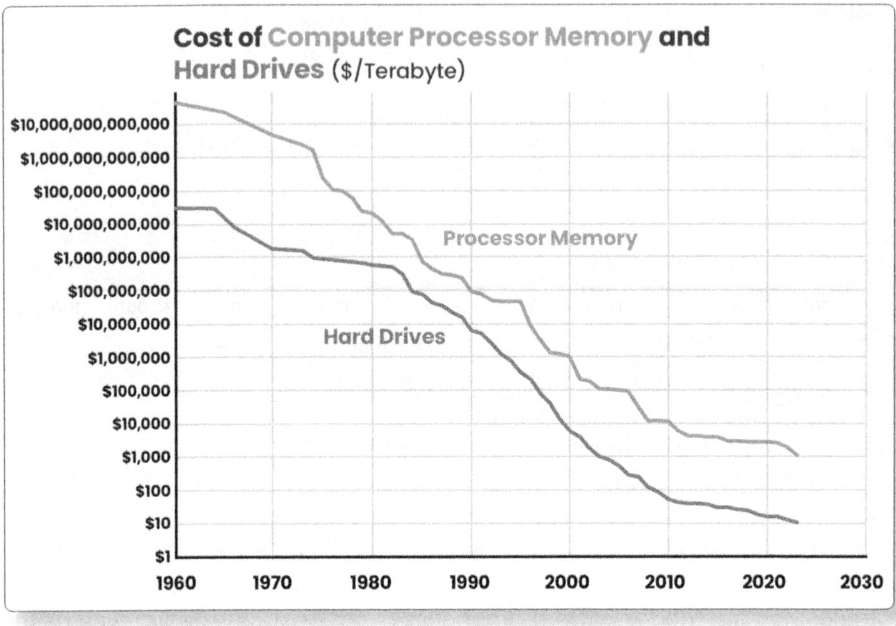

Figure 12 – Cost of Computer Memory and Hard Drives. Figures are in constant 2020 US dollars. *Source: Our World in Data*

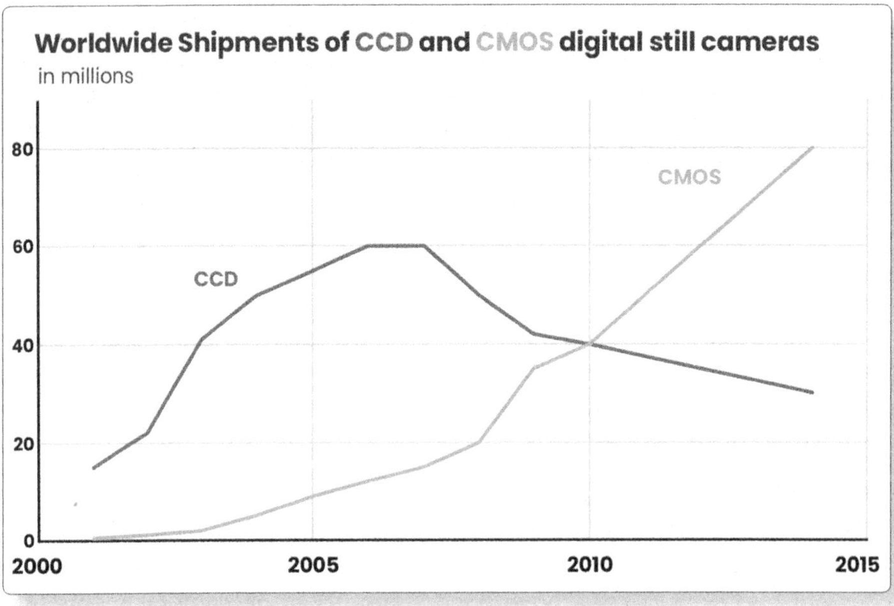

Figure 13 – Charge-coupled Devices (CCD) and Complementary Metal-Oxide Semiconductor (CMOS) sensors. *(Roy, 2018)*

The decrease in cost of hard disk drives was less dramatic, but by 2010 they were still more than five *million* times cheaper than they had been in 1970, and in 2020 they were about 1/3rd of the price of ten years prior.

Cheap computer processing and cheap data storage have become dramatically more useful as sensors like microphones and cameras have grown cheaper. So-called CMOS (Complementary Metal-Oxide Semiconductor) image sensors, the kind found in mobile phone cameras, took over the market from Charge-Coupled Device (CCD) sensors in under ten years. Now, cameras are available on many common electronic consumer devices and taking photos is essentially free.

Laser-based scanners (lidar), which are used on some autonomous vehicles, have likewise become dramatically cheaper in just the past few years. Speaking at the Downtown Seattle Association's 2016 'State of Downtown' convention, Tony Seba said that in 2012, when Google revealed some facts about its self-driving car, they said the lidar was "seventy thousand dollars... Within a year, the cost of the next generation lidar went down to $10,000." Seba continued, "a Silicon Valley company announced a $1,000 lidar. But wait, there's more! Last month, that same company announced a $250 lidar. So essentially lidar will have gone from $70,000 to 250 bucks within four or five years. The next-gen from this company is going to be the size of a postage stamp and it's going to cost $90. You can put it on your iPhone. I don't know what you're going to do with it, but you could." In late 2020, sure enough, Apple announced that the new iPhone 12 Pro would have a built-in lidar.

One implication of vehicles that drive themselves could be the end of car ownership for many people, as taxis hailed by phone and paid for by distance would eliminate the expense of personal car ownership and much of the cost of current taxis. Sensors and software that allow a machine to drive like a person also threaten all forms of labor if packaged in the form of a humanoid robot.

A side effect of the revolution in computer technology, including automation in gene analysis, has been cheaper sequencing and manipulation of DNA. Using genetically-modified microorganisms to 'brew' the food, medicines, and materials we desire could not only make production 10X more efficient, but could open the door to new possibilities that are technologically or economically unfeasible today.

Precision Fermentation and Cellular Agriculture (PFCA) will Radically Improve Nutrition and Materials and Reduce Their Cost

One of the newest technologies shaping the future of food and materials is also one of the oldest: fermentation. Just as brewer's yeast can convert some sugars into alcohol, scientists are now intentionally modifying microbes to produce desired molecules – proteins that taste extremely sweet, dye molecules, the key proteins found in milk and eggs, and even heme, the oxygen-carrying blood protein that helps make meat taste like meat. We can now 'brew' milk proteins without cows and egg proteins without chickens, using the technology of *precision fermentation*. We can also brew the hormones that signal to animal cells in a vat that they should divide. Instead of using animal agriculture to get meat, we could use *cellular agriculture*.

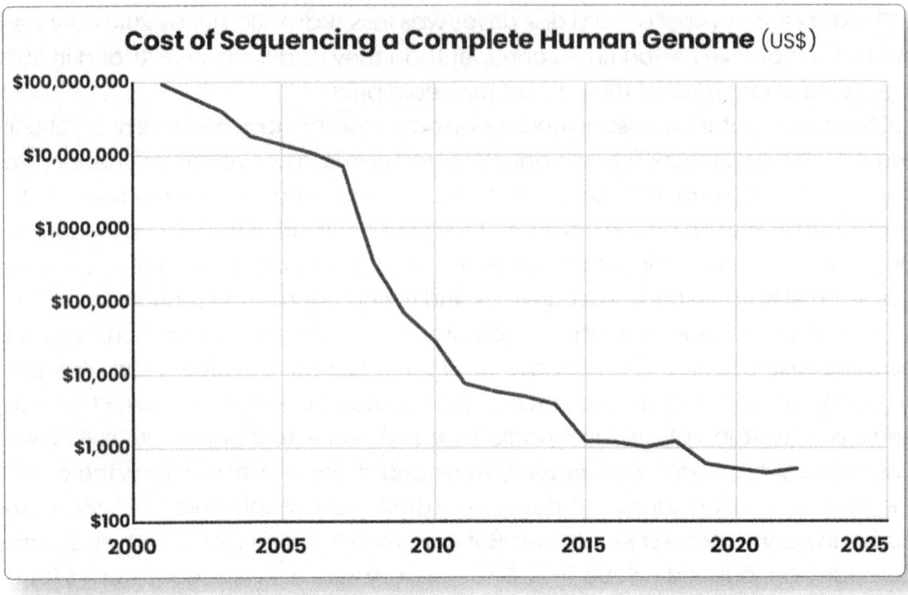

Figure 14 – Cost of Sequencing a Complete Human Genome. *Source: Our World in Data*

Both of these technologies – precision fermentation and cellular agriculture – are being unlocked by the rapidly declining costs of DNA sequencing ('reading' the genetic material from existing organisms), DNA synthesis (making sequences of genetic material that are either identical to what is found in nature or, in some cases, that are entirely new), and gene editing (inserting genetic material into an organism).

According to *Our World in Data*, the cost of sequencing a complete human genome in 2003 was more than $40 million. By April of that year, the total number of human genomes that had ever been sequenced, the culmination of a thirteen-year-long project begun in 1990, was exactly one. (Actually, some particularly difficult parts of the genome took many more years to sequence, with the last parts only being completed in 2022.) By 2022, the cost had fallen to about $500.

As the cost of gene sequencing has come down, its popularity has gone up dramatically, with profound scientific and medical implications.

Sequencing DNA from dozens or even hundreds of ancient individuals is becoming increasingly commonplace in scientific research, helping to re-write our understanding of our own species and our domesticated animals. A 2022 study of the rapid migration of the Avar people from Central Asia to Europe roughly 1500 years ago involved sequencing sixty-six genomes. Harvard Medical School professor David Reich led a study, published in 2021, which sequenced the genomes of hundreds of individuals who lived between 4000 BC and 300 AD in the UK, finding a sudden change in genetic composition of the population around 2400 BC. Now is the first time that this kind of science of ancient people is possible.

But it is not limited to ancient, dead people. In 2014, the UK government funded the '100,000 Genomes Project' with roughly £300 million (about $500 million at the time) to sequence the genomes of 100,000 people in Britain. Less than ten years later, genomics company 54gene sequenced the genomes of more than 100,000 Nigerians at roughly 10% of that cost. The British government has recently announced funding for a project to sequence the genomes of 100,000 infants to screen for two hundred rare genetic disorders. The total cost of the project will be about one-third what the original 100,000 Genomes Project cost.

All these technologies – in areas like energy, agriculture, materials production, transportation, and communications – are influencing and accelerating the development of the others. These advances are not just making expensive things cheaper, they are unlocking new spaces of possibility in what we can do. Boston-based company Ginkgo Bioworks took DNA from flowers from now-extinct species and inserted the fragrance-creating sequences into yeast to create aromas that were chemically identical to those produced by the original flowers. For the first time, we were able to smell the aromas of extinct species. Using this approach, perhaps we could create aromas from species that have never existed.

Revolutions in energy, materials, and information have the power to change virtually every aspect of our lives, over a short period of time. The only limitation is what we choose to do with these technologies.

"Faster Than Ever Before"

The idea that the pace of technological change is faster now than ever before is not a new one. In 1591, Robert Barret wrote in *The Theoretike and Practicke of Moderne Warres*, "Time altereth the order of warre, with many new inventions daily." Just a few years later, in 1600, a Barnaby Rich wrote, "one of the great diseases of this age is the multitude of books that doth so overcharge the world that it is not able to digest the abundance of idle matter that is every day hatched and brought into the world." (Durant, 1935) It seems that, at every time, people have felt overwhelmed by the pace of change of new technologies and the rate of flow of new ideas.

And they might have been justified in feeling that way – if technological change keeps going faster, then every time is the time of fastest change to date. Why does technology seem to be changing faster than ever before?

Figure 15 (overleaf) shows how much time it took to go from 10% to 80% market adoption for more than one hundred and fifty examples of disruptions from RethinkX's dataset. As with the experience curves we showed earlier, this graph is plotted on a logarithmic axis, in order to make the trend clearer.

(Each technology is plotted according to when it reached 10% market adoption, which means that for recent technologies that have reached 10% adoption, but not yet reached 80% adoption, it is not possible to plot a point for them. The region where it is physically impossible to see points because not enough time has passed is marked with a solid, black line.)

In the latter half of the 1800s, we found few examples of disruptions that took less than ten years to go from 10% to 80% adoption. Synthetic alizarin (a red fabric dye) is one example, as is synthetic indigo a couple of decades later. Even the Bessemer process took less than a decade to replace the puddling furnace as

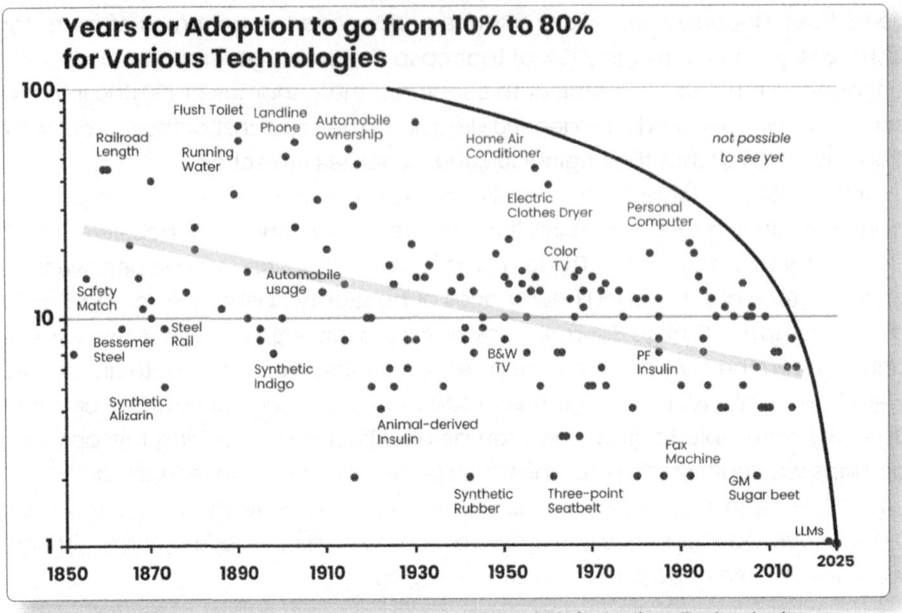

Figure 15 – Years for Adoption to go from 10% to 80% for Various Technologies.

the primary means of making steel. But railroad tracks, the flush toilet, and running water all took decades to reach 80%.

In the 1900s, automobile ownership took more than a half-century to go from 10% of US households owning a car in 1915 to 80% in 1970. But the usage of automobiles - as a fraction of vehicle-miles traveled - changed much faster. Cars travel much further per year than horses, six to seven times the distance per year, so the fraction of vehicle distance traveled per year in the US in automobiles went from under 10% in 1909, the year after Henry Ford's assembly line became operational, to 80% by 1922, just thirteen years later.

By the latter half of the 1900s, we commonly find technologies that are adopted in only a few years and few that take multiple decades. Home appliances like the air conditioner, electric clothes dryer, and even the personal computer took a relatively long time to be adopted, while ones adopted by businesses, like the genetically-modified sugar beet, fax machines, the three-point seatbelt, and synthetic rubber, only took a few years.

Overall, the trend in adoption times has been downward. In the mid-1800s, it took about twenty-five years for the technologies in RethinkX's dataset to be adopted, while by one hundred and fifty years later, it took ten or fewer years. That is, for every decade in the past one hundred and seventy years, the time it has taken for a technology to get adopted by the marketplace has fallen by about one year, on average. The world really is changing faster than it ever has before.

A 2018 study by Bryan Kelly and co-authors at the National Bureau of Economic Research looked through US patents from 1840 to 2010 to identify what they call "breakthrough innovations", inventions like the telegraph, elevator, airplane, and television that were both "distinct from prior patents" and also influential on later patents.

Many patents are issued each year by the US Patent and Trademark Office (USPTO), but some inventions have much, much greater impact in the marketplace and on future inventions than others. Kelly and his co-authors analyzed the text of millions of patents to find ones that used words and phrases that did not appear often in previous patents, but which did appear often (relatively speaking) in subsequent patents.

For example, Nikola Tesla's 1888 description of an electric motor "was among the first patents to use the phrase 'alternating current,' a phrase used with great frequency throughout the 20th century", even among patents that do not directly cite Tesla's work.

According to Kelly, "the USPTO categorizes patents... based on the nature of the technology represented by the patent," labelling those related to 'agriculture/food' distinctly from those that are 'weapons', those involved in 'chemistry and metallurgy', those that are related to 'electricity and electronics', and other subjects. This allowed Kelly and his co-authors to analyze how the subject areas of breakthrough patents changed over time.

In the 1840-70 period, they claim that "the most important inventions" took place in "engineering and construction" and "consumer goods". In the 1870s and 1880s, they saw "the beginning of the second industrial revolution" with breakthrough patents largely in electricity and transportation. In the early 1900s it was agriculture, and in the 1920s and 1930s they saw breakthroughs "in manufacturing, particularly plastics and chemicals".

Again, this analysis is concerned only with 'breakthrough' patents, the distinctively influential ones that shaped the direction of future innovations, not simply all patents, as "the composition of technology classes among all inventions has remained relatively stable over time."

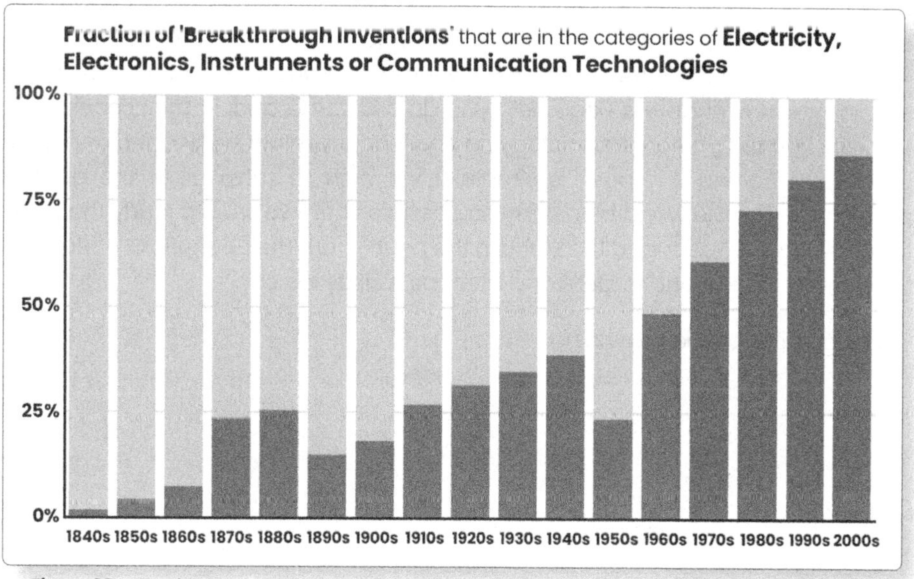

Figure 16 – The fraction of 'Breakthrough Inventions' that are in the categories of Electricity, Electronics, Instruments, and Communication (EEIC). *(Kelly, 2018)*

Four of these subject areas stand out among the others in terms of their importance to the later part of the 1900s: electricity, electronics, instruments, and communication technologies, or 'EEIC'. The graph in Figure 16 shows the percentage of breakthrough inventions that were in one of these four subject areas by decade, starting in the 1840s.

Inventions in these four categories constituted less than 25% of all 'breakthrough' patents before the early 1900s, and again dipped below that level even as late as the 1950s. But since then, EEIC inventions were about half of all breakthrough inventions in the 1960s, nearly 75% by the 1980s, and by the 2000s were over 85% of all breakthrough inventions.

One explanation for why technology seems to be changing 'faster than ever before' might simply be that most breakthrough inventions today are all things that contain transistors – electricity, electronics, instruments, and communications devices. We have largely stopped innovating in all other areas – construction, transportation, agriculture, and manufacturing – except if the invention contains electronics. It's a common refrain for technologists to claim that 'the pace of innovation is speeding up' or 'things are changing faster now than ever before'. But another way of looking at Kelly's analysis is simply to say that *we are making major innovations in fewer subject areas than ever before.*

There remains one bright spot that Kelly and co-authors identified in the portion of recent breakthrough inventions that are not EEIC-related, one area that in fact might become the start of a new wave of major breakthroughs: molecular biology and genetics.

A great idea is more than just a singular insight. It is a lens through which to view the world and to go back and re-evaluate everything you thought you already knew. And by that metric, the concept of 'disruption' as I have outlined it in this Chapter is a great idea. It has caused me to rethink much of what I had known about the history of technology and to investigate new areas that previously did not seem particularly interesting to me when I first started looking at them. I hope that in next few chapters I can help you do the same and, in the process, gain some insight as to where technology and our society might be headed as a result.

The idea behind this book is simple: if we wish to solve all of the greatest problems ailing the world as quickly as possible, then we should study the things that have rapidly changed the world in the past - both their beneficial effects and the harmful ones - and understand them as well as we can.

Let's go back to the beginning.

Chapter 2:
Early History

◆ ◆ ◆

"We can learn about our future from our past because, regardless of technology or the speed of innovation, people are still people."
– Simon Sinek, author

◆ ◆ ◆

Disruption is as old as fire, maybe older. But if you listen to many technologists, forecasters, and business people, they make it sound as if rapid and dramatic changes in technology are a feature only of the modern world, particularly the part centered around San Francisco Bay. The flip-style cellphone replaced the landline phone. The smartphone then replaced the flip phone. Digital cameras replaced film cameras. Computers replaced typewriters. Streaming music services replaced digital downloads which replaced compact discs which replaced cassette tapes which replaced eight-tracks which replaced vinyl records. These people often do not talk about technologies much older than those ones, except perhaps as far back as the car disrupting the horse.

Historical times were periods of long stability, they argue. A son often had the same occupation as his father (farmer, most likely), used the same basic tools and techniques, and lived the same way as him, even on the same piece of land. The father's father had lived the same way and the son's sons would too. Everything is different now, they claim, since today we live in an Age of Disruption.

These people are right, but not quite. There of course have been long eras of technological stability, and there are tools from ancient eras that look and work exactly as their modern day equivalents, but dramatic technology driven disruptions are also nothing new. "I've looked at technologies going back 10,000 years," Tony Seba says, and what he has seen is what ecologists call 'punctuated equilibrium'. There can be periods of stability that last for a very long time, but when a dramatically better technology or idea comes along, the change to the new way of doing things happens quickly, and the world becomes different from how it was before the disruption. In fact, it is these relatively short periods of rapid change that divide history into 'eras' and 'ages' in the first place.

Cooking, Without Fire and With

> *Gorillas spend six or more hours a day eating. For humans, that is only about one hour per day, because we cook our food. This disruption unlocked one-third of our waking day for other, more-productive tasks*

Our technology shapes how we live and how we live has shaped our bodies. The first episode of a 2021 four-part *BBC Sounds* series by journalist Justin Rowlatt called 'A Pyrotechnic History of Humanity' concerns the invention of fire. Dr. Richard Wrangham, a professor at Harvard University, says that cooking, starting with *Homo erectus*, "completely transformed our biology".

As a result of cooking our food, we have much smaller guts than other primates, so our rib cages do not flare out at the bottom like apes' do. Spending less energy digesting food means more for brain development, which means that our heads are larger at birth. And cooking our food instead of eating it raw also simply meant less time chewing.

Cooking Food Takes Less Effort and Requires Much Less Time

Essentially, cooking moves part of digestion outside of the body. But cooking also simply softens food, making it physically easier to chew and swallow. The softening of food by itself leads to a radically faster and better way of eating. "Humans devote between a fifth and a tenth as much time to chewing as do the great apes", wrote Dr. Wrangham in his 2009 book *Catching Fire: How Cooking Made Us Human*. Wrangham said in his interview with *BBC Sounds*, "that's how big an effect cooking has in softening the food and just making it easier to chew."

Famed chef Ferran Adrià's book *The Origins of Cooking: Paleolithic and Neolithic Cooking* explores the earliest ways hominids discovered to prepare food, from roasting to boiling. "*Origins* is adamant," a *Bloomberg* news article about the book reads, "that, in terms of cuisine, the raw and the cooked is a false dichotomy. Adrià ... has come up with a third term: the 'unraw' - food that's been physically transformed without heat." Identifying and selecting naturally fermenting fruits, soaking foods in water and slicing them into smaller, more easily eaten bits all constitute some of the ways that we manipulated and modified our food to make digestion easier for ourselves, both in the amount of energy used by our bodies for digesting, and the amount of time and effort devoted to chewing.

Processing Food Expanded the Range of Food Available to Us

Processing food, in all its forms, did not result in us eating the same things in just a softer format. Processing makes some foods safer, not only meats but also some kinds of plants, increasing the quantity of food available to us and also expanding the range of our diets and our own range of travel as we "went out to the savannah to hunt", as Rowlatt said.

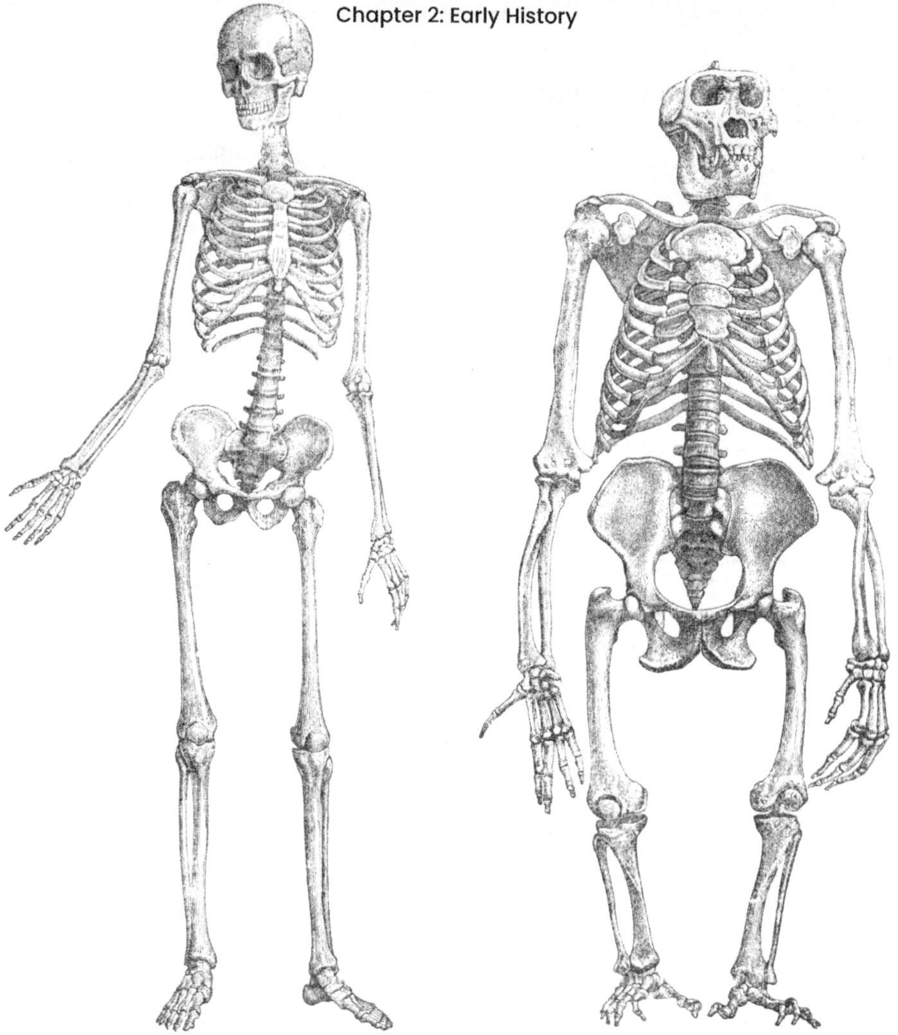

Figure 17 – Human and gorilla skeletons. The gorilla's ribcage is flared at the bottom compared to the human's ribcage to hold a much larger intestinal tract, as required by a roughage-based diet. We also have larger skulls and more-rigid feet, among other differences

This had a profound effect on our brains. Each increase in the number of neurons increases a brain's demand for energy linearly. That is, a brain with 10% more neurons requires about 10% more energy. But the *connectivity* of neurons, and thus the processing capability of the brain, increases exponentially. A brain with 10% more neurons is more than 10% smarter than a smaller one.

This created a self-reinforcing feedback loop: the smarter we became, the better we got at acquiring sustenance, which made us even smarter. Fire obviously has many other applications besides cooking. We have used it to heat our bodies, keep predators at bay, deliberately burn the landscape to flush out animals and to modify our stone tools, among other things. Fire even eventually led to the discovery of the smelting of metal, which replaced stone tools. But the mere act of softening food with fire was a disruption as profound as any other.

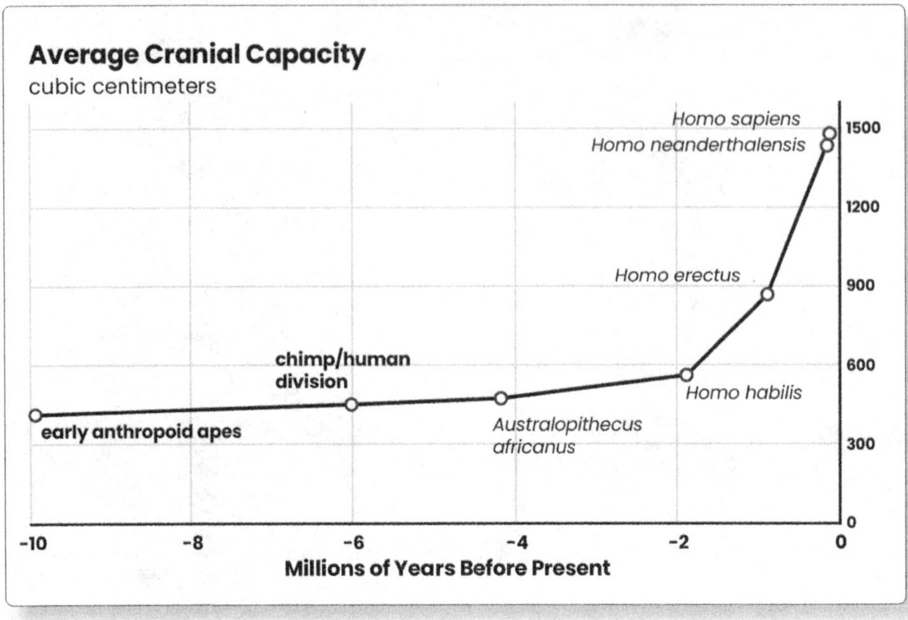

Figure 18 – Average Brain Size of Hominins over time. *(Wilson, 2021)*

In an interview with *Smithsonian Magazine*, Dr. Briana Pobiner, a research scientist and educator at the Smithsonian National Museum of Natural History, said "by about two and a half million years ago, early humans started to occasionally eat meat. By about 2 million years ago, this happened more regularly. By probably about a million and a half years ago, humans started to get the better parts of animals. They shifted from just scavenging the leftovers to maybe getting earlier access to carcasses."

"It really was the invention of technology and stone tools [that made meat-eating possible] — like using rounded rocks to bash open bones to get at the marrow inside and sharp-edged rocks to slice meat off bones." (Joosse, 2021)

Disruption is in Your Bones

Disruption has literally shaped your skeleton. Cooking foods to soften them resulted in smaller guts and so smaller rib cages. Eating meat contributed to a larger brain and thus a larger skull volume. Dairy products and other soft foods led to an overbite, affecting which sounds you can make. And walking as a biped meant narrower hips, which resulted in children being born at an earlier stage in their development, and thus helpless after birth for far longer than other species. (To accommodate an early birth, even our shoulder bones develop less prior to birth relative to other primates.) The greater parenting burden humans thus have relative to other animals drove the need for social development and social structure, which fortunately could be supported by our larger brains. The entire early history of humanity is written in your bones.

Put those opposable thumbs of yours into your armpits and then trace your magnificent hands slowly down the sides of your rib cage. Note that it is relatively straight instead of flared out at the bottom, as a gorilla's rib cage is. (If you have any doubts, go find a gorilla and tickle his ribs.) Now put your hands around your head and feel how voluminous it is, how large the region behind your forehead. Gorillas don't have much of that in comparison to us. What you lack in rib flare, you make up for with prefrontal cortex.

Now put your hands on your hip bones. They too are much narrower than a gorilla's, making walking bipedally easier. In his book *First Steps: How Upright Walking Made Us Human*, author and Dartmouth College professor Dr. Jeremy DeSilva claims there is evidence of terrestrial bipedalism in human ancestors dating back six and maybe even seven million years. "Bipedalism is the most ancient human adaptation. It's what happened first," DeSilva said in a 2022 interview with Wisconsin Public Radio. "It set the stage for all the things that happened later, from our language and our dietary flexibility, to our reliance on technology and tools, to the way we give birth and raise our children. All these things really have their roots in how we move." The issue of why our ancestors began walking upright remains a matter of scientific debate. Perhaps it was to free the hands to collect and carry food, although it was likely not related to toolmaking as we know bipedalism came millions of years before the first evidence we have of tools, even ones made of wood or bone. Biologist Richard Dawkins suggests that walking on two legs might have been the first 'meme' – that is, it was not done for any specific purpose other than as a form of showing off.

This was more than just a behavioral change, it reshaped our feet, knees, and lower backs. A December 2021 article in the *Journal of Human Evolution* by Alisha Anaya of Duke University and co- workers found that "a rigid lateral foot may have evolved as early as" 3.2 million years ago. A 2021 article in the journal *Evolutionary Biology* by Dr. Scott Williams of the Department of Anthropology of New York University showed that a specimen of *Australopithecus sediba* from about two million years ago "possessed a lower back consistent with lumbar lordosis and other adaptations to bipedalism." Less flexibility in the feet and knees makes climbing in trees more difficult, but it makes walking upright on flat ground far easier. In a study of movement in humans and chimpanzees, Dr. Michael Sockol of the University of California at Davis found that human bipedal locomotion has roughly one-quarter the energy cost of either bipedal or quadrupedal locomotion in chimpanzees. That is, walking on two legs is roughly four times as efficient for us as for other primates whose greater limb flexibility, and lack of the angulation in the toe that gives us a little 'spring' in our step, requires them to expend much more energy to move the same distance.

Being three or four times as efficient at walking has enormous implications for gathering resources. If one organism is three or four times as efficient at walking than another, then the area of a circle it can reach for a given expenditure of energy is nine to sixteen times larger. Roughly speaking, the earliest bipedal humans had access *to ten times* as much land as chimps, allowing us to live in savannahs and other low-food-density places they cannot. A 10X disruption! (This is perhaps part of the reason why humans have ten times the density of sweat glands compared to chimpanzees and macaques, allowing us to better regulate our body temperature.)

It is not just in walking that we are several times more efficient than primates. In chewing too, due to cooking and eating softer, less-fibrous foods. In her book *Four Lost Cities: A Secret History of the Urban Age*, Annalee Newitz says, "Domestication appears to be a self-reinforcing process: humans were drawn to domestic foods [like milk, grain, sheep, and goat], which transformed our bodies, and over time our bodies were better suited to those foods, which made them all the more alluring."

Now feel your jaw and how your teeth align. Milk may have played a role in that, and in our ability to make new sounds. Foods from domesticated plants and animals, Newitz writes, "made human faces more delicate and thinned out our body hair. Our jaws became shorter and more rounded, which may have allowed new sounds to enter our languages. Specifically, the "v" and "f" sounds, produced when our top teeth press against our lower lips, are only possible in mouths where the lower jaw bites behind the upper."

All these things combined – walking upright, using tools, making fire, cooking food, and getting more protein from meat and dairy – resulted in ever larger brains, and larger brains equal larger societies. In a now-famous 1992 research paper also published in the *Journal of Human Evolution*, British anthropologist Robin Dunbar studied 38 genera (the biological classification that can contain multiple closely related species) of primates and found that "the ratio of neocortex volume to the rest of the brain is the best predictor of group size". That is, the larger our brain, the more combinations of interpersonal relationships we can understand, and therefore the larger a group we can live in. Humans, having the largest brains of all primates, also live in the largest groups.

As Mark Moffett wrote in the 2019 book *The Human Swarm: How Our Societies Arise, Thrive, and Fall*, "societies had already reached populations of several hundred individuals by the time the genus *Homo* emerged 2.8 million years ago – when the human diet first begun to include more flesh." And that innovation, living in large groups, driven by larger brains and more eating of meat, might have been our most-profound innovation ever. DeSilva recounts that famed anthropologist Richard Leakey and his team in Kenya in the 1970s found a hominin upper leg bone that contained a new feature never seen before in bones – a healed fracture. To fracture an upper leg bone and not die from that injury, to be able to rest long enough for it to heal while being cared for by friends and family, is realistically the first glimmer we have of what we could call humanity. And if you have ever experienced a bone fracture and had it heal, then that part of humanity is written in your bones as well.

When the Sun Set on European Neanderthals

> *The population of Neanderthals in Europe collapsed over just a few thousand years, starting about 45,000 years ago. Computer simulations suggest that an invasive species – us – may have been at least four times better at acquiring resources as them*

Neanderthals *(Homo neanderthalensis)* were the main hominin species in Europe for hundreds of thousands of years, but they disappeared over a period of just a few thousand years. Were they wiped out by competition with us, interbreeding with us, climate change, a pandemic, volcanoes? Did we cause their decline, or did their decline simply enable us to move into areas they formerly occupied? This topic is hotly debated among scientists.

It is not only *what* killed off the Neanderthal that is uncertain, but also *when* they died out. Recent findings from a cave in Belgium indicate that the last of their species lived in the area around 40,000 years ago.

Dr. Axel Timmermann, a professor at Pusan National University in South Korea, thinks he might have the answer. As he described in a peer-reviewed scientific paper, Dr. Timmermann simulated the spread of an invasive species (us) into lands populated by Neanderthal over the past one hundred thousand years. Over this time scale, the climate of Europe changed dramatically, with changing regional plant productivity and ice sheet extent, even the coastline of Europe changed. These things are all included in his calculations.

He divided up the land into a checkerboard-like grid and at each location *Homo sapiens* and Neanderthal competed for food and other resources. But we did not just compete, we also cooperated. We even interbred. (It is well-established that *Homo sapiens* have some Neanderthal DNA. But also teeth found on the island of Jersey, in the English Channel between Great Britain and mainland France, dated to less than 48,000 years, right at the time *Homo sapiens* and Neanderthal were both present in Europe, suggest "a possible shared Neanderthal and *Homo sapiens* ancestry", according to a recent scientific study.) (Compton, 2021) Dr. Timmermann conducted a series of analyses to test parameters that represented how good each species was at harvesting resources, how quickly we spread to new areas, how intensely we competed or cooperated, and other factors, finding a sudden S-shaped collapse of the Neanderthal population starting about 45,000 years ago, as the Homo sapiens population rose, just as scientific evidence suggests.

"A realistic extinction of the Neanderthal population can only be simulated when *Homo sapiens* is chosen to be considerably more effective in exploiting scarce glacial food resources as compared to Neanderthals," Dr. Timmermann wrote. By 'considerably more effective', his simulations suggest that we were at least *four times better* at acquiring resources as them and that each parcel of land could support about twice as many of us as Neanderthals. We might have likely killed off the Neanderthal through "differences in hunting strategies, social behavior, technological and physiological characteristics." For example, recent discoveries in a cave in France from 54,000 years ago by Dr. Laure Metz of the University of Connecticut and co-workers suggest that *Homo sapiens* were using bows and arrows at that time, the oldest evidence of this technology outside Africa. According to Dr. Timmermann, "climatic effects alone would not have been sufficient to cause a [Neanderthal] extinction". In other words, it wasn't the climate that did Neanderthals in, it was us.

The graph below shows Dr. Timmermann's findings for the population density of *Homo sapiens* and Neanderthals in central Europe (the area corresponding roughly to modern Germany). The number of Neanderthal per area is roughly stable until *Homo sapiens* began to grow dramatically starting around 45,000 years ago.

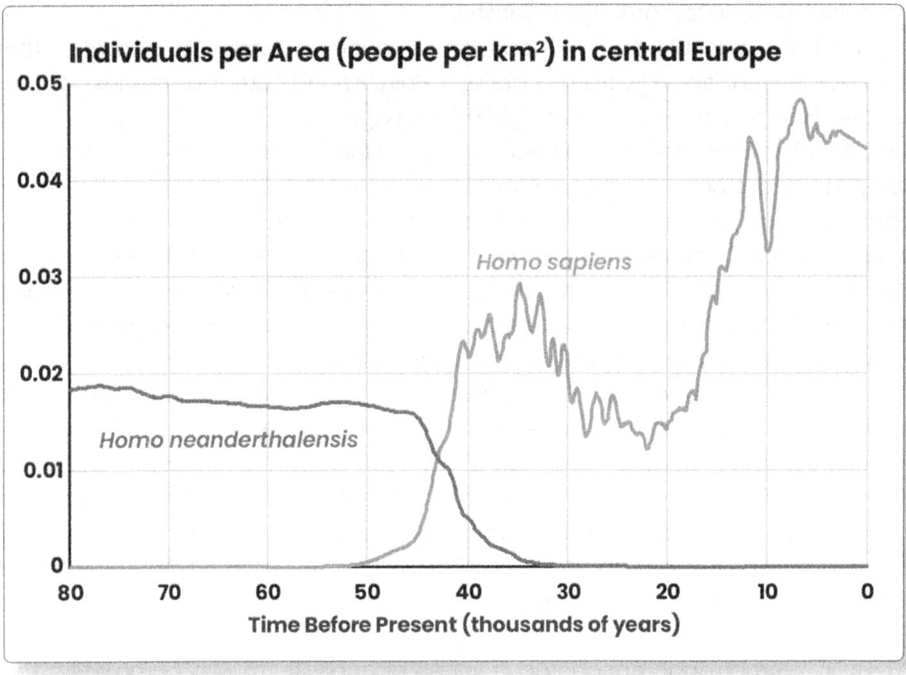

Figure 19 – Population density of Neanderthals and Homo sapiens in central Europe.
(Timmerman, 2020)

Within two thousand years, the number of *Homo sapiens* had surpassed the number of Neanderthal and, by two thousand years after that, the number of *Homo sapiens* surpassed the highest level Neanderthals had seen in at least tens of thousands of years. That is, this was not a one-for-one disruption. By 10,000 years ago, the number of *Homo sapiens* was more than double the number that Neanderthal had ever been and, obviously, in the past couple of hundred years has grown dramatically beyond that.

Social Learning: Animals Disrupt Their Behavior Too

> *It is not just humans who develop and share innovative tools and rapidly disrupt their behavior, animals do too*

In 2011, members of a group of thirty chimpanzees in Uganda were seen using new techniques for drinking water: dipping moss or leaves into divots to soak up water that was otherwise difficult to scoop by hand and then putting the moss into their mouths to suck the water out. These behaviors spread throughout much of the troop over a period of just five days. (Hobaiter, 2014)

But these were not chimpanzees in a zoo they were wild. No human had taught them to do this. They developed these techniques themselves. It was the first time ever that wild chimpanzees had been seen inventing a new tool and then 'culturally transmitting' its usage, without any outside influence. For the study, a team of scientists from universities in Canada, the UK, and Switzerland took video of the entire sequence of events, so that the scientists know not only which chimps engaged in the new behavior, but also who in the group they each learned it from, and which yet-to-be-converted chimps were near them at the time.

Innovation in behavior has also been noted in whales. An analysis published in 2021 of American whaling ship logbooks from the mid-1800s in the north Pacific Ocean found that "the rate at which whalers succeeded in harpooning whales fell by about 58% over the first few years of exploitation in a region." (Whitehead, 2021)

The scientists' rigorous analysis showed that this decline could not be explained by a change in the competence level of the whalers, nor by the initial killing of particularly vulnerable individuals. "It appears", the authors wrote, "that whales swiftly learned effective defensive behavior. The whalers themselves wrote of defensive methods that they believed the whales were adopting, including communicating danger within the social group, fleeing - especially upwind - or attacking the whalers."

Astonishingly, these behaviors were also adopted by whale pods that the whalers had not previously encountered. That is, the scientists think that the attacks changed the behavior of groups who were not yet attacked 'through echolocation and communication' or by whales from the attacked groups joining other whale pods. "Social learning of defensive measures between social units is the best-supported explanation for the rapid decline in strike rate following the first sperm whale sighting within a region," they concluded.

But one of the best-known examples of the rapid spread of a new behavior throughout a population of animals was observed in the UK. Starting in 1921, tits (a kind of small bird) in Swaythling, in the far southern part of England, were first seen opening the tops of milk bottles delivered to peoples' doorsteps to get at the cream inside. Within twenty years, the behavior had spread throughout the country.

The examples so far in this chapter, dealing with cooking, Neanderthals, and animals, show that rapid and dramatic innovations in technology and behavior are not limited to our own species nor to the present time. Hominin species who lived before us invented, and taught each other about, stone tools and fire. These were not the inventions of *Homo sapiens*, and we have no right to claim them as our own. Hominins who lived alongside us, like the Neanderthal, made glue, string, clothing, and perhaps art and flutes. They stored food and, like us, they cared for injured individuals and buried their dead. And the studies described above show at least chimps, sperm whales, and birds engage in and spread the use of new behaviors in ways similar to the ways we do.

Admittedly, the animal examples pale in comparison to our own. The ape, whale, and bird cases do not seem to represent 10X disruptions compared to previous behavior. But they do point to the tantalizing idea that disruption is a feature of sentient behavior that transcends us. If intelligent aliens were to ever come to Earth and we had to find a way to communicate with them, we might try to do so using mathematics, since they would have likely discovered many of the same properties of numbers that we have discovered. And we might try to communicate with them about chemistry, since they would likely be comprised of the same basic chemical elements we know about. And if we were to find a way to do so, we might also try to show them what we know about technology disruptions, since it is possible that they have experienced the same pattern of disruption themselves.

The Disruption of Wild Animals by People and Our Livestock

> *96% of the mass of mammals on Earth are humans and our livestock. More than half of birds are chickens. As far as wild animals go, the disruption is already over, and they have lost*

Homo sapiens have been disrupters from the beginning. In a 2018 study published in the *Proceedings of the National Academy of Sciences* (PNAS), a team of scientists from the Weizmann Institute of Science in Israel and the California Institute of Technology estimated, among other things, the total mass of all mammals on Earth now and also the 'pre-human biomass', the mass before humans had a major impact on the diversity of life. (Y. M. Bar-On, 2018)

They found that the mass of all mammals on Earth today is more than four times as large as it was 100,000 years ago. However, wild mammals are now only about 4% of the mass of all the mammals on Earth, while people are about 36% and our livestock about 60%. So, while the mass of all mammals quadrupled, the mass of wild mammals 100,000 years ago was about six times larger than it is today. Humans and our livestock went from essentially 0% of all the mammals on Earth to 96% now.

Hannah Ritchie of the website *Our World in Data* wrote, "it wasn't just that we lost a lot of mammals. It was almost exclusively the world's largest mammals that vanished... Travel back 100,000 years and the planet was rich with a wide array of wild mammals. Mammoths roamed across North America; lions across Europe; 200-kilogram wombats in Australasia; and the ground sloth lounged around South America. They're now gone."

Ritchie describes how the mass of land mammals declined by about 25% in the period from 100,000 years ago to 10,000 years ago, about the time that the current period of relatively warm, stable climactic conditions known as the Holocene period started and about the time that we domesticated the first livestock, gave up hunter-gatherer lifestyles, began agriculture and settled down into semi-permanent settlements. So, it is only in the last 10,000 years that people and our livestock have come to dominate the mass of mammals on Earth.

The benefits of livestock in a warm, stable climate are obvious. From them we get not only the 'primary' products like meat, bone, hide and blood (those things that involve killing or harming the animal to obtain) but also 'secondary' products, like wool, dried dung for fuel, muscle power, and perhaps most importantly, milk. Recent studies have found milk proteins in the dental calculus encrusted on the teeth of people who lived in Britain at least six thousand years ago and also in eastern Africa from about the same time. (Charlton, 2019) (Bleasdale, 2021)

In the study of ancient east Africans there is even evidence of the consumption of dairy products before people developed the genetic mutation that allowed some people to continue into adulthood to digest lactose, the sugar found in milk, perhaps by making cheese or by fermentation into yogurt, both of which reduce the lactose content. In some parts of the world, milk is fermented into alcoholic beverages, which also reduces the amount of sugar.

Young mammals can digest milk sugar, though this ability generally does not persist into adulthood. But genetic mutations in some people, especially in northwest Europe, though found in many other parts of the world as well, allow them to digest milk sugar throughout their lives. Obviously, this can provide an enormous survival advantage to people who have access to milk from domesticated animals. In 'The Documentary Podcast' by the *BBC World Service,* Mark Thomas, a professor of evolutionary genetics at University College of London, says that the ability to digest milk sugar is "the most-strongly selected single gene trait that has evolved in humans in the past 10,000 years".

A scientific study of the genes of people who died at a Bronze Age battle near the river Tollense in Germany in around 1200 BCE found that only 7.1% of the warriors had a genetic variant that enabled them to break down lactose. "When we look at other European genetic data from the early Medieval period less than 2000 years later, we find that more than 60 percent of individuals had the ability to drink milk as adults, close to what we observe in modern Central European countries, which ranges from 70 to 90 percent" said Krishna Veeramah of Stony Brook University, one of the study's participants. "This is actually an incredibly fast rate of change for the gene that controls milk digestion." (Burger, 2020)

There is serious scientific consideration that the switch from wild-gathered foods to foods from domesticated plants and animals literally reshaped our heads and our languages. Linguist Charles Hockett noted in the 1980s that the languages of hunter-gatherer societies involve fewer labiodental sounds, those that involve touching the lower lip to the upper teeth, like 'v' and 'f'. "On average," scientists from the University of Zurich in Switzerland wrote in a 2019 study published in the journal Science, "hunter-gatherer societies have only about 27% the number of labiodentals exhibited by food-producing societies." (Blasi, 2019)

The scientists believe that the softer foods like cheeses and those made from ground grains caused a shift in the human population to favor people with an overbite, as opposed to edge-to-edge alignment of the top and bottom teeth, and having a slight overbite makes production of labiodental sounds easier. As a result, the "probability of labiodental sounds" in Indo-European languages rose "from a median estimate of about 3% in the proto-language (6000 to 8000 years ago)" to over half of the languages spoken around the world today and 76% in Indo-European languages.

From Stone Tools to Metal

> *Good tool-quality stone, like any limited resource like gold, tin, or petroleum, is not evenly distributed around the world. Scientific research suggests that stone tools changed suddenly, like modern-day technologies do, and that the decline of incumbent industries had many of the same characteristics of the declines in modern incumbents*

There was not one Stone Age but several. The oldest known flaked stone tools come from Kenya and date to 3.4 million years ago. (Pargeter, 2019) In the 1998 textbook *People of the Earth: An Introduction to World Prehistory, 9th Edition*, author Brian Fagan estimates that flint worked in the way of the oldest tools would only produce about three inches of cutting edge per pound (or about 17 centimeters per kilogram). From about 1.7 million years to 100,000 years ago, the Acheulean period, we made pear-shaped tools which were flaked on opposite sides, yielding about one foot of cutting edge per pound of flint (or 66 cm per kg). And only more recently, in the Neolithic era, did we make stone-tipped spears, bows and arrows, and other more sophisticated tools and weapons, some of which used birch tar pitch to glue stone parts to wooden ones.

"The Neanderthals were far more efficient stone artisans than their predecessors," Fagan wrote. The flaked tools they made resulted in about thirty inches of cutting edge per pound (168 cm per kg). By the late Paleolithic (starting 40,000 years ago, when *Homo sapiens* replaced Neanderthals in Europe), our blades yielded thirty *feet* of cutting edge per pound (20 meters per kg).

That is, the pear-shaped tools *Homo erectus* made had 4X the cutting edge per pound of material compared to the simple 'Oldowan' tools *Homo habilis* made. The Neanderthals made flaked tools with 2.5X the cutting edge per pound of Acheulean tools. And *Homo sapiens* made tools that had 12X as much cutting edge as those made by the Neanderthal.

Actually, the Stone Age Did End Because We Ran Out of Stones

Former Saudi oil minister Sheikh Ahmed Zaki Yamani is quoted as having said about our modern world's use of petroleum that "the Stone Age came to an end not for a lack of stones, and the Oil Age will end, but not for a lack of oil." (Ratcliffe, 2016) That is, the Stone Age did not end because we ran out of stones. But, in fact, there is good evidence to suggest that a lack (per person) of good-quality stone was a key factor in the end of last era of the Stone Age and a driving force in the search for better materials for making tools.

Not every type of stone is equally common in the world nor equally good for making tools. Stone toolmakers favored materials like flint and obsidian, chert, quartz, jasper, chalcedony, and some others. Flint, found in chalky areas and along riverbeds, was the workhorse of the Stone Age, yielding an edge that is both sharp and durable. Obsidian, which author Matt Ridley, a former member of the UK House of Lords, calls "the 'rich man's flint' of the stone-age world ... was once one of humankind's most sought-after materials." While flint is a sedimentary rock, obsidian is a form of volcanic glass which fragments into extraordinarily sharp blades that kept it in high demand even after the first metal tools were introduced. (In fact, according to journalist Peter Shadbolt, some surgeons still use these stone blades to carry out fine incisions. "Obsidian - a type of volcanic glass - can produce cutting edges many times finer than even the best steel scalpels.")

And not every *size* of stone is equally good for making tools. Stoneworking is reductive, which means that stone tools are essentially a 'flow', like a conveyor belt that moves only in one direction: from larger in size to smaller. A large stone can be chipped and flaked to reveal a cutting edge, but as that edge gets dulled from use, the stone must be chipped and flaked again to reveal a fresh edge. Over time, the stone gets smaller until it is finally discarded. This means that nomadic Stone Age hunter-gatherers needed to move through or live in places with access to a lot of good-quality stone, or trade with people who did.

But the invention of agriculture put an end to that way of life. Suddenly (relatively speaking), people were living at a much higher density than ever before and remaining in the same area. Professor Vaclav Smil in his 2004 book *Enriching the Earth: Fritz Haber, Carl Bosch, and the Transformation of World Food Production* wrote that even "the earliest practitioners of the least intensive form of agriculture... could support population densities between 0.1 and 0.6 people/ha [about 25 to 150 people per square mile] — an order of magnitude higher than even the relatively most affluent hunting and gathering societies".

A high density of people, and the need for specialized farming tools that required large amounts of cutting edge, like sickles and threshing boards, meant large local demand for stones that were fundamentally unevenly distributed and difficult to move long distances. What the earliest farmers really needed were 'recyclable rocks' — material that could be melted down when it broke and sharpened when it got dull. That is exactly what metal provided.

Unlike stone, refined metal is a 'stockpile'. Once extracted from the ground and purified, it can be used over and over again. If a bronze axehead splits during use, it can be re-cast as good as new, or even better. An axehead can be melted down and made into knives, or knives can be melted together to make a sword. Using and re-using the metal does not use it up (for the most part).

But, for the millions of years that the Stone Ages lasted, people saw many of the same sorts of technology disruptions that we have seen since. "One of the oldest manufacturing sites is Olorgesailie, Kenya," wrote Christoph Roser in *Faster, Better, Cheaper in the History of Manufacturing.* Stone "was carried from two different mountains about 20 km [12 miles] apart to a manufacturing site between these two mountains. The production of an Acheulean tool easily took hours. It also appears that the manufacturing site was organized. For example, based on stone fragments there seems to have been one area for making new axes and another for sharpening worn-out axes. Hence, this site may be one of the first examples of specialization, where multiple individuals focused on different tasks. In any case, the site was in operation for 1 million years, making it easily the longest-running manufacturing site in the world." Perhaps the moral of the story here is that even one million continuous years in business is no protection against disruption. And the Stone Age was full of disruptions.

Experimental archaeologist Dr. James Dilley, who makes replica prehistoric artifacts, noted that during the Neolithic Era in Britain a certain style of arrow was used and then that style disappeared as a new design emerged, with little overlap in usage between the two styles. For example, 'leaf' type arrow heads were used from approximately 4000-3400 BCE and 'chisel' style arrow heads from approximately 3400-2600 BCE. Revolutions like these in arrowhead design happened multiple times in Britain.

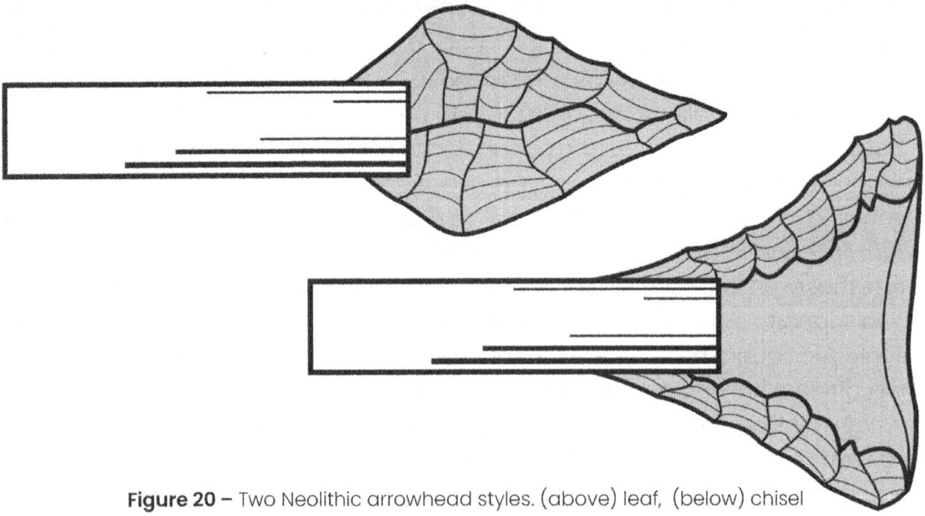

Figure 20 – Two Neolithic arrowhead styles. (above) leaf, (below) chisel

The 'leaf' arrowhead was mounted on the arrow shaft so that its single sharp tip faced toward the prey. But some Neolithic genius realized that by turning a triangular arrowhead around and burying its tip into the shaft of the arrow, a sharp edge of the stone would face the prey and do much more damage when it struck its target.

Dr. Dilley does not just make replica stone tools and weapons, he tests them too. In one of his experiments Dr. Dilley fired a chisel-head arrow he had made into a block of ballistics gel, a material that realistically mimics flesh. He found that the 'leaf' arrowhead makes a wound that tends to close around the shaft of the arrow, reducing blood loss. The wide chisel-shaped tip, on the other hand, opens a channel around the arrow shaft, holding the wound open and inducing what would clearly be devastating blood loss in any living creature it was used against. Additionally, the two backward-facing edges of the chisel-shaped arrowhead themselves are very sharp, so removing the arrow does additional damage, exacerbating an already bad situation.

Not only is the chisel-shaped tip more dangerous, it is also easier to make. Dr. Dilley said, "leaf arrowheads require more platform prep as general flaking. At least double the time investment." So, twice as much time to make an arrowhead that does not work as well. No wonder the switch was rapid and total.

For reasons we do not fully understand, the predominant design of arrowhead remained stable for centuries, and then was quickly replaced by a different design. If specific arrowhead designs were used for certain prey, we might expect the design to have stayed constant, or a variety of arrow heads to be produced contemporaneously for different purposes, but this is not the case. And this is not just true in Britain, but elsewhere in the world. Similar arrowhead sequences are seen, according to Dr. Dilley, throughout Europe and in Africa as well.

Not only did technologies rapidly replace each other in the Neolithic age just as they do today, but incumbent industries collapsed in similar ways, too.

A scientific report published in 2002 by archaeologists from universities in Tehran and Bradford (UK) looked at sites in Iran that were inhabited in the late Stone Age and early in the age of metals. They found that, as metalworking became widespread in northern Iran after 4500 BCE, the production system of stone tools throughout the region collapsed. That part is not so surprising. What is surprising is that this collapse appears to have many of the same characteristics as the collapse of modern incumbent industries. (Fazeli, 2002)

Northern Iran was probably not the first place in the world to make copper tools. Evidence of earlier copperworking has been discovered at Tell Halaf (Syria), Çatalhöyük (Türkiye), Jarmo (Mesopotamia), and other places. But this study is interesting because the researchers were able to look at several sites that are physically close to each other.

At a site called Fakrabad, south of Tehran, which was occupied only from about 4800 - 4500 BCE, more than 60% of the stones the scientists found were a type of rock called grey chert. Few source stones were found there, leading the scientists to write, "it is hypothesized that grey chert is being processed at another locality and transported mostly as prepared cores to Fakrabad."

But at Chakmak Tepe, about 50 miles (80 km) away and occupied only after 4500 BCE, less than 10% of the stones found were grey chert. More than 50% were red chert. (At Fakrabad, less than 3% of the stones were red chert.) "At Chakmak Tepe," they write, "there is much evidence for the preparation and processing of cores as well as the manufacture and use of formal tools." They also found obsidian, which is "most likely not local", and which was only found at Chakmak Tepe and only from after 4500 BCE, not in any other place they looked and not from any other time.

So, comparing Fakrabad to Chakmak Tepe, we see "a dramatic shift from the use of good quality material" at the earlier site, in the words of the Iranian and British scientists, "to use of primarily poor quality local material and limited quantities of very high quality imported material" at the later site. In Tony Seba's parlance, this sounds a lot like a 'disruption from above': Just as Tesla electric cars initially competed against high-end luxury cars, metal tools first displaced tools made from high-quality stone. Only very-high-quality obsidian, which was not found in the region prior to this, was able to compete with metal.

As a result, "the lithic production system, including the role of local lithic craft specialists, appears to collapse", the scientists write. "Instead, it appears that poor quality local stone is procured and manufactured in tools as a part-time activity primarily within the domestic household. This suggests that there may be an increasing division of the population between elites who have access to quality goods, and subordinate groups who find that they must depend on themselves to produce their required technological equipment."

So, the introduction of metal did not just change the kinds of stone used in making stone tools, it also changed where people lived, who made tools, and how tools were made.

Bronze and its Daughters

> *Making metal alloys, and long-distance trading to get the raw materials, redirected the course of humanity and unlocked powerful new technologies like the potter's wheel and the wagon*

Only a few metals can be found in essentially their pure forms or obtained by heating rocks that contain them: including gold, silver, copper, and tin. Even lead (though rare) and iron-rich meteorites (even rarer) can also be found sometimes by themselves. It is easy to understand why for much of ancient times these metals were largely used for decorative purposes, as all except meteoric iron are too soft to hold a good edge for long.

In the Near East about 3500 BCE, it was discovered that mixing a small amount of impurities into copper made a superior metal, bronze, the first manufactured alloy of metals that was not found in nature. It had a color like copper, but held a sharp edge better and melted at a lower temperature, so it could be more easily cast (and re-cast) into a variety of shapes. The native people of the Lake Superior region of North America also made copper tools, but had abandoned the practice by 1000 BCE. Professor Michelle Bebber of Kent State University, writing in the scientific journal *Nature*, believes that "the unusual level of purity in the North American copper" negatively affected the performance of tools made from that material. "After an equal number of blunting events," her experiments found, "copper and stone knives possess the same sharpness." Simply put, copper that is *too pure* yields a tool that quickly becomes no better than tool made from chert, a common toolstone, even though the copper material takes much more effort to mine, smelt, and sharpen in the first place.

Adding just a few percent of impurities, people in the Middle East discovered, made the metal both harder than pure copper tools, and easier melt. According to *Perry's Chemical Engineers' Handbook*, for bronze made with ten percent tin, the yield strength, that is the maximum stress a material can withstand before permanently bending or deforming, is more than twice as high as for pure copper. As well, this harder metal is easier to melt. Energy radiates away from a hot material at a rate that depends on the *fourth* power of the temperature (the temperature multiplied by itself four times). Metal at the melting temperature of bronze (1740°F / 950°C) loses heat at only about two-thirds the rate of metal at the melting temperature of pure copper (1985°F / 1085°C). This means that, realistically speaking, metalsmiths working with bronze can make a product that is more than twice as hard as pure copper, using only about two-thirds of the fuel.

It is understandable why the possession of bronze, knowledge of how to make it, and control of trade routes for obtaining its raw materials became the basis of new empires. British academic Carol Bell, quoted in Eric Cline's book *1177BC: The Year Civilization Collapsed*, says that "the strategic importance of tin" in the Late Bronze Age "was probably not far different from that of crude oil today." At that time, tin was available in quantity only from Afghanistan, Cornwall in the UK, and a few other places. Bell continues, "the availability of enough tin to produce ... weapons grade bronze must have exercised the minds of the Great King in Hattusa and the Pharaoh in Thebes in the same way that supplying gasoline to the American SUV driver at reasonable cost preoccupies an American President today!"

Metal Tools Opened New Possibilities

The invention of bronze casting did not just result in the replacement of stone tools with sharper and more durable metal ones on a one-for-one basis. It also opened new possibilities to make things that could not be made economically, or perhaps at all, with stone. Over just a couple of centuries around five and a half millennia ago, a cascade of interrelated technologies, perhaps all indirectly deriving from the invention of casting bronze tools, revolutionized life in Mesopotamia. In a comparatively brief time, people living in the land between the Tigris and Euphrates rivers invented or were introduced to the axle, which gave rise to both the foot-powered potter's wheel and the wagon, and also domesticated oxen, using them to pull plows. Oxford University professor Amy Bogaard believes that the use of the plow itself was about a 10X disruption. "A farmer with an ox team could prepare the same land for cultivation that would have required anywhere from 2–15 hoe-farmers", she wrote in the journal *Antiquity* in 2019. The late Oxford academic Andrew Sherratt championed the idea of a contemporaneous 'Secondary Products Revolution', where metal tools enabled us for the first time to not only use animals for meat, but also other outputs, "secondary products" like milk, wool, and labor.

Individually these were each disruptive innovations. But in combination they were instrumental to the rise of agriculture, a population boom, cities, income inequality, taxation, writing, and organized warfare as well.

◆ ◆ ◆

"The greatest shortcoming of the human race is our inability to understand exponential growth"
— Albert Bartlett, physicist

◆ ◆ ◆

Pottery, Fast and Slow

> *The foot-powered wheel made pottery ten times faster and easier than before. But, paradoxically, this resulted in a sharp drop in the quality of goods produced, as pottery went from a luxury for the rich to something almost disposable*

Here are two pots from Mesopotamia. The one on the left was made in the Ubaid Period, roughly seven thousand years ago (5000 BCE), while the one on the right is about two thousand years newer (3000 BCE).

Figure 21 – Two pots from Mesopotamia. The one on the left is from the Ubaid Period and is roughly two thousand years older than the one on the right, from the Uruk Period.

Though the two are similar in size and basic style, they are remarkably different in how they were produced. Ubaid Period pottery, on the left, was made on a 'slow wheel', a flat stone slab laid on top of another flat stone slab, where the top slab can be turned with one hand while shaping the clay with the other. The 'Fall of Civilizations' podcast episode 'The Sumerians - Fall of the First Cities' said: "the ancient pottery of the Ubaid Period was exceptionally beautiful... It was a luxury item for the select few."

But an invention by some ancient genius, the axle, dramatically changed that, and much else. The oldest axle ever discovered was not used for transportation, but instead as the heart of a potter's wheel. Pushing a rotating table by foot freed both hands to shape and style the clay, dramatically increasing a potter's productivity. The 'Fall of Civilizations' podcast continues: "Jars and pots could now be made in great numbers by workmen in intensive workshops. This was the first era of mass production." While only a lucky few could own the decorated Ubaid pot, pottery pieces in the Uruk Period became commonplace, and so they were typically undecorated and simple in design.

This had enormous implications for daily peoples' lives. Pottery is useful not only as cookware for food preparation, but also a means of storing, transporting, and preserving foods. Potting was used regularly up until the invention of canning in the early 1900s to store meat and vegetables, which can be cooked in a pot and then covered in clarified butter or fat, which solidify when they cool, resulting in food that is isolated from the air. Done properly, potted meat and fish can keep for months, even up to a year.

But how much faster was the foot-powered potter's wheel compared to the 'slow wheel' that preceded it? In the 1959 article 'The Potter's Wheel: An Analysis of Idea and Artifact in Invention', George Foster, himself quoting Harry Holbert Turney-High's 1949 book *General Anthropology*, says: "The invention of the potter's wheel enabled the artisan to make from *ten* to *twenty vessels* in the time formerly required to produce one, thus introducing mass production in ceramics. Pottery thenceforth became a cheap and easily manufactured commodity and the standard of living consequently rose for the bulk of the people." (italics added)

Foster's claim of an order-of-magnitude increase in the rate of pottery-making is supported by actual demonstrations. In one video by the 'Make it Primitive' channel on YouTube, it takes about six minutes to shape a small pot using a slow wheel. For comparison, in 2016, professional potter Joel Cherrico set the Guinness World Records title for "most pots thrown in one hour by an individual" by throwing 159 pots on a traditional Japanese kick-wheel, one every 24 seconds, or more than ten times faster than the slow wheel.

The two styles of pots, Cherrico's and the one from 'Make it Primitive', are not identical, even though they appear to be similar volume and use a similar amount of clay. The pot made on the slow wheel was more like a large cup without a handle, while Cherrico's pots had straight sides and a distinctive lip. And the person doing the 'slow wheel' demonstration did not seem to be working as quickly as possible, while Cherrico was clearly shaping clay at what looked like the fastest pace he could sustain for an hour. Nevertheless, it is reasonable to think that any experienced person working a foot-powered wheel at even half of Cherrico's speed must be at least several times more productive than a similarly skilled person with just a hand-turned wheel.

◆ ◆ ◆

"The flow of history is accelerated when two rivers converge."

- Walter Isaacson, The Code Breaker: Jennifer Doudna, Gene Editing, and the Future of the Human Race

◆ ◆ ◆

The Wagon, the First "Viral" Invention, May Have Disrupted Uruk

> *The rich farmland of Uruk was a great place to build the world's first city. But when metal tools enabled the wheeled wagon, the advantage of living on a muddy floodplain suddenly became a disadvantage*

Ribbons of flame turned into a towering column of thick, black smoke as they rose into the air from the center of the city of Uruk in what is now Iraq. Perhaps fifty thousand people lived there at the time and, though just a town by modern standards, five millennia ago when the fire happened, this was not only the largest city in the world but perhaps the first and only one. Its position on a floodplain on the banks of the Euphrates River meant the city was blessed with rich farmland and plentiful water that its inhabitants used to turn it into a 'Venice of the desert', laced with canals and greenery and dominated at its center by the Eanna temple complex that seems to have acted as a religious institution, government center, and a business district all rolled up together.

According to David Graeber in his 2021 book *The Dawn of Everything: A New History of Humanity*, "around 3200 BC the original public buildings of the Eanna sanctuary were razed and covered with debris, and its sacred landscape redesigned". While the fire did not spell the end of Uruk as a place of residence, it did mark the end of Uruk as the predominant center of power in the area. "By 2900 BC," Graeber says, "we have evidence for local kings of rival city-states battling it out for supremacy over Uruk".

Five thousand years after the fire, in 1936, German archaeologist Adam Falkenstein excavated the site, finding thousands of clay tablets covered in proto-cuneiform writing. David Anthony wrote in his 2007 book *The Horse, the Wheel, and Language* that "about thirty-nine hundred tablets were recovered from level IVa, the end of Late Uruk. In these texts, among the oldest documents in the world, a pictograph shows a four-wheeled wagon with some kind of canopy or superstructure. The 'wagon' sign occurred just three times in thirty-nine hundred texts, whereas the sign for 'sledge' — a similar kind of transport, but dragged on runners not rolled on wheels — occurred thirty-eight times. Wagons were not yet common... Sledges still were far more common than wagons in the city of Uruk at that date."

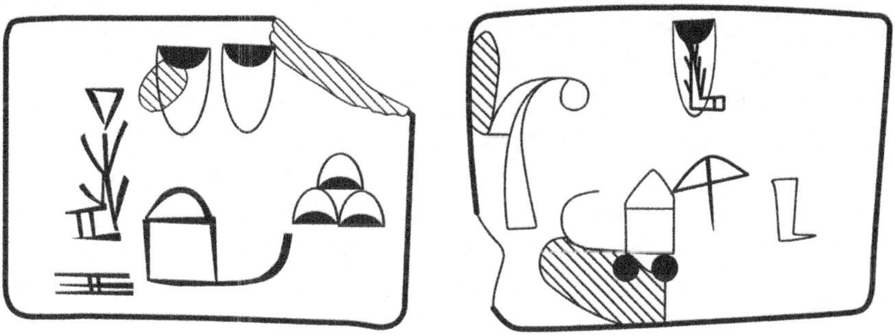

Figure 22 – An illustration of tablets from Uruk IVa. The tablet on the left depicts, among other things, a sledge, while tablet on the right depicts, among other things, a wagon.
illustrations: Stan Tucker

It is understandable why sledges were common in an agricultural area. The threshing sledge, or *tribulum*, is a wide board fitted with flint stones or metal nails on the underside. Harvested stalks were placed on flat, hard ground and animals would drag the sledge over the stalks, with a driver standing or sitting on the board, to separate out the grain kernels. People threshed grain this way so recently that there are *photographs* of people operating sledges like these. The image below shows a threshing sledge like those seen on five-thousand-year-old tablets, with a person sitting in a seat affixed to it, in Heliopolis, Egypt, in the late-1880s.

Figure 23 – A threshing sledge in operation in Heliopolis, Egypt, in 1889

Mounting axles and wheels to a sledge to make a wagon was one of the most profound inventions in human history, in part because the wheel has no precedent in nature to serve as an inspiration. The wagon was enabled by a convergence of technologies and ideas. One of the reasons the wheel was invented only at this point is because historians believe that metal tools, particularly ones made of bronze, which was only just becoming common, made fabricating wheels and axles significantly easier.

The technology spread so quickly that it is difficult to determine where the wagon was actually invented. According to Roland Ennos in *The Age of Wood: Our Most Useful Material and the Construction of Civilization*, "wheels appeared almost simultaneously in the Sumerian civilization of Mesopotamia, in the Caucasus, and in Eastern Europe. Evidence from written texts suggests that the first vehicles were four-wheeled wagons, whose design had been converted from that of sleds... The earliest actual wheel found to date is the Ljubljana Marshes Wheel, which dates from around 3150 BC."

In the *Digital Atlas of Innovations*, Florian Klimscha writes: "It spread so quickly that its diffusion can only partly be traced by archaeological means. Wheeled vehicles were so interesting for prehistoric societies that within a short time, perhaps even in the course of a single century, different, locally adapted versions of the wagon spread between the North Sea and the Baltic sea in the north, and the Alps and the Black Sea in the south."

From Two Hundred Pounds to Four Thousand

The wagon was disruptive, at least a 10X disruption. William J. Bernstein wrote in *A Splendid Exchange: How Trade Shaped the World* that "a draft horse can carry about two hundred pounds on its back. With the help of a wagon and a good road, it can pull four thousand pounds." Thus, even in the absence of a 'good road', it is reasonable to think that a draft horse with a wagon might be able to haul ten times the load that one could carry without a wagon.

Perhaps the rise of the wagon is why the Late Uruk Period was, in fact, the late Uruk period. Uruk's flat floodplain (the original 'silicon valley', which made significant contributions to irrigation, metalsmithing, urbanism, theology, and writing, among many others) was excellent ground for agriculture and threshing sledges, but muddy soil is poor ground for wheeled wagons. Land that was previously an asset in a world of sledges quickly became a liability in a world of wagons. What a coincidence it was that just as the symbol for wagon was making its first appearances in the written texts of an agrarian society, that society was on the cusp of being radically reshaped by that very technology, isn't it?

A Pattern We Have Seen Before

We have seen equally radical transformations in technology and society in our own time. Language can change faster than technology. Google has digitized millions of modern books and allows you to search, using the Ngram Viewer tool, to see how often a given term appeared, sorted by the year the books were published. So we can find that, among American English books published in the year 1890, the word 'horse' accounted for about ninety out of every million words published in American English-language books in that year.

The automobile was invented in 1885 and the word 'automobile' became more popular in American books than the word 'horse' by 1920. However, the total references to 'horse' and 'automobile' combined only go up only a little during this period. That is, as automobiles got more popular, there was largely just a switch from talking about horses to talking about automobiles.

According to Google's data, the number of instances of the word 'automobile' in American English-language books published the year 1900 was about 5% as much as the word 'horse', even though in 1900 there were only eight thousand registered automobiles in the US (compared to over eighteen *million* horses).

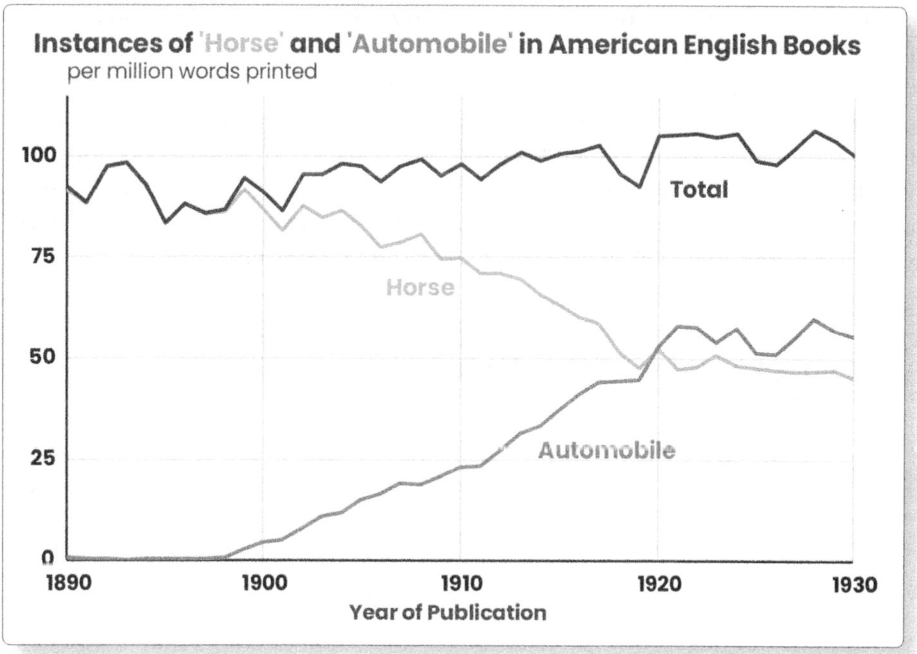

Figure 24 – Instances of 'horse' and 'automobile' in American English-language books, per million words printed, by year of book publication. *Data from Google Books Ngram Viewer*

In 1900, horses in the US outnumbered automobiles by a ratio of more than two thousand to one. If an automobile travelled ten times as much per year as a horse, then miles travelled by horses outnumbered miles travelled by cars by more than two hundred to one. There were probably around one million horses born in the US in the year 1900 while the number of automobiles on the road was only about five thousand higher than in 1899, so horse births outnumbered automobile 'births' also by about two hundred to one. But the use of the word 'horse' in American English-language books published in 1900 was only about twenty times as much as the use of the word 'automobile'.

New technologies and developments command a disproportionate share of our attention, including in our writing. This is not restricted to transportation. The ratio of references to 'radio' in American English-language books in 1918 was about 7% of the peak level that was seen in 1945, even though the first regular commercial radio broadcasts did not begin in the US until 1919. Before we had radios, we were writing about them.

Television too. The word 'television' was coined in the first decade of the 1900s, but the first functional electronic television system was not demonstrated until 1928, more than twenty years later, by 22-year-old Philo Farnsworth in San Francisco. By 1938, according to Google's Ngram data, the number of references to 'radio' in American books was about thirteen for every reference to 'television', even though the first regular TV broadcast in the US would not begin until the spring of the following year, 1939. It would take eleven more years after that, until 1950, for even 9% of American homes to have a television set.

During all this time, the use of the word 'television' was rising rapidly in American English-language books, according to Google's data. From the mid-1950s until after the year 2000, the word 'television' accounted for a consistent 40 to 50 words out of every million words written in American books. By the time 9% of American homes had a television set, in 1950, the word 'television' was already about 30 words per million. Televisions themselves went on to see, as Robert J. Gordon wrote in *The Rise and Fall of American Growth*, "the fastest diffusion of any appliance or device in history". By 1955, two-thirds of American homes had a TV.

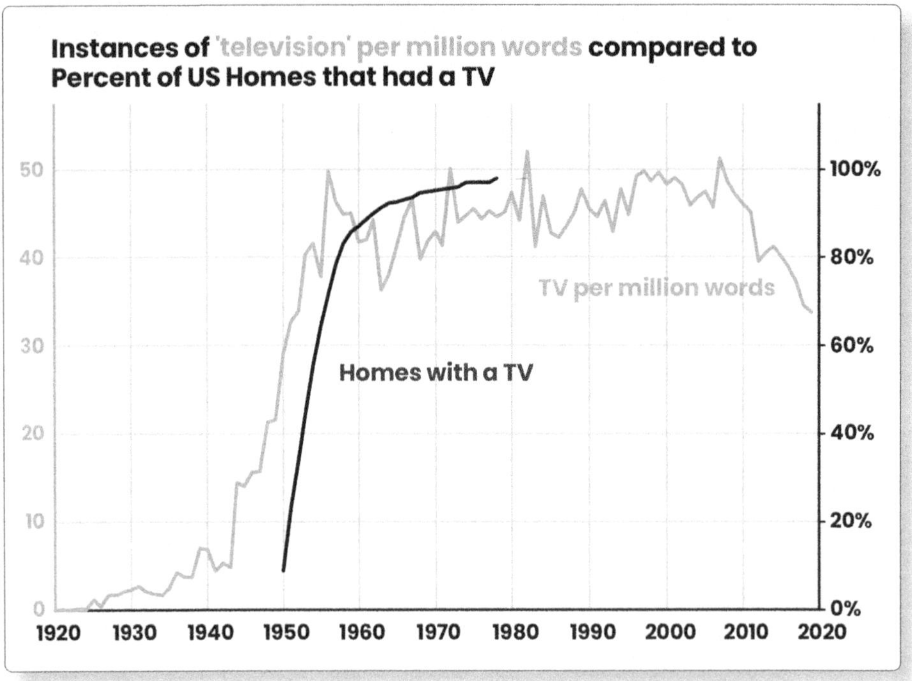

Figure 25 – Instances of 'television' per million words in English-language American books, compared to the percent of US homes that had a TV set.
Language data from Google Books Ngram Viewer. TV adoption data from Our World in Data.

Isn't that strange? In the 1890s and early 1900s we were talking about the automobile more and more even though almost nobody owned one. In 1910s we were talking about radio more and more even though almost nobody owned one. In the 1930s and 1940s we were talking about television more and more even though almost nobody owned one. Isn't it reasonable to think that mentions of wagons in ancient Sumerian tablets would precede the actual adoption of wagons too?

The point is: our written references to new technologies often precede and far surpass the rate of adoption of that technology itself, especially in the earliest days. In the November 3, 1899, edition of the five-cent weekly magazine 'Secret Service', a New York City father-and-son detective team travel to San Francisco, but before they departed they "purchased and shipped to San Francisco a light automobile. It was one of the latest patents, with an inexhaustible battery, which it was not necessary to recharge." At this point there were only about three thousand automobiles in the United States (and electric automobiles certainly did not have 'inexhaustible' batteries), but that did not stop the author from writing about this new contraption.

In the finds from Eanna temple in Uruk, we saw thirty-eight symbols for 'sledge' along with three symbols for 'wagon', a ratio of about thirteen to one. For comparison, the ratio of 'horse' to 'automobile' references in American English books was also about thirteen-to-one in the year 1902, six years before the Model T was released. It might be reasonable to think that the scribes who lived in Uruk in about the year 3100 BCE pressing the symbols for 'wagon' into wet clay were literally recording the start of a rapid transformation of their culture that would prove to be every bit as dramatic as what the automobile would do to American culture five thousand years later.

◆ ◆ ◆

> "Pharaoh had a dream: He was standing by the Nile, when out of the river there came up seven cows, sleek and fat, and they grazed among the reeds. After them, seven other cows, ugly and gaunt, came up out of the Nile and stood beside those on the riverbank. And the cows that were ugly and gaunt ate up the seven sleek, fat cows. Then Pharaoh woke up."
>
> **— Genesis 41**

◆ ◆ ◆

The Camel Disrupted the Donkey

The camel was domesticated about 3500 to 4000 years ago and, according to William J. Bernstein in *A Splendid Exchange*, it "revolutionized trade over the sands of the Middle East and through the steppes of Asia."

It is easy to understand why. In a hot, arid environment, the key metric by which to judge a pack animal is the amount of water it requires to move a given load a certain distance. Prior to the camel, the main pack animal in the Middle East and northern Africa was the donkey. "If the donkey was a family sedan," Bernstein wrote, "able to convey light loads over smooth hard surfaces, then the camel was a Land Rover".

In 'Comparative Physiology of Desert Mammals', Knut Schmidt-Nielsen estimated that the summertime daily water requirement for a 175-pound man is about 1.5 gallons per day. Even though camels are significantly larger than people, "the water expenditure in the camel is of the same magnitude as in man." That is, camels consume the same amount of water per day as us even though they are about six times heavier than a person.

Donkeys, in comparison, need about 1 gallon per 175 pounds of body weight according to Schmidt-Nielsen, much closer to what a person requires than to what a camel needs. As a result of their lower water requirement, camels can go a longer distance between drinking than other animals, upwards of five days of travel, or about one hundred miles. Considering that a camel can carry a total load of about 35% of its own body weight, while a donkey only can only carry about 20% of its own weight, the water requirement for a camel is about seven times lower per unit of load carried than for a donkey. It is no wonder that the camel has so thoroughly displaced the donkey as the preferred pack animal in sandy desert environments. (U.S. Army, June 2004)

This is a conservative estimate because, on an extended journey, the animals would need to carry water as part of their load, reducing the amount of actual cargo they could bear.

Let's say that on a trip we would like to bring enough water for our animals for five days. A 715-pound (325 kilogram) camel would need about five gallons (18 liters), which would be about 15% of the total load it can sustainably carry. In comparison, a 220-pound (100 kg) donkey would need over six gallons (23 liters), which is essentially all the load it can sustainably bear.

If you want to bring a donkey on a desert trip with five days of water, the donkey could bring nothing else. So, the domestication of the camel did not just make transport in arid environments more efficient, it enabled trips (and therefore, connections) that were simply not possible before.

There was no ancient Egyptian word for 'camel'. Because they lived along a river, where water was plentiful, the Egyptians used donkeys as beasts of burden, and their world thus extended up and down the Nile River. But the widespread use of the camel before and at the end of the dynastic age, when the Persians took control around 525 BCE, enabled travel by land away from the river. This was the time when an oracle to Ammon was founded at the Siwa Oasis, near what is now the western border of the modern country of Egypt, and reachable by traversing a series of oases, each accessible from the previous by camels, but not by donkeys. When Alexander the Great set foot in Egypt, he insisted on using camels to reach the oasis and it was there at the oracle, in the desert west of the Nile, that he was declared to be the divine descendant of the god Ammon and thus the legitimate Pharaoh, allowing him to take control of Egypt from the Persians without a fight.

◆ ◆ ◆

"Sooner or later, everyone sits down to a banquet of consequences"

— Robert Louis Stevenson

◆ ◆ ◆

Alexander and his Macedonian Startup

> *Over just thirteen years, Alexander the Great defeated the Achaemenid Empire because his father had designed the Macedonian army around one principle: take Persepolis, the enemy capital*

When Persian King Darius III ascended to the throne of the Achaemenid Empire in 336 BCE, the lands under his control stretched from his capital at Parsa ('Persepolis') in what is now southern Iran, east to the Indus River and as far west as Egypt and the Aegean Sea, a span of 2500 miles, a distance as long as the width of North America. Effectively, his Empire was the western world's only superpower.

Greece at the time was a loose assemblage of city-states just beyond the edge of his Empire, frequently in conflict with each other but united by their hatred and fear of their enemies just across the Dardanelles who had burned their sacred temples during the Second Persian War a century before. Macedon was at the edge of the Greek world and the Greeks were at the edge of the Persian world.

In the very same year that Darius III became king of the Persians, Alexander, age twenty, inherited the throne of the Greek kingdom of Macedon from his father, King Philip II, and with it an army like none the world had ever seen before. Philip had spent much of his twenty three years as king reforming his army and had been authorized by the League of Corinth to invade Asia Minor. Shortly before his death, he sent 10,000 men to what is now Türkiye to prepare for a war to liberate the Greek peoples there from Persian rule and to exact revenge on their enemies. But Philip realized that the short-distance military logistics that were adequate for local clashes were not suitable for the longer distances required to operate in Asia Minor and beyond.

While king, Philip had made "sweeping reforms to the Macedonian logistics system", including replacing slow, heavy ox-drawn carts with smaller, lighter horse-drawn wagons, "the first time a Western commander had done this", according to the 'Kings and Generals' channel on YouTube.

He also imposed new rules on who could tag along with the army. Previously, Greek armies would march with a retinue of "thousands of non-combatants – attendants, women and slaves... sometimes outnumbering the soldiers." But "the Macedonian king drastically reduced the number of non-combatants accompanying the army. Women were forbidden while the number of servants was drastically reduced. Each cavalryman would have one servant, while for the infantry there would be one servant for every ten Macedonians."

It was not any one change Philip made that prepared his army for conquest, but rather the *convergence* of several innovations, all dictated by the overarching goal of being able to move long distances sustainably and to defeat the Persians on their own land, an aging Empire facing internal struggles and areas already on the brink of rebellion.

Effectively, Philip's reform regarding non-combatants alone was a nearly-10X disruption. As the infantry constitutes the bulk of any army, reducing non-combatants from 'outnumbering the soldiers' (or roughly one non-combatant for every soldier) to 'one servant for every ten Macedonians' meant the Greeks could move further, faster, and could devote more resources to siege weapons and cavalry. Much of Philip's reforms are credited to his general, Iphicrates, who chose torso protection made from quilted linen instead of bronze plate. Metal armor is heavy, requires a lot of maintenance, and is difficult to put on and take off without the help of attendants, so it is understandable why quilted linen might be preferable for an army about to make a very long expedition. The general even designed lightweight and durable boots to replace the sandals Greek soldiers typically wore, the first time soldiers were expected to wear boots. The main weapon of the infantry soldier was chosen to be the *sarissa,* a long spear which could be dismantled into two pieces for easier transport. And the Macedonians drilled relentlessly in order to act as a unit.

Alexander spent his first two years as king solidifying his rule by crushing rebellions in his own territory. Then, in the spring of 334 BCE, he crossed the Hellespont into Asia with 48,000 infantry and 6,000 cavalry and won his first major battle against the Persians, at a location along the Granicus River near Biga, Türkiye, that scientists announced they had positively identified only in early 2025.

Over the next eighteen months, the Macedonians pushed the Persians back across what is now the western half of Türkiye. At first, Darius did not take their actions in Asia Minor very seriously, as there were many other areas of instability and unrest in his empire besides this one. Just a couple of years before, he had put down a major revolt in Egypt. In fact, the Greeks themselves had been active in Asia Minor since the very start of Darius's rule, when Phillip II sent 10,000 men.

But as they crept eastward, Darius must have grown more concerned about the fate of Asia Minor, if not lands closer to home, as he decided to personally lead his army to face Alexander, which he did at the very north-eastern corner of the Mediterranean Sea, at a place called Issus.

Jacob Abbott wrote in his 1848 account of the life of Alexander: "King Darius took his wife and all his family with him, and a large portion of his treasures, on this expedition against Alexander. His mother... and his wife came, each in her own chariot, immediately after the king. Then there were fifteen carriages filled with the children and their attendants, and three or four hundred ladies of the court, all dressed like queens. After the family there came a train of many hundreds of camels and mules, carrying the royal treasures."

He continues: "It was in this style that Darius set out upon his expedition, and he *advanced by a slow progress* toward the westward, until at length he approached the shores of the Mediterranean Sea. He left his treasures in the city of Damascus, where they were deposited under the charge of a sufficient force to protect them, as he supposed. He then advanced to meet Alexander." (italics added)

By the time the two great armies engaged, the Greeks had conquered only about 8% of the land area that had been the Persian Empire less than two years before, which means that it only required Alexander taking a few percent of the Persians' land (a sort of 'market trauma' in Tony Seba's parlance) for Darius to decide that a personal response was necessary. Darius lost the battle at Issus and fled the field, leaving his camp unguarded. His mother, wife, and two daughters were taken hostage.

Over the next two and a half years, Alexander and his men racked up a series of unparalleled victories: at Issus they defeated an army perhaps twice their size. Over a matter of just a few months, they successfully laid siege to Tyre, wealthiest of all Phoenician settlements and famous center of purple dye production, claiming the entire eastern coast of the Mediterranean in the process. (In early 332 BCE, Alexander laid siege to the island city of New Tyre, Lebanon, building a causeway of rocks and debris out from the city of Old Tyre, which he had ransacked. That causeway still exists and today is the site of a Starbucks coffee shop.)

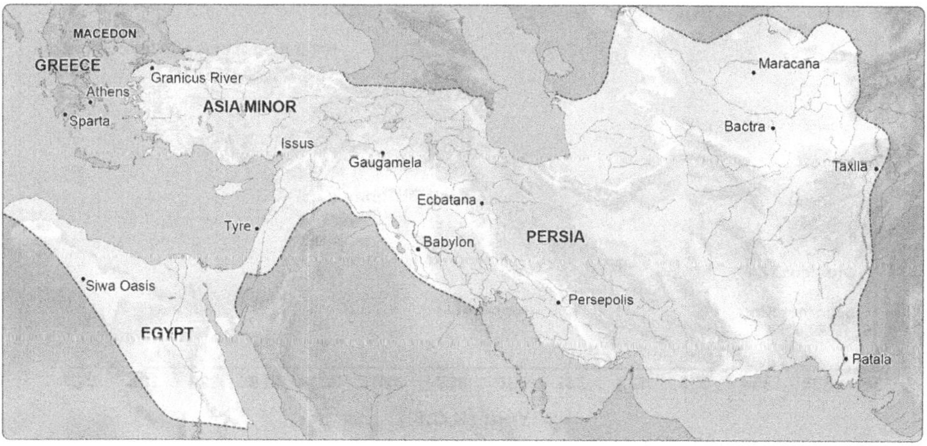

Figure 26 – The Lands Alexander Conquered

Less than six months later, in late 332 BCE, Alexander was declared Pharaoh of Egypt at the oasis of Siwa after traveling from oasis to oasis from the Nile River. News of that event must have stung for Darius III, as the contact between the Siwa Oasis and the Egyptian civilization had been established about three hundred years prior due to the Persian importation of camels. Darius's namesake Darius I (also known as 'Darius the Great') had been Pharaoh of Egypt, as was Darius III himself. And now Alexander, officially a man-god, turned back east and north to ride toward the heart of the Persian Empire.

At Gaugamela, in 331 BCE, he defeated Darius a second time and won control of what is now Iraq. And in early 330 BCE, less than four years after they had left Greece, the Macedonians entered Persepolis, looted its riches, and burned it to the ground. It was then that Alexander earned the nickname by which he would ever be known in Persian hearts, 'Alexander the Accursed'.

King Darius III, defeated three times in battle by Alexander, fled his city and died a few months later, killed by one of his own men, effectively ending the Persian empire for good. But at that point, Alexander's work was still only half done. Over the next few years, he conquered areas that are now in Iran, Turkmenistan, Uzbekistan, and Afghanistan, and then pushed down the Indus River in what is now Pakistan, before returning to Babylon, the city he chose as his capital.

In June 323 BCE, Alexander the Great died in his new capital at the age of thirty-two. Over the thirteen years of his rule, his empire had grown in an S-shaped curve to nearly twenty times the size of his Greek homeland. On his deathbed, Alexander passed his signet ring to Perdiccas, a calvary leader, whose assassination the following year kicked off forty years of civil war that carved into pieces the kingdom Alexander had fought so hard to build.

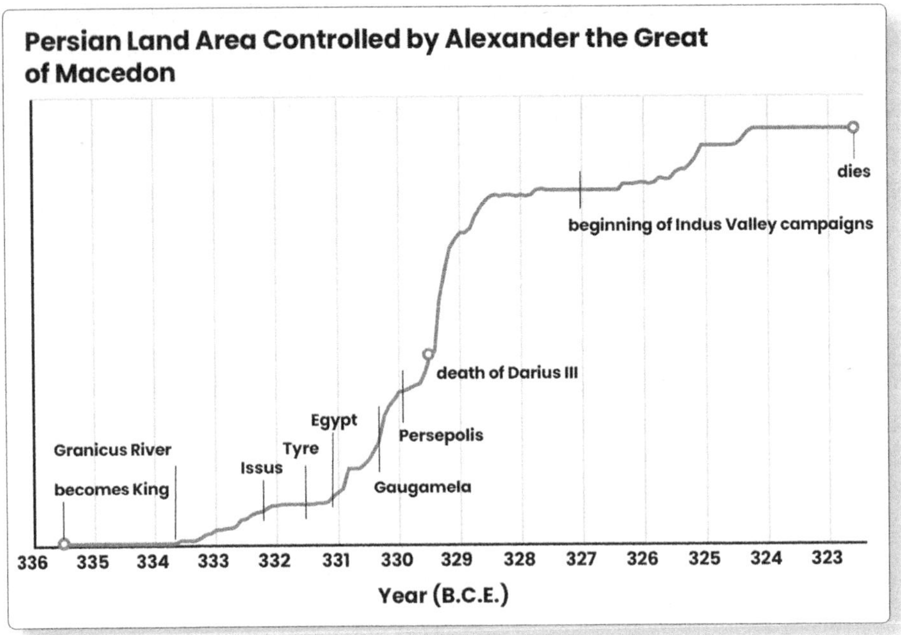

Figure 27 – Land Area Conquered by Alexander the Great

For all intents and purposes, the Macedonian army that King Philip II built was a startup company. It had a BHAG (a 'big, hairy, audacious goal') – to use a term popularized by James Collins and Jerry Porras in their book *Built to Last: Successful Habits of Visionary Companies* – namely, to destroy the Persian Empire. And when you can state your goal clearly, it becomes a standard against which you weigh all your decisions.

In Philips' case, he needed an army that could move a very long distance, quickly and sustainably. Every reform of his was made with the BHAG of defeating the Persians in mind. Wives and mothers were kicked out of the army. So were oxen. It is fine to carry a large shield if you are making moves against another Greek city-state, as it is not a very long distance between the Greek cities, and the armies would meet halfway between them anyway. But to go to Persepolis meant casting aside the large 'hoplon' shield Philip himself had carried as a younger man and replacing it with the 'pelta', a smaller, lighter, and thus, more portable, piece of equipment.

When Alexander became king of Macedon, he inherited an army well-suited to the task at hand, and rode off to face a man perhaps twenty-five years his senior, who led an army that was just as ill-prepared as its leader for what was coming.

A City of Marble

> *The death of Julius Caesar kicked off seventeen years of civil war and strife between his opponents and supporters. In the end, one man was left standing, Octavian, who took the name Caesar Augustus and led Rome into two centuries of peace and prosperity. But it was not just the beneficence of Augustus that undergirded the Pax Romana*

In March 44 BC, as Julius Caesar lay in a pool of his own blood on the floor of Pompey's Theatre in Rome and exhaled his last breath, his great-nephew Octavian, the man named in Caesar's will as his principal heir and the man with the greatest claim to succeed him, was just nineteen years old.

Caesar's death kicked off seventeen years of conflict, first as Octavian and his allies Mark Antony and Marcus Lepidus united as the "triumvirate" to defeat Caesar's assassins, generals Brutus and Cassius (as well as the ample armies who supported them), at the Battle of Philippi in Greece in 42 BC, leaving 40,000 men dead. After the battle, fearing capture, Brutus and Cassius fled to nearby hills where Brutus reportedly said, "by all means must we fly, but with our hands, not our feet", before grabbing his sword and falling onto it to commit suicide. Cassius cut off his own head with the help of a servant.

The triumvirate split the Roman world amongst themselves, but then they turned against each other. In 36 BC, Marcus Lepidus sent fourteen legions to Sicily – a region under Octavian's dominion – to help put down a revolt. But Octavian claimed that these troops were thenceforth under his own command, and so Marcus Lepidus fled into exile and spent the rest of his life peacefully in relative obscurity at his seaside home about halfway between Rome and Naples. Still officially a consul, Lepidus returned to the capital only occasionally for important government matters to allow Octavian to humiliate him by calling him last in the voting rolls.

In 30 BC, Octavian invaded Egypt. To avert a civil war, Mark Antony had married Octavian's sister, Octavia, but then carried on an extended and public affair with his lover, Queen Cleopatra VII of Egypt. When Octavian won the Battle of Alexandria, Mark Antony stabbed himself and was carried by his men to Cleopatra, where he died in her arms. Shortly after, Cleopatra was either murdered or committed suicide, perhaps by a snakebite or by poisoning herself. Either way, her death marked the end of the last vestige of any opposition to Octavian's rule.

In 27 BC, Octavian declared the end of the Roman Republic and the founding of the Empire, taking the name for himself of Caesar Augustus. He united the Italian peninsula administratively and politically as a single district, the first time that anyone had called that land 'Italia'. He would rule Rome and its territories as emperor for the next forty-one years until his death in 14 AD, quietly at age seventy-five, while visiting the town near Naples where his own biological father had died.

The first two centuries of the Empire saw unprecedented peace and economic prosperity, the so-called *Pax Romana.*

In the January 2010 article 'Stone Sawing Machines of Roman and Early Byzantine Times in the Anatolian Mediterranean', Paul Kessener of Radboud University in the Netherlands wrote: "wealth increased throughout the empire, about every town gained its aqueducts supplying public and private *thermae* that became standard elements of city life. The walls and floors of these bath houses and of many other buildings were adorned with marble slabs". On his deathbed Caesar Augustus is recorded to have said: "I found Rome a city of brick and left it a city of marble", attesting to his own greatness.

But Rome owed its transformation to more than just Caesar Augustus's beneficence.

Kessener continues: "A relief on the cover of a 3rd c. AD sarcophagus at the north necropolis at Hierapolis in Phrygia [or Pamukkale, in modern-day Türkiye] shows … a water-powered twin stone saw mill. The mill is equipped with the crank and connecting rod system, previously thought to have been a medieval invention, by which the rotary movement of the water wheel is transformed into a reciprocal linear movement enabling mass production of marble slabs."

That is, the Romans were able to use waterpower to turn a wheel that moved wooden frames back and forth – frames that held steel saw blades.

It was only around the start of Augustus's reign that the Romans figured out how to temper steel. Making steel itself had always been a sort of hit-or-miss enterprise. Steel is 98% iron, but for many centuries we did not understand what was the magical other 2% that transformed it into steel and as a result, steel was rare and expensive right up until the mid-1800s. (It turns out, the other 2% is carbon.)

Red-hot steel plunged into water or oil will quickly quench, leaving a hard metal that can be brittle. But a second heating at a lower temperature, followed by letting the steel slowly cool over a long time, leaves a product that can take a durable edge. It was this second tempering step, which made the steel more durable, that the Romans discovered around the start of Pax Romana.

Roman saw blades were not like modern saws with a row of sharp, hard teeth. Instead, they were simply a thin slab of steel which could be dragged back and forth across a surface as wet sand was pushed into the gap (or 'kerf'). Slowly, the hard blade and abrasive sand would grind their way through almost anything, including stone.

Kessener wrote that in 2009 Fritz Mangartz, an archaeologist at the Römisch-Germanisches Zentralmuseum in Cologne, Germany "conducted experiments on a 1 to 1 reconstruction" of a Roman sawmill. A "saw frame was moved to and fro for about 20 cm [8 inches], to cut a block of 'Jura marmor' similarly sized as the Ephesos marble blocks. A mixture of sand and water was applied continuously by hand onto the saw slit while the sawing process was monitored." The experiments showed that one person "would produce about 330 square meters [or 3500 square feet] of slabs in a year... From this Mangartz concluded that in the best case the machine would produce twelve fold compared to manual sawing."

To repeat, experiments showed that a water-driven mill could saw marble at twelve times the rate of manual sawing. A 10X disruption!

Aqueducts brought ample fresh water to Roman cities, but by the use of water wheels, aqueducts were more than just sources of water, they were sources of labor. Connected to giant round mill stones, they could mill grain and press oil from olives (itself a source of energy used for both food and lighting). Connected to tempered steel blades, waterwheels could cut stone or timber, materials used to not only build buildings but also to build more aqueducts, creating a self-reinforcing cycle of growing economic output.

Stone blocks could be carved into statues or the sections used to make columns. Slabs were used as veneers for the façades of buildings and for floors. And slabs could also be split up into small square chips used to make mosaics. No wonder the Empire saw unprecedented peace and prosperity under Augustus! No wonder he was able to leave his beloved Rome 'a city of marble'.

The *Pax Romana* lasted for two centuries – undergirded by a convergence of technologies: aqueducts, water-driven mills, tempered steel, stoneworking, and concrete – but some of its effects have endured far longer. The Pantheon was built on the site of a temple commissioned during the reign of Augustus and to this day remains the world's largest unreinforced concrete dome. And decades after Augustus's death the Romans began construction of the Colosseum, which was built of concrete and faced with sawn stone slabs.

Christianity and Islam Disrupted Pagan Faiths

> *At the time of the death of the Prophet Mohammed, the entirety of the Arabian Peninsula was Muslim, and Islam continued to quickly spread to lands far beyond. In Rome, Christians went from being persecuted by the state to having their faith be the Empire's de-facto religion in a period of four years. There must be powerful, disruptive reasons why change happened this quickly for so many people*

Few innovations have spread faster, further, or affected more people more deeply than rises of Islam and Christianity. Today, more than half of all people on Earth consider themselves to be a member, at least nominally, of one of these two faiths.

Christians now outnumber adherents to Judaism, from which Christianity was derived, by a factor of roughly 100-to-one. In the Roman Empire, Christianity replaced the pagan pantheon, a religion that had existed for about one thousand years, over about a century. Emperor Diocletian's persecutions of the years 303 to 312 AD (the 'Great Persecution', the Empire's largest and bloodiest official anti-Christian pogrom) gave way to Christianity becoming the official state religion under Emperor Theodosius I in the year 380 AD, less than a lifetime later. In fact, Christianity became the empire's preferred religion under Constantine in the year 324, just twelve years after Diocletian's purge ended and just four years after Emperor Licinius reneged on his pledge to tolerate Christianity.

Islam spread faster than almost any idea ever. The Prophet Mohammed began receiving revelations in 609 AD in a cave near Mecca. Fifteen years later, in 624 AD, he led three hundred warriors in a raid on a Meccan merchant caravan at Badr. And in 629 or 630 AD with ten thousand warriors he conquered Mecca itself. By the time he died in 632 AD, most of the Arabian Peninsula had converted to Islam over a period of just eight years. According to William J. Bernstein in *A Splendid Exchange*, "Chinese sources suggest that Islam arrived in Canton in about 620 [AD], a full dozen years before the death of the Prophet." In Iran, Islam replaced Zoroastrianism, the state religion of various Persian empires spanning more than a millennium, with about 70% of the conversion taking place "in just a little over one century", according to Columbia University professor Richard W. Bulliet.

Bulliet's 1979 book *Conversion to Islam in the Medieval Period: An Essay in Quantitative History analyzed* "changes over time in the patterns of first-name giving" for more than six thousand biographies listed in the book *Shadharāt adh-dhahab fī akhbār man dhahab* ["Nuggets of gold in the stories of those who have passed away"] by Ibn al-'Imād, who died in the year 1687 AD. Bulliet used the popularity of 'distinctive' Muslim first names like Mohammed, Ahmad, Ali, Hasan, and Hussein to estimate the portion of people who adhered to Islam over time.

The biographical listings include people who died between the years of the Islamic calendar AH 100 (719 AD) to AH 1000 (1592 AD) in places ranging from Iran, Iraq, Syria, Egypt, and Tunisia to Spain. As with many other disruptions, the

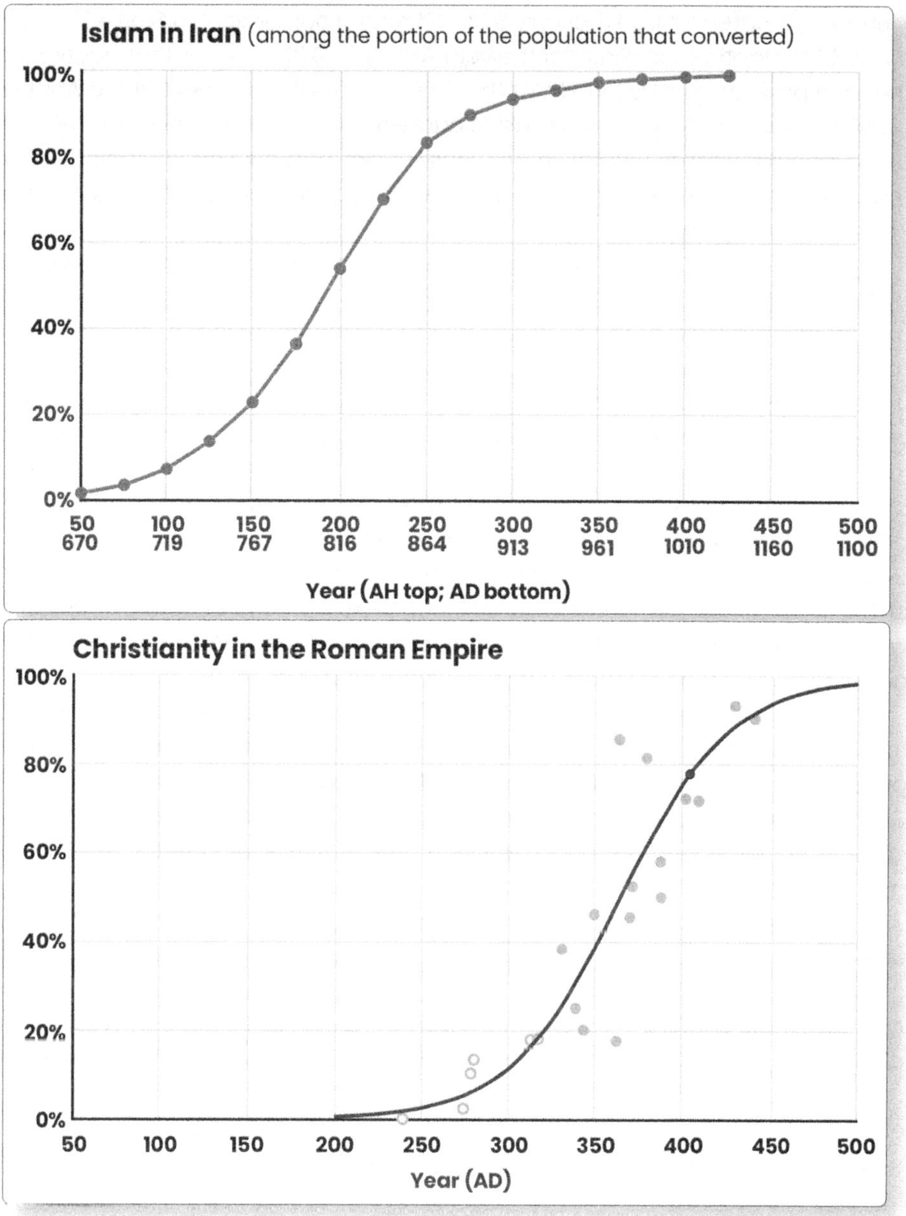

Figure 28 – Adoption of Islam in Iran and Christianity in Rome. Islam data from Bulliet, Table 2. Christianity data from Turchin. Black dot marks the year of the abolition of gladiatorial combat by Honorius I.

adoption of Islam in these lands each followed an S-shaped pattern. In Persia, adoption rose from about 10% in AH 132 (750 AD) to 80% in about AH 250 (864 AD). These years are important in Persian history because the first (AH 132 / AD 750), corresponds to the revolution of the Abbasids, the Arab caliphate based in Iraq, while the latter year (AH 250 / AD 864) corresponds to the establishment of the Saffarid dynasty, an indigenous Persian movement that drove the Abbasids out of power in Iran.

Bulliet estimated for Iran that only 80% of the total population converted to Islam, so his '10% adoption' estimate for the year AH 132 / AD 750 means that about 8% of the total population was Muslim at the time of the establishment of the Abbasid caliphate. That is, the Muslim Abbasids came to power in Iran when less than 10% of the population was Muslim. In this sense, Islam had caused Iranian society to reach a point of 'market trauma', where this new, disruptive idea (Islam) became irreversibly ascendant.

As is the case with many other disruptions, the spread of Islam in Iran was driven by literal outsiders (Arabs) but also by 'outsiders' within Persian society: artisans, merchants, and even prisoners of war. While rural landowners, the "*dihqans*, a kind of landed gentry who constituted the backbone of the Iranian upper class in pre-Islamic times" were initially reluctant to convert, the early adopters of Islam in Iran were, according to Bulliet, "people of very low social status who had no status to lose". Christianity, too, was developed by outsiders. It arose in a distant province of the Roman Empire and its earliest followers were not Roman. Even among the people of the Holy Land, Jesus was an outsider. Luke 22:59 identifies Jesus as a Galilean and Matthew 26:73 that he spoke with a distinctive accent. Eruvin 53b of the Talmud, the ancient Judaic text, suggests that the people of the region were 'bumpkins', known for their imprecise way of speaking. Historians suggest his native language was Aramaic, which originated in Syria.

Estimates of the portion of the Roman population that was Christian come from historian Peter Turchin, a professor at the University of Connecticut. For years prior to 300 AD, Roger Bagnall assessed the number of people with identifiable Christian names mentioned in Egyptian papyri. The later estimates on the graph come from the 1978 dissertation by R. von Haehling that included a list of Pagan and Christian officeholders in ancient Rome between 324 and 455 AD.

Because the initial spread of Islam was an urban phenomenon, it resulted in the unforeseen consequence of promoting urbanization. "It was extremely difficult to convert to Islam and lead a normal life," Bulliet writes, "without moving to one of the Arab governing centers." It is known from work by scientists like Geoffrey West, author of the book *Scale: The Universal Laws of Life and Death in Organisms, Cities and Companies*, and physicist Luis Bettencourt that urbanization has strongly positive effects on economic growth, human well-being, and even the rate of technological and scientific progress. Urbanization in the Islamic world had a profound effect, turning the cities of the Middle East into centers of trade, culture and learning that would outshine surrounding non-Muslim areas for centuries to come.

During the period of the most rapid growth of Islam in Iran, the non-Muslim incumbents, still in a dominant position but quickly losing 'market share', responded with a series of anti-Islamization revolts which started first in the urban areas where Muslims lived and then spread to more rural areas. Bulliet said, "the first revolts took place around major cities in eastern Iran and the last ones in the mountainous and sparsely settled region of Azerbaijan" but that "non-Muslim revolts die out as the middle point in the conversion process is reached. From the assassination in 754 [AD] of Abu Muslim, the great Iranian leader of the Abbasid revolution, until the capture and execution of Babak, the last leader of a major non-Muslim revolt, in 838 [AD], Iran experienced an almost unbroken series of insurrections."

Bulliet's quantitative analysis found similar dynamics of the adoption of Islam in other lands, characterized by an S-shaped adoption profile. In Iraq, Egypt and Tunisia, the conversion process took place largely between 750 AD and 960 AD, going from 10% to about 80%. In Spain it took from 750 AD until about 1075 AD to go from 10% to 80%.

In each case, the spread of the Muslim faith was met with anti-Islamization rebellions. "Iraq was visited by a steady succession of Khariji revolts that almost exactly coincides with the period of early adopters" of Islam, Bulliet writes. It is perhaps the recognition by the incumbents that they are still dominant, but that a disruptive competitor is rapidly gaining in popularity, that caused the incumbents to begin their reactionary response. "The correlation between events in Egypt and Tunisia and events in Iran discussed earlier is quite striking," Bulliet continues. "In all three cases the outbreak of non-Muslim rebellions occurs at the same point on the respective curves of conversion".

Reaching the point of 'market trauma' caused turmoil in the Roman Empire as well. According to Charles Freeman in *A New History of Early Christianity*, "most authorities agree that by 300 AD, between seven and ten percent of the population of the Roman Empire were Christian." The following couple of decades would be tumultuous ones. The 'Great Persecution' saw a series of edicts in 303 and 304 AD that, among other things, prohibited Christians from assembling for worship and from responding to actions brought against them in court, ordered the destruction of Christian scriptures, and stripped Christian senators, soldiers, and others of their ranks.

Constantine became Emperor in 306 AD, but needed eighteen years of civil war to eliminate his rivals and be declared sole Emperor in 324 AD. The following year, 325 AD, he convened a council of Christian bishops in Nicaea (now the city of İznik, Türkiye) who produced a statement of the Christian faith now called the Nicene Creed, the statement which begins "I believe in one God, the Father Almighty, maker of heaven and earth, of all things visible and invisible..." With his rival Licinius, Constantine had issued the Edict of Milan in 313, granting Christians a status of tolerance within the Empire, but in 320 AD, Licinius reneged and resumed persecution of Christians. In 324, Constantine defeated Licinius's men in a battle near Edirne, Türkiye, and in two further battles near what is now Istanbul, becoming the sole Emperor of Rome and ending the civil wars. This war was explicitly Christian versus pagan, with Licinius's men aided by Goths and Constantine's men marching adorned with the chi-ro symbol denoting Christ. From that point, Christianity was the preferred religion of the Roman Empire and a little more than a half-century later, in 380, Theodosius I declared Nicaean Christianity to be the official state religion. There were no frieze sarcophagi with Christian iconography before 280. In turn, those with pagan iconography disappeared after Constantine. A complete transformation within a generation.

Constantine built himself an imperial residence in the city of Byzantium, a Greek settlement established one thousand years before, in the seventh century BC, and renamed the city Constantinople. It became the capital of the eastern Roman Empire for the next thousand years and beyond, until the Ottoman sultan Mehmed II ("Fatih Sultan Mehmed") conquered the city in 1453.

In retrospect, it is easy to see why Christianity and Islam were disruptive to Judaism and Zoroastrianism. The Jewish faith has more than six hundred *'mitzvot'*, or rules, concerning everything from proper worship to sexual relations to food and drink. The laws of the Jewish faith (the *'halakha'*) only consider a person to be Jewish if their mother is Jewish, or if they convert and adhere to all the mitzvot. In Zoroastrianism, the dominant religion in Persia before Islam, conversion to the faith is generally not allowed, and someone (like rock legend Freddie Mercury) is typically only considered to be a member if both of their parents are.

In comparison, both Christianity and Islam welcome converts with open arms and membership in these faiths is trivial to obtain: to become Christian, Acts 16:31 says simply, "Believe on the Lord Jesus Christ, and thou shalt be saved" and John 3:16 says, "whosoever believeth in him [Jesus] should not perish, but have everlasting life." To become Muslim, one must only make an honest and sincere recitation of the *shahada*, a single short sentence recognizing Allah as the one true God and acknowledging Mohammed as his prophet. For these religions, unlike their predecessors, it does not matter who your parents are, only what you believe.

Christianity and Islam Replaced Traditional Religions in Sub-Saharan Africa

The replacement of traditional religions by Christianity and Islam continues to this day, more than one thousand years after those faiths conquered Rome, the Arabian Peninsula, Persia, and beyond.

In 2010, the non-partisan think tank Pew Research Center published a study called 'Tolerance and Tension: Islam and Christianity in Sub-Saharan Africa' that included information about the profound change in religious adherence that has occurred in much of Africa over the past century.

Using the World Religions Database, which publisher Brill claims "contains detailed statistics on religious affiliation for every country of the world", Pew showed that the percentage of sub-Saharan Africans who identified themselves as adherents of traditional African religions fell from about 75% of the population in 1910 to under 25% by 1970.

Likewise, the percentage who identified as either Muslim or Christian (or other) rose in that time from about 25% of the population to more than 75%, representing a complete reversal of the population, from three-quarters adherents to traditional religions to three-quarters not, in a period of about seventy-five years.

Astonishingly, Peter Turchin and Richard W. Bulliet found very similar patterns in their studies of the adoption of Christianity in Rome and of Islam in Iran (and other places). Turchin found that Christians went from about 25% of the

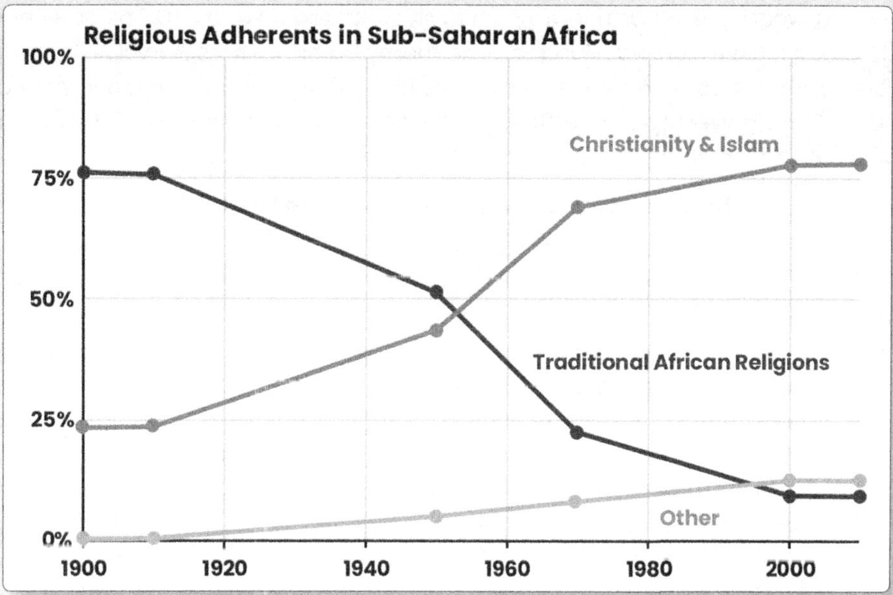

Figure 29 – Religious Adherents in Sub-Saharan Africa. *(Pew Research Center, 2010)*

population of Rome in 325 AD to 75% in 400 AD, seventy-five years later. Similarly, in Iran, Bulliet found that the population went from 25% Muslim in about 155 AH (772 AD) to 75% in about 230 AH, seventy-five years later.

A survey in England and Wales found that the portion of people who identify as Christian fell by about one-quarter of the total population over a period of twenty years from 2001 to 2021. Most of this decline was due to people who formerly called themselves Christian identifying instead as having 'no religion', not because of the growth of other religions. (Booth, 2022) Interestingly, this is as quickly as traditional African religions declined in popularity from 1950 to 1970, and also as quickly as the portion of the population of the Roman Empire who professed Christianity changed in the thirty years from 360 to 390 AD.

The Cannon Disrupted the Castle

> *Why did Constantinople fall to Fatih Sultan Mehmed in 1453 and not 1443 or 1463? Mehmed's father had also tried to take the city, as did his grandfather before that and his great-grandfather before that. But a revolution in artillery starting in the 1420s and 1430s meant that the young sultan could roll up to the Theodosian Walls, the edge of the last bastion of the Roman Empire, with a cannon that could deliver at least 10X the force of his forebear's best tool, the counterweight trebuchet*

Prior to the invention of the bombard, an early form of cannon, the most powerful device for attacking castle walls was the counterweight trebuchet, which consisted of a long, wooden lever arm that was heavily weighted on one end. The other end held a projectile which was flung toward enemy barricades. As Daniel A. Bertrand wrote in 'Building the Medieval Trebuchet', his undergraduate capstone project at Utah State University in 2019, "trebuchets were the heavy artillery of the Middle Ages, dominating siege warfare for almost 300 years."

The Incumbent Technology and the Disruptor

Warwick Castle, southeast of Birmingham, England, boasts one of the world's largest, working, historically accurate recreations of a medieval trebuchet. Standing nearly sixty feet high when its arm is fully upright, about the height of a six-story building, it can hurl a mass of 330 pounds the length of three football fields at a speed of up to 120 miles per hour.

In comparison, the bombard 'Mons Meg', now located at Edinburgh Castle in Scotland, is one of the largest-calibre cannons ever built, with a diameter of nineteen inches. Made of wrought iron, it was constructed in 1449 in Mons, in what is now modern-day Belgium, and was presented to King James II of Scotland in 1454, the year after Constantinople fell to the Ottoman Turks. Like the Warwick Castle trebuchet, it could launch stone balls of approximately 330 pounds. (Mons remains pivotal in the defense world nearly six centuries later, as it is the home of the military headquarters of NATO.)

Figure 30 – Warwick's trebuchet (left) and Edinburgh's bombard Mons Meg (right)

A 2015 study by Ian Lewtas and co-workers published in *Defence Technology* made a sophisticated computer simulation that estimated the muzzle velocity of the Mons Meg to be just over 700 miles per hour and the speed of its projectile at the time of impact after flying over a range of about two miles to be about 465 miles per hour. This means that the speed of the bombard's ball at the time of impact is almost four times as fast as the launch speed of the trebuchet.

Even the Earliest Cannons Could Deliver 10X the Force of a Trebuchet

The amount of energy a projectile will impart onto its target upon impact increases as the *square* of the speed of travel. So, the energy that would be delivered to a target by one of the Mons Meg's projectiles from a range of two miles is more than *fifteen* times the energy the Warwick Castle trebuchet would deliver using the same projectile to a target placed immediately in front of it. If the trebuchet's ball slows down during its flight due to air resistance, even a little bit, then the trebuchet would be even less effective compared to the cannon. That is, the bombard cannon is at least a 10X disruption in terms of energy delivered to castle walls compared to a trebuchet, and perhaps much more.

In April of 1453, the 21-year-old sultan Mehmed II, the leader of the Ottoman Empire, positioned his bombard cannons northwest of the walls of Constantinople. The Ottomans at the time ruled portions of what is now Türkiye and parts of southeast Europe, but not this critical city, the last remnant of the Roman Empire, in the geographic center of their lands. The young man had only been sultan since February 1451 and had spent much of the time since then preparing for this attack, including crafting cannons and stockpiling gunpowder.

The guns had been cast from bronze in Edirne, the Ottoman capital, in the autumn of 1452, test-fired in January 1453, and then dragged the 140 miles to within sight of the walls that had defended Constantinople for more than one thousand years. It is said that the Prophet Mohammed himself in one of his *hadiths* had prophesied that "One day Constantinople will be conquered. Great is the commander who will conquer it. Great are his soldiers." By April of 1453, more than eight hundred years later, that prophesy had still not yet come true. But that was about to change.

According to Roger Crowley in the September 2007 issue of *Military History* magazine, "Mehmed probably had about 69 cannons, a huge artillery force by the standards of the day. They were augmented by more traditional technologies for hurling stones, such as the trebuchet. The latter had been effective in the Muslim capture of crusader castles 300 years earlier, but now it looked like a device from another age."

On April 12, "lighted tapers were put to the touchholes of the sultan's guns along a four-mile sector of the front line, and the world's first concerted artillery barrage exploded to life." According to eyewitness reports: "nowhere was the wall strong enough or sturdy enough or thick enough to withstand it, or to hold out totally against such a force or the velocity of the stone ball." The Ottomans employed a "bear and cubs" strategy, using smaller cannons to keep the defenders on edge while the larger guns were repositioned and reloaded.

On May 29, the walls of the city were breached. It had only taken about as long to knock down the walls as it had taken to move the cannon from Edirne to Constantinople. Crowley says that when the walls fell, people from inside the city "ran out of their houses, beating their chests and crossing themselves. Women fainted in the streets. Churches were thronged with people voicing petitions and prayers." After a fierce battle, Mehmed II took the city for himself and, by extension, the titles of "Fatih" ("Conqueror") and also of Caesar of the Roman Empire, a claim all of Europe rejected.

The attack in 1453 was not the only siege in the recent history of Constantinople. Mehmed's father, sultan Murad II, had laid siege unsuccessfully in 1422, his grandfather, sultan Mehmed I, unsuccessfully in 1411, and his great-grandfather sultan Bayezid I laid siege unsuccessfully twice, once in 1391, after which he built one of the largest armies in the known world at the time, and then again from 1394 to 1402.

Figure 31 – This scene is from a mural of Fatih Sultan Mehmed's siege of Constantinople. On the top, there is one cannon among several trebuchets. On the bottom, one trebuchet among several cannons. *Photos by Bradd Libby.*

Meanwhile, on the Other Side of Europe...

By the time Henry V started to lay siege to the city of Harfleur, on the northwest French coast, in August 1415, England and France had been at war for nearly eighty years, but with several recent decades of relative peace. The French city took longer to fall than expected, so rather than march on to Paris that autumn, Henry instead headed east toward Calais.

South of Calais, in October 1415, the English army, comprised mostly of longbowmen, met a numerically superior French army at Agincourt. To call the result a *désastre* for the French would be an understatement. The English lost six hundred men compared to France's six *thousand*. (A similar thing had happened nearly seventy years prior, at the battle of Crécy, where the French lost two thousand knights and soldiers to longbow archers, while the English lost just fifty men.) Henry then took Caen in 1417, Rouen in 1419, and Paris (by treaty, not by force) in 1420. He laid siege to Meaux, east of Paris, for seven months starting in October 1421 before it finally fell in May 1422. Henry V himself died a few months later at the age of thirty-five, perhaps from dysentery contracted during this long winter siege, but the English continued the war. They laid siege to Orléans, southeast of Paris, starting in October 1428. This siege also lasted nearly seven months, but this time with a very different outcome. In late April, Joan of Arc, a sixteen-year-old peasant girl who claimed to have foreseen divine visions portending French victory, snuck into the city under the cover of darkness and within days a series of French assaults had broken the siege.

At Patay, northwest of Orléans, in June 1429, the French army inflicted on the English one of the most lopsided victories in history, with the English counting two thousand dead (out of five thousand deployed) and the French only three. Not three thousand. Three.

Figure 32 – The 1428 Siege of Orléans. English cannons could not get the French to surrender.
Vigiles de Charles VII, by Martial de Paris, d'Auvergne, 1484

An attempted French siege of English-held Paris lasted only a few days in September 1429. Joan of Arc was captured and then executed by the English in 1431. As Christopher Lee wrote in *This Sceptred Isle*, she "perished on May 30, 1431, and thereafter the tides of war flowed remorselessly against the English... The

French artillery now became the finest in the world. Seven hundred engineers [fired] gigantic stone balls against the numberless castles which the English still held. Places which, in the days of Henry V, could be reduced only by famine, now fell in a few days to smashing bombardment."

A Convergence of Technologies Enabled the Disruption

A convergence of technologies worked in favor of the French. A discovery in the 1420s that the addition of limestone during refinement made the ore easier to work reduced the loss of iron and thus reduced the cost of iron. Greater purity and better consistency of metal meant that less labor was needed to turn it into a final product. And the lower cost of iron and greater ease of working resulted in longer gun barrels, which increased both the accuracy and the muzzle velocity of shots. Improved recipes for gunpowder approached the now-known chemically ideal ratio of saltpeter, sulfur, and charcoal, and exploded better due to our enhanced understanding of how grain size affects powder performance. As gunpowder got better, guns could be improved, allowing for better powder recipes to be used, in a self-reinforcing feedback loop. And eventually, we figured out to make the cannonball itself from iron, rather than stone, making both the ball and cannon smaller, and the projectile unlikely to fracture on impact.

By 1449, the French had retaken Rouen. At the Battle of Formigny in April 1450, the destruction of England's last army in Normandy allowed the French to retake all the northwest portion of their country. Bordeaux, the capital of Gascony, the last province held by the English, was surrounded, and surrendered without a fight on June 30, 1451. The English were decisively defeated at the Battle of Castillon, the last battle of the Hundred Years' War, in July 1453, taking severe casualties from French cannon fire. While the French lost about one hundred men, dead or wounded, the English lost about four thousand, a complete reversal of the ratio from Crécy.

The Disruption Played Out Over Just a Couple of Decades

This is how quickly the world of artillery changed. In 1422, Fatih Sultan Mehmed's father laid siege unsuccessfully to Constantinople though he had bombards. In 1428, the English could not take Orléans in seven months, though they had bombards as well. But in 1453, Constantinople fell after six weeks of cannon attack, and at about the same time, Bordeaux fell to the French without a fight at all.

Much of the improvement in artillery capability happened over just the decade of the 1440s. As Clifford Rogers wrote in 'The Military Revolutions of the Hundred Years' War', "as late as 1440 Harfleur was able to resist an English siege for over three months [but] made terms after a seventeen-day bombardment in 1449... As early as 1442, the French government was spending more than twice as much on its artillery train as it spent on more traditional war materiel - arrows, lances, bows, etc."

As with other technologies we have seen, artillery underwent a long period of stability, punctuated by a time of sudden change. "Cannon evolved for a full century", Rogers writes, "before they were able dramatically to change the European way of war, and they continued to improve steadily (if slowly) for centuries afterward."

In 1429, the French could not lay siege to Paris. But by 1449, they had the English on the run. Rogers continues, "A contemporary English document lists one hundred strong points taken by the French in 1450, including Chateau Gaillard, St.-Saveur-le-Vicomte, Cherbourg, Roche-Guyon, and Rouen, all of which had earlier required long sieges to capture... Based on the above accounts, it seems fair to say that a revolution occurred in the art of war around the 1420s to 1430s, as gunpowder artillery overturned the centuries-old dominance of the defensive in siege warfare." It took less than two decades for artillery to switch from trebuchets to cannons and for thick castle walls to go from being an advantage to being a burden, or at the very least, obsolete.

A similar dynamic played out in other parts of Europe. According to Rogers, "thanks to their command of a siege-train of some 180 guns, the 'Catholic Kings' Ferdinand and Isabella were able to reduce within ten years (1482-92) the Moorish strongholds in the kingdom of Granada that had defied their forbears for centuries... When the French marched into Italy in 1494, their artillery could do in a few hours what in Italy used to take days."

Rogers continues, "Central governments of large states could afford artillery trains and large armies [which] enabled the central governments to increase their control over outlying areas of their realms, or to expand at the expense of their weaker neighbors. This increased their tax revenues, enabling them to support bigger artillery trains and armies, enabling them to increase their centralization of control and their tax revenues still further, and so on. One scholar has estimated that the tax revenues of central governments in Western Europe doubled in real, per capita terms between 1450 and 1500; this feedback loop between military capability and economic mobilization ability helps account for that phenomenon."

Back in *Kostantīnīye,* the new capital of the Ottoman Empire, it was just the beginning of three decades of success for Mehmed the Conqueror and his giant guns. Over the course of his life, he would gain territories in what are now Serbia, Greece, Bosnia, Albania, Moldova, and Türkiye. In Wallachia, in what is now Romania, he defeated Vlad the Impaler (also known as Dracula) in 1477. By expanding his empire and solidifying his hold on his territories, Fatih Sultan Mehmed's Ottoman Empire posed an obstacle to European trade routes to Asian markets. In the latter half of the 1400s, Europeans suddenly found themselves in need of ships that could survive on open seas for long distances to carry them to far-off lands.

In Spain, cannons had given King Ferdinand and Queen Isabella formerly Muslim lands in their own south, but cannons in Muslim hands on the other side of the Mediterranean blocked them from accessing Asia. An Italian visitor to their court, Cristopher Columbus, had earlier traveled to Iceland to examine Viking maps of the North Atlantic. And now he had an idea to reach Asia by sailing west.

It is fair to say that the bombard cannon not only revolutionized siege warfare, but brought about the end of the era of castles altogether. But the rise of the cannon did not spell the end of fortified cities or of sieges. In fact, some of the longest in European history happened in the 1600s as people rethought their defenses and built fortified towns, often with walls in a star-shaped layout, with cannons on top.

The Age of Discovery

> *With Constantinople in Muslim hands, Europeans would need to find a new way to get to Asia. Building larger ships to sail around Africa required new forms of shipbuilding techniques. But Columbus's expeditions westward opened a new possibility: re-create the Asian export economy in an entirely new place*

In March 2021 the cargo ship *Ever Given*, en route from Malaysia to the Netherlands, careened to a halt in the Suez Canal, lodged into the sandy banks of the canal, where it remained aground for nearly a week blocking all traffic. She is one of the world's largest container ships, capable of carrying over 220,000 tons, or more than 20,000 twenty-foot-unit-equivalent (TEU) containers. Yuval Noah Harari, author of such best-selling books as *Sapiens* and *Homo Deus*, wrote shortly before the accident: "in 1582, the English merchant fleet had a total carrying capacity of 68,000 tons and required about 16,000 sailors." That is, the *Ever Given* on its own had a total carrying capacity more than three times the entire English merchant fleet in the late 1500s, while employing only twenty-five men. Put another way, the world of international maritime shipping went from transporting a little over four tons per sailor per voyage to nearly 9000 tons per sailor per voyage, a disruption of more than 2000X in four hundred years – and that's not even accounting for difference in sailing speed or cost.

Harari is not the first person to remark on how large new merchant ships are. Writing in 'England in 1685', the third chapter of his 1848 book *The History of England*, historian and politician Thomas Babington Macaulay wrote "The shipping [registered in Liverpool in 1685] was about fourteen hundred tons, less than the tonnage of a single modern Indiaman of the first class, and the whole number of seamen belonging to the port cannot be estimated at more than two hundred." That is, for at least 170 years we have been remarking that our ships today are larger than entire fleets were 170 years ago. That drive for Europeans to build ever-larger ships was propelled by the need to find routes to eastern Asia that avoided the Muslim-controlled Middle East.

The fall of Constantinople in 1453 laid the way for three centuries of ascendant power by the Muslim world, the so-called Gunpowder Empires – the Ottomans, the Safavids, and the Mughals – whose lands stretched from north Africa and southeastern Europe, through the Middle East, and throughout much of modern-day India. And their power was built on the confluence of their knowledge of metalworking and gunpowder to make cannons and firearms.

As the cannon disrupted the trebuchet and the castle, firearms disrupted the longbow and a thousand years of the use of metal armor. As we learned earlier with the bombard cannon replacing the trebuchet, the amount of kinetic energy delivered by a projectile to a target depends on the mass of the projectile and the square of its speed. By that standard, even the firearms of the early 1500s were ten times superior to longbows and crossbows. In *The Knight and the Blast Furnace: A History of the Metallurgy of Armour in the Middle Ages & the Early Modern Period*, Alan Williams estimates the initial energy of a lead ball fired from a 1500s arquebus, one of the earliest practical firearms, to be about twenty times the initial energy of a longbow's arrow and more than ten times the initial energy of a medieval crossbow's bolt.

With their access to Asian spices, sugar, silk, and other goods restricted, Europeans needed to find an alternative route to Asia, or some other means to acquire what they desired. In Europe, spices were much more than just condiments, they were medicine. As William J. Bernstein wrote in *A Splendid Exchange: How Trade Shaped the World*, "the contents of a medieval French spice store and a nineteenth-century American pharmacy would have been nearly identical." By the late-1400s, iron cannonballs and better metalworking techniques allowed cannons to get small enough to be carried aboard ships. All that was needed to reach distant lands were vessels that could endure the voyage.

From Clinker to Carvel Construction

Sailing across oceans requires more than just an adventurous spirit, it also takes vessels capable of surviving extended periods in open water. Early European cargo ships used so-called 'clinker' construction, in which a wooden frame was outfitted with overlapping boards. Viking ships like the longship and the knarr (the longship's less-stylish, cargo-bearing cousin) used this technique and over several centuries regularly made trans-Atlantic voyages from the northern parts of the European continent to Iceland, Greenland, and the north American mainland. But ships like this were clearly not going to get people halfway around the world and back again.

Starting in the 1450s, the Portuguese developed the caravel, with a displacement of about sixty tons, and advanced the design of the carrack (or *nau*), with a displacement of about one hundred tons, which they used to trade and explore for decades along the Atlantic coast of West Africa.

In 1470, Italian astronomer Paolo dal Pozzo Toscanelli suggested sailing west across the Atlantic to reach India and east Asia. Christopher Columbus, backed by Spain, set off on a journey in 1492 that made this crossing in seventy days. (In comparison, the first steamship to cross the Atlantic did so in 1819 in 207 hours, about eight and a half days, more than an 8X difference.)

Vasco da Gama, backed by Portugal, had another plan. In 1497 he set off to reach India by rounding the southern tip of Africa, which Bartolomeu Dias had found in 1488 while sailing in a couple of fifty-ton caravels. da Gama's largest ship, the *São Gabriel*, at 120 tons was one of the world's largest carracks at the time and only a little larger than Columbus's flagship, the *Santa María*.

By sailing around Africa, da Gama did something Columbus never achieved - he reached India. With the route to the East proven, there was a sudden market demand for ships that maximized the amount of cargo they could carry. In the year 1500 the second major Portuguese expedition to India, the Cabral armada (which famously reached Brazil along the way), had some ships between 240 tons and 300 tons. The *Flor de la Mar*, built in 1502 to make the India run, was four hundred tons, while at least one of the ships of the third major Portuguese expedition, the Albuquerque armada of 1503, is reported to have been as large as six hundred tons. That is, over a span of just *four* years after da Gama's return to Portugal in 1499, newly built merchant ships grew five times larger, from 120 tons to nearly six hundred tons. And that was not the end of it. The largest vessels built by the Portuguese for the India run, "the largest seagoing vessels of the era" according to Bernstein, included the 1600-ton *Madre de Deus,* built in 1589, and the *Cinco Chagas,* which made its maiden voyage in 1593, was two thousand tons, more than sixteen times larger than da Gama's flagship from one century prior.

The increase in the amount of actual cargo that these ships could carry was just as dramatic as the increase in the size of the ships themselves. A man might require about one gallon (four liters) of beverage per day, along with a pound of salted meat or dried fish, a pound of hard biscuit bread, some pickled vegetables, cooking oil, salt, honey, and other assorted provisions, adding up to about ten pounds (five kilograms) per sailor per day in total. A three-month voyage across the Atlantic would then require about a half ton of goods per person and a one-year voyage to India would require about two tons of provisions per person, taking away from the space available for actual cargo.

The Viking knarr had a cargo limit of about twenty-four tons and the crew would require about three or four of those tons just for their own provisions for a trans-Atlantic trip, leaving about twenty tons of space for actual cargo, or about 2.5 tons per sailor per voyage if there were eight people onboard. A ship the size of da Gama's *São Gabriel* or Columbus's *Santa María* loaded with enough provisions for forty people for three months would have had no more than one hundred tons available for actual cargo, again a ratio of about 2.5 tons per sailor per voyage.

Small ships like these were acceptable for exploration and for resupplying small colonies but supporting higher-volume trade with distant continents would require larger vessels. Much larger. The *Jesus of Lübeck*, launched in 1520, the first English ship to carry enslaved people from Africa to the Americas, used eighty men for a 700-ton ship, which meant that the capacity for actual cargo, after crew provisions were loaded, was about eight tons per sailor. Had the *Jesus of Lübeck*'s cargo been inanimate, rather than people, the ratio of cargo to sailors might have been about ten tons, very similar to the *Madre de Deus.*

Of course, sending enormous ships to India and beyond did not simply replace the overland spice trade on a one-for-one basis. In *A Splendid Exchange*, Bernstein writes, "When da Gama left Lisbon on his first voyage in 1497, Europeans consumed less than two million pounds of pepper [spice] per year. By 1560, this amount had grown to between six and seven million pounds."

While da Gama enabled the Portuguese to sail to Asia, Columbus allowed the goods from Asia – spices, cotton, silk – to be produced in an entirely new location, one far closer to home. In 1505 the first sugar cane was grown in the New World, cotton had been purposely grown there by native people for thousands of years, silkworms were imported to Mexico in the 1530s, entirely new products like tobacco, chocolate, vanilla, corn (maize), and the potato were exported for the first time, and in this period the regular commercial trade began in importing the labor, in the form of enslaved people, from Brazil and the Caribbean to the North American mainland.

Mechanized Production of Pulley Blocks

In his book *Exactly: How Precision Engineers Created the Modern World*, Simon Winchester describes how the Portsmouth Block Mills were able to supply the British Royal Navy with the 130,000 pulley blocks it needed annually, by replacing skilled craftsmen with machines: "Where more than a hundred skilled craftsmen had once worked, and had filled, just, the navy's insatiable appetite, now this thundering factory could feed it with ease, without ever breaking a sweat: the Portsmouth Block Mills would turn out the required one hundred thirty thousand blocks each year, one finished block every minute of every working day, and yet it required a crew of just ten men to operate it." Another 10X disruption.

Chapter 3:

The Industrial Era

◆ ◆ ◆

"From barbarism to civilization requires a century; from civilization to barbarism needs but a day."

Will Durant, The Story of Civilization

◆ ◆ ◆

The Steam Engine, Copper-Bottom Ships, and the Atlantic Slave Trade

> *Watt's steam engines revolutionized mining, making slave-trading ships faster (and more profitable) by giving them copper bottoms. But steam power also disrupted the desire to enslave people in the first place*

Transatlantic slavery started with Christopher Columbus. On his second trip to the Americas in 1493, Columbus himself sent "several hundred" Taino people from the island of Hispaniola to Spain. In 1518, King Charles I issued an edict authorizing the shipment of enslaved people directly from Africa to the Americas, effectively launching the commercial Atlantic slave trade.

One of the primary reasons for bringing people from Africa to the Americas was obvious. In demographic terms, the effect of the first century of European voyages and colonization on the indigenous population of the Americas was nothing short of devastating. According to a study by Dutch researchers published in the scientific journal *The Holocene*, the population of what is now called Latin America fell by almost 80% between the years 1500 and 1600, and it would not surpass its pre-Colombian level until the 1860s, more than three-and-a-half centuries after first contact. (Klein Goldewijk, 2010) Much of this drop happened in the first half of the century. While the population of Mexico was more than twenty million at the start of the 1500s, a smallpox outbreak in around 1520 killed about eight million people and the so-called Cocoliztli epidemic, of unknown cause, killed at least twelve million more around 1545. By 1550, the population of Mexico had decreased about 90% compared to its pre-contact level.

The website *SlaveVoyages.org*, a project funded by the U.S. National Endowment for the Humanities, offers a freely downloadable database which records Atlantic slave-carrying trips departing from Africa and landing in the Americas, the Caribbean, Europe, and elsewhere over a period of more than three hundred years.

This is a disturbing data set, but one that illustrates some of the key features we see in the adoption of innovations. Below is a graph of the number of people transported by year in the Atlantic slave trade. In total, it contains the records of nearly ten million individuals, more than half of whom were sent to the Caribbean, 35% to Brazil, 3.75% to mainland North America, and 10% to other places. Of the people transported across the Atlantic, more than 96% went aboard European ships. (In December 2022, Dutch prime minister Mark Rutte issued an apology for the Netherlands' role in the Atlantic slave trade, the first European country to do so.)

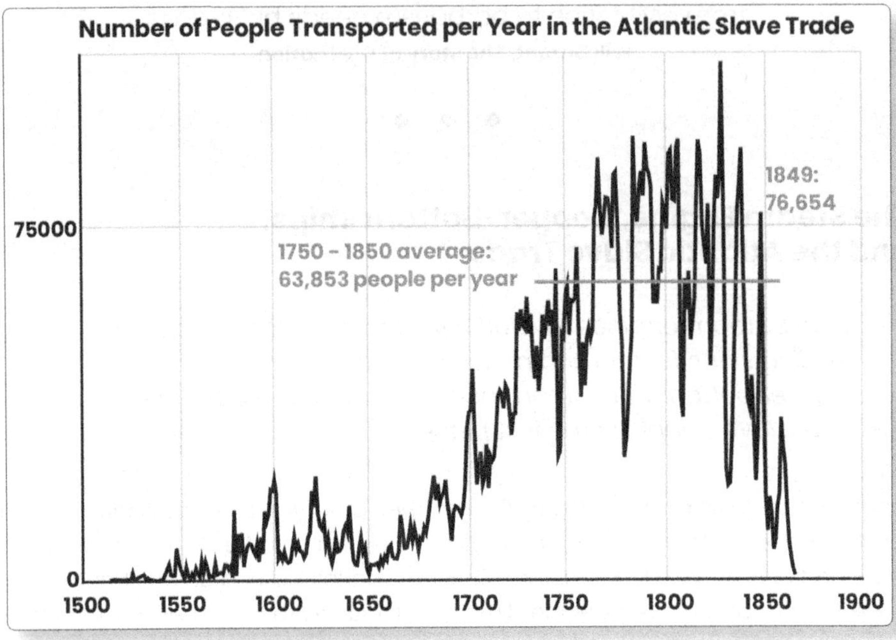

Figure 33 – Number of People Transported per Year in the Atlantic Slave Trade.
Data from SlaveVoyages.org

The total story of the trade in enslaved people forms a hump-like curve – with an S-shaped rise over a couple of centuries, a peak era that lasted roughly from 1750 to 1850, and then a decline in the mid-1800s that was relatively rapid. In 1849, an above-average number of people were brought across the Atlantic. But in 1850, the number of people transported was half of 1849, and 1851 was half of 1850. Within fifteen years, the trade had ended completely.

We already saw in the previous chapter how the fall of Constantinople in 1453 launched the Age of Discovery, where European nations raced to find ocean routes to the eastern parts of Asia. The new travels to the Americas opened an alternative possibility: the creation of exotic goods on other continents, ones that were much closer to Europe. But reliable trans-Atlantic travel would require eliminating a problem that seemingly had no good solution.

A Problem with Seemingly No Good Solution

The wood-eating shipworm, *Teredo navalis* (a type of long-necked clam), plagued seafarers for centuries. The ancient Norse text *The Saga of Erik the Red* tells of an Icelander, Bjarni Grímólfsson, who voyaged onboard his merchant ship to Greenland and spent a winter there with Erik the Red, the father of Leif Eiriksson, who was as the first European to set foot in the Americas. On Bjarni's return trip from Greenland to Iceland, his ship was blown off-course and, as the Saga says, "came into a part where the sea was infested by ship-worms... They had a ship's boat which was smeared with tar made of seal-fat. It is said that the ship-worm will not bore into the wood which has been smeared with the seal-tar." (Sephton, 1880)

But the lifeboat "held not more than half the men." They cast lots to choose who would get a place onboard, but Bjarni decided to stay behind so that a younger man could take a place. "It is said that Bjarni perished there in the Worm-sea," the Saga continues, "and they who were with him in the ship; but the [life]boat and those who were in it went on their journey until they reached land, and told this story afterwards."

It seems unlikely that a ship would have been eaten so quickly just from passing through worm-infested waters. However, worms are known to begin attacking wood literally within weeks of a ship being launched, so if Bjarni's ship had passed through those waters on the way to Greenland, then it is possible the planks could have been rotten through by the time of the return voyage.

These events described in the Saga happened around the year 1006, but five centuries later, in 1503, it is said that Columbus lost two ships to worms in his fourth voyage to the Caribbean. As the sailing routes to Africa and the Americas grew more popular and ships spent more time in tropical waters where worms act faster, the need for worm-fighting techniques grew more pressing.

Shipmakers tried a variety of possible solutions. Tar made of seal-fat, of course, was in limited supply outside Arctic waters, so sacrificial wooden cladding was added to ships so when the wood got infested, the cladding could be removed and replaced. Lead plates did not hold up well in salt water and taoking many flathead nails all over a wooden surface was expensive, as nails were made individually by hand until the 1800s.

For centuries shipwrights filled in the gaps between the boards of their carvel-built vessels with hemp or other materials and slathered the surface with tar or pitch to protect the wood from being eaten. But this ultimately could not stop the worms, nor keep barnacles and weeds from growing below the waterline, weighing the ship down and slowing its travel. So, periodically, ships would need to be hauled out of the water, scraped clean, and re-tarred.

An Outsider Finds a Solution

But a Sunday stroll in May 1765 changed that, and much else, forever. Newcomen steam engines, what one early account called "The Invention for Raising Water by Fire", had been around for decades, but they were not very efficient. Steam would fill a cylindrical chamber and then cold water was sprayed in to cause the steam to condense, creating a vacuum that would pull down a piston head. In this way, the heat from a fire would boil water, which would power a pump that could raise water out of mines, making mining possible in places where it otherwise would not be.

The only problem was that these steam engines were not very good. They took a lot of energy to power a pump that was both slow and unreliable. But on that walk in May in Scotland, twenty-nine-year-old James Watt got the idea to separate the condenser, which would always by kept cold, from the steam cylinder itself, which was always warm, and to use a valve to connect them.

This arrangement was about five times more efficient than Newcomen's engines, taking only about 20% as much fuel to do the same amount of work. But this caused new problems for Watt and his business partner, Matthew Boulton. How much should they charge for this device? This required Watt and Boulton to make a business model innovation. They would make key parts of the steam engine, and then take one-third of the savings in the expenditure for coal, leaving the remainder for the buyer.

The invention of this new technology also required new metrics to describe it. Potential customers wanted to know how much work the new steam engine could do, but there were no good terms to describe this concept. In a 1936 article 'The Origin of the Horsepower Unit' in the *American Journal of Physics*, Harry L. Smith said that James Watt "kept 'Blotting and Calculation' books, wherein he made many notes and calculations from time to time. It is in such a book for the years 1782 to 1783 that we find a record of the steps by which Watt arrived at the unit of horsepower now generally accepted and used."

So, to bring his steam engine to customers, Watt not only needed to invent the device and build a facility to make its key parts, but he also needed to invent the 'we get one-third of the savings, you get two-thirds' business model and invent the mathematical equation to describe how much work his engines could do compared to the standard Newcomen engines (or, conversely, how much fuel his engines need to do the same amount of work as a Newcomen engine).

But Watts' engines did not do the same amount of work as Newcomen's. More-powerful, more-efficient engines kicked off an industrial revolution, first being employed in mines that supplied the metals to make machines (including metals for steam engines, a self-reinforcing feedback loop). And their steady power enabled mechanized factories to spin thread, weave fabric, carve gun stocks or furniture parts, and all sorts of other processes that Newcomen's unreliable and underpowered engine was unable to do.

In northwest Wales, on the island of Anglesey, Parys Mountain was one of the largest known copper deposits in the world. It had been exploited since the Bronze Age because the deposits were near the surface, where they were easy to access. But unfortunately, the ore was not very high grade which meant that it took a lot of energy, and therefore a lot of coal, to refine it into metal product. So much coal was needed that it was easier to bring the ore to the coal than the coal to the ore.

Watt's engines did not just make metals easier to obtain and fabric and woodwork easier to produce, they also made it easier to mine for the coal that fuelled those machines and, in later years, steam-powered trains and ships. They were a revolution in materials, energy, and transportation all wrapped up in one. They would also turn out to be the solution to the shipworm problem.

Industrialized mining began at Parys in 1775 and within fifteen years, it was the largest copper mine in the world. Ore was loaded onto ships and brought south to Swansea, where there were coal reserves that could be exploited due to Watt's steam engines. Combined with the output from Cornwall in the southwest of England, by 1790 British mines were producing more than 75% of the world's copper. (Symons, 2003)

Ten years after Watt's patent, over a period of just two years from 1779 to 1781, the entire British Navy was clad with copper bottoms. According to Rees: "copper sheathing not only solved the problems of worm and hull fouling [like barnacles and weeds], but actually improved sailing speed as an unexpected and welcome by-product."

Copper-clad ships travelled about 15% faster, meaning that an 80-day Atlantic crossing could be cut by more than ten days. So, copper cladding did not end with the Navy. All ships that went to tropical waters and that needed to move quickly benefited from copper bottoms, and there was one kind of merchant ship that benefited from greater speed more than any other - slave traders. They too went from less than 10% having copper bottoms to more than 70% having copper bottoms over a period of just two or three years, starting in the late 1770s.

The glory days of the Parys mine did not last long. By 1800 the output was down by more half from its peak. When Thomas Williams, the "Copper King" of Anglesey, died in 1802, he was the richest man in Wales and 1200 people were employed in his Parys and nearby Mona mines. Five years later, that number had dropped to just 120.

Copper sheathing had a substantial effect upon death rates in the Atlantic slave trade. The *SlaveVoyages.org* database suggests that the death rate was about half on a copper-bottom ship compared to one without sheathing. The *SlaveVoyages.org* records show that, prior to 1780, the proportion of enslaved people lost during the voyage was approximately 20% and, after the early 1780s, decreased to about 10%.

The Collapse of the Atlantic Slave Trade

What brought about the end of the trans-Atlantic transport of enslaved people? Speaking before the UK Parliament in 2019, environmentalist and TV presenter David Attenborough said, "there was a time in the 19th century when it was perfectly acceptable for civilised human beings to think that it was morally acceptable to actually own another human being for a slave. And somehow or other, in the space of twenty or thirty years, the public perception of that totally transformed."

Why did the trade collapse so suddenly? Perhaps it was a mass spiritual awakening, like the one by English clergyman John Newton, who in 1772 wrote the hymn that begins "Amazing Grace how sweet the sound, that saved a wretch like me", based on his experiences a couple of decades before as a slave trader. The 1850s were a time of growing abolition movements in the US, spurred on by books like Harriet Beecher Stowe's *Uncle Tom's Cabin*, published in 1852. But they were also time of rapid development of key technologies, particularly concerning coal-fired steam power, with which muscle power could no longer compete.

The Barbados Bubble: An Island of Sugar

> *Sugar was once at least as important to Europe, economists estimate, as the Internet is today. So important that a few small islands became prized possessions and the sites where much slave labor was used*

When John Russell, the 4th Duke of Bedford, sat down in Paris in 1763 to sign the treaty that sealed British victory against France in the Seven Years' War, there were many who thought he was agreeing to a bad deal. During the war, the British had captured from France the Caribbean island of Guadeloupe, named by Columbus himself on his second voyage in 1493, and now Russell was giving this fertile land back in exchange for France's North American colonies east of the Mississippi River (except New Orleans). That's how valuable sugar was: it was a genuine uncertainty whether it was a good deal to give up an island about the same size as Greater London in return for the land between the Appalachian Mountains and the Mississippi River plus most of eastern Canada.

On Columbus's first voyage, he brought cuttings of sugar cane with him, for the chance to grow a new supply of sugar if an established source could not be found. Muslims had done the same many centuries before as their faith spread throughout the Middle East and around the edges of the Mediterranean. According to historian J. H. Galloway, "sugar followed the Koran." Even the word 'sugar' comes from Arabic, *al-sukar.*

The Crusaders discovered those white crystals, what they called "sweet salt", during the Middle Ages but even by the time Constantinople fell to the Ottomans in 1453, sugar was still a rarity in Europe. The voyages by Columbus across the Atlantic and by da Gama around the southern tip of Africa, what economist Adam Smith called "the two greatest events recorded in the history of mankind", would change that forever.

According to James Rogers in 'A History of Agriculture and Prices in England', in the first half of the 1400s, sugar in England cost '24 shillings per dozen pounds', but sugar was so rare that this estimate was based on only 'three entries from opulent persons'. "Between 1490 and 1510", he wrote, the price of sugar "rapidly declines. In 1495 and 1503 it is about 2 shillings 9 pence [for] the dozen [pounds]," a drop of nearly 90% in less than a century. By 1572, Fernand Braudel wrote in *The Structures of Everyday Life*, a French observer commented that "people devour it out of gluttony... What used to be a medicine is nowadays eaten as a food."

Sugar was more than just a sweetener, it was also a source of calories, a food preservative, and an ingredient for making alcohol. A 2009 economic study by Jonathan Hersh of the Wharton School of Business at the University of Pennsylvania estimated that by 1850, the average Englishperson should have "been willing to forego" about 8% of their income "to maintain access to sugar" (which is not to say that is how much they spent, only how much economic benefit they got from having sugar in their lives). In comparison, a similar study found that economic benefit of the Internet in 2005 was about 3%. Sugar had more than double the impact on Englishpeople a couple of centuries ago that the Internet had on us by 2005. It's no wonder that sugar cane revolutionized the economy of every land where it could grow.

In 1637, sugar was introduced to Barbados by Dutch traders from Brazil. According to Willian J. Bernstein in *A Splendid Exchange*, "by 1645, 40 percent of the island was planted in sugar... By 1660, it had more settlers than Virginia or Massachusetts... four times the population density of England. It had become the world's largest sugar producer, supplying almost two-thirds of England's consumption." By 1660, Barbados generated more trade than all the other English colonies combined. So much of the island had been taken over by sugar cane that its inhabitants imported food, to give themselves more land to grow cane. They not only tore up their gardens, they cut down their forests. According to J. R. McNeil in his 2010 book *Mosquito Empires: Ecology and War in the Greater Caribbean, 1620-1914*, "surviving sources make no mention of forest after 1665, and by 1666 Barbados imported timber and fuelwood from Surinam."

In his 1990 book *The Rise and Fall of the Plantation Complex: Essays in Atlantic History*, Philip Curtin asked "why did a sugar revolution occur in the eastern Caribbean in the 1640s, rather than earlier or much later?" He attributes this local revolution to a confluence of factors including the Dutch bringing sugarmaking technology from Brazil and also starting to bring enslaved people from Africa, who were newly available in the Caribbean as part of a recent shift in the slave trade away from Spanish Mainland Americas to the Caribbean and Brazil.

The production of sugar requires a great deal of labor. Sugar plantations were the reason for 84% of all enslaved people being brought from Africa to the New World, and about 86% of all enslaved people transported in the Atlantic slave trade went to either Brazil or the Caribbean. In 1644, the population of Barbados was less than 3% of African descent. By 1660, just sixteen years later, it was over 50%.

The effect this blossoming production had on the price of sugar in England was profound. Hersh estimates that the price of sugar in England was about twenty-seven pence per pound in 1630, 1640 and 1650, but by 1660 had fallen to sixteen pence per pound, a drop of about 40% in ten years, and it continued generally downward in later decades. Lower prices, in turn, meant greater consumption. "While medieval Cyprus produced no more than an estimated 50-100 tons of sugar per year," Hersh wrote, "England in 1700 imported approximately 10,000 tons."

One of the unintended consequences of greater sugar availability was a rapid rise in tea consumption in the 1700s and coffee in the early 1800s. English breakfast, Hersh wrote, "changed from porridge and cold meat consumed with beer, to the morning meal we still consume today – bread or pastry, combined with a sugary hot beverage full of caffeine."

But Barbados's place of importance in the English sugar trade would not last long. "After 1680," Bernstein wrote in *A Splendid Exchange*, "falling sugar prices, English tariffs, and depleted soil and forests ruined its plantations, and many planters fled to greener pastures in the New World."

The Seeds of the U.S. Civil War

> *Radical improvements in spinning and weaving in the late 1700s and early 1800s greatly increased the demand for cotton fiber. A hybrid breed of Mexican cotton revolutionized the American industry in the early 1800s, opened new areas of the South for cotton production, and brought in a lot of money in the export trade, making some Southerners think an attempt at secession could be to their benefit*

By 1860, the eve of the U.S. Civil War, the United States accounted for about three-quarters of world cotton production and cotton was more than 50% of US exports. For a Federal government which did not levy personal income taxes or Social Security taxes, and which made most of its money from trade tariffs, cotton was king. But it was not always that way.

When Abraham Lincoln was born in 1809, cotton accounted for less than about 15% of US exports of goods and services combined. By the time he was first elected to the Illinois State Legislature in 1834 at the age of twenty-five, cotton exports had surpassed 60% of US combined exports of goods and services, and cotton would continue to generally account for about half of US exports until the war began.

Cotton exports did not just grow as a percentage of US exports, they grew in absolute terms as well. By 1860, exports of cotton were about fourteen times larger in terms of the number of bales and ten times larger in terms of dollar revenue than they were in the year of Lincoln's birth. The American cotton industry, and with it the contentiousness of slavery as a moral issue, had grown up largely within Lincoln's lifetime.

Growing cotton outside the United States, of course, had a long history. In her 2020 book, *The Fabric of Civilization*, Virginia Postrel wrote that "the Islamic caliphate spread the cultivation of early-blooming cotton along with its new faith. Islam promised the faithful silks in heaven but forbade them to Muslim men in this world. Wearing cotton became a mark of devotion, and the demand for cotton grew with each new convert." "By the 10th century AD," she continued, "cotton was found growing in nearly every region of the Muslim world, from Mesopotamia and Syria to Asia Minor, and from Egypt and the Maghreb to Spain."

But it took an outsider to invent a device that revolutionized the spinning of cotton fiber, and in doing so, to help launch the Industrial Revolution. "In 1764, an illiterate weaver from Lancashire [England] named James Hargreaves produced an ingeniously simple device known as the spinning jenny, which did the work of ten spinners by incorporating multiple spindles," author Bill Bryson wrote in *At Home: A Short History of Private Life.*

The impetus for Hargreaves' invention was a simple imbalance in the output of weavers compared to spinners. "Spinning was the bottleneck in fabric production," Postrel wrote, "and a bottleneck is a problem waiting to be solved." A report from the International Institute for Applied Systems Analysis (IIASA) in Austria says, "it took three or four spinners to supply one full-time weaver by traditional methods... An even more severe imbalance was created by the invention by John Kay in 1733 of the flying shuttle, which essentially doubled the output of hand looms when it was adopted widely in the 1750s". (Ayers, 1989)

So, the widespread adoption of the flying shuttle in the 1750s meant that about eight spinners were needed for each weaver, which created the demand for a more-efficient way to spin cotton thread. "Hargreaves's patent specified 16 spindles," correcting this imbalance, the IIASA report continues. However, "the number rapidly increased; by 1784 jennies could handle 80 spindles, and by 1800 the number had risen from 100 to 120. It reached 1,200 by 1832... In effect, the productivity of a single operator was multiplied by 100-fold in the space of a generation and 1,000-fold in two generations."

The effect on British cotton thread production was profound. "Before his invention," Bryson wrote, "homeworkers spun 500,000 pounds of cotton in England every year by hand. By 1785, thanks to Hargreaves's machine and the refined versions that followed, that figure had leapt to 16 million pounds." That is, in just twenty years, annual cotton thread production went from about one ounce per person, to about two pounds per person. "Thread output soared so much", Postrel wrote, "that weavers became the new bottleneck."

"Those within the handloom weaving trade enjoyed a golden heyday," Postrel wrote, "with as much work as they could want and at high rates of pay... The golden heyday was not to last. Power looms arrived at the turn of the century and with them the fabled Luddite movement, as yesterday's winners became the new economic losers. In one of history's ironies, the hand weavers who broke the looms that threatened their jobs — and became synonymous with resistance to new technology — owed their imperiled livelihoods to earlier, and far more disruptive, advances.

According to IIASA, the productivity of power looms increased dramatically during the early part of the 1800s, as cheaper cast iron made it possible to build looms from materials that were sturdier and more precisely made than wooden frames. By 1800, an operator of a power loom could produce roughly double the amount of fabric as a hand loom operator from just a few years prior. By 1810, the power loom operator could produce about four times as much, and by 1830, nearly twenty times as much. "Within a few years, water-powered spinning mills had spread throughout northern England, turning out previously inconceivable amounts of low-cost cotton yarn," Postrel wrote.

A New Business Model of Fabric Production

Power looms and spinning jennies did not just change how quickly and easily cotton fabric could be produced, they also caused a radical change in the business model of where and how people worked. Power equipment not only required a large building near fast-running water, it also required consistent quality of thread. Textile production went from something done at home, in small amounts, for direct domestic consumption to a massive, world-moving industry run in large factories. This also changed who did the work. Hand looms were typically operated by men and spinning wheels by women, but powered weaving and powered spinning could be done by children.

"Abundant yarn created ripple effects within nearly every aspect of life," Postrel wrote. As the first business to be industrialized, textiles were therefore also responsible for many of the non-tangible innovations associated with the Industrial Revolution: standardized sizes, quality inspectors, financiers, letters of credit, wholesalers, middlemen, brokers, and so on.

The effects of low-cost fabric radically transformed life for nations like Britain and the United States, not only by creating of new export industries, but also by literally providing the sails to carry those exports around the globe.

For most of history, sail-making was laborious and, therefore, expensive. In *The Fabric of Civilization*, Virginia Postrel says, "a Viking Age sail 100 meters square took 154 kilometers (60 miles) of yarn. Working eight hours a day with a heavy spindle whorl to produce relatively coarse yarn, a spinner would toil 385 days to make enough for the sail. Plucking the sheep and preparing the wool for spinning required another 600 days. From start to finish, Viking sails took longer to make than the ships they powered."

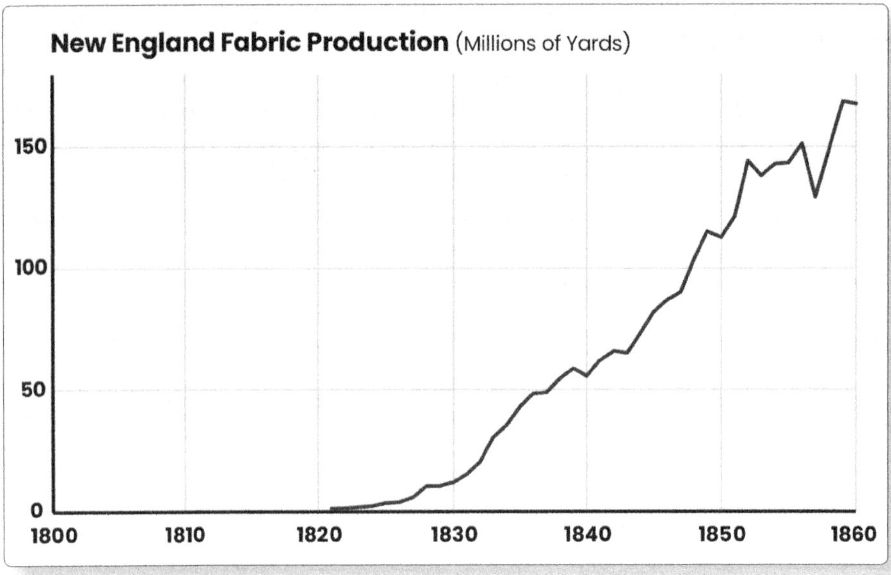

Figure 34 – New England Fabric Production (Millions of Yards)

One thousand years later, cheaper sail fabric not only reduced the cost of shipping, it also sparked a revolution in ship design, with the mid-1800s seeing the rise of clipper ships, which were fast because of their large sail area. Sail power drove the expansion of the British Empire, boosted the China trade, and enabled the whaling boom and the guano (fertilizer) trade, among other things.

"The story of thread illustrates how, despite immediate dislocations, labor-saving technologies can create abundance and free people's time for more economically valuable and personally satisfying purposes," Postrel wrote. But for some people, these new technologies did not do any of those things. While the mechanization of spinning cotton fiber into thread and weaving cotton thread into fabric saw tenfold (or greater) advancements over a short period of time in the early 1800s, the agricultural activity of producing spinnable cotton fiber struggled to expand accordingly.

Eli Whitney had patented a device for removing seeds from cotton bolls in 1794, long before the North had industrialized textile production and before the South was even beginning its development to becoming the cotton-growing center of the world. A person could pick the seeds from about one pound of cotton per day by hand. But Whitney's cotton gin was able to process fifty pounds per day.

With revolutions in fabric weaving and thread spinning underway, supplying the textile mills would require a revolution in cotton growing, one whose seeds were provided (quite literally) in 1806 by an American named Walter Burling on a trip to Mexico, when he found a variety of cotton that, according to Postrel, "proved ideal for the Mississippi frontier". "It ripened early, avoiding the frost. The bolls all appeared at about the same time, allowing an efficient harvest, and they were

Figure 35 – Percent of US Exports of Goods and Services that are Cotton. *(Francis, 2021)*

large and opened especially wide, which made them much easier to pick... The ratio of fiber to seed was significantly better, yielding about a third more usable cotton after ginning. And the Mexican cotton was immune to a disease called the rot, which threatened to wipe out the area's cotton production. By the 1820s, farmers in the lower Mississippi valley had widely adopted the new breed."

She continues to say that Mississippi valley cotton farmers "also improved it, both accidentally and deliberately. By carelessly allowing it to cross-pollinate with Georgia Green Seed, they inadvertently created a hybrid that preserved most of the advantages of the Mexican variety while eliminating its biggest drawback: the tendency of bolls to drop on the ground if not immediately harvested. Seed breeders then worked intentionally to refine the stock. By the early 1830s, a new Mexican-based hybrid called Petit Gulf dominated the Mississippi valley and was flourishing in the red clay soils farther east."

Agricultural historian John Hebron Moore wrote, "Mexican cotton... produced more abundant yields than its predecessors and could also be harvested more easily. Ordinary field hands were able to pick about sixty pounds of seed cotton a day in fields of Siamese or Georgia green seed cotton; in fields of Mexican cotton, they could gather 250 pounds [more than four times as much] with no more exertion."

US cotton production began to take off with the introduction of Burling's Mexican breed and then accelerated in the 1830s with the Petit Gulf hybrid. It doubled from 1820 to 1835 and then doubled again in the next fifteen years to 1850. In just the years from 1850 to 1859, cotton production nearly doubled one more time. Something had to give.

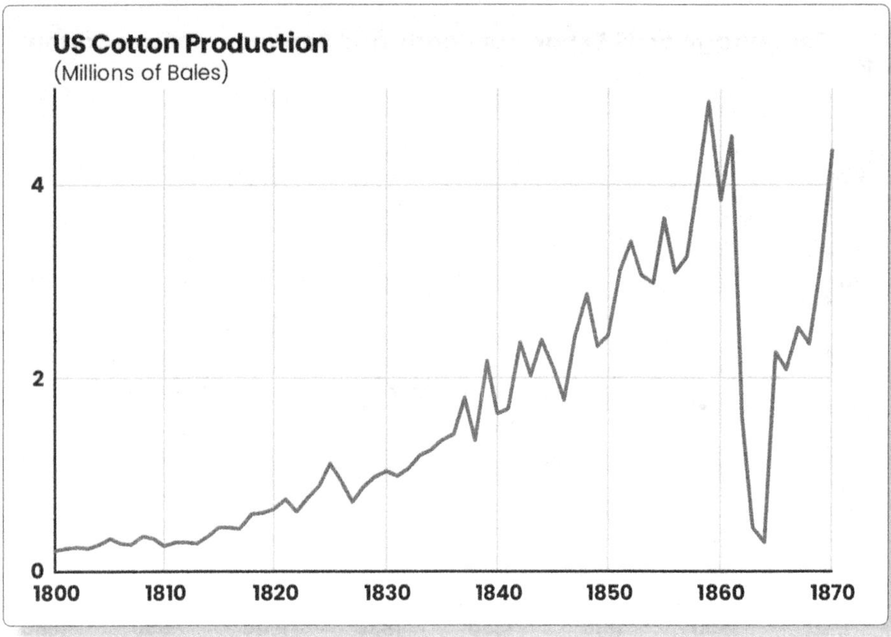

Figure 36 – US Cotton Production (Millions of Bales). *(Watkins, 1908)*

Petit Gulf did not just replace acreage of other cotton breeds on a one-for-one basis, it opened entire new areas to cultivation. While in 1810, 93% of cotton production was in the four Atlantic states that were part of the original thirteen colonies (Virginia, Georgia, and the Carolinas) and only 7% in the remainder of the South, by 1830, production was evenly split between those 'Original South' and the 'Extended South': Alabama, Mississippi, Florida, Tennessee, Louisiana, Arkansas, Missouri, Oklahoma, and Texas. By 1860, more than three-quarters of US cotton production was outside of the four Southern states that were part of the original thirteen colonies.

As the cotton industry grew immensely in the decades leading up to the war, the Federal government tried to maintain a balance of power between Free and Slave states; new states were admitted to the expanding nation such that the Senate remained split evenly for most of the period from 1810 to 1850. However, population growth meant that the percentage of seats in the House of Representatives (and therefore, the Electoral College) that were from Slave states fell almost continuously through the early 1800s.

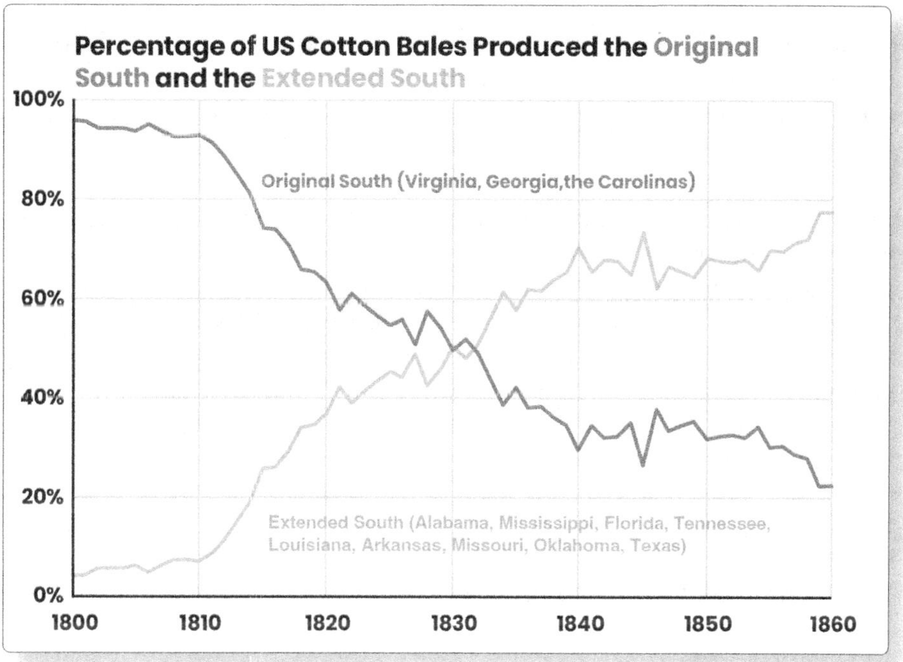

Figure 37 – US Cotton Production in the Original South and the Extended South. *(Watkins, 1908)*

A series of admissions of Free states in the fifteen-year period prior to the start of the war put the South into the position of delivering the bulk of US exports while also being outnumbered in both the House of Representatives and the Senate. Texas was the last Slave state admitted before the war, in 1845. But the admission of the Free states of Iowa in 1846, Wisconsin in 1848, California in 1850, Minnesota in 1858, Oregon in 1859, and then Kansas in January 1861, meant that the South had little chance of seeing balance in the Senate ever again.

Researcher Joseph A. Francis wrote, "Southern slaveholders recognised that their cotton exports were being used to finance their own disempowerment in Congress, so they seceded. The North would not let them go, however, because they threatened to take with them the customs revenues that the Federal Government had relied upon up to then. The result was the Civil War".

The five southeastern-most states – South Carolina, Georgia, Alabama, Mississippi, and Florida – which were the five with the highest percentage of the population that were enslaved, were also the first five to secede.

From 'the Scramble for Africa' to Asian Shipbuilding

> *Steam power made it as easy to travel north-south as east-west, making the coast of Africa more alluring to European colonizers. But when diesel-powered ships disrupted coal-powered ones in the 1920s, the expansion of African colonies ended, and the industrialization of northeast Asia took off*

Enslaved muscles were disrupted more by coal-fired steam power than any hymn or book. If we wish to understand the decline of the trans-Atlantic trade in enslaved people, we should study the rise of coal-fired steam power. On the ocean, the rise of steam ships spelled the start of the long decline of sail ships.

The graphs below show both the number of sailing ships and steam ships, as well as the tonnage of sailing ships and steam ships, for the United Kingdom by year from the year 1800 to the eve of World War II. The number of sailing ships peaked in the early 1860s, when the number of steam ships was only about 5% of the number of sailing ships. By the early 1900s, steam ships outnumbered sail ships, but this was not a one-for-one disruption: the peak number of steam ships was only about half of the number of sail ships.

When we look at the tonnage of ships, however, it was a very different disruption. Steam ships in the 1850s were about the size of sail ships, but as time went on, steam ships grew larger because they were not restricted by sail size and could be made of steel instead of wood.

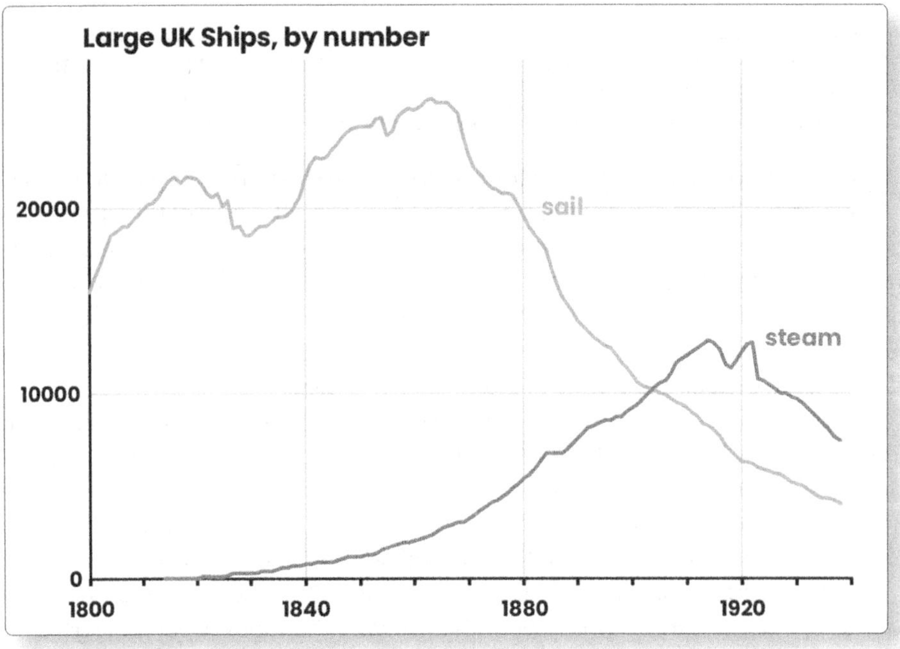

Figure 38a – Large UK ships powered by sail and steam, by number and by tonnage. *Data from National Bureau for Economic Research (NBER) Cross-country Historical Adoption of Technology (CHAT) data set (Comin, 2009)*

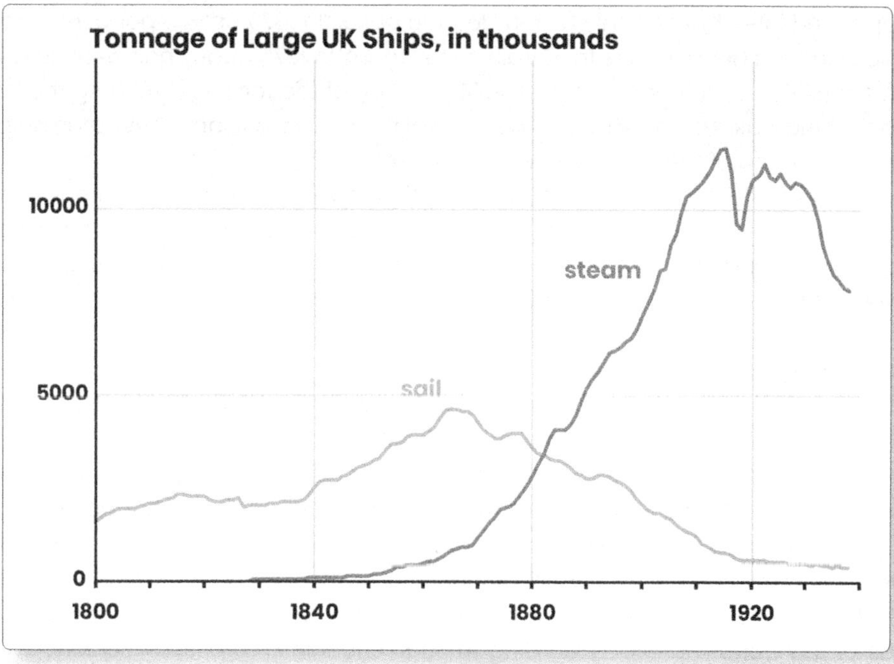

Figure 38b

In 1870, the tonnage of UK steam ships constituted about 20% of the total tonnage of ships in the water. Just fifteen years later in 1885, less than the normal operational lifetime of one ship, steam ships were 55% of the tonnage.

As with some other disruptions, steam-powered ships started as a 'disruption from above'. That is, steam-powered travel was initially more expensive than sail-powered travel, but steam ships are also 'higher performing' than sailing ships in that they offer opportunities that sailing ships did not. Steam ships have a greater flexibility in which directions they can travel and at which times, being able to go almost as quickly against the wind as with it. This was important for expensive, perishable goods like tea leaves, "whose price stood in direct proportion to their freshness, galvanizing the shift from sailing ships to steamers," according to On Barak, author of *Powering Empire: How Coal Made the Middle East and Sparked Global Carbonization.*

Unlike sailing ships, steam ships required fuel, which made them expensive to operate. Charles Harley in the book *Essays on a Mature Economy: Britain After 1840* estimated the cost of cargo travel from UK to New Orleans, USA, in 1872 at £1230 for 1000 tonnes by sail and £1400 for the same cargo by steam ship. For UK to Kolkata (Calcutta), India, in the same year, £2610 by sail and £2640 by steam. So, even in the early 1870s, steam travel for cargo was still more expensive than sail.

As a result, according to Harley, "most of the uses of steamships prior to 1865 involved premium freights for speed and regularity". "In 1856, 96 per cent of the passengers arriving in New York from foreign ports arrived in sailing ships. By 1862 the steamers carried 32 per cent of the westbound passengers to and almost all the eastbound from New York." For products like tea and cotton the switch happened later, but still only took a decade or so. "In 1870 only 10 per cent of the ships clearing New Orleans for Britain were steam, in 1875 nearly half were steam, and in 1880 over 70 per cent... In 1869, 14 per cent of the tonnage entering the U.K. from China was steam. By 1873, 70 per cent of the trade was carried by steamships and by the end of the decade over 90 per cent."

This steam revolution required coal and therefore coaling stations around the world. Unbound by the direction of prevailing winds, steam ships were able to move north and south as easily as east or west. This set off what historians call 'The Scramble for Africa'. "In 1880 the continent of Africa was largely unexplored by Europeans. Less than thirty years later, only Liberia and Ethiopia remained unconquered by them." (Pakenham, 1992)

Steam power brought about the end of the reliance on muscle power for labor and sails for transport. But steam ships themselves got disrupted by diesel engines after about 1920. The propulsion for ships got disrupted in the years before World War II and, after World War II, the locations where ships were built changed as well.

Though today eastern Asia is the global center of manufacturing for "textiles, steel, consumer electronics, and automotive products... the first major industry in which the East Asian countries became globally 'dominant' - with more than half of the world market," according to Stig Tenold in *Enterprise & Society*, "was shipbuilding. Shipbuilding increased rapidly in Japan in the first postwar decades, and in 1956," just eleven years after the war ended, "the country replaced the United Kingdom as the leading nation in terms of tonnage launched... Around 1970 the market for new ships was dominated by Japan and European countries, and the tonnage exported from South Korea was negligible. Fifteen years later, South Korea was the world's largest shipbuilding nation."

It is both surprising and understandable that this revolution happened. Surprising because, when Japan became the world's largest shipbuilding nation in 1956, the post-war Allied occupation had ended only four years prior. And in the early 1970s, when South Korea began to disrupt European shipbuilders, they were economically far behind any industrialized country. In disruption parlance, Asian shipbuilders were outsiders to an industry dominated by northern Europeans.

But this revolution was also understandable from an economic perspective. According to the Organization for Economic Co-operation and Development (OECD), in 1970 South Korea was last among its members in terms of GDP per hour worked (in inflation-adjusted, 2015-dollar equivalents). While the GDP per hour worked in Sweden in 1970 was $30 and in the UK was $23, in South Korea it was only $3, less than 10% of the United States' value of nearly $34. For comparison, the second-lowest country in the OECD's dataset was Türkiye, with a GDP per hour worked in 1970 of $11, and Japan was the third from the bottom at nearly $14.

Industries like textiles, steel, automotive manufacturing, and shipbuilding have a strong incentive to move to lower-cost locations, particularly those places that can deliver results. While South Korea and Japan were the lowest and third-lowest OECD countries in terms of GDP per hour worked, they were the highest and third-highest countries in terms of annual hours worked per person employed, with

South Korea averaging 2919 and Japan 2243. (Number two was Ireland, at 2334 hours.) In comparison, the US was 1857 hours, the UK 1775, and Sweden was the lowest in the OECD dataset at just 1565 hours per worker per year.

The Kockums Crane in Malmö, Sweden, was the largest shipyard crane in the world when it was installed in 1974, just when South Korea began dramatically expanding its shipbuilding capacity. It went largely unused for most of the 1970s and 1980s and was eventually dismantled in the summer of 2002 and shipped to Korea. According to Port News, the Koreans dubbed it the "Tears of Malmö" since "residents of the city are said to have cried as they watched it depart". In 2006 China overtook Japan as the world's second-biggest shipbuilding nation. Today, more than 90% of all ocean-going ships are built in South Korea, Japan, or China.

From the Lords of Guano to the Elders of Zion

> *A booming trade in guano fertilizer from South America to Europe brought the potato blight pathogen with it. The resulting social unrest spread throughout Europe in the late 1840s, providing fertile soil for the later development of Communism, modern anti-Semitism, and Naziism. Nitrate-bearing soils quickly displaced guano in the late 1800s, and then chemically synthesized ammonia replaced the natural deposits as sources of nitrates. By the time the Nazis started their war, both Germany and the Allies could produce both fertilizers and explosives in unprecedented quantities*

In the spring of 1854, thirty-seven-year-old George Washington Peck, an American, published a book called *Melbourne and the Chincha Islands* describing a voyage he had taken the previous year onboard a sailing ship from Melbourne, Australia, to Peru. Passing the North Cape of New Zealand, he wrote in his diary, "today the wind is still light; the ocean wide and blue... I feel the poetry of the ocean..." But upon reaching the Chincha Islands on the other side of the Pacific, just off the coast of South America, his attitude changed dramatically.

On these islands it rains seldomly and the surrounding waters are rich with fish, so droppings from birds who live there have accumulated over centuries into a hardened layer as deep as 150 feet or more and loaded with nitrates and phosphates that make *guano*, one of the best natural fertilizers in the world. "Guano," Peck wrote, "lies upon all the islands in the form of high, smooth, rounded hills, and covers nearly every part of all of them".

For at least four thousand years people have harvested these deposits and brought them to the mainland. "The use of guano as a fertilizer was fundamental to sustain the agricultural development" of the Inca Empire, wrote Pedro Rodrigues of the Universidad Austral de Chile, "and, it has been argued, was the basis for its rapid growth." Garcilaso de la Vega, born in Peru to a Spanish ship captain and an Inca noblewoman, in his 1609 book *Comentarios Reales de los Incas* described how the Inca protected the islands by outlawing killing or disturbing birds, landing on the islands during breeding season, and stealing eggs, among other prohibitions, violations of which were punishable by death. Rodrigues wrote, "these measures were among, or actually were, the first habitat conservation measures implemented by humans to achieve a sustainable use of seabird guano as a natural resource."

But a roaring international trade in Peruvian guano ran from about 1840 to 1880, when a war between Chile, Peru, and Bolivia brought most of the business to a rapid halt. In his book *The Latin American Economies*, Roberto Condo wrote, "Europe received its first shipments of Peruvian guano in 1841, and within a few years the guano boom was on. Only six years later, in 1847, guano had become Peru's most important export".

A convergence of factors helped to propel the guano trade. Bruce Farcau explains in his 2000 book *The Ten Cents War. Chile, Peru, and Bolivia in the War of the Pacific, 1879-1884* that "the Industrial Revolution had created an urgent need for vastly expanded agricultural production on a decreasing land base in Europe, and the market for American cotton" which was growing rapidly due to the introduction of Mexican hybrid strains, "made necessary a reliable means of renewing the exhausted soil of southern plantations."

The cotton industry not only needed fertilizer but also whale oil as a lubricant for cotton looms, which brought a booming number of American and British clipper ships to Pacific waters, some of which sometimes had spare shipping capacity. In the *Journal of Latin American Studies*, Juan Stemmer estimated that the cost of shipping guano and nitrates from the west coast of South America by sail to the UK fell from over £5 per ton of weight in 1842 to £3 per ton ten years later. Freight rates only declined further over time with the rise of 'steamers', that is, steam-powered ships. Graeme Milne in his 2000 book *Trade and Traders in Mid-Victorian Liverpool* wrote, "the building of steamers ... created a buoyant market in used sailing ships, which were put to good use in the South America guano trade."

Guano quickly grew indispensable to American and European agriculture. "The first shipment of guano arrived in the United States in 1844 and consisted of 700 tons of the stuff," wrote Columbia University's Christina Burnett. "A decade later, guano imports into the United States peaked at 175,849 tons." President Millard Fillmore during his 1850 State of the Union address, just six years after the first shipment of guano had reached American shores, said the US should "employ all the means properly in its power" to secure its supply. In his 2005 book *A Most Damnable Invention: Dynamite, Nitrates, and the Making of the Modern World,* Stephen Bown wrote "by 1860 imported guano made up nearly 45 percent of the commercial fertilizer sold in the United States," just sixteen years after it had first been introduced.

When George Washington Peck arrived at the Chincha Islands in 1853, he found what economic historian Jimmy Skaggs in his book *The Great Guano Rush: Entrepreneurs and American Overseas Expansion* described as "a hell on earth." "The guano is dug from the hills", Peck wrote, "by coolies [workers], who are brought to Peru by English ships from the free ports of China... Their labor is much more severe and injurious than railroad digging; they have no liberty days, no protecting laws, no power to obtain even the pittance said to be paid them".

"Working and living conditions on the Chincha Islands were ghastly," Bown wrote. "Men choked on the acrid chemical dust when the winds picked up. They were remorselessly prodded by pitiless overseers to chip and shovel the valuable powder from the earth and load it into wheelbarrows. The barrows were pushed across the bleak, rugged surface of the island by other workers who were chained to the handles because their hands were so cracked and bleeding that they were useless for other work such as wielding heavy tools ... Disease and sickness were common and included histoplasmosis and shigellosis from accidentally ingesting the feces, as well as a host of bewildering ailments brought on by constant exposure to the ammonia dust: respiratory illness, internal hemorrhaging, and severe gastrointestinal irritations."

Alf Hornborg, in *Ecology and Power*, reports that "some workers were branded on the cheeks with hot irons." Peck noted, "almost every week some of them commit suicide by throwing themselves from the cliff". Cemeteries, Bown wrote, "reportedly overflowed with the rotting corpses of those who failed to survive and who had been buried in graves too shallow to protect remains from scavenging dogs, human bones consequently being liberally scattered about."

Speaking on the floor of the House of Commons in London, Sir Charles John Wingfield, a member of Parliament, said "in 1854, a number of British ship-masters who had taken coolies to the Chincha Islands in Peru, horrified at the cruelties they saw inflicted on the Chinese, whose dead bodies they described as floating round the islands".

"The Coolies who dig guano are brought here in English vessels," Peck wrote, "their transporters and those who transfer them to the Peruvian government, know that they are engaged in a slave trade as much as if they brought slaves from Africa... It is, moreover, the worst and most cruel slavery of all forms of slavery that exist among civilized nations. It was universally said to be so by captains who had visited every quarter of the globe; I am sure I never saw anything like it on the plantations, when I once resided for several months in the country adjoining that where the scene of the popular tale of 'Uncle Tom's Cabin' is laid."

Departing for Lima, Peck wrote, "the islands seem to me to be a kind of human *abattoir*, or slaughterhouse of men; and I feel a relief in being away from them, as one feels who has escaped out of some gloomy dream." The government of Peru largely turned a blind eye to the conditions on the islands as the revenue proved indispensable to the nation. As W.M. Mathew wrote in *The Economic History Review*, "by 1861-2 almost four-fifths of the government's income was coming from guano".

A resource this valuable would not go uncontested. In 1864, Spain seized the islands. Peru got them back by paying a ransom. When a three-century-old stone slab was dug out of Isla Chincha Norte in 1847, that engraved coat of arms of an indigenous Peruvian guano lord was nothing but a relic of a bygone era when the islands were revered and protected by the Inca. But something else came out of the Peruvian export trade that began in the early 1840s. Something whose effects would be felt long after guano exports collapsed.

By the start of the international guano trade, the potato, itself a native of Peru, was a staple crop throughout Europe. It is understandable why – the potato yields about three times the calories per area of land planted than wheat or oats. Peruvian guano proved to be an excellent fertilizer for potato crops in Europe. But recent research by scientists at North Carolina State University and the Norwegian University of Science and Technology, using genetic analysis of historic samples of potato plants, suggests that the blight that swept through Europe in the mid-1840s originated in South America and then spread to North America and Europe through the potato and fertilizer trade.

"The guano ships brought South American potatoes to Europe", Austin Bourke wrote in 'The Emergence of Potato Blight 1843–1846' in the scientific journal *Nature* in 1964, and with that trade came a disease that would irrevocably alter the course of European history.

Few places were hit as hard as Ireland. "Initially, reports describing the appearance of a mysterious disease on the potato crops in various parts of Europe in 1845 were regarded with curiosity rather than alarm within Ireland," Christine Kinealy wrote in her 1994 book *This Great Calamity*. "In the previous year, a number of Irish newspapers had carried reports from American journals and newspapers concerning a disease which had attacked the potato crop there for the second consecutive year." By the middle of 1845, the disease had been detected throughout much of northern and central Europe, Belgium, the Netherlands, France, and England.

Over the next few years, catastrophic failures of Ireland's main food crop claimed the lives of about one million people there and caused more than a million others to flee (ironically, to places like Boston and New York, from which the blight might have been brought to Ireland in the first place), bringing down the island's population by about one-quarter. It was not until 2023, one hundred and seventy-five years later, that Ireland's population climbed above five million people for the first time since the blight struck.

Similar, though less dramatic, crop failures throughout the rest of Europe led to social and political disruptions in 1848 in Sicily, France, Germany, Italy, the Austrian Empire, and many other places that nourished both strong left-wing and strong right-wing reactions. The Chincha Islands did not escape the attention of Karl Marx. In an essay called 'English Ferocity in China' published in the *New York Daily Tribune* on April 10, 1857, Marx called conditions on the coast of Peru "worse than slavery". The late German historian Reinhard Rürup described the 1848 Revolutions as 'a turning point' in the development of modern anti-Semitism, due in part to the prominence of openly anti-Semitic writers like Hermann Goedsche. A few decades after his death in 1878, Goedsche's writings would help inspire a new and devastating wave of European anti-Semitism that culminated in World War II. But global production and trade of nitrogen would contribute to a few wars before then.

The Peruvian guano trade would collapse starting in 1879 with the invasion of southern Bolivia and southern Peru by Chilean forces, starting the War of the Pacific. But Chile's victory consolidated ownership of the Atacama Desert into the hands of a single nation, facilitating foreign commercial investment in exploiting its resources. The Atacama is one of the driest places on Earth and geologists suggest that nitrogen-laden groundwater that has slowly percolated through the desert soil and evaporated is responsible for a layer of soil infused with sodium nitrate.

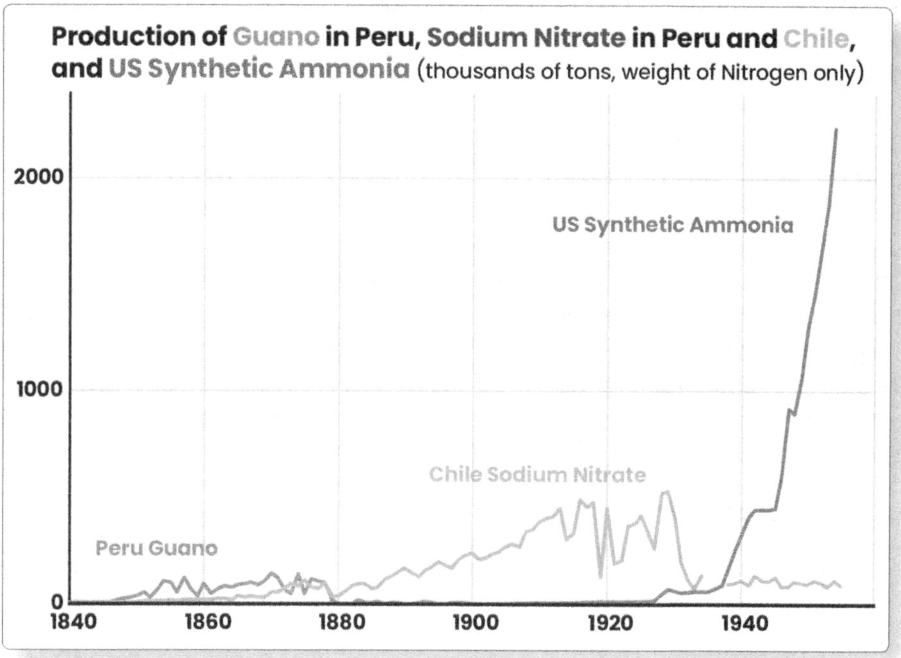

Figure 39 – Production of Guano in Peru, Sodium Nitrate in Peru and Chile and US Synthetic Ammonia. *(Condo, 1985)*

Guano is an unparalleled natural fertilizer because it contains not just nitrogen, but also phosphorus, iron, calcium, magnesium, and other chemicals. Sodium nitrate, on the other hand, contains only the elements sodium, nitrogen, and oxygen. Though this sounds like a limitation, it is actually the opposite. Purified sodium nitrate can be used for other applications besides simply fertilizer, for example the production of gunpowder and explosives. While phosphorus-bearing rocks have been found and exploited elsewhere in the world, like Morocco and the South Pacific island of Nauru, the Atacama is the only known natural source of sodium nitrate.

Unlike Peru's guano deposits, which are located exposed on islands, the nitrate-bearing soils of the Atacama are more difficult to access. They are located further inland and buried under higher layers of desert soil which must be removed to access the nitrate-bearing layer, with no rivers running down to the shore to utilize for transport. There are also 1000-foot (300-meter) cliffs between the sea and the high desert plateau.

The ability to economically exploit these resources in the latter part of the 1800s and early part of the 1900s was due to a convergence of factors. Dynamite, which was invented in the late 1860s and became widely available throughout the 1870s, made accessing the soils easier. "Holes are drilled through the overlying material," R.H. Whitbeck wrote in 1931, "and charges of powder or dynamite are shot off at the bottom. The explosion shatters the mass for some distance around. In the past the largest pieces of *caliche* [nitrate-bearing soil] had to be broken up with sledge hammers." (Interestingly, sodium nitrate is one of the ingredients in dynamite. So, the greater availability of sodium nitrate made it cheaper to produce dynamite, which in turn made it easier to access additional sodium nitrate in a self-reinforcing feedback loop.)

Steam power not only made shipping cheaper, especially in the later part of the 1800s, but also steam-powered machinery made mining the nitrate-bearing soil layer easier and steam-powered railroads made it easier to transport from the high plateau to the shore. As the sodium nitrate mining industry expanded, the length of functional railroads increased almost in lockstep.

By the 1890's, Whitbeck reports, "Chilean nitrate supplied about four-fifths of the inorganic nitrogen used in the world." As with Peru a few decades before, the government of Chile became dangerously dependent on a single resource. Nitrates accounted for less than 10% of government revenue in 1880, but by 1890, just ten years later, it was over 50%. Just a few years later, it approached 70% and remained above 50% until the dawn of the First World War. (Marr, 2007)

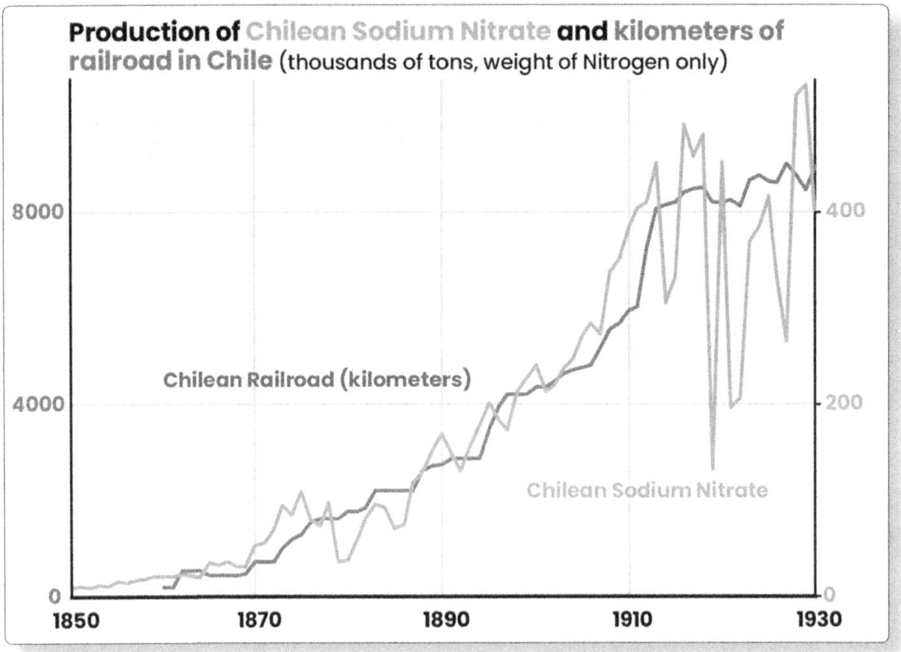

Figure 40 – Nitrate production and kilometers of railroad in Chile, 1850-1930

The irony of a monopolistic, global trade in sodium nitrate is that nitrogen surrounds us. It constitutes about 80% of the air we breathe, but because it is locked up in the form of pairs of nitrogen atoms, most living things cannot access it directly, ourselves included. In 1909, chemist Fritz Haber demonstrated that ammonia, a single atom of nitrogen bound to three hydrogen atoms, could be produced at conditions of high temperature and high pressure from air and methane (natural gas), which Thomas Hager, author of *The Alchemy of Air. A Jewish Genius, a Doomed Tycoon, and the Scientific Discovery that Fed the World but Fueled the Rise of Hitler*, called "the most important discovery ever made". The BASF company, founded in the 1860s to produce artificial dyes, bought the rights to Haber's process and BASF chemist Carl Bosch began to develop its industrial-scale production. (Haber would be awarded the Nobel Prize in Chemistry for 1918 for his work. The Nobel Committee initially overlooked Bosch, but then recognized him with the same prize in 1931.)

"By 1918 synthetic nitrogen accounted for 23% of total production," Simon Hanson wrote in his 1951 book *Economic Development in Latin America*. "From 1920 to 1933 a veritable revolution took place in the nitrogen industry. The leading industrial countries rushed mass production of synthetic ammonia from the air; world capacity to produce nitrogen rose more than threefold from 1.6 million to nearly 5 million tons. While it is true that world consumption was also rising, more than doubling in the period, Chilean production dropped from 430,000 tons per year to 76,000 tons."

Initially, Chilean nitrate producers downplayed the threat that synthetic ammonia posed to their business. "For a time the nitrate producers and government officials had taken the ground that there really was no danger from synthetic nitrogen," R.H. Whitbeck wrote in 1931. "After all, the world must have nitrate, and Chile was the only place to get the real 'natural' article. However, things have looked increasingly different since the [First] World War. Germany, the largest pre-war buyer, is making all the nitrogen it needs and is exporting a surplus."

It was not just Germany turning from a customer to a competitor that threatened Chilean nitrate exports. Chilean miners thought 'natural' nitrate was superior to synthetic products. In fact, the opposite was true, as mechanized farming machinery required uniform inputs. Paul Marr of Shippensburg University writes, "American agriculture had become mechanized and it was impractical to use natural nitrate fertilizers with mechanized agriculture at the turn of the twentieth century".

Starting in the 1910s, the Chilean nitrate mining industry entered a period of wild oscillations in production. Marr continues, "ultimately British nitrate companies in Chile found themselves with ever increasing quantities of unsold nitrates which would then be dumped on the market at a discount. These periods of oversupply resulted in the familiar boom and bust cycles". Sodium nitrate production decreased 75% from 1929 to 1932 while prices dropped 40%.

Economically, nitrogen mined from South America could not compete with nitrogen taken from the air. Nitrate-bearing soils contain up to 20% sodium nitrate, but even pure sodium nitrate itself is only about one-sixth nitrogen, by weight. So, every load of nitrate-bearing soils are only about 3% nitrogen. Synthetic ammonia, by contrast, is more than 80% nitrogen, by weight.

In *Statistics on Fertilizers and Liming Materials in the United States*, Arnon Mehring reports the prices of Chilean sodium nitrate and synthetic ammonia by year in the US from the mid-1920s though the early 1950s. In 1926, for example, sodium nitrate sold for $3.27 for twenty pounds, but only about 16% of this weight is the desired nitrogen. Synthetic ammonia sold for $1.75 for twenty pounds, less than half the price per pound for a product that is more than 80% nitrogen. That is, on a dollar-for-dollar basis, buyers got more than nine times as much nitrogen from synthetic ammonia as from sodium nitrate. But refinements helped make the Haber Process even more efficient. By the years after World War II, ammonia sold for less than $1 for twenty pounds, and buyers got more than fifteen times as much nitrogen from synthetic ammonia as from sodium nitrate. As a result, the growth of the synthetic ammonia industry was staggering. The graph above only shows U.S. production, and world production according to the United States Geological Survey, a government agency tasked with compiling such statistics, was three to five times the level of U.S. production.

The disruption of sodium nitrate by synthetic ammonia also changed the way that synthetic ammonia was produced. As late as 1935, less than 10% of the ammonia made in the world was from oil or natural gas. Most ammonia synthesis used coal as the feedstock. But by 1950, just fifteen years later, more than 90% of the world's synthetic ammonia was from oil or natural gas.

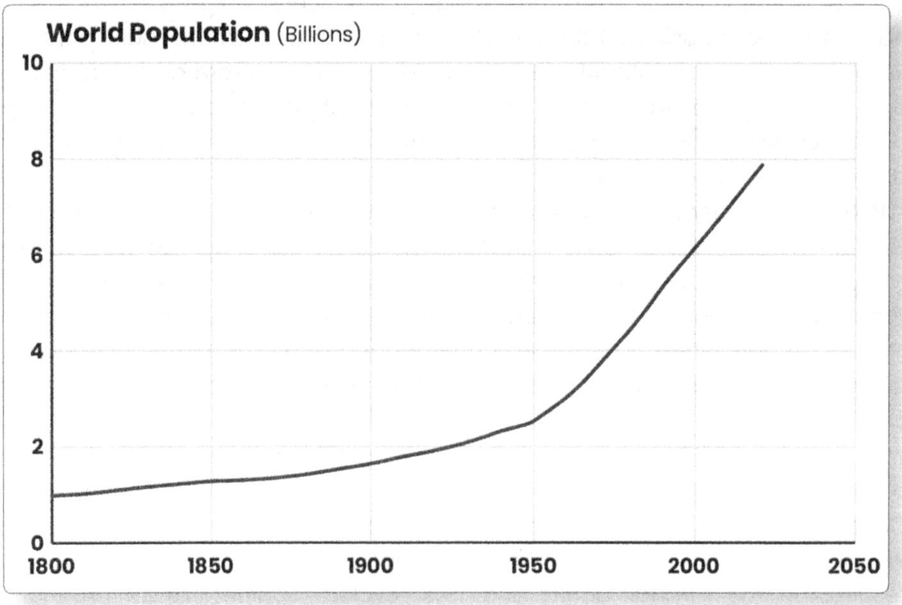

Figure 41 – World Population, in billions

Ammonia is useful as a fertilizer, but it is also a key ingredient in explosives, so the Haber Process not only enabled the enormous food (and population) boom of the past century, but also the two biggest wars in world history. It is bitterly ironic, then, that the devastation of the First World War provided fertile ground for the rise of a new and exceptionally virulent strain of anti-Semitism that contributed to the Second World War. The notorious *Protocols of the Learned Elders of Zion*, which inspired a generation of enthusiastic anti-Semites, was produced in the early 1900s from plagiarized and re-worked passages of text that historians have shown contains parts from as early as the 1848 and 1849 works by Hermann Goedsche and others, written during the potato famines and their resulting social unrest that were ultimately caused by the guano trade. In short, Naziism literally came from shit.

After inventing the process for ammonia synthesis that bears his name, Haber turned his attention to developing chemical weapons during the First World War, which he greatly supported. But this put him at odds with his wife, Clara Immerwahr, the first German woman to be awarded a doctorate in chemistry in Germany, a women's rights activist, and a pacifist.

In April of 1915, Haber personally oversaw the first successful use of chemical weapons in combat, a release of chlorine gas against French troops during the Second Battle of Ypres, resulting in about 67,000 casualties, earning himself the title of "father of chemical warfare." On the first of May, Haber held a gathering to celebrate both the attack and his promotion to the rank of captain in the German army.

Figure 42 – Clara Immerwahr, born 1870, the first German woman to earn a doctorate in chemistry in Germany. *Photo circa 1890*

Sometime late that evening, after the party was over, Clara retrieved her husband's army pistol, walked out into the garden at the director's mansion at the Kaiser Wilhelm Institute where they lived, and shot herself in the heart. She died on the second of May, and on the following day her widowed husband left her body behind to go observe the first gas attack against Russians on the Eastern Front.

Dynamite

Swedish chemist Alfred Nobel revolutionized the explosives industry over just a few years in the 1860s. Twenty-seven-year-old Nobel learned to make nitroglycerin in 1860 from thirty-five-year-old Italian chemist Ascanio Sobrero. The Italian had discovered the oily chemical a few years prior, but refused to work with it. Roughly seven times more powerful than gunpowder and prone to explode from mishandling, nitroglycerin is so dangerous that Sobrero had kept its discovery a secret after he first made it.

But Nobel saw nitroglycerin's great power in a positive light. The Nobel family's armaments business, which had supplied explosives to Russia during the Crimean War that had ended a few years prior, became defunct and the young Alfred was investigating non-military applications of explosives.

The rock-breaking part of the mining process had not made any significant advancements in centuries. Since antiquity people had been breaking large rocks and rock walls by lighting fires near them, then splashing them with cold water. Gunpowder had completely revolutionized that part of mining, but gunpowder is prone to exploding without warning.

Despite a similar shortcoming, 'blasting oil', a mixture of nitroglycerin and gunpowder, helped Nobel rebuild his family business as he experimented with improved formulations. In 1867, Nobel patented a mix of 75% nitroglycerine and 25% diatomaceous earth which he called 'dynamite'. As Stephen Bown wrote in his 2005 book *A Most Damnable Invention,* "although it was weaker than pure nitroglycerin, it was still five times as powerful as gunpowder." It also was safe to store and easy to put into place for blasting. It was a perfect combination of being more powerful than gunpower, but also much safer than either gunpowder or blasting oil.

"Dynamite ushered in a new era of industry and mining. It caused a profound sea change in the way civil engineering projects could be conceived and executed," Bown continued. "Anything gunpowder could do, dynamite did far more safely, far more efficiently, and far more cheaply. Dynamite rapidly became the most sought-after explosive in the world." Nobel did not just disrupt the world of explosives, he also disrupted his own nitroglycerin business – as soon as dynamite was in production in 1867, Nobel stopped shipping blasting oil entirely.

According to Bown, Nobel's factories made eleven tons of dynamite that first year, seventy-eight tons in 1868, and 185 tons in 1869. "By 1874", fourteen years after he learned how to make nitroglycerin, Bown writes, Nobel's "factories were producing an astonishing 3,120 tons of dynamite".

According to American journalist Henry Dam, who visited one of Nobel's facilities in Ayrshire, Scotland in 1897, one of the key ingredients in dynamite was nitric acid, made from sodium nitrate imported from Chile, and "absolutely essential to the process was the one legged stool. This innovation of Nobel's prevented the operator of the nitroglycerin reaction vat from falling asleep on the job." Producing nitroglycerin releases heat which speeds up the reaction further in a self-reinforcing cycle. Failing to monitor the reaction closely could have fatal results, not just for the operator, but also for anyone else in the general vicinity.

Of course, dynamite did not simply replace gunpowder on a one-for-one basis, it also opened new possibilities in mining and tunnelling. Bown writes "easy access to coal deposits fueled industrial growth. Cement and concrete became more common construction materials when gypsum and lime could be blasted. Fields could be cleared for agriculture much quicker as a carefully placed stick of dynamite demolished stumps and shattered boulders... Oil could be located by setting off dynamite blasts and tracking the vibrations of the earth. Ruined buildings could be demolished in moments. Irrigation and transportation canals could be dug in a fraction of the time they had taken with gunpowder."

As a child, Nobel had lived in Saint Petersburg, Russia, and by the time he had gotten his dynamite business off the ground, his brothers Robert and Ludvig owned a small oil refinery in Baku, on the Caspian Sea. Dynamite revolutionized the industry in the oil-rich region. By the 1880s, oil production in the Baku region amounted to about one-third of the level of oil production in the US, and Russia's position as one of the world's top oil nations was secure.

An unfavorable 1888 obituary of Ludvig Nobel, which confused him with his brother, saying, "Dr. Alfred Nobel, who became rich by finding ways to kill more people faster than ever before, died yesterday", caused Alfred to re-evaluate his life's legacy.

In 1895, Nobel signed his last will and testament, donating the bulk of his assets after his death to the creation of a memorial prize to be given each year to worthy individuals who have made extraordinary contributions to the fields of physics, chemistry, medical science or physiology, literature, and the furtherance of peace.

The Cigarette Won its War in Europe, Then Lost the Battle at Home

> *The cigarette rolling machine revolutionized how we consume tobacco. Electronic vaporizers are doing it again*

The cigarette has been both disruptive and disrupted. In 1880 James Albert Bonsack, still not even twenty-one years old, won $75,000 in prize money (nearly $2 million today) for inventing a rolling machine which could produce two hundred cigarettes per minute instead of the just four or so that a skilled person could make when rolling by hand. (His first working prototype, ironically enough, was destroyed by a fire and went up in smoke, but he rebuilt it.) According to Patrick Porter in a 1969 article in *Business History Review*, "the machines brought about a tremendous reduction in the cost of manufacturing... In 1876, labor costs were about 96 cents per thousand; by 1895, labor costs for the same cigarette were slightly over 8 cents per thousand," a decline of more than 90%. A 10X disruption! Mr. Bonsack's machine quickly displaced human labor in rolling cigarettes.

But in the late 1800s cigarettes were simply not a popular form of tobacco. Even by 1900, cigarettes were only about three percent of US tobacco sales, by weight. Pipes, cigars, and chewing tobacco were all far more common. "Cigarettes were for 'dandies and sissies,'" wrote Joel Bius in *Smoke 'Em if You Got 'Em: The Rise and Fall of the Military Cigarette Ration*. It would take a second revolution, a social one, for cigarettes to become the most-popular means of consuming tobacco in America.

"The commercial success of manufactured cigarettes occurred because of the convergence of several technological and social innovations," wrote Michael Thun of the American Cancer Society and colleagues in the scientific journal Oncogene in 2002, "including the development of safe and portable matches, the creation of equipment to mass-produce and package cigarettes inexpensively, the introduction of mass media campaigns to promote special brands, and the distribution of free cigarettes in military rations during World Wars I and II."

During the First World War, US military officials promoted cigarette smoking among the troops, according to Peter Andreas, author of *Killer High: A History of War*

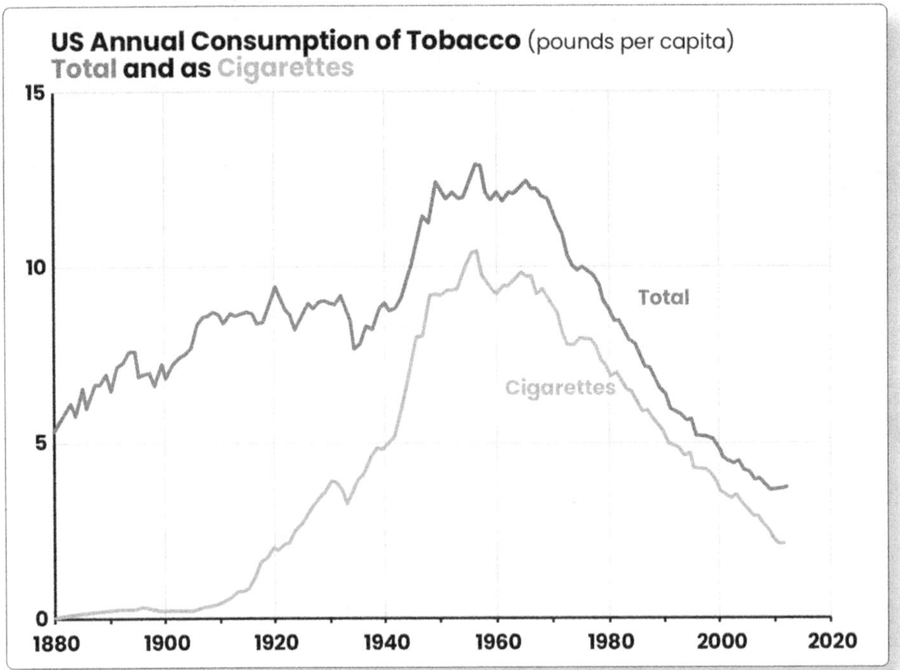

Figure 43 – U.S. Annual Consumption of Tobacco Products (pounds per capita) Total and in the form of Cigarettes. *(U.S. Department of Health and Human Services, 2014)*

in Six Drugs, "to sustain morale and discipline, calm nerves, boost alertness, foster camaraderie and suppress hunger", going so far as to include them in soldiers' rations. Tobacco consumption per person in the US in the form of cigarettes more than doubled in just the five years from 1915 to 1920.

Again quoting Bius: "In 1920, responding to the increase in cigarette smoking, one observer concluded that WWI had changed the way Americans saw cigarettes: 'Ten or fifteen years ago, cigarettes didn't have much of a standing in the community. There was a neat distinction between the man who smoked cigarettes and the man who smoked cigars or a pipe. That distinction seems to have disappeared today.'" In 1910, only about 5% of tobacco sold (by weight) in the US was in the form of machine-rolled cigarettes. By 1930, it was over 40%, 50% by 1935, and by 1950 had reached nearly 75%.

A third major disruption in cigarettes occurred over just a few years in the mid-1950s with the filtered tip. We know from Tony Seba that a 10X improvement in performance relative to cost will cause consumers to adopt a new product. Sometimes, however, it is the mere perception of a significant improvement that sparks a disruption, which is exactly what filters did for cigarettes.

In terms of functionality, filters on early brands of cigarettes were "essentially cosmetic", according to a 2001 report by the National Cancer Institute, part of the U.S. federal government's National Institutes of Health. Cigarette manufacturers had offered various kinds of filtered cigarettes since they were patented by Hungarian inventor Boris Aivaz in 1925, but they were always a niche product, never amounting to more than a few percent of the US market before the 1950s.

A spate of medical reports in the early 1950s suggested that smoking caused lung diseases or other health problems. Although there was no conclusive evidence at the time that smoking was harmful, marketing studies suggested that consumers would respond very favorably to the idea of cigarettes that were 'low tar' or that had a 'smooth' or 'mild' flavor.

In 1953, Claude Teague of the R. J. Reynolds Tobacco Company devised a tip that contained chemicals that darkened when exposed to smoke, but essentially did nothing else. They were an instant hit. Now that people could see a change in the tip, they assumed that the new cigarettes must be better for you. In the five years from 1954 to 1959, filtered cigarettes went from about 5% of the US market to 50% and continued growing rapidly, today being nearly 100%.

In fact, Teague's cigarette filters did little except change color. U.S. Congressional investigations in 1958 found "some firms' filtered products delivered even more tar and nicotine than their unfiltered traditional products. Reversals occurred even within brand families, with Brand X filtered versions yielding higher tar and nicotine than the unfiltered Brand X products that they ostensibly improved upon," according to the National Cancer Institute. It turns out that the filter caused people to inhale more deeply for any individual cigarette and, thinking that filtered cigarettes were safer, to simply smoke more.

Now we are amid another disruption in smoking. In 2003, Chinese pharmacist Hon Lik patented a replacement for the cigarette, a lithium-ion battery heated pen that vaporizes a solution of liquid nicotine and propylene glycol. In *Rethink: The Surprising History of New Ideas*, Steven Poole says that in 2015 "the UK research body Public Health England declared that, according to the 'best estimate so far', vaping was 95 percent less dangerous than smoking", although this finding is controversial. Data from the National Youth Tobacco Survey, a collaboration between the Centers for Disease Control and Prevention, the Food and Drug Administration, and other US federal government organizations, indicates that vaping with e-cigarettes has already become more popular among youth than smoking cigarettes.

Vaporizers (or 'e-cigarettes') were only introduced to the US market in 2009, but in just the five years to 2014 they became the most popular form of nicotine consumption for US high school students, with 13.4% of students reporting having used vaping cigarettes at least once in the prior thirty days, according to the National Youth Tobacco Survey (NYTS), a long-running survey conducted by the federal government of the nicotine usage of middle- and high-schoolers. By 2019, that figure had more than doubled to 27.5%, meaning that nearly 90% of students who had consumed nicotine in any form in the prior thirty days (cigarettes, cigars, chewing tobacco, electronic cigarettes or other) had used electronic cigarettes. (Some people consume nicotine in multiple forms, so 20% of high school nicotine users also had smoked at least one combustible cigarette in the past thirty days, too.)

Also in 2019, for the first time in more than a decade, more than 30% of high school students reported trying a nicotine product of any form in the past thirty

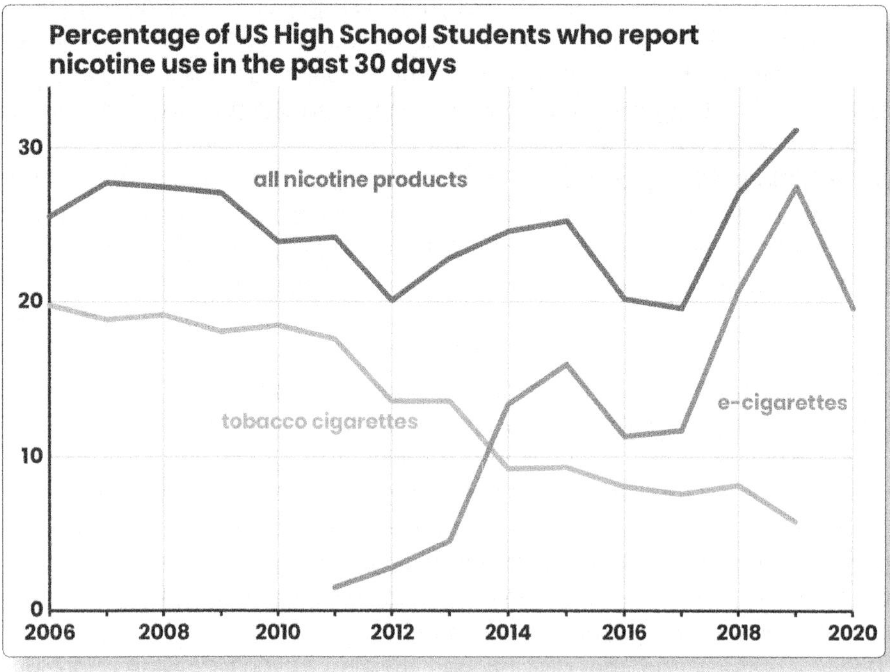

Figure 44 – Percentage of US high school students who reported using tobacco cigarettes and e-cigarettes in the past thirty days

days. So, as we have seen with many other technologies, vaporizers are not simply a one-for-one replacement for other nicotine products: they have replaced tobacco cigarettes, are used in conjunction with tobacco cigarettes, and seem to have gotten some high schoolers to consume nicotine who would otherwise not have done so. All of this is despite growing efforts by governments in the later part of the 2010s to restrict vaporizer sales, limit advertising, and make them less attractive to consumers by eliminating flavoring additives in some areas.

Electronic cigarettes not only changed who smoked and how, but also which brands dominated the market. Marlboro, Newport, and Camel split most of the US cigarette market. But these brands have zero share in the vaping business, where the top five brands – Juul, Vuse, MarkTen, Blu and Logic – own 97% of the market. Incumbent companies, rather than diluting their existing brands with vaping products, chose new brands to lead the way into vaping. Marlboro (Altria) invested in Juul while Vuse is owned by Camel parent company British American Tobacco.

Electronic cigarettes also changed how suppliers obtain nicotine. The nicotine used in vaping cigarettes is extracted from *Rustica*, a member of the tobacco family also known as Aztec tobacco or strong tobacco, and it contains up to nine times more nicotine than common tobacco. Since consumers do not smoke it directly, producers grow the strongest plant they can.

Smoking Bans

It seems crazy to think about now, but people used to be allowed to smoke cigarettes in restaurants, during university lectures, and even on airplanes. In fact, bans on cigarette smoking are surprisingly recent. (Airlines would *give* you

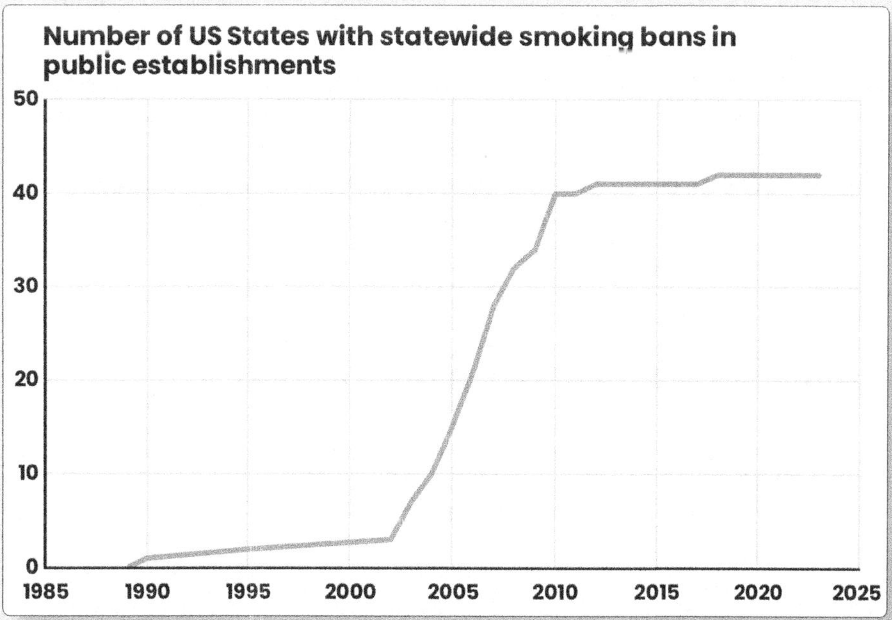

Figure 45 – Number of US States with state-wide smoking bans

cigarettes! If you did not have any, just ask a flight attendant and they would bring you a small pack.) The U.S. ban on in-flight smoking began in April 1988 with domestic flights of two hours or less and extended to all domestic and international flights in the year 2000.

That is, it took only twelve years from there being no smoking bans on flights, to there being bans on domestic flights in some countries, to there being bans on all domestic and international flights.

The advertising world began rejecting tobacco advertisements at the same time. The *New York Times* banned cigarette advertisements in 1999. "More than a dozen other American newspapers refuse cigarette advertising, among them *The Seattle Times, The Deseret News of Salt Lake City* and *The Christian Science Monitor*," Doreen Carvajal wrote in the *New York Times*. "They are already barred from TV and radio ads."

It was just after the time that advertisers began rejecting tobacco advertisements that US states began banning smoking in public establishments *en masse*. As of the year 2000, only two US states had state-wide smoking bans: South Carolina and California. But over the next decade, the number would rise to forty of the fifty US states.

◆ ◆ ◆

"Automobiles will be cheaper than horses are today. Farmers will own automobile hay-wagons, automobile truck-wagons, plows, harrows and hay-rakes. A one-pound motor in one of these vehicles will do the work of a pair of horses or more. Automobiles will have been substituted for every horse vehicle now known. There will be, as already exist today, automobile hearses, automobile police patrols, automobile ambulances, automobile street sweepers. The horse in harness will be as scarce, if, indeed, not even scarcer, than as the yoked ox is today."

— John Elfreth Watkins Jr., "What May Happen in the Next Hundred Years",
***The Ladies Home Journal*, 1900**

◆ ◆ ◆

The Bumpy Road from the Horse to the Car

> *Few inventions have shaped our world more than the automobile, and with it the irrevocable switch from muscle to motor as the basis of transportation*

On June 4, 1896, thirty-two-year-old Henry Ford was in his garage on Bagley Street, in the heart of Detroit, with his wife Clara and his friend and co-worker Jim Bishop. Ford had not slept for the past couple of days and finally, early in the morning hours, he grabbed an axe and began breaking out the frame and bricks around the garage's main doorway.

Ford was a gifted mechanic, having already worked for a company that made steam engines and was now working for the Edison Illuminating Company, where he had met Bishop. A couple of years before, Ford had started making a gasoline engine from plans he found in a magazine, testing it in his kitchen sink. The engine needed an electric spark to start the ignition, and fortunately because of Henry's employment, the Fords were early adopters of household electricity. Henry ran an electric wire from the ceiling light into the sink and spun the flywheel while Clara hand-fed in gasoline as the engine sputtered to life.

Now, a couple of years later, with the door from the garage removed so that his 'Quadricycle' automobile could be rolled out, Ford spun the flywheel again and took off down the street with Bishop on a bicycle ahead of him. After a short run, according to the Ford company's website, "the two men returned triumphantly to the Ford home, got a few hours of rest and then reported for work at Edison." The Quadricycle was Ford's first car, but far from his last.

The gasoline engine was first unveiled in 1876 by Nicolaus Otto and Eugen Langen as part of a series made by the company they had founded fourteen years prior. But a motor by itself is not useful as transportation. It took a convergence of technologies and ideas for the internal combustion engine to disrupt muscle power: the coach, steel, petroleum fuels, and pneumatic tires, to name a few. But the first of these ideas, of course, was the very concept of using a motor to move a wheeled vehicle.

At the time of the invention of the automobile, wheeled vehicles were not common in much of the world. According to Tom Standage in his book *A Brief History of Motion*, "five thousand years after the invention of the wheel, it was still not in widespread use. In North Africa and much of the Middle East it had been abandoned altogether; medieval Arabic and Persian do not even have words for wheeled vehicles, and travelers' accounts as late as the eighteenth century remark on their total absence in the Arab world."

In Europe, things were not much different. There, according to Standage, "wheeled vehicles had come to be seen as socially unacceptable by elite men. But then, during the sixteenth century, European attitudes toward wheeled vehicles underwent a sudden turnaround. You might call it a revolution."

What caused that horse-drawn wagon revolution more than anything else, according to Standage, was handheld firearms. He writes, "coaches came to be held in high regard because of their military connotations – and, in particular, a novel military use for wagons that emerged in the late fifteenth century. Facing the threat of the expanding Ottoman Empire, Hungarian commanders adopted a new tactic: arranging wagons on the battlefield in a ring and chaining them together to form a wagon fort, a mobile defensive fortification that could resist cavalry charges. The wagons, equipped with gunports, also acted as protected platforms from which men could fire a small cannon or an early form of gun called an arquebus. This cutting-edge combination of wagons and gunpowder weapons made armored knights on horseback look suddenly old-fashioned."

Though Nicolaus Otto wanted to make larger and more powerful engines for stationary applications, his factory manager, Gottlieb Daimler, set out to develop a half-horsepower engine small enough fit on a bicycle frame. 1885 became a banner year for mechanized transportation. Daimler and his business partner, Wilhelm Maybach, debuted their 'Reitwagen', the first motorcycle and also the first vehicle of any kind to use a gasoline-powered internal combustion engine. In the UK, John Kemp Starley invented the first commercially successful "safety" bicycle, the now-ubiquitous style that has two same-sized wheels and a chain-and-gear drive, and named both his bicycle and then later, his company, Rover. (The company would later get into car manufacturing and has now become Jaguar Land Rover.) And in Germany, Karl Benz built his Patent-Motorwagen, the first vehicle ever designed to be a car (rather than just a horse carriage with a

Figure 46 – Karl Benz's 1885 Patent-Motorwagen

motor mounted in it) and also the first commercially available automobile.

The car arose from bicycles. Most of these early automobiles were bricolage, that is, devices built from available parts. The Benz Patent-Motorwagen used bicycle parts – tubular steel, wire-frame tires, gears, and chains. It's not surprising that Benz did so, since he himself was a known bicycle lover. Brothers Charles and Frank Duryea, founders of the first American gasoline car company, started as bicycle makers. Ford's first car, the Quadricycle, used pneumatic bicycle tires, chains, and gears too. Ford himself was a bicycle nut. George Foss, maker of Canada's first-ever car, the Fossmobile, was a blacksmith who ran a bicycle repair shop in Québec. The Wright Brothers, too, inventors of the airplane, ran a bicycle shop. Škoda, Peugeot, and Kia all got started as bicycle companies. Michelin originally made tires for bicycles, not cars.

The automobile was literally the convergence of key technologies developed for other applications: the horse-drawn wagon combined with Otto and Daimler's gasoline engine, built with parts developed for bicycles.

As with other industries we have seen, it took about fifteen years for the automobile to go from single-unit prototypes built by hobbyists to a mass-production industry. The Duryea brothers built the first US gasoline automobile for themselves in 1893 and sold the first American-made car three years later, in 1896, the same year Henry Ford built his Quadricycle. Just twelve years later, Ford launched the Model T, bringing an automobile within reach of every worker.

All of this happened despite the conditions at the time, not because of them. At the start of the 1900s, the automobile did not have much in its favor. First, few people knew how to drive or how to maintain a piece of internal combustion machinery, or even knew someone who could possibly teach them these things. If you bought a car in 1901, there were only 650 miles of paved roads in the entire US to drive on. (Kaszynski, 2012) There were no fuel stations. (Those were not invented until 1913. Until then, people bought their fuel at hardware stores.) And cars were not very user-friendly. In early 1908, automobile manufacturer Byron Carter, who had previously run a bicycle shop, died from injuries he sustained while hand-cranking a car to start it. His death prompted Charles Kettering to invent the electric self-starter. Early gasoline cars were so hard to start that, according to Tom Standage in *A Brief History of Motion*, "Henry Ford bought his wife, Clara, a Detroit Electric rather than one of his own Model Ts." The electric self-starter became a major factor in the widespread adoption of gasoline automobiles.

Other people had made cars, but Ford's great innovation, of course, was applying the moving assembly line, a concept he learned from the Midwest meat processing industry, where animal carcasses moved along an elevated track and each worker performed the same task repeatedly. Ford built his Quadricycle in a shed, but putting Ford cars beside every American home would mean that they could not be built in sheds. Ford incorporated his company in 1903 and produced the models A, B, C, F, and K before the Model T was launched in 1908. Even then, his assembly line did not begin operation until 1913, five years after the Model T started production and a decade into his company's existence.

It is no surprise that it took so long. In an interview reported in *Business Insider,* the CEO of Tesla, speaking about his own personal experience with starting a car company in the 21st Century, is quoted as saying that "for cars it's maybe one hundred times harder to design the manufacturing system than the car itself." (Rapier, 2019)

For Ford, that effort was worth it as the time it took to assemble a Model T vehicle went from twelve hours without an assembly line to just ninety-three minutes with one, an 8X difference. Though an assembly line might seem simple, it is actually the convergence of a number of ideas. The first was the concept of the general production line itself, pioneered by Thomas Blanchard in 1819, nearly a century before Ford put it to use.

The second important idea was to precisely manufacture parts so that they were interchangeable. In his book *Exactly: How Precision Engineers Created the Modern World*, Simon Winchester describes how at Ford before the assembly line, "parts of the car were all brought to one location on a factory floor, and a jostling gang of men welded and hammered and soldered and bolted and snapped and levered and turned screws and *filed*, always they would *file* to achieve the proper fit". (italics in original)

According to Winchester, when designing his assembly line Ford "was insistent that no filing ever be done in his motor making factories". In 1896, Swede Carl Edvard Johansson had developed a means to manufacture precision-milled steel blocks that could be stacked together to calibrate measuring equipment. He used his wife's sewing machine to make the first ones. Johansson's gauge blocks (or "Jo blocks") are still the standard way to ensure the precision production of machine parts. Gauge blocks first came to the United States in 1908 just as Ford was starting production of the Model T, and Ford later bought Johansson's company.

Thus, precision manufacturing forms a self-reinforcing feedback loop in which more precise production and measurement has led to even greater precision in measurement and production.

Manufacturing automobile parts very precisely turned the production of cars from work that required skilled craftsmen to one that was primarily less-skilled assembly work. It also dramatically decreased the time need to make an automobile. "Where once it had taken workers ten hours to construct an entire engine," Winchester wrote, "it now took just four". He continues, "it used to take two and a half hours to put one axle together; on the new line, twenty-six minutes. Another production line then cut by exactly half the time needed to put together a transmission". "Before the advent of the assembly line, it would take a worker 20 minutes to assemble a magneto", a "device that would produce the spark for the ignition of the fuel in the engine", "with the assembly line running ... it would take just five minutes."

The moving assembly line made the resulting low-cost Ford car almost irresistible.

We could directly compare the price of a car to the price of a horse, but that would not be fair, since the performance of the two is not the same. If you were in the market for transportation in the year 1908, the Model T cost $825 (in 1908 currency), about the cost of two very nice horses. But the car produced twenty horsepower, not two, and had a top speed of forty-five miles per hour. Racehorses unburdened by a carriage could perhaps compete with the automobile over short distances, but if you added in the cost of a carriage and only ran the horses at a sustainable pace, then James Arbib and Tony Seba in their book *Rethinking Humanity* estimate that "the Model T price/performance was 10x" better.

Over the next fifteen years, the car displaced the horse as the primary means of transportation (besides walking), a position the horse had held for more than four thousand years. The graph on the next page shows the number of horses, the total number of equines (including horses, donkeys, mules, and burros), and the number of cars in the US from 1850 to 1990.

There are several interesting things to see here. First, is that the number of

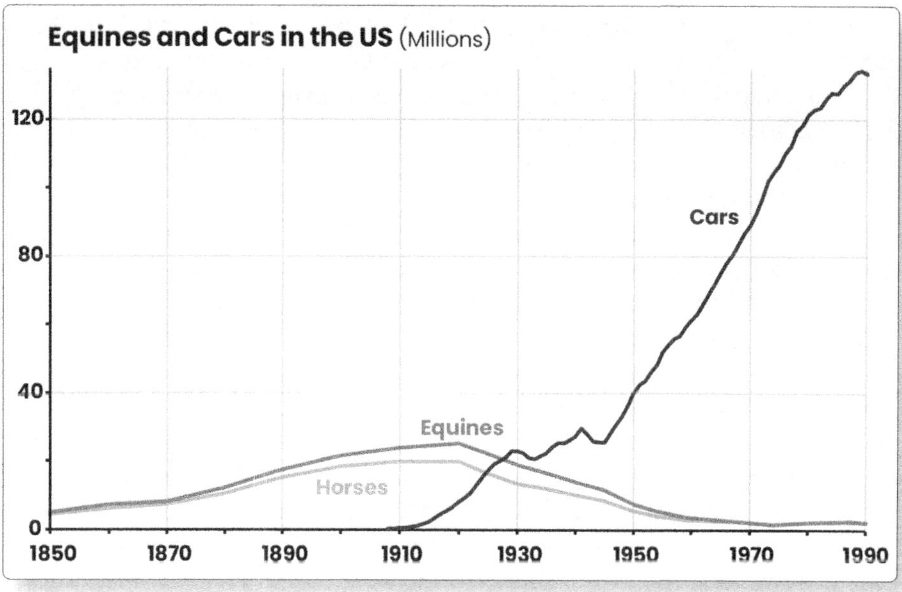

Figure 47 – Number of Equines (Horses, Donkeys, Mules, and Burros) and Cars in the US

horses and other equines grew throughout the latter half of the 1800s and into the early 1900s. However, the human population of the US grew at a similar rate in this period. In 1860, on the cusp of the Civil War, there was one horse for about every five people. By 1900, when the United States became the world's largest economy, there was still only one horse for about every four people. If America was going to get moving in the 1900s, it would need an entirely new way to do it.

And that entirely new way was the automobile. For each year from their introduction until the start of World War I, the registered number of cars on the road increased by at least one-third. This resulted in a doubling in the number of cars about every two years.

In 1910, there were nearly twenty million horses in the US and just under one-half million cars. That is, the number of horses peaked when the number of cars was just over 2% of the number of horses. This is what Tony Seba refers to as 'market trauma'. When a new product constitutes just a few percent of the market, it 'crosses the chasm' (to use Geoffrey Moore's term) from the earliest adopters to the 'early majority' of customers. Seeing that this new technology has the potential to be more than a niche product serving a limited market, attention and investment rapidly switches away from the old and toward the new.

The iPhone of the Steppes

The modern domesticated horse got disrupted for transportation purposes by the automobile, but thousands of years ago, starting on the steppes near the Volga River, north of the Caspian Sea, the horse itself was the disruptor.

Professor Ludovic Orlando of the University Paul Sabatier in Toulouse, France, and over 150 co-workers used genetic analysis to show that horses were domesticated twice, first about five-and-a-half thousand years ago in the Botai region of Kazakhstan and a second time about four to five thousand years ago in the area north of and between the Black Sea and the Caspian Sea. (Librado, 2021) These were separate lineages of horses, domesticated by two genetically distinct groups of people.

Today, there are two groups of horses in the world: the endangered Przewalski's horse, which is wild and has never been domesticated by people, and all other horses, which are descendants of that second domestication of horses four to five thousand years ago.

In an interview with author David Miano, Orlando said that this new breed of horse "was indeed extremely successful, because they managed to spread out from [the area between the Black Sea and Caspian Sea] to the rest of the world. And what our paper shows is that [the people who domesticated the horse] managed to spread [horses] out in almost a few centuries only. So, the success was not only massive but it was immediate. It was almost overnight that when we invented this new type of horse, it became really fashionable. Everyone wanted to have that horse. Probably because it gave you a very strong advantage in terms of how you could move, how you could actually share work with the animal, and also share goods and trade with your neighbors, in fact. So probably this was the equivalent of a sort of iPhone, but back four thousands of years ago. When it was invented, everyone wanted to have this technology."

Tom Standage in *A Brief History of Motion* wrote, "the number of horses on American roads went into rapid decline, falling by 60 percent in the biggest cities between 1910 and 1920. By the mid-1920s only 3–6 percent of vehicles were still being drawn by horses." It is difficult to see from the graph, but according to the 'Fifteenth Census of the United States, Census of Agriculture, The Farm Horse', published in 1933, "the high point of the horse and mule population of the country appears to have been reached in 1913 or 1914."

This is remarkable. In the year 1895, there were four cars in the US. Not four per person. Just four, in total. In 1900, there were eight thousand. Thirteen years later, when there were about 1.1 million cars, the number of horses peaked at nearly twenty million. If you owned a horse at that time, you were in for a rough ride.

Economists Olmstead and Rhode estimate that in 1912, the price of a horse was about 90% of the price it had been in 1910. By 1917, just five years later, horses averaged just 44% of their 1910 price. Horses can live for twenty-five years or more, so if you had purchased a three-year-old in 1910, just seven years later you would have lost more than half of your investment, and the horse would still not even be middle aged.

The price of the Model T fell dramatically in that time as well. From $900 in 1910 to $500 in 1917 and $260 by 1925. But the price of horses and the price of Model T's were falling for completely different reasons. The car's price fell because Ford's business grew dramatically. From fewer than 20,000 units in 1909, his business was ten times larger in 1914 and then grew ten times larger again over the next decade, averaging almost two million units per year in 1923, 1924, and 1925. As Tom Standage wrote in *A Brief History of Motion*, "the automobile was the computer industry of its time... There were 4000 registered cars in the US in 1900 and 4,000,000 in 1920. That's 10 doublings in 20 years, or one doubling every two years – which is Moore's Law, long before computer chips."

Part of the decrease in the cost of making a Model T was due to greater economies of scale – making things in larger quantities tends to mean a lower price per unit. But part of the decrease was also due to Wright's Law, the relationship we saw in Chapter 1 where the marginal cost of making an item tends to decrease by the same percentage with each doubling in the cumulative number of units ever produced. Ford's obsession with vertical integration (owning every part of the supply chain right back to the mines and forests that supplied raw materials) and with cost-cutting also played a role. The first six years of the production of the Model T, versions were produced that were green, blue, gray, or red. But starting in 1914, the Model T was available in "any color so long as it is black." Due to these price decreases, from $900 in 1910 to $260 in 1925, people were simply able to buy a lot more horsepower for their dollar in 1920 or 1930 compared to 1890 or 1900.

As a result, the disruption of the horse by the gasoline automobile was not simply a one-for-one replacement. The number of cars on the road in the US first exceeded the number of horses in 1925, seventeen years after Ford's Model T debuted, and just two years later by 1927, the last year the Model T was produced, the number of cars on the road was greater than the number of horses had ever been. Today, the number of passenger cars is more than ten times the peak number of horses from 1910, meaning that there is one car for about every 1.33 people. And since each car goes much further per year than a horse, the number of vehicle-miles travelled has increased astronomically.

By the 1930s, one in every seven Americans was in employment linked to the automobile industry, an industry that basically did not exist twenty years prior.

No More Donkey Milk

One other interesting thing to see in the graph of the number of 'Equines and Cars in the US' is the virtual extinction of the donkey, mule, and burro after World War II. These animals were used primarily for hauling goods, not the transport of people, and early engines were not powerful enough for use pulling heavy loads, so the donkey, mule, and burro took longer to disrupt than horses. Their peak number did not come until about 1925, around ten or fifteen years after Peak Horse, when the non-horse equines numbered about 5.68 million. Their numbers were down nearly half by 1945 and by 1969 were less than 1% of their peak value.

The decline of the donkey, mule, and burro had some unexpected side effects. One was the elimination of the supply of donkey milk (or 'jenny milk'). According to Wikipedia, "donkey's milk was sold until the twentieth century to feed orphaned infants and to cure delicate children, the sick and the elderly" due to "closer resemblance to [human] breast milk" than cows, sheep, or goats.

As with many other disruptions, the switch from horses to cars changed key metrics of interest. Horses are easy to count. Miles less so. But it is readily apparent to a car owner that his or her vehicle can travel a far greater distance per day or per year than a horse. Benz's 1885 engine is believed to have been two to three horsepower while both George Foss's Fossmobile and Henry Ford's Quadricycle, each built about ten years later, were about four horsepower.

If we assume each horse in the early 1900s travelled one thousand five hundred (1500) miles per year (that's five miles per day for the six working days per week, fifty weeks per year) and each car was run about 10,000 miles per year, then each car travelled about 6.67 times as far as a horse. (Those are rough estimates, but coincidentally are similar to the ratio that Tesla described in an April 2019 presentation where they suggested that autonomous electric taxis might cover 90,000 miles per year each, about 6.67 times greater than the roughly 13,500 miles a human-driven privately owned car travels in the US.)

As cars got more popular, it quickly became apparent that what people primarily wanted was not a horse carriage or even a car, but instead the ability to travel a distance quickly. If a horse travels 1500 miles per year and a car 10,000, then we can calculate the proportion of all horse or car travelling done by cars solely from the number of cars and horses each year. In 1909, over 9% of all miles travelled by horses or cars in the US were done by cars. Just thirteen years later, in 1922, 80% of all miles were by car even though, at that point, horses still outnumbered cars.

Looking at the 'Portion of Car and Horse Travel in the US Conducted by Car', we can see that the bulk of this disruption in traveling ability happened over just ten to fifteen years. But in a sense, in the 1920s, 30s, and even 40s, the disruption was still only just beginning. As we saw earlier, the number of cars by 1930 already exceeded the highest-ever number of horses, and the number of cars kept growing after that.

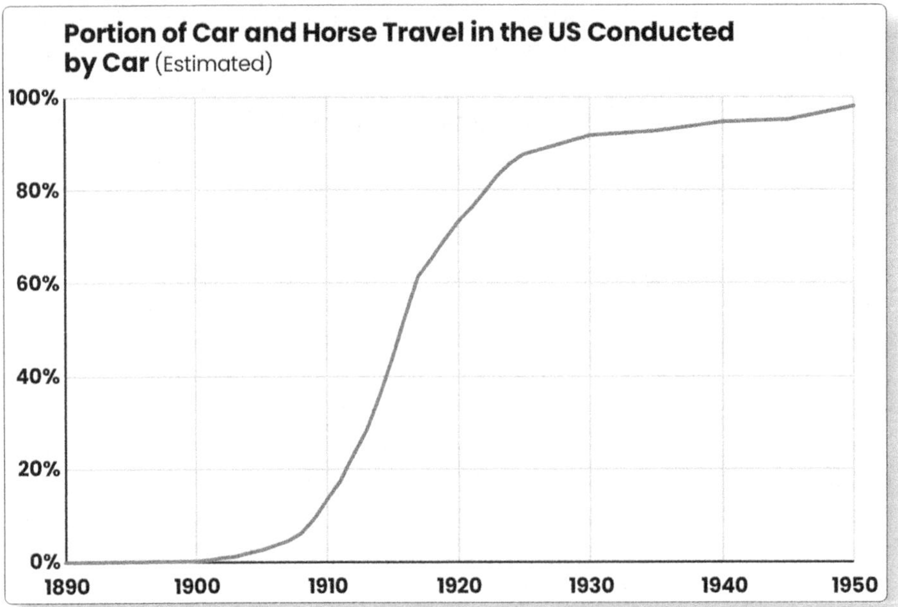

Figure 48 – Portion of Car and Horse Travel in the US Conducted by Car

If we look from the perspective of the number of annual vehicle-miles of travel per person in the US, the disruption that started with the Model T is still underway. In 1850 there was one horse for each 5.3 people in the US. Again, assuming each horse travelled 1500 miles per year means that there was 280 miles of horse travel available per year per American. By 1900, economic growth had made horses more prevalent, but there was still only 360 miles of horse travel available per American, not even 30% more than what was available fifty years prior.

But motorized transport changed that forever. By 1930 in the US there was one car for every 5.3 people, the same as the ratio between horses and people in 1850. But since cars travel more per year than horses, the number of vehicle-miles available per person, on average, went up by about sevenfold over that time, with most of that change happening in the years from 1910 to 1930.

The number of cars per person has only increased since then. By 1950 there were fewer than four people per car in the US. By 1960 there were fewer than three people per car and, by 1980, fewer than two. Today, with about 340 million people and about 285 million registered passenger cars, there are only about 1.2 people per car. Even if each car in the US now only travels about 10,000 miles per year, there would still be over seven thousand vehicle-miles of personal transport available, on average, to each person in the US, more than twenty times what was available when we relied on horses to move ourselves around.

The automotive industry is a hungry beast. According to David Landes in the 1969 book *The Unbound Prometheus: Technological Change and Industrial Development in Western Europe from 1750 to the Present*, by 1929, only a couple of decades into the development of the industry, the American automobile production was responsible for about one-fifth of US steel consumption, a quarter of tin and nickel, three-eighths of aluminum, more than half of malleable iron, nearly three fourths of plate glass and almost 85% of rubber.

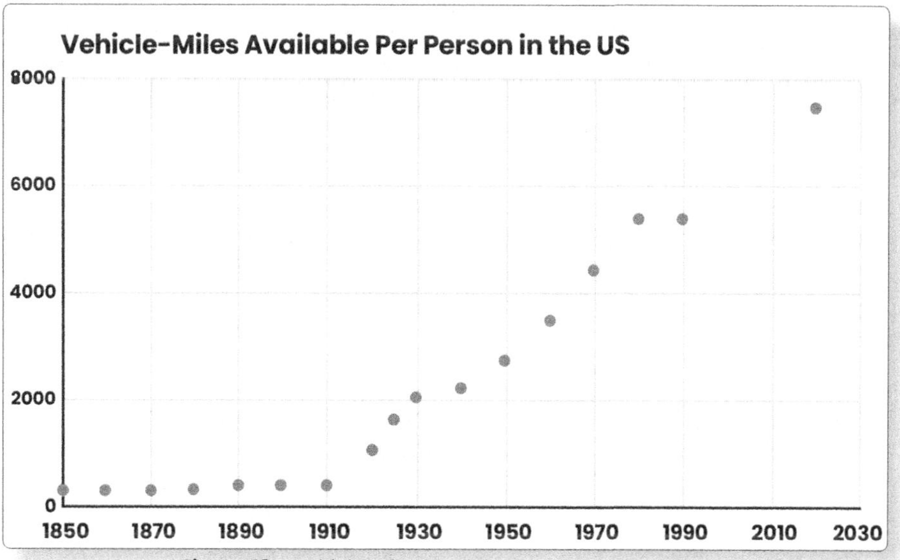

Figure 49 – Vehicle-Miles Available per Person in the US

The adoption of the automobile unleashed a multitude of disruptions in how cars were designed and purchased. Early automobile designs left the passengers exposed to the elements, but as cars got cheaper and appealed to a broader range of people, closed-body designs became more popular. As Tom Standage recounts in *A Brief History of Motion*: "Closed bodies were unusual before the First World War, but the proportion of American cars with closed-body designs rose from 10 percent in 1919 to 85 percent in 1927."

General Motors was founded in 1908 by William C. Durant, who made a fortune in carriagemaking and then moved into cars, one of the few examples of an incumbent transitioning to the new disruptive industry. Founded in 1886, the Durant-Dort Carriage Company was the leading manufacturer of horse-drawn vehicles in the United States by 1900, just fourteen years later. Durant sold out of this business in 1914, at the age of fifty-three, and the company stopped manufacturing carriages in 1917.

In many ways, General Motors was the opposite of Ford. Unlike Ford, GM offered loans, annual models, and a range of options. Ford's Model T's were available only in black, but starting in 1923, according to Standage, GM "collaborated with DuPont, a chemicals giant that owned a large stake in GM, to develop DuCo, a fast-drying and durable lacquer resin that could be made in a range of colors including blues, greens, brown, and orange."

The "color revolution" that GM started had ripple effects that went outside the family garage and straight into the home. Durant had also founded Frigidaire, the manufacturer of household refrigerators and automobile air conditioning systems, and so DuPont began offering colorful and durable DuCo paints to home appliance manufacturers. A *Fortune* magazine article from 1930 describes this new "suddenly kaleidoscopic world" of household appliances that the automobile had brought about.

General Motors also pioneered the concept of financing, a business model disruption in which the automobile seller lends the money (at interest) to buy the vehicle. According to Standage, GM "embraced the idea of allowing customers to pay for cars in installments, a practice that emerged around 1910 and accounted for 65 percent of new car sales by 1925." In contrast, at that time, the Model T's price was so low that Ford made only $2 on each unit he sold. More than 95 percent of Ford's profits came from spare parts and accessories.

GM was the world's largest manufacturer of automobiles from 1931 to 2007. At its peak, it sold 50% of the cars in the US.

◆ ◆ ◆

"In times of rapid change, experience could be your worst enemy."

— J. Paul Getty

◆ ◆ ◆

Early Visions of the Future of Cars and Computers

Otto and Langen's 1876 gasoline engine was not the first internal combustion engine.

In 1859, thirty-seven-year-old French engineer Étienne Lenoir had built a stationary engine that burned a mixture of coal gas and air ignited by a spark and, over the next few years, built various powered carriages, including a one-cylinder model that ran on hydrogen.

In 1863, fellow Frenchman Jules Verne wrote *Paris in the Twentieth Century*, a science-fiction account about what life might be like in one hundred years, in the 1960s. In his imagination, Paris would be overtaken by "countless carriages clogging the boulevards", each hydrogen-powered with a Lenoir engine, and they would be quiet, *"simple and maneuverable"*, with a steering wheel and a brake pedal.

Each vehicle "with the power of several horses, did not cost, per day, one eighth the price of a horse", he wrote, and cost "nothing when it was not in use". Hydrogen "gas hydrants" would be "set up at the various carriage parking places" to refuel vehicles. To get horse-drawn vehicles off the street, Verne wrote, "the Ministry of Police forbade any cart, dray, or wagon to pass through the streets after ten in the morning".

Verne also described electric street lighting resulting in "boulevards lit as brightly as by the sun" and wood-pulp paper displacing linen rag about fifteen years before these happened. And he mentioned the electronic transmission of documents (which he called "photographic telegraphy") and calculating machines "which indeed did resemble huge pianos: by operating a sort of keyboard, sums were instantaneously produced", things that would not be commonplace for more than a century.

All these new technologies would revolutionize our languages, Verne suspected. "A 1960 dictionary that wants to include all terms in use," he wrote, "is twice the size of an 1800 dictionary".

In this prediction, Verne was not bold enough. The original 1828 Webster's dictionary contained roughly 70,000 entries, Merriam's 1860 dictionary was about 114,000 entries, Webster's 1890 edition contained about 175,000 entries, and the 1909 edition was more than 400,000 entries. That is, the dictionary was already more than twice the size in 1890 it had been seventy years prior, and by 1909 it was nearly six times the size of the original.

It Got Harder to Not Own a Car

The adoption of the automobile was driven not only by the fact that automobiles got better over time, but also because not owning a car got more difficult, in a self-reinforcing cycle.

The higher average speed of automobiles compared to horse-drawn wagons, pedestrians, and bicyclists required a rethink of what constituted acceptable road usage. Before automobiles, pedestrians, bicyclists, horse-drawn wagons, and people on horseback all shared the streets, each traveling at about the same speed. But cars often go about ten times faster than these other forms of transport and motorists in their expensive new contraptions were not willing to wait for others to slowly bumble out of the way.

The introduction of automobiles was a technological disruption which led to a social one, what Peter Norton in his 2008 book *Fighting Traffic* calls "the violent revolution in street use circa 1915 – 1930". Over a period of only about fifteen years, a "social reconstruction of the street" destroyed the idea of roads as shared spaces. "By 1930", Norton writes, "most street users agreed that most streets were chiefly motor thoroughfares." Pedestrians who walked in the street were now branded 'jaywalkers'. "Most jaywalkers after 1930 would agree that they were jaywalking," Norton said, "though in 1920 most would have objected to the term."

Tom Standage in *A Brief History of Motion* agrees that "behavior had shifted entirely by 1930, and the default was that streets were for cars, and pedestrians should limit themselves to crosswalks. The [car] industry had successfully changed attitudes... that cars, not people, had the first claim on the roads." Walking in the street became a finable offense.

The rapid attitude shift in the 1920s was not only to the detriment of pedestrians, but also to horses and horse-drawn wagons. This is a dynamic we have seen with many other disruptions: as the old system becomes less popular, it quickly loses its 'social licence to operate' and sees new restrictions put in place, or supports taken away, which make the old technology even less popular, in a self-reinforcing vicious cycle.

Standage recounts: "The Los Angeles Traffic Commission, set up with the backing of the local automobile club and chaired by the head of a local car dealership, funded studies of traffic and made a series of recommendations that were adopted by the city in January 1925. 'The old common law rule that every person, whether on foot or driving, has equal rights in all parts of the roadway must give way before the requirements of modern transportation,' declared the code's author. The code excluded horse-drawn vehicles from the central business district in the evening rush hour, to speed the flow of traffic." Who in a large city like L.A. would purchase a horse or horse-drawn wagon in 1923 knowing they might soon be banned from the roads entirely?

As horses became less popular, urban areas began removing watering troughs, making owning a horse in a city more difficult. What were once necessary pieces of infrastructure were quickly branded public health hazards due to their role in the transmission of infectious diseases, and became liabilities.

Public water troughs became less popular in the US throughout the 1920s, essentially disappearing by the 1930s. Europe saw a similar trend. In London, England, the Metropolitan Drinking Fountain and Cattle Trough Association, had been building troughs in the city since the 1860s. But according to journalist Ian Molyneaux, "by the 1930s the association stopped building troughs as using horses was being phased out by the growth of cars".

Over time old roadways were reshaped, and new roadways were built, with the motor vehicle in mind. Sidewalks, when built, were often poorly maintained. Bicycle

Figure 50 – Local residents Eugenia Dunbar and Mary Moose (right) with their pet ducks at a horse watering trough in Washington, D.C., April 1927

lanes were often not built at all. All of these helped make not owning a car more difficult than owning one.

But the rise of electric vehicles and the possibility of robo-taxis and delivery pods will require us to rethink our roads once again. Why punish jaywalkers when autonomous vehicles will be able to stop safely for pedestrians? Cheap lithium-ion batteries will not only make electric cars more affordable, but electric bicycles too. And electric cars and trucks should make cities quieter and their air cleaner, making bicycling more pleasant. Another self-reinforcing feedback loop will be underway – the more popular electric cars get, the more appealing alternatives to cars will become.

Tetraethyl Lead

In October 1924, thirty-five-year-old engineer Thomas Midgley Jr. walked into a press conference in the Manhattan offices of Standard Oil. In *The Poisoner's Handbook: Murder and the Birth of Forensic Medicine in Jazz Age New York*, Deborah Blum wrote, "he assured reporters that handled properly there was nothing dangerous about his prize discovery", tetraethyl lead, or TEL. "To prove it, he washed his hands in a bowl filled with TEL. 'I'm taking no chances whatever,' he said."

But in fact, tetraethyl lead absorbs readily through the skin. A few months after his press conference," Blum continues, "Thomas Midgley Jr. left for an extended European vacation, seeking treatment for the effects of lead poisoning."

Just a few years prior, in 1921, Midgley had discovered that tetraethyl lead would boost the octane rating of gasoline, making it burn better. A 20% mix of ethanol into gasoline had the same effect, but Standard Oil supported the idea of adding a few grams of TEL per gallon rather than replacing one-fifth of the gallon with ethanol, so TEL quickly became the standard 'anti-knock' agent. In February 1923, TEL went on sale for the first time in America.

According to Dr. William Kovarik, writing in the *International Journal of Occupational and Environmental Health*, "workers in a Standard [Oil] refinery in Bayway, New Jersey, went violently insane after making leaded gasoline. Seven men died and 33 were hospitalized there; meanwhile, ten more were killed at a DuPont facility, and at least two died and 40 were hospitalized in Dayton, Ohio." Thus, Midgley's Manhattan press conference to reassure the public that TEL was harmless.

"Leaded gasoline was the subject of a federal inquiry," Kovarik continues, "but in 1926 the Public Health Service concluded that the dilute additive in gasoline posed no immediate threat to the public." Author Tim Harford wrote that, "Midgley breezily assured the surgeon general that 'the average street will probably be so free from lead that it will be impossible to detect it or its absorption'".

By 1929, "approximately 10% of the total gasoline sold... was leaded", according to Gil Oudijk of Triassic Technology in Hopewell, New Jersey. "By 1936", thirteen years after TEL's first sale, "90 percent of all gasoline sold in America was leaded," Tom Standage reports in *A Brief History of Motion.* "Alcohol was abandoned as a fuel additive in America, though it continued to be used in Europe, at concentrations up to 25 percent, until the Second World War." Oudijk claims that "by 1969, 97.5% of gasoline worldwide contained lead additives."

Though tetraethyl lead became wildly popular, Midgely himself moved on to other research topics, helping the Frigidaire division of General Motors in the late 1920s to develop dichlorodifluoromethane, also called "Freon", which had a devastating impact on the Earth's ozone layer.

For his work on both poisonous tetraethyl lead and ozone-depleting Freon, environmental historian John McNeill in his 2000 book *Something New Under the Sun: An Environmental History of the Twentieth-Century World* said that Thomas Midgley "had more impact on the atmosphere than any other single organism in Earth's history".

In 1940, at the age of fifty-one, Midgley contracted polio and invented a system of ropes to help lift himself out of bed. But a few years later, in 1944, he became entangled in his own invention and accidentally strangled himself to death.

But starting in the late 1960s and early 1970s, a convergence of forces began to doom TEL to the dustbin of history. Kovarik noted that as least as far back as Roman engineer Vitruvius, who lived in the century before Christ, people have seen that lead fumes and handling or injesting lead had adverse health effects, and that "printers were also frequent victims of occupational lead poisoning from handling and heating lead alloy type." But decades of scientific investigation and campaigning in the mid-1900s by people like chemist Dr. Clair Cameron Patterson proved that lead poisoning, from both leaded gasoline and lead in paint, was more than just an anecdote. Higher-grade metal parts eliminated the need for the lubrication of engine valve seats that tetraethyl lead provided. And starting in the mid-1970, the inclusion of catalytic converters in new cars required the use of unleaded gasoline, as lead fouled the catalyst materials.

US lead consumption in gasoline peaked in 1976, and it was down more than 90% by 1986, but it took until 1996 for lead to be banned entirely from road vehicles. But the US was not the first to fully ban lead in gasoline. Both in terms of the number of countries banning leaded gasoline and the percent of the world's population living in countries that allowed leaded gasoline, the end of leaded gas came quickly.

In 1985, no country had fully banned leaded gasoline. Brazil, which has always used a large amount of ethanol in its fuels, banned lead in 1989. But the biggest impact happened in the year 2000, when both China and India (among other countries) banned leaded gasoline.

At the very end of 1995, 92% of the world's population lived in areas that permitted at least some leaded gasoline. By the end of 2006, little more than a decade later, that number was 3%. The last country to ban leaded gasoline, Algeria, did so in 2021, one hundred years after Midgley discovered tetraethyl lead to be an anti-knock agent.

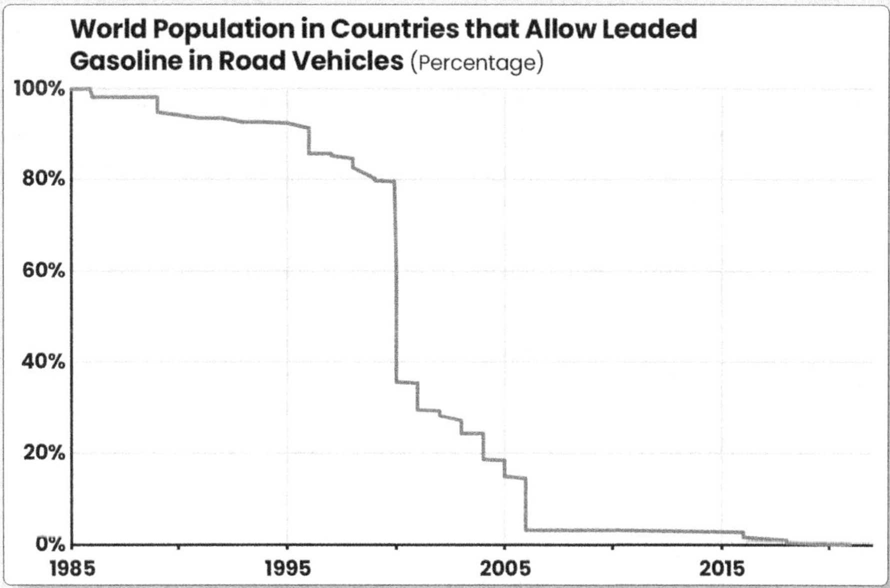

Figure 51 – Percentage of World Population in Countries that Allow Leaded Gasoline in Road Vehicles. *(Ritchie, 2022)*

The social impacts of eliminating tetraethyl lead, even just in the US, have likely been enormous, as lead exposure is known to lower IQ, and promote behavioral problems and learning disabilities. According to *Mother Jones* magazine, researcher Rick Nevin, who summarized his work in the 2016 book *Lucifer Curves: The Legacy of Lead Poisoning*, claims that "gasoline lead may explain as much as 90 percent of the rise and fall of violent crime over the past half century."

◆ ◆ ◆

"Americans are a horse-loving nation... the widespread adoption of the motor-driven vehicle in this country is open to serious doubt."

— *Lippincott's* magazine, 1903

◆ ◆ ◆

The Diesel Engine: Tractors, Trucks and, Unfortunately, Passenger Cars

Otto and Langen unveiled their gasoline engine in the mid-1870s, but basic thermodynamic calculations suggested it was not the most-efficient design possible. So, twenty-seven-year-old German mechanical engineer Rudolf Diesel started developing a simpler engine that ignited its fuel purely through pressure. If you have ever pumped up a bicycle tire, you will know that compressing a gas makes it warmer. Diesel's idea was that, if you compress air enough, it will get hot enough to ignite fuel sprayed into a chamber. He began his work in 1885, which led to the first successful demonstration of the engine that bears his name twelve years later, in 1897.

The resulting design was not a clear victor over the gasoline approach. Diesel's engine was inherently more efficient, but it was also prone to vibration, which required a heavier, more robust construction than a gasoline engine, reducing its efficiency advantage. It was also louder. But fuel for Diesel's engine was cheaper to refine from crude oil than gasoline and also gave off less fumes, so it was easier to store and less likely to explode, which made it attractive for military purposes. By 1904, Diesel had gotten his engines into French submarines and started to penetrate other markets where powerful, simple engines were needed.

The 1910s turned into a good time to buy tractors and a bad time to sell horses. By the middle of the decade, diesel-powered tractors were taking over the market. Tractors fell in cost over the decade of the 1910s to less than 40% of their price at the start of the decade. With mechanical horsepower so cheap, the value of horses followed quickly. As Tony Seba said, "if you had invested in a horse barn in 1900, basically you lost all your money."

But Rudolf Diesel would not live to see the success of his life's work. In 1913, while crossing the Channel from France to England, Diesel disappeared during the night. With Europe on the verge of war, some suspect that he was murdered and thrown overboard to avoid revealing secrets to the British. Others suspect that, depressed his work had not found the success he thought it deserved, he jumped of his own free will. It was only a few years later, after the First World War, that Diesel's invention began to realize its commercial potential. The first diesel-powered trucks started gaining popularity in the 1920s and trains in the 1930s. By 1939 a quarter of global maritime trade was fuelled by diesel-powered engines.

Not everyone was happy with this new system. In his 2019 book *Food 5.0: How We Feed the Future*, Robert Saik wrote, "believe it or not, people protested the tractor

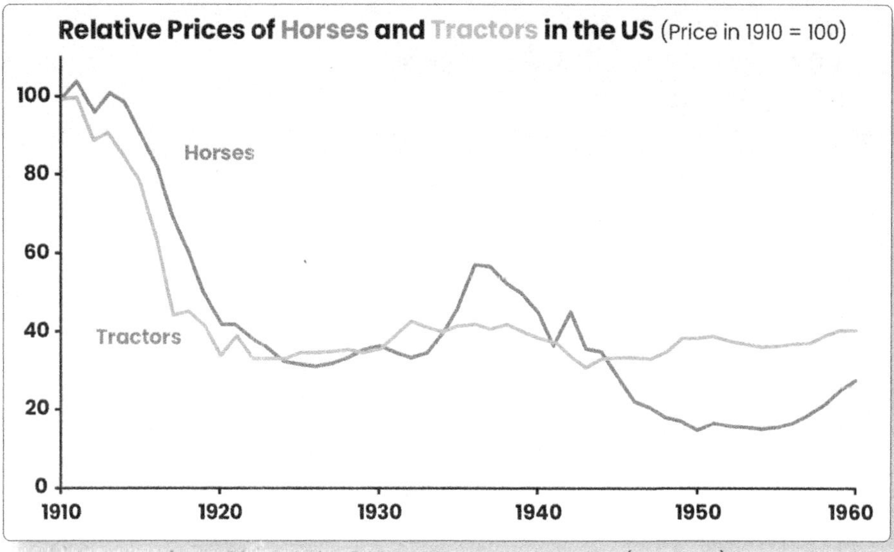

Figure 52 – Relative Prices of horses and tractors (1910 = 100).
(Olmstead & Rhode, 2000)

much like people today protest and resist GMO seed. The horse industry put up a huge resistance to the tractor. At one point, the Horse Association held rallies across the country to support horses and wrote legislation outlawing tractors."

But the rise of the mechanical tractor was unstoppable. The horsepower of tractors in the US, which was essentially zero in 1915, exceeded the number of work equines just fifteen years later and was five times higher than the number of equines when the Second World War ended. This was not a one-for-one replacement of equine power by tractor power. By 1935, according to Yujiro Hayami's 1971 book *Agricultural Development: An International Perspective*, the combined horsepower of all tractors in the US was at about the same level that the peak number of work equines had been in 1915, twenty years prior. By 1945, the combined horsepower of tractors was more than double what the peak number of work equines had been in 1915, and by 1955, it was nearly five times the peak.

Even more dramatically, the population of farm workers in the US was declining during these years, so the amount of horsepower available per worker increased even faster. In 1915, there were a little more than two work equines per farm worker in the US. But by 1955, forty years later, the population of farm workers in the US had fallen by half even though the total US population had grown, so there were then more than twenty horsepower of tractors per farm worker, a tenfold increase in the availability of work per person in four decades.

It was not just tractors that found the diesel engine useful. In 1910, two years after the Model T was launched, there were roughly 460,000 registered automobiles in the US, but only just over 10,000 trucks, roughly one truck for every forty-six cars. From 1910 through the start of World War II, the number of trucks per car kept rising, reaching one truck for every ten cars just after World War I and one for every six cars in 1933. By the period from 1945 to 1970, there was roughly one commercial truck in the US for every five cars.

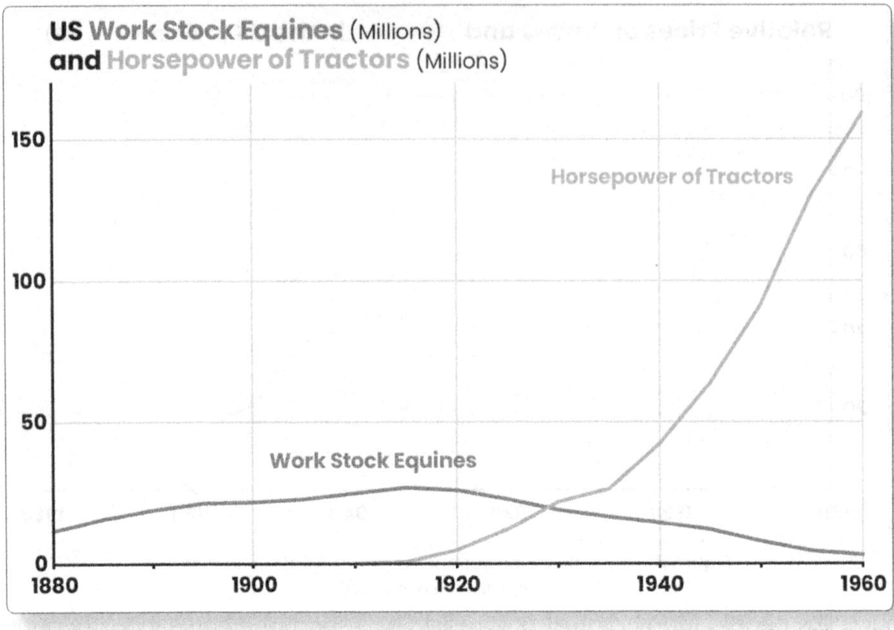

Figure 53 – US number of Work Stock Equines and total Horsepower of Tractors, in millions.
(Hayami, 1971)

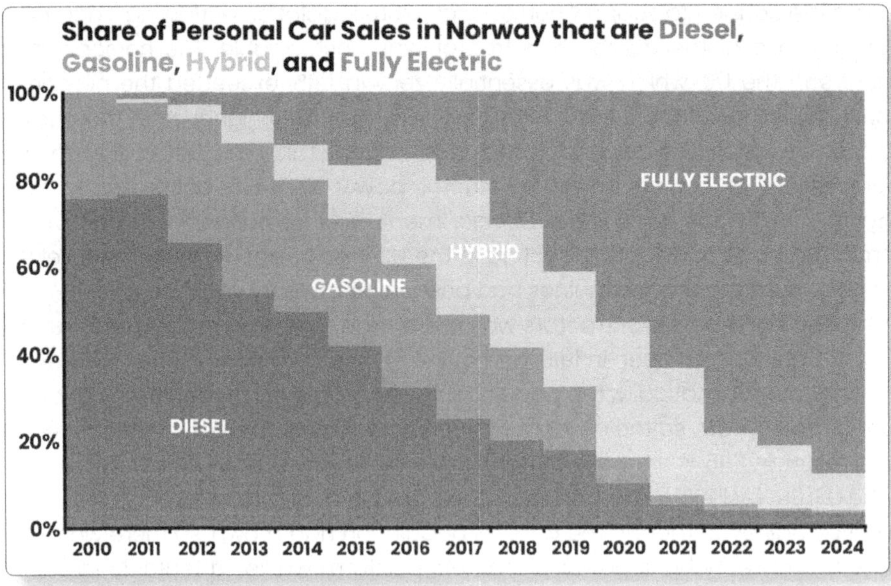

Figure 54 – Market Share of Diesel, Gasoline, Hybrid, and Electric cars in Norway

It took decades more for Diesel's invention to come into its own in Europe. Unfortunately, European authorities introduced tax incentives that favored the use of diesel engines for passenger cars so that European automakers could dominate that market. By the early 2000s, diesel engines represented the majority of sales of personal passenger cars in places like Norway. But, unable to get their 'clean diesel' technology to work properly, companies like Volkswagen, the world's then-largest automobile manufacturer by unit sales, resorted to installing cheat devices on their cars to deceive consumers into thinking diesel vehicles were better for the environment than other vehicles. However, the rapid rise of electric vehicles in the early 2010s quickly displaced all sorts of combustion vehicles – diesel, gasoline, and hybrid. In 2015, researchers at the University of West Virginia discovered that Volkswagen's vehicles were using software to cheat emissions tests. The collapse of combustion vehicles followed swiftly. By the early 2020s, about five years later, electric cars, which constituted less than one-in-five new vehicle sales as late as 2016, were the majority of new car sales. This is how quickly a market can change – in just five years a century-old technology can be wiped off the map forever.

◆ ◆ ◆

"Any man who can afford to keep a road horse will keep one, regardless of all the bicycle and horseless-carriages fads"
— *The Kansas Sunflower* (Clyde, Kansas), March 2, 1899

◆ ◆ ◆

Synthetic Rubber: The Other Manhattan Project

> *The sudden stop to imports of Asian latex rubber at the start of World War II required the US to build an entire new industry from nothing, the manufacture of synthetic rubber. While building this new industry, synthetic rubber manufacture itself underwent a disruption, with 5X-cheaper petroleum-based ingredients replacing alcohol-derived ingredients. Synthetic rubber manufacturing became the second-biggest wartime project after the development of nuclear weapons*

The entrance of the United States into World War II suddenly cut off America's supplies of some key materials from Asia, including silk and rubber. In *Paths of Innovation: Technological Change in 20th-Century America*, David Mowery writes that "the federal government invested approximately $700 million in the construction of fifty-one plants that produced the essential monomer and polymer intermediates needed for the manufacture of synthetic rubber... The synthetic rubber program was second only to the Manhattan Project in terms of rapid and extensive mobilization of human resources in order to achieve an urgent wartime goal."

Wallace Carothers, the inventor of nylon, had also invented neoprene, the first synthetic polymer, in the early 1930s, but high costs of production inhibited its adoption. "Applications of neoprene were limited by its price of $1.05 per pound at a time when natural rubber, which had superior performance characteristics as a general-purpose product, was priced at less than 5 cents per pound," Mowery wrote. "In 1940, not surprisingly, natural rubber accounted for 99.6% of the U.S. rubber market and synthetic rubber a mere 0.4%."

The transformation of the U.S. rubber industry during the war was nothing short of breath-taking. Rubber production in the US went in 1941 from approximately 750,000 tons, all of which were from natural sources, to over one million tons five years later, more than 85% of which were synthetic. As with many other disruptions, this was not a one-for-one replacement of natural rubber with synthetic alternatives, but instead resulted in a significant growth of the total market.

The process of synthetic rubber production itself underwent a radical transition during the wartime build-up of this new industry. According to Frank Howard's 1947 book *Buna Rubber. The Birth of an Industry*, while the production of butadiene, a key ingredient used in making synthetic rubber, in 1943 came largely from alcohol (made by fermenting corn), by the end of 1944 butadiene was derived primarily from petroleum. It is understandable why the switch from alcohol-derived butadiene to petroleum-derived butadiene was so rapid. According to Paul Wendt, writing in 1947, "the butadiene produced in these [alcohol-based] plants cost roughly five times that produced in the petroleum plants."

We had already tried to make synthetic rubber from fermented chemicals a few decades prior. In 1912, gifted thirty-eight-year-old biochemist Chaim Weizmann, born in what is now Belarus, isolated a bacterium that can produce acetone, butanol, and ethanol from starch and glucose. Weizmann's attempts to make rubber commercially proved fruitless, so he pivoted to producing acetone for use in making smokeless gunpowder. Weizmann's work is recognized to be start of modern biomanufacturing and his ideas were quickly adopted by others to produce amino acids, organic acids, vitamin C, and other chemicals. Weizmann himself was later elected the first president of Israel.

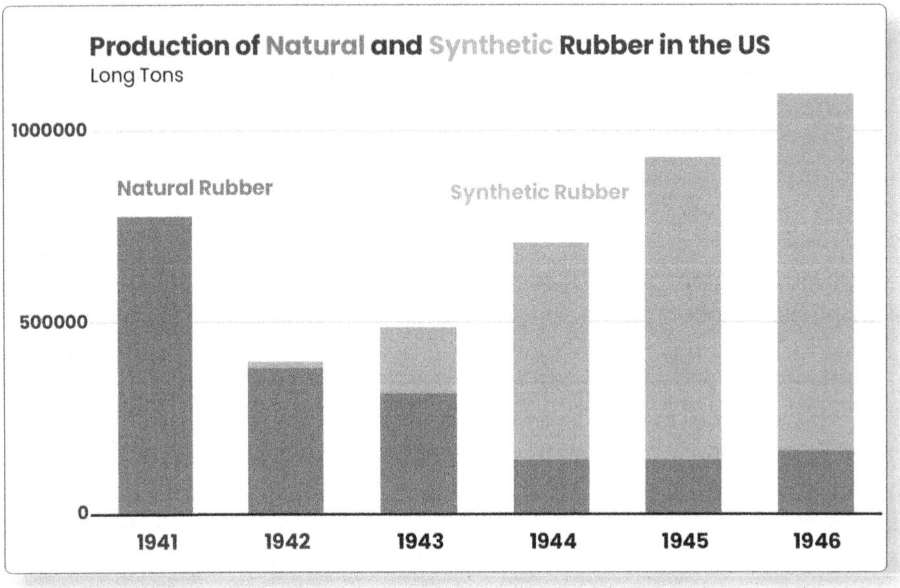

Figure 55 – Production of Natural and Synthetic Rubber in the US. *(Mowery, 1998)*

According to Katrina Cornish of the USDA, global production of natural rubber dropped precipitously during World War II, but bounced back after the war ended, temporarily overtaking synthetic production for the fifteen years until 1960. From that point, synthetic rubber production consistently exceeded natural rubber production worldwide for the remainder of the century.

The Radial Tire Revolution

> *Radial-ply tires last 5X longer than bias-ply ones, but they require a different design of the car chassis. Getting manufacturers and tire manufacturers to convert was a tall order. Even so, these new tires disrupted the old ones over a period of about fifteen years. And autonomous cars might require tires to get disrupted again*

In a world of change and disruption, one thing has been constant for more than a century: the Michelin Man. His appearance has been largely unchanged throughout his life for a guy who is more than 130 years old. Originally he was made from bicycle tires because making bicycle tires was how his creators, the Michelin brothers, got their start. But as cars grew more popular in the 1910s and 20s and the brothers switched to car tires, he switched too.

Figure 56 – The Michelin Man throughout the ages

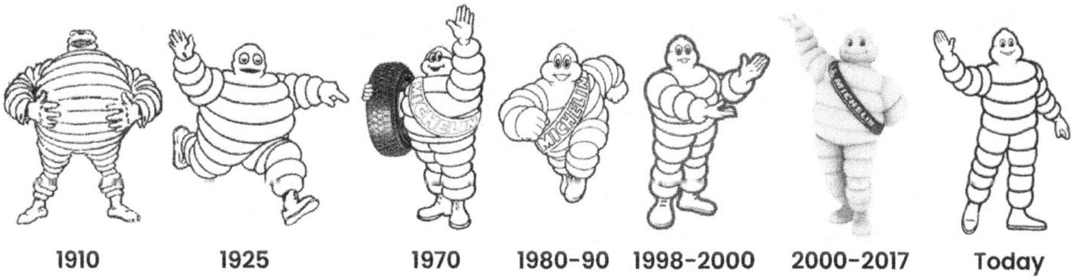

Why is the Michelin Man white even though tires are black? Because tires have not always been black. Latex is naturally light in color, and carbon black was only added to rubber as a strengthener as cars got more powerful and faster after the early 1910s. "Over 90% of the carbon black manufactured is consumed as reinforcing and compounding agents for rubber, mainly for motor vehicle and aircraft tires," it says in *Van Nostrand's Scientific Encyclopedia*, "these agents extend the life of a tire in normal usage by eight to ten times." (Considine, 2008) Ford's Model T, first released in 1908, originally came with white tires and in 2020, when Harley-Davidson introduced their Serial 1 electric bicycle based on their oldest product, their Serial Number One motorcycle from 1903, they chose white tires for the new cycle, just like the original.

Just after World War II, the Michelin company commercialized the 'radial-ply' tire, which was more durable than the standard 'bias-ply' tire. As with many disruptive products, the radial tire had a long incubation period from the time of its invention to its commercialization. The first radial designs were patented in 1914 in the UK and 1916 in the US, but it took thirty more years for Michelin to introduce the first radial-ply tire for sale to the marketplace.

By Seba's classification system, radial tires constitute a 'disruption from above', a more-expensive but better-performing technology that ultimately offers a lower cost-per-mile to the consumer. Radials lasted about 40,000 miles instead of the 12,000 miles typical of bias-ply tires, and they were known for their longer tread life, better steering characteristics, and lower rolling resistance, which improves gas mileage. However, they were also stiffer, which required car manufacturers to redesign their car suspensions, as cars that take bias-ply tires typically cannot take radials, and vice versa. Richard Tedlow said in his 2010 book *Why Business Leaders Fail to Look Facts in the Face* that in 1968 the product review magazine "*Consumer Reports* awarded five of its top six ratings to radials even though they made up a mere 2 percent of the American market that year."

Convincing a car manufacturer to design for the stiffer tire was not an issue for Michelin in France, as the Michelin tire company owned the Citroën car company. Over the couple of decades, radial tires became standard in Europe and Asia. By 1970, according to the National Bureau of Economic Research (NBER), 98 percent of tires sold in France were radials. (Rajan, 2000)

U.S. manufacturers made radial tires in their European subsidiaries, but the technology did not catch on at home, with American tire manufacturers and car manufacturers 'locked in' to bias-ply. Car manufacturers did not want to design their cars for tires that were not popular, and tire manufacturers did not want to make the massive investments necessary to switch to radial-ply production if few cars could use them.

But in less than five years during the early 1970s, tires sold on new cars in the US

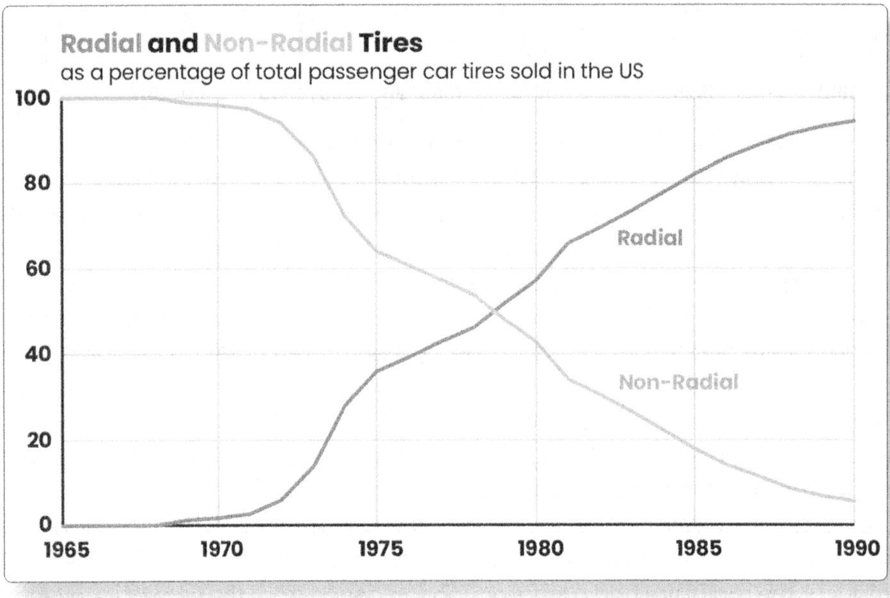

Figure 57 – Radial Tires as a percentage of total passenger car tires sold in the US. *(Rajan, 2000)*

went from essentially nothing to more than 60% of tires shipped. Because a car can use several sets of tires in its lifetime, sales of replacement tires are typically about three times the sales of original equipment tires. As a result, it took about fifteen years for radial tires to go from just a couple of percent of the total US tire market to over 80%.

"What changed in the United States in the early 1970s to propel adoption?", the NBER report asks. "An important factor in this decision was Michelin's decision in 1970 to reenter the U.S. market. Another factor was the growing volume of automobile imports from Europe and Japan that convinced manufacturers of changing consumer preferences."

Though foreign automobile imports into the US in the decade after World War II were negligible and sales of imported cars were never more than 10% (by value) of US car sales prior to 1965, records from the US Bureau of Economic Analysis show that the portion of the US car sales market (in dollar terms) taken by imported vehicles rose steadily afterwards. From 1965 to 1970, imports went from about 4% of the market to 20%, the level where they stayed for most of the 1970s. US auto manufacturers could no longer ignore radials.

US tire manufacturers did not emerge from the radial revolution unscathed, quite the opposite. "By the end of the decade" of the 1980s, NBER says, "traditional American firms like Firestone, Uniroyal, Goodrich, Armstrong, and General Tire belonged to foreign companies. Consequently large U.S.-owned tire manufacturers, who in 1971 represented 59 percent of the world production and included four out of the top five producers, in 1991 represented only 17 percent of world production with only one of the top five producers," Goodyear.

Radial tires were not the only automobile innovation to quickly conquer the market. According to the website *Our World in Data*, only 6% of new cars had power steering in 1953. By 1968, just thirteen years later, it was 80%. "In 1923," Nebojsa Nakicenovic wrote in *The Automobile Road to Technological Change*, "a little more than 1% of all new cars were offered with four-wheel brakes, by 1927 they were standard equipment on 90% of new cars." Disk brakes were on less than 1% of new cars in 1966 and 100% thirteen years later. In 1951, 36% of new cars had an automatic transmission. In 1957, just six years later, it was 80%. One of the most dramatic of all car-related innovations was the electronic ignition, found on 11% of new cars in 1980 and 100% just four years later. In 1990, more than 10% of new US passenger cars (not including light trucks like SUVs) had at least a driver-side airbag. By 1995, just five years later, more than 80% of new cars had dual airbags in the front.

Radical disruption in the car industry is the norm, not the exception. Tesla made their first car in 2008 and they became the number one car company in the world in terms of market capitalization just thirteen years later.

We soon might need another major disruption in car tire technology, due to the rise of autonomous taxis. A September 2020 article on CleanTechnica's website says, the "Tesla Model 3 Standard Range Plus (SR+) comes with Michelin Primacy MXM4 tires... You should get around 40,000 miles out of those tires." But if a robotaxi lasts for a million miles, then "24 tire changes would be needed and that would add up to $32,592!" (Fosse, 2020) This amounts to a little over $0.03 per mile just for the tires for a taxi, which is a significant portion of the $0.20 per mile that Tesla estimated in the October 2024 launch presentation their taxi would cost to operate.

To bring down this cost, tire manufacturers will need to offer products that are literally ten times more durable or cheaper, or a combination of the two. Some electric car manufacturers have suggested that small air thrusters could improve acceleration, braking and cornering. Much of the wear that tires experience is from when they are subjected to stress – namely, accelerating, braking, and cornering. Using thrusters on a car to assist it in only these times, combined with a software-based driving system that accelerates and brakes more smoothly than a human driver, could conceivably both reduce tire wear and extend their lifetimes enough to cut the effective tire cost significantly.

This could have enormous environmental implications. Small 'microplastic' particles from worn tires and brake discs are polluting our waterways, and our own bodies. A December 2020 article in *The Los Angeles Times* reported on a promising explanation for mysterious mass die-offs of salmon, "an alarming phenomenon that has been reported from Northern California to British Columbia". Researchers on the West Coast of the US now believe that a chemical in car tires when exposed to the air turns into another chemical known to be 'very poisonous', even at concentrations of just a few micrograms per gallon of water. When it rains, car tire dust accumulated on and alongside roads washes down into the local waterways. Better tires, smarter driving, or even unusual solutions like air thrusters might help reduce or eliminate this problem.

Automotive Safety and Security Features Have Been Disruptive... and May Get Disrupted

> *Some of the fastest-adopted technologies in the 1900s were automotive safety and security features. But if vehicles can drive themselves, most of these features might soon become obsolete*

The story of the automobile is full of examples of safety and security features that have rapidly been adopted, either due to consumer demand or to mandated changes, and driven by the automobile's inherently high potential for loss of life or valuable property.

Disk brakes and the electronic ignition were among the fastest-adopted technologies ever, faster even than television or radio. Even at a time when households were slowly adopting things like electricity and the vacuum cleaner, automotive technologies like the car ignition key and door lock were being rapidly introduced. According to Tom Standage in *A Brief History of Motion*, "manufacturers added ignition keys and door locks to cars around 1912, letting owners leave them unattended on the street during the working day or overnight, something that had not been possible with horse-drawn vehicles. Parked cars, a rarity in photos of American streets before 1910, are commonplace after 1915."

It might seem staggering that a feature added 'around 1912' would make on-street parking 'commonplace' by 1915, but in fact the number of registered cars in the US in 1912 was just over 900,000 and the number in 1915 was over 2.3 million. So, by 1915, more than half the cars on the road were less than three years old. (In any rapidly growing system, many of the members of the group will have been newly added and therefore causing a change to the inflow of new members will rapidly change the composition of the group.)

Over the decades, sales of new cars have seen similar, rapid adoptions of a variety of security and safety innovations, including airbags, antilock brakes, car alarms, child locks, and numerous other improvements. Perhaps the most impactful among them were seat belts. Other designs existed, but in 1959, Volvo introduced the three-point seat belt and then freely allowed other car manufacturers to use it. According to Harold Linstone in his 1976 book *Technological Substitution*, fewer than 10% of new cars in 1963 and only about 15% of new cars in 1964 included seat belts, but after the publication of Ralph Nader's 'Unsafe at Any Speed' in 1965, seat belts jumped to being offered in more than 95% of new cars in 1966.

While the presence of seat belts in new cars changed dramatically in just two years, the actual use of the belts lagged behind. An August 2008 report from the National Highway Traffic Safety Administration (NHTSA), 'How States Achieve High Seat Belt Use Rates', estimated seat belt use nationwide for drivers and right-front passengers by observing vehicles on the road. Even as late as 1984, US front seat belt use was observed nationwide to be under 20%. But then "seat belt use increased rapidly from 1983 to 1993, a period when most States enacted seat belt use laws." Their usage went from fewer than 20% of front-row people to over 60% in just six years.

Like seat belts, many of the major safety developments in cars in the past decades have focused on maximizing the chance that passengers will survive an accident or minimizing the amount of harm that will come to them in an accident. General Motors began offering airbags in the 1973 Chevrolet Impala, initially in government fleet-owned vehicles. In Europe, airbags were almost unheard of until the early 1990s. By 1991, four manufacturers – BMW, Honda, Mercedes-Benz, and Volvo – offered the airbag on some of their higher-end models. By 1999, finding a new mass-market car without an airbag at least as optional equipment was difficult.

Automobile safety and security equipment is now poised on the cusp of what might be its final disruption: artificial intelligence resulting in the replacement of individually owned, human-driven cars with robo-taxis, thus possibly eliminating most car crashes as well as the need for most anti-theft devices. Emergency braking and autonomous driving offer the possibility of dramatically reducing the number of crashes that happen in the first place, eliminating the need for (or, at least, use of) most safety equipment. In an interview, automobile industry analyst Sandy Munro said, "the easiest accident to survive is the one that never happened." And in the event of an actual crash or other medical emergency during your travel, the robo-taxi or an autonomous ambulance would be able to take you to a hospital faster than any human could, saving even more lives.

◆ ◆ ◆

"History shows us that disruptions are not linear. They always follow an S-curve, starting off slowly, before exploding exponentially, and then eventually slowing down again as they reach saturation. As the new technologies grow, they wipe out the older ones that can no longer compete. Because they are nonlinear, disruptions tend to sneak up on us, and then suddenly they unfold much, much faster than we expect – within just a decade or two."

— Adam Dorr, Brighter: Optimism, Progress, and the Future of Environmentalism

◆ ◆ ◆

Ripple Effects: Rethinking Retail... and Demographics

> *The car reshaped retailing, giving rise to the supermarket, the suburb, interstate highways, the shopping mall, the big box store, the drive-in theater, and the recognition of a new group of people who were not quite adults, but also no longer children, and who had both money and driver's licences in their pockets*

Motor vehicles transformed almost all areas of life, reshaping our cities and our societies in unforeseeable ways. They affected where we lived, how we shopped, and what we bought.

"The suburb is not an American or even Western invention," Amanda Kolson Hurley wrote in her 2019 book *Radical Suburbs*. "Suburbs have been around as long as cities have. In the third millennium BCE, the suburbs of Ur stretched miles beyond the city," But the end of World War II brought with it a flood of returning veterans. As Tom Standage wrote in *A Brief History of Motion*, "one of those returning veterans was William Levitt. He had served in the navy and saw how mass-production techniques, honed in the car industry, had been applied to military housing. Returning to civilian life at his family's building firm, he set out to do the same for civilian housing." Levitt's process was, Standage wrote, "the inverse of a Detroit assembly line... Levitt called his company 'the General Motors of the housing industry,' and the comparison was apt: it offered houses at different prices to suit different buyers and even adopted the idea of annual model updates, with improved versions of its houses announced each year. Prices started at $7,990 with 5 percent down (0 percent for veterans), and each house included modern features such as a television and hi-fi system."

And each house also had mouths to feed, which meant the need for new food markets – supermarkets – accessible by car. Standage quotes the *Journal of Marketing* in 1941 as saying: "Perhaps no phase of retail distribution has so captured the imagination of the American public as has the supermarket... Within a period of less than a decade, approximately 50% of the American consuming public has changed its food-shopping habits." Supermarkets came to western Europe a little later but were embraced just as quickly. Tony Judt recounts in his 2011 book *Postwar: A History of Europe Since 1945* that "supermarkets began to appear, notably in the 1960s, the decade when the impact of the increase in purchasing power was felt most dramatically. In the Netherlands, which boasted just seven supermarkets in 1961, there were 520 ten years later. In the same decade, the number of supermarkets in neighboring Belgium rose from 19 to 456; in France from 49 to 1,833."

The rise of the supermarket was enabled largely by the ownership of automobiles, which had reached more than half of American households as early as the mid-1920s, and which also drove the ownership of refrigerators. William C. Durant started the Frigidaire Company, which was owned by General Motors from 1919 to 1979. Judt says, "in 1957 most west European households still did not possess a fridge (the figure ranging from 12 percent in West Germany to less than two percent in Italy). The reason was not so much technical (by the mid-1950s virtually all of western Europe had full electricity service, with the exception of parts of rural Norway and southern and upland regions of Italy) as logistic: until housewives could afford to buy a lot of perishable food at one outing, and could transport it home, there was not much point in spending large sums of money on a fridge."

But less than twenty years later, the picture had changed dramatically. "By 1974," Judt wrote, "in Belgium and the UK, 82 percent of households had one; in France, 88 percent; in the Netherlands and West Germany, 93 percent. Most remarkable of all, 94 percent of Italian households now owned a fridge, the highest ratio in Europe."

The development of the US Interstate Highway System, constructed largely over a fifteen-year period starting in the late 1950s, and of containerized shipping over roughly the same time, enabled the rise of shopping malls and 'big box' stores, like Walmart and its Supercenter grocery stores, often situated in more-rural areas at the intersections of major thoroughfares. Car ownership changed not only retail, but also entertainment as well. In *A Brief History of Motion*, Tom Standage writes that "America had only fifteen drive-ins by 1939, and ninety-six by 1945" but "the number of drive-ins jumped to two thousand by 1950 and more than four thousand by 1958."

Standage also says that a tax law called 'accelerated depreciation' "introduced in 1954 in an effort to boost investment in manufacturing, had the side effect of triggering a surge in construction of large-scale shopping centers." Developers of large buildings were able to claim that the property lost its entire value over a period of just five years, rather than the normal thirty or forty, greatly reducing their tax burdens. Any retail outlet located in a shopping mall effectively paid much less tax than any similar facility located in a town center. "In 1970, when the 1954 tax loophole started to be closed, America had more than thirteen thousand shopping centers (including enclosed malls), nearly all of which had been constructed in the previous fifteen years in suburban areas, accelerating the decline of traditional downtowns."

Supermarkets and box stores themselves were platforms for further disruptions.

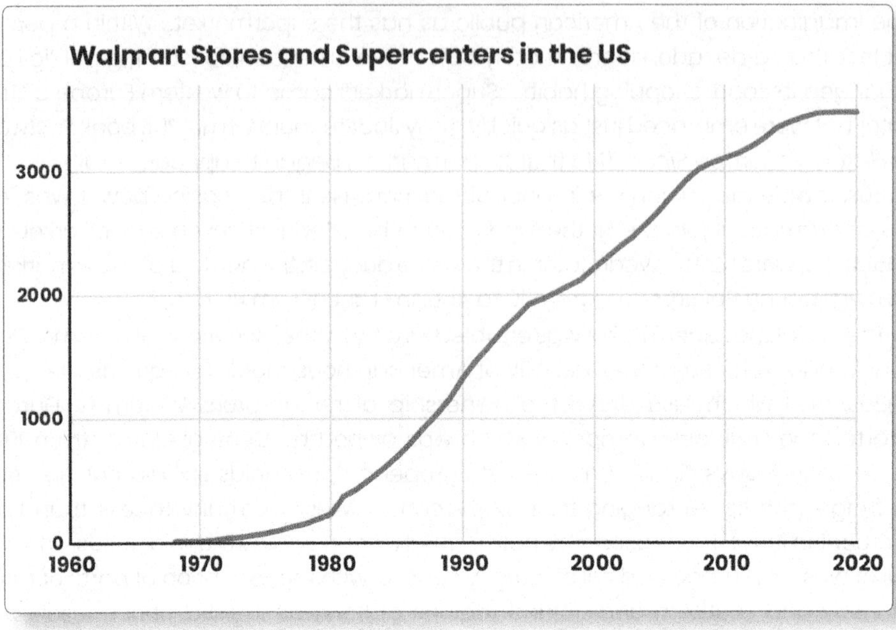

Figure 58 – Walmart Stores and Supercenters in the US. *Data from walmartnearby.com*

In 1960 physicist Theodore Maiman fired the world's first laser. Fourteen years later, in 1974, a pack of chewing gum now on display in the Smithsonian Institution became the first item scanned by a laser-powered UPC code reader. There were scanning machines in fewer than two hundred grocery stores nationwide by 1977. But, according to Emek Basker of the National Bureau of Economic Research, "once scanners commercialized, they caught on quickly: about a third of U.S. supermarkets adopted scanners within ten years of the first commercial installation."

The convergence of the post-War economic boom, increasing suburban living, rising rates of automobile ownership, and everything from the rise of shopping malls to drive-in theaters, created a new class of people that actor Jon Bernthal, portraying automobile executive Lee Iacocca in the 2019 film 'Ford vs. Ferrari' described as "the first group of 17-year-olds in history with money in their pockets" and the ability to spend it on movies, music, clothing, and more.

For the first time ever, there was a group of people young enough to not be adults but old enough to be able to drive, to work and to earn money that they could keep for themselves, and no good term to describe them. Google's Ngram Viewer tool, which records the occurrences of different words published in books, by year, notes that the use of 'teenage' and 'teenager' suddenly began increasing in popularity starting in the 1940s and 1950s, with essentially no use prior to then.

Europeans were not far behind. As Tony Judt wrote, "around 1957, for the first time in European history, young people started buying things themselves", like records and clothing. Google's scans show similar rises of these words in British books and even German-language books. When The Who sang in 1971 about a "teenage wasteland", the term 'teenage' was only about as old as the band members.

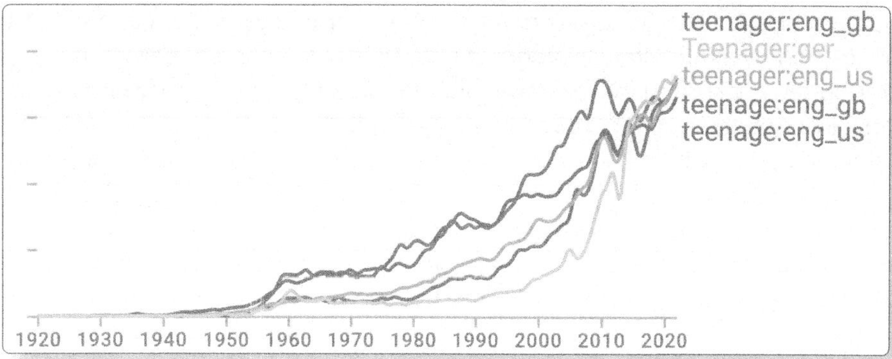

Figure 59 – Frequency of use of 'teenage' and 'teenager' in US English ('us'), British English ('gb'), and German ('ger') books. *Graph from Google Books Ngram Viewer*

◆ ◆ ◆

"Can we expect that a government will be well modelled by a people who know not how to make a spinning-wheel, or to employ a loom to advantage?"

— David Hume, 1742

◆ ◆ ◆

Fabrics, Bleach, and the Sewing Machine

> *For tens of thousands of years, fabric-making and sewing were done in essentially the same way. In fact, even Neanderthals made thread. But in the 1700s and 1800s, various aspects of producing fabric and clothing were revolutionized over short periods of time*

Fabric production is one of the cornerstones of human civilization. It drove long-distance trade in antiquity, spurred the Industrial Revolution, and contributed to modern computing. In her 1991 book *Prehistoric Textiles*, Elizabeth Barber wrote "the textile industry, in fact, is older than pottery or metallurgy and perhaps even than agriculture and stock-breeding, and it probably consumed far more hours of labor per year, in the temperature climates, than pottery and food production put together." "The spindle was the first wheel," she is quoted as saying in an interview. "It wasn't yet load bearing, but the principle of rotation is there."

Virginia Postrel, the author of *The Fabric of Civilization: How Textiles Made the World*, suggests that the Stone Age should instead be called the String Age. "The two prehistoric technologies were literally intertwined. Early humans used string to attach stone blades to handles, creating axes and spears." She continues, "the story of textiles is the story of human ingenuity. Agriculture developed in pursuit of fiber as well as food. Labor-saving machines, including those of the Industrial Revolution, came out of the need for thread. The origins of chemistry lie in the coloring and finishing of cloth; the beginning of binary code — and aspects of mathematics itself — in weaving. As much as for spices or gold, the quest for fabrics and dyestuffs drew merchants to cross continents and sailors to explore strange seas... It spread the alphabet and double-entry bookkeeping, gave rise to financial institutions, and nurtured the slave trade."

Postrel theorizes that one of the reasons the ancient Greeks were apparently so interested in mathematical issues like which numbers are prime is precisely because this knowledge is key to developing weaving patterns. Mathematics arose wherever the weaving loom was present and people from Greece to India to China were masters of both. Ensuring a good harvest of crops, not only for food but also for fibers, propelled astronomy as a means to measure annual cycles.

Fabric production not only drove developments of mathematics and astronomy, but also chemistry and computing. In his 2002 book *Salt: A World History*, Mark Kurlansky wrote, "until the late eighteenth century bleaching was accomplished by soaking fabric in buttermilk and then laying it out on the ground to be whitened in sunlight for weeks." But in 1786, just ten years after Swedish chemist Carl Wilhelm Sheele discovered the element chlorine, French chemist Claude-Louis Berthollet showed that it, when absorbed in potash, created a liquid bleach. "In a little more than a year," according to Wikipedia, "industrial bleaching became a major activity in the British textile industry."

Starting in the late 1790s, according to David Landes in his 1969 book *The Unbound Prometheus*, industrial bleaching powder was made from "chlorine and limestone, which was cheaper than chlorine and potash. The price of chlorine [bleach] fell to one-tenth its original level – from £140 to £14 per ton."

Industrial bleaches not only revolutionized the textile industry, but also land use, by freeing up large areas of urban land that were previously used as 'bleach fields', and revolutionized communications, by enabling wood pulp to become the primary material for making paper in the late 1800s.

Making paper is not the only way that the fabric industry affected communications. The Jacquard loom punch card system, according to Virginia Postrel, "was an enormous boost in productivity, especially for fancy patterns. A single master weaver could now turn out two feet of brocade a day compared to a mere inch working with an assistant on the old drawloom." "When Jacquard's device was patented in 1804", she writes, "it was unquestionably the most complex mechanism in the world."

Jacquard's punchcard system revolutionized many other fields as well. Postrel says that "shipbuilders designed similar systems to control the automatic riveting machines used to build the era's new ironclads," punch cards enabled automatic player pianos, and every computer science student knows that the punchcard system was instrumental to Charles Babbage's Analytical Engine, the precursor to the computer, and to early computer programming.

"By 1812," Postrel notes, "incremental improvements boosted the loom's speed and cut the cost in half. Far from devastating [Lyon, France's] employment, Jacquard's invention ushered in a new golden age of Lyonnais silk... Jacquard's breakthrough made weavers' work easier, improved the quality of their fabrics, and expanded the market for their products to middle-class shoppers. The card-driven system spread to makers of ribbons, woolens, and carpets."

But making fabrics is only a portion of the textile-working process – making clothes also means a lot of stitching and for tens of thousands of years that was done entirely by hand. How marvellous it must have been in American homes in the late 1860s and early 1870s when they first carried a sewing machine into the house. The new availability of cheaper, high-quality steel enabled the strong needles and other precision parts, lubricated with whale oil.

Compared to everything else in the household of the time, there was nothing remotely like it. In a typical home around 1870, heating and cooking were done by a cast iron stove fired by either wood or coal. According to Clive Thompson in the July/August 2022 issue of *Smithsonian* magazine "by the 1860s, American households were finally – and rapidly – shifting to coal." By 1870, less than 10% of homes had a flush toilet. Obviously, none had a telephone, electric refrigerator, electric lighting, or electric anything. Preserving food was done in Mason or Ball jars, whose revolutionary two-part steel lids that allowed them to be unscrewed before breaking the vacuum seal had only been invented twelve years prior. Tools like hammers, saws, and chisels had not changed for thousands of years. Transportation was largely by horse-drawn wagon, as it had been for many centuries.

The handheld needle and thread pre-date them all by far. In a 2020 article in the journal *Scientific Reports*, Bruce Hardy and co-workers at Kenyon College in Ohio describe identifying in a French cave where Neanderthals lived "a 3-ply cord fragment made from inner bark fibres". At 41,000 to 52,000 years old, that piece of thread is the oldest ever discovered. Sewing needles roughly as old have been found throughout the world from South Africa to Asia to Europe.

But sewing, which had been done in roughly the same way for tens of thousands of years, began to change dramatically in the 1850s with the introduction of sewing machines like Isaac Singer's "Family" model, the first machine ever targeted at the home market. Although at first sewing machines were expensive, after a key patent by Elias Howe expired Singer introduced the cheaper "New Family" model in 1865, the first household sewing machine that became popular.

And as with many other new technologies, the sewing machine spawned new business models. According to 'Atlantic Threads: Singer in Spain and Mexico, 1860-1940', the 2013 Ph.D. dissertation by Paula de la Cruz-Fernández at Florida International University, when the New Family model was introduced, Isaac Singer's business partner and lawyer Edward Clark developed the idea of the "hire purchase system", or 'installment plan', "in which consumers could acquire the appliance by making weekly or monthly payments", rather than by making the full payment up front. As a result, sales of Singer's machines boomed.

In the ten years after the Civil War ended in 1865, more than 4.5 million sewing machines were sold in the US, a country of about forty-five million people, or roughly one machine for every ten people. Considering that the average household size in 1876 was just over five, this means that by the mid-1870s there were sewing machines in roughly half of all households.

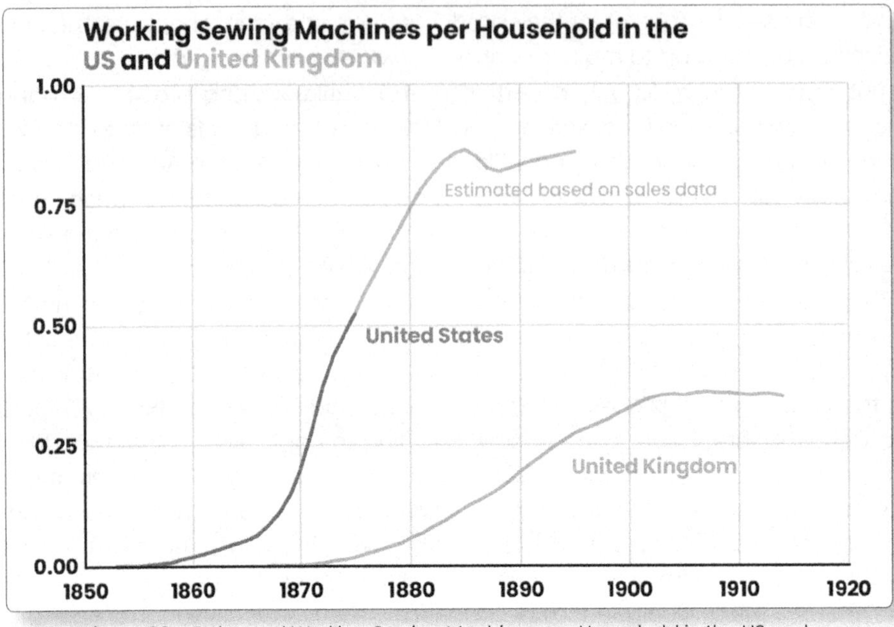

Figure 60 – Estimated Working Sewing Machines per Household in the US and the United Kingdom. Sales figures from *(Cooper, 1968)*

The appeal of the new machine was obvious: according to Yaffa Claire Draznin in *Victorian London's Middle-Class Housewife: What She Did All Day*, with a sewing machine, a person could produce a man's shirt in about one hour, compared to 141/2 hours by hand, or a chemise in less than an hour instead of ten hours by hand. That is, the sewing machine was a 10X disruption compared to hand stitching.

It was not just in the home but also in factories that sewing machines had an impact. Ruth Brandon wrote in *Singer and the Sewing Machine*, that "in 1860, Oliver F. Winchester, a shirt manufacturer of New Haven, Connecticut, stated that his factory turned out 800 dozen shirts per week, using 400 sewing machines and operators to do the work of 2,000 hand sewers. The price for hand sewing was then $3 per week, which made labor costs $6000 per week. The 400 machine operators received $4 per week, making the labor cost $1600 per week." That is, industrial sewing machines reduced the cost of producing Winchester's shirts by almost three-quarters, while increasing employee wages by about one-third.

Just as Gutenberg's printing press required a new kind of ink, the sewing machine required a rethink of the needle and the thread. Sewing machine needles pierce fabric, but do not pass fully through it, so the eye of the needle must be placed at the tip instead of the back, where it had been for tens of thousands of years. Unlike hand sewing, where a needle pulls one thread, machine sewing uses two threads, one pulled from one side of the fabric by the needle and the other from the underside of the fabric by a bobbin. Machine-sewing also puts more tension on the thread than hand sewing, so the three-ply twisted thread we had made since Neanderthal times was not sufficient either. According to Grace Rogers Cooper in *The Sewing Machine: Its Invention and Development*, "it was not until the six-cord cabled cotton thread, which was suitable for both machine and hand sewing, was perfected that the industry progressed into full operation."

In the UK, the sewing machine revolution started later, took longer, and never reached American levels of adoption. Cheaper, faster, easier production of clothing meant that people could increasingly buy ready-made clothes, in standard sizes, from clothing retailers. According to Nancy Green in her 1997 book *Ready-to-Wear and Ready-to-Work: A Century of Industry and Immigrants in Paris and New York*, "in the late 1860s, twenty-five percent of garments produced in the US were ready-made, but by 1890, the portion had risen to sixty percent." So, by the time the British were in the steepest part of the adoption curve, they could just buy clothes from a shop instead and the number of machines per capita never reached American levels.

Rose Madder and Alizarin

> *Synthetic red dye revolutionized fabrics, and kicked off the pharmaceutical industry*

Madder plants have been cultivated in places like Egypt and India since as early as 1500 BCE for their unforgettable red-colored dye. Cloth dyed with madder has been found in the tomb of Tutankhamun in Egypt, and in Pompeii, Italy. In China madder was the primary source of red dye before the Han Era. Virginia Postrel, author of *The Fabric of Civilization: How Textiles Made the World*, says that "textile fragments excavated from Masada, the desert palace fortress famous for the mass suicide of Jewish rebels in 73 CE, include bright red, salmon pink, deep burgundy, purple, purplish black, and reddish brown — all dyed with madder." According to Dr. Judith Lopez, whose doctoral thesis concerned the commercial adoption of artificial dyestuffs in New England, madder was "the world's most widely and continuously used red dyestuff" for many centuries.

A particular shade of madder dye called 'Turkey red', first imported to Europe in the 1740s, became popular in England, the Netherlands and France (where it was known as *rouge d'Adrinople*, after the Turkish city of Edirne). Bright red cloaks, sometimes called 'cardinals', dyed with Turkey red or red cochineal dye from Mexico, became especially popular with English country women and, by extension, with American colonists. So much so that bright red cloaks are now *de rigueur* in any story set in the era, from M. Night Shyamalan's *The Village* to Margaret Atwood's *A Handmaid's Tale*. Turkey red (or cochineal) even appears in the flags of Britain and many of its former colonies, along with blue from woad.

Natural madder contains a mixture of color-bearing chemicals, including munjistin, which is orange, purpurin, which can be red or yellow, and the magnificent red alizarin. (The purpurin and alizarin chemicals were first isolated from natural madder in the 1820s by Pierre Jean Robiquet, the French chemist who also identified the first amino acid.) Dried madder roots can be ground and treated to produce a range of colors, including red, pink, orange, and brown, though up to 98 percent of the plant material ends up as waste.

Treating the roots with sulphuric acid produces a dye called garancine which, though more expensive by weight than the natural madder dyestuff, requires less dye per batch of fabric and is therefore actually cheaper than madder per pound of finished textile product. As well, the colors are brighter and more consistent, although the range of colors available from garancine is less than from natural, untreated madder.

As a means to get red coloration, madder and garancine had competition from a variety of other sources, including kermes and cochineal. Kermes, from the Arabic *qirmiz*, is of ancient origin; jars of kermes have been found in a Neolithic cave-burial at Adaouste, northeast of Aix-en-Provence, France. The early Egyptians made use of it too. Cochineal came from Mexico, so it was unknown elsewhere before Columbus. After Europeans arrived in the Americas, the market for red dyes changed quickly. "In 1500," Postrel writes, "Europe's best and most valuable source of red dye was kermes, made from a tiny insect that lives on European oak trees. Fifty years later, it was Mexican cochineal, derived from a similar parasitic insect that grows on the nopal, or prickly pear, cactus."

"Containing ten times as much colorant as kermes," she continues, "cochineal was one of the New World's greatest gifts—a tribute to Mexico's native farmers. While kermes grew wild, indigenous people had cultivated cochineal for centuries. Like the Chinese who domesticated silkworms, they devoted careful attention to both the insect and its host plant. By the time the Spanish arrived, their selective breeding had produced 'the closest thing Europe had ever seen to a perfect red,' writes Amy Butler Greenfield in her history of the dye."

A Synthetic Dye Disrupts its Natural Equivalent

In 1868, alizarin became the first naturally occurring dye to be produced artificially when German chemists Carl Graebe and Carl Liebermann, working for BASF, found a way to make the exact same chemical that is found in nature from a chemical found in coal tar. The product was not only more consistent, as it was the single chemical alizarin rather than a mix of chemicals, and it was also more concentrated. Thus, while dyeing requires fifty pounds of natural madder per one hundred pounds of fabric or thirteen pounds of garancine, it takes just two pounds of pure, synthesized alizarin. Put another way, garancine is about four times as potent as natural madder and alizarin is about twenty-five times.

Using historical business records from US textile factories, including the Hamilton Manufacturing Company of Lowell, Massachusetts ("the largest of the companies studied according to the amount of its madder and alizarin consumption"), Dr. Lopez estimated that the price of a pound of synthetic alizarin imported from Europe in the early 1870s was a little more than twenty-five times more expensive than madder, but since twenty-five times more natural madder is required for a given amount of textile, the price for dyeing a given batch of fabric was essentially the same for using either natural madder or synthetic alizarin. (Garancine, in comparison, was thus cheaper for a given amount of textile than either natural madder or artificial alizarin, though again, more limited in its potential uses. So, garancine was an 'inferior' product, as it was both *cheaper* and *less useful* than natural madder.)

Ten Times Cheaper

Artificial alizarin, on the other hand, had the potential to disrupt natural madder. In 1871 chemists discovered an easier way to obtain from coal tar the main ingredient for synthesizing alizarin, making the artificial chemical significantly cheaper. According to Dr. Lopez, "the synthetic alizarin could be produced for a fraction of the cost of the natural product, and the market for madder collapsed virtually

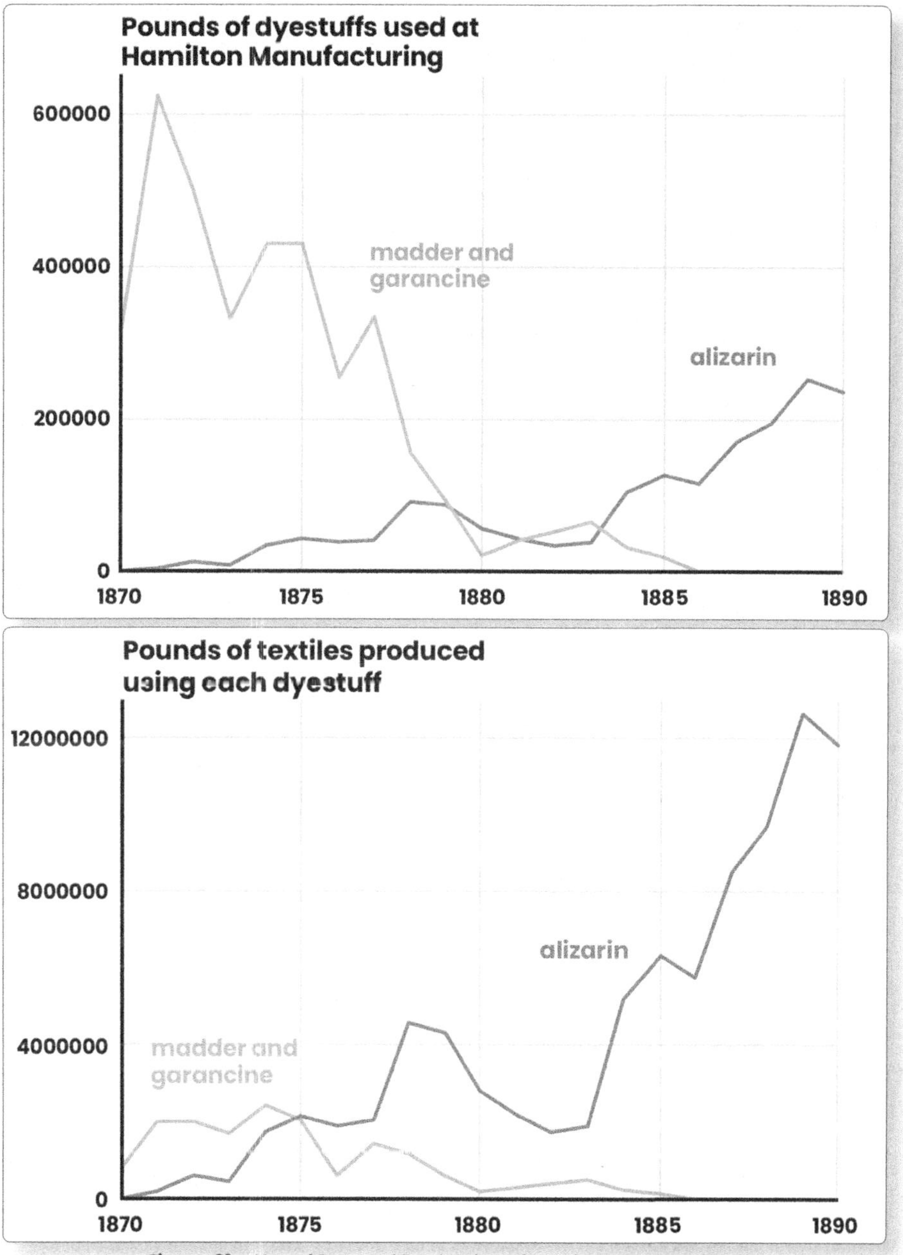

Figure 61 – Use of Dye and Production of Fabric at the Hamilton Manufacturing Company of Lowell, Massachusetts. *(Lopez, 1989)*

overnight." In 1874, alizarin was 60% cheaper than it had been two years before, making it about the same price per one hundred pounds of fabric as garancine, and two years later, in 1876, it was 60% cheaper still. The following year, the price fell another 40%, making alizarin 10X cheaper in 1877 than it had been just five years before.

In the three years from 1877 to 1880, the Hamilton company's use of natural madder and garancine combined fell nearly 94%. The company's total expenditures on dyestuffs fell from $131,000 in 1872 to $54,000 in 1877, though the output (in pounds of fabric produced) in the two years was nearly identical. Even the factory's expenditure on just artificial alizarin alone fell from $40,000 to under $15,000 over those same five years. The Germans were not just disrupting the madder market, the rapid collapse in the cost of producing alizarin was leaving money on the table. Facing a catastrophic fall in revenue from their customers, Dr. Lopez says, "evidence suggests Germany deliberately controlled alizarine prices from 1879 to 1883, when the German patents expired."

Figure 62 – Cost to Dye 100 Pounds of Textile with each kind of dyestuff: madder, garancine and alizarin. *(Lopez, 1989)*

By 1883, the price of dyeing one hundred pounds of fabric with alizarin was about $4, a little below that of natural madder. But as the chemical went off patent, its price again collapsed more than 90% over the next three years to 1886, the year that artificial alizarin became the only dyestuff among the three sources still used at Hamilton.

In short, natural madder dye along with garancine dye from the madder plant were 99% of the dyestuffs used at Hamilton Manufacturing company in 1871, and 0% fifteen years later in 1886. Dr. Lopez notes that "alizarin itself has been in turn largely replaced today by the more light-resistant quinacridone pigments developed at DuPont in 1958."

There are several interesting lessons in the story of how alizarin disrupted natural madder and garancine dyes:

First, this disruption was driven by a 10X change in price. The average price that Hamilton paid for dyes in 1885 (per one hundred pounds of fabric produced) was less than 1/10th what they had paid thirteen years before, in 1872. And in 1886, the last year that madder and garancine dyes were used there, the price of alizarin was also about 10X cheaper than the natural alternatives.

A second observation is that this was a business-to-business (B2B) disruption. The sellers and the buyers were both companies, as opposed to individuals. BASF did not need to convince consumers that their new product was better than the existing product. Consumers were willing to buy Hamilton's dyed products regardless of the ingredient used to make a specific color.

This example shows that a fifteen-year disruption can happen not only to an entire industry, but also to a single (albeit very large) company.

Another lesson was that several years of price collusion in the late 1870s and early 1880s by producers who held exclusive patent protection did not prevent the disruption. In fact, it might legitimately have been their best option to maximize their own revenue by keeping the price just below that of the next best alternative dyestuff, knowing that their patents would expire in just a few years thus making alizarin a commodity.

As we have seen with other technologies that are both better and cheaper than the market-leading alternative, this was not a one-for-one disruption in output. Hamilton's output of fabrics (in weight) was larger in the six calendar years in the period 1885-1890 combined than in the fifteen years from 1870-1884 combined. In fact, Hamilton's production of fabrics was almost five times as much (again, in weight) in 1889 as in 1872, but even so, their expenditures on dyestuffs in the three years from 1888-1890 combined were less than for the single year of 1871.

A New 'Possibility Space'

The discovery of synthetic dyes also opened a new 'possibility space' of color-related chemistry. The synthesis of naturally occurring dye molecules enabled a variety of similar chemicals to be fabricated, some of which are not found in nature, that had a wide range of colors and even chemicals that had uses beyond fabric dyeing, in fields such as food, materials, and science. Amaranth, for example, is used in the United Kingdom to give Glacé cherries their distinctive color. Sudan Yellow 3G is used today in the inks in inkjet printers and nigrosin in black shoe polish. Amido black 10B is used in biochemical research and in criminal investigations to detect blood present with latent fingerprints.

Red-colored madder was not the only dyestuff to be replaced by artificial alternatives – far from it. Today, virtually all dyes commercially used in large quantities are synthetic. In fact, virtually the same story that happened to madder happened again, a decade or two later, to purple-colored dyes like indigo, raising a very interesting question: why didn't producers learn their lesson the first time around?

◆ ◆ ◆

"Natural indigo dyes give a more durable color than artificial dyes - the latter will not displace natural dyes in the long term."

— W. B. Hudson, President of the Bihar Indigo Planters Association

◆ ◆ ◆

Indigo and Tyrian Purple

> *Tony Seba asks, 'why do smart people in smart organizations fail to anticipate disruption?' Just a decade after madder and garancine dyes were disrupted, indigo and woad dyes were disrupted in essentially the same fashion. Smart people saw it coming. And yet no one could stop it*

Indigo, the dye that makes blue jeans blue, used to come primarily from natural sources, in particular *Indigofera tinctoria*, a member of the pea family that grows well in south Asia. A chemically identical compound is found in Europe in the wood of *Isatis tinctoria*, a relative of mustard commonly known as wood. But a couple of men in Germany overturned the entire indigo market in the late 1800s.

Indigo dye is not easy to use. The chemical itself does not dissolve in water, so it can only be used by first converting it into a yellowish form by using sodium hydroxide (lye) and sodium dithionite, soaking the fabric, and then exposing the fabric to the air, causing the purple color to return. Sodium dithionite can be made from sulfur dioxide, which can be produced by burning elemental sulfur or by roasting pyrite, both of which can sometimes be found in volcanic areas. It's really a wonder anyone ever figured out the process in the first place.

To make it worse, harvesting indigo dyes from nature is as complicated as the dyeing process itself. The Phoenician people of the eastern Mediterranean for centuries made a fortune off of 'Tyrian purple' or 'royal purple', a natural mix of indigo dye (which is blue) and 6,6'-dibromoindigo (which is red), which they harvested from *Murex* snails. The pigment was complex to produce and expensive, so fabrics colored with it became associated with royalty and wealth. Pliny the Elder described the production of the dye in his *Natural History*:

> "After the rising of the Dog-star [Sirius] or else before spring ... the vein is extracted [from the snail] ... to which it is requisite to add salt, a sextarius [about 2 and half cups] to every hundred pounds of juice... leave them to steep for a period of three days... then set to boil in vessels of tin [or lead]. About the tenth day, generally, the whole contents of the cauldron are in a liquefied state, upon which a fleece, from which the grease has been cleansed, is plunged into it by way of making trial; but until such

> time as the color is found to satisfy the wishes of those preparing it, the liquor is still kept on the boil. The tint that inclines to red is looked upon as inferior to that which is of a blackish hue. The wool is left to lie in soak for five hours, and then, after carding it, it is thrown in again, until it has fully imbibed the color."

It takes upward of one hundred kilograms (220 pounds) of snails to get one gram (1/30th of an ounce) of dye, which can color seven or eight grams (about one-quarter of an ounce) of silk or wool. This means that literally 99.999% of the mass of snails becomes waste. Indigo derived from plants is not much better. As much as twenty tonnes of leaves are needed to produce one hundred pounds of pigment, so about 99.75% of the plant goes to waste. In the end, it takes an incredible amount of time and effort to make natural indigo, so today the resulting dye can cost up to $4,000 per gram.

The Phoenicians produced Tyrian purple from before 1000 BCE to the fall of Constantinople in 1453, a total of at least 2500 years, before the knowledge of how to produce the dye was lost for centuries. In recent years people have rediscovered the process, but for many hundreds of years Europeans resorted to other sources of blue dyes, like woad, the mustard relative mentioned earlier. But anyone who had access to indigo would use that instead. "Woad," E.J.W. Barber wrote in the 1991 book *Prehistoric Textiles*, "was doomed to take a back seat to the indigo plant wherever the latter was known (if only by importation), since the indigo plant contains ten times the concentration of indigo colorant as woad."

In 1878, with the recipe for Tyrian purple still unknown and Indian plantations the primary source of indigo dye, Adolf von Baeyer discovered a method to synthesize indigo dye chemically, for which he won the Nobel Prize in Chemistry. But it took until 1897 for the German company BASF to find a commercially viable route to produce the first synthetic indigo for sale. The product was initially met with scepticism. W. B. Hudson, President of the Bihar Indigo Planters Association, wrote a letter to the government of India in March 1900, saying "natural indigo dyes give a more durable color than artificial dyes - the latter will not displace natural dyes in the long term."

On the contrary, artificial indigo did displace the natural dye in the long term. In the short term, too. When BASF began making indigo dye in 1897, annually 19,000 tons of indigo were being produced from plant sources. By 1914, this had dropped to 1,000 tons, a decline of almost 95% in seventeen years. "In India," Virginia Postrel wrote, "the transition was particularly abrupt. In its peak year, ending in March 1895, British India exported more than nine thousand tons of indigo dye. A decade later, that volume had plummeted by 74 percent while the revenue received dropped by 85 percent."

But some people did foresee the impending disruption. An 1894 article in *McClure's Magazine*, published just a couple of years before BASF commenced large-scale production, quoted the French Professor of Chemistry Marcellin Berthelot as saying, "madder fields, so far as Europe concerned, have practically ceased to exist. So with indigo. The chemists have now succeeded in making pure indigo direct from its elements, and it will soon be a commercial product. Then the indigo fields, like the madder fields, will be abandoned, industrial laboratories having usurped their place."

Many incumbent producers did not heed the Professor's warnings. F. Mollwo Perkin, in an article in the year 1900 in the scientific journal *Nature*, wrote that "one would have supposed that the indigo producers would have taken warning from the extinction of the artificial alizarin industry, and called to their aid experienced agriculturists to see if it were not possible to increase the yield and quality of the indigo plant, and chemical experts to endeavour to improve the process of manufacture. This, however, has not been done."

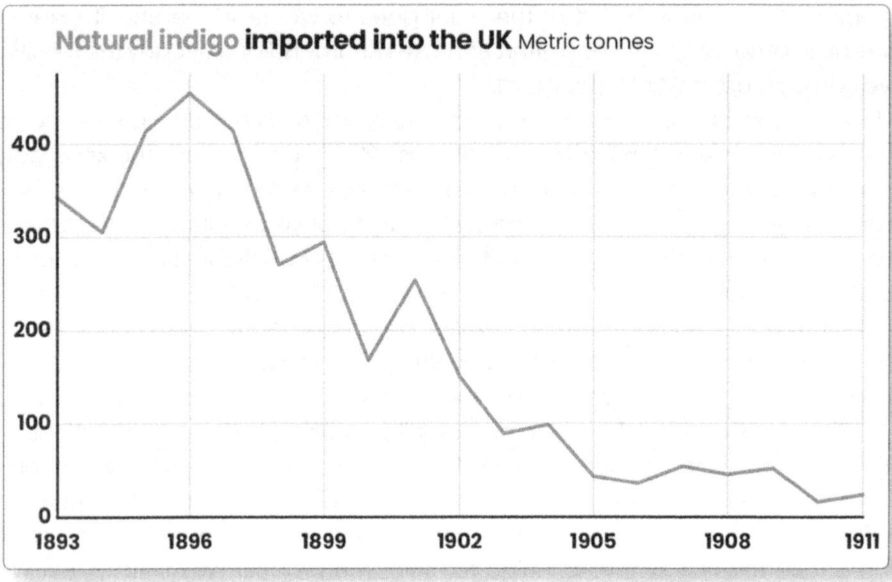

Figure 63 – Tons of Natural Indigo imported into the UK.
Source: various editions of the Statistical Abstract for the United Kingdom

The development of the first synthetic dyes unlocked a 'combinatorial explosion' in possibilities for other dyes. As Johann Murmann explains in an article in the *Journal of Evolutionary Economics*, "A firm that made many intermediates and dyes could make an even larger number of dyes because a particular dye often served as an intermediate for other dyes. Making hundreds of dyes and intermediates gave rise to the possibility of making thousands of different dyes. As a result, the number of different dyes increased exponentially with the number of different intermediates available."

In just over thirty years, German ingenuity in the form of chemically synthesized indigo disrupted the production of the natural, plant-derived dye. But even though synthetically produced indigo was cheaper and more consistent in quality than the natural product, it is also damaging to the environment. The ingredients for Baeyer's indigo come from petroleum and use chemicals like benzene and formaldehyde. "A kilogram of standard indigo dye", an article in *Fast Company* says, "requires 100 kilograms of petroleum to produce." (Peters, 2020) A 100-to-1 ratio of input to product is far better than *Murex* snails, yet still leaves room for greater efficiency.

Soon, though, this century-old method might get disrupted again. Huue, a California startup company founded by biologists Tammy Hsu and Michelle Zhu is seeking to commercially produce indigo from genetically modified bacteria: "The way that we're resolving these problems is that we are using a biological process to produce the indigo," Zhu said. "So, instead of needing the petroleum base for it, we can literally program bacteria, microbes to grow and secrete the indigo without the need for toxic chemicals."

The Boy Who Invented Everything

> *An accidental discovery by teenage student William Perkin produced a purple-colored chemical never before found in nature, and opened the door to synthesizing pharmaceuticals, plastics, and many of the other products of our modern world*

When in 1453 the Ottoman ruler Fatih Sultan Mehmed was finished conquering Constantinople so thoroughly that even its name no longer remained intact, let's just say that Europeans would need to set out around the world to find another route east to get Chinese herbs and spices.

And set out they did, in two great branches. Some, like da Gama, headed south down around Africa and over to Asia. Others, like Columbus, went west. Of course, those who crossed the Atlantic did not find Asia, but in the lands they did reach they sought to create the products of the Middle Eastern and Asian economy on new soils: silk, cotton, and sugar.

There they also discovered many delightful new things previously unknown outside the Americas: potatoes and peanuts, tomatoes and tobacco, cacao and coca ...

... and the cinchona tree.

If you are not familiar with the cinchona tree, then you have probably never had malaria, or looked at the official coat of arms of the nation of Peru, which is where the cinchona tree grows.

Malaria had plagued humanity basically since the invention of agriculture. Some historians say that outbreaks might have contributed to the fall of the Roman Empire. Basically anywhere warm enough that there is wet marshy land, there can be mosquitos that carry the malaria parasite.

For most of history, we did not know any of that. In medieval times, people thought that maybe it spread by 'bad airs', which in Latin is where the word 'malaria' comes from. But by the early 1600s we had discovered that the bark of the cinchona tree could treat malaria, though we did not know what it was in the bark that had medicinal effects until much later.

It turns out that it is the chemical quinine. (In fact, treatment of malaria with quinine was the first known use of a specific chemical compound to treat an infectious disease.) The first isolation of quinine from cinchona bark is credited to the French chemists Pierre Joseph Pelletier and Joseph Bienaimé Caventou in 1820. Its molecular formula was determined in 1854 by German chemist Adolph Strecker.

A Marvellous Mistake

The previous year, in 1853, exactly four hundred years after Constantinople fell, fifteen-year-old William Henry Perkin had enrolled in the Royal College of Chemistry (now the Department of Chemistry at Imperial College London) to study under August Wilhelm von Hofmann.

von Hofmann had some ideas on how to synthesize quinine from aniline, a chemical found in coal tar, which London had loads of in the mid-1800s due to the extensive use of coal gas lighting. He gave the task to his student, Perkin. If successful, humanity would no longer rely on tree bark to get quinine and could make as much of the medicine as we wanted.

It was an excellent task for a professor to foist onto a student, as aniline is toxic to humans, ignites readily, and stinks of rotten fish. During the Easter holiday in 1856 while working in his home laboratory, one of Perkin's experiments failed marvellously.

Instead of making quinine, he instead produced a brilliant purple dye, which he called 'aniline purple' and was later named mauveine. It was the first-ever synthetic dye, a chemical that does not exist anywhere in nature (as far as we know). And it quickly made young Mr. Perkin famous - and a lot of money.

A Convergence of Economies

But there was one problem: aniline was expensive, which meant that Perkin's aniline purple (mauveine) was too.

"When I discovered aniline purple," Perkin recounted in an 1861 lecture, aniline was "only to be met with in the laboratory; in fact, half-a-pound of aniline was then esteemed quite a treasure, and it was not until a great deal of time and money had been expended that I succeeded in obtaining this substance in large quantities, and at a price sufficiently low for commercial purposes."

Fortunately, in 1854, a couple of years before Perkin made his discovery, just as Strecker was determining the molecular formula of quinine, French chemist Antoine Béchamp had figured out a way to make aniline from nitrobenzene, another chemical in coal tar which was much more abundant.

It took a couple of years for Perkin to begin commercial production of both his purple dye and the nitrobenzene-derived aniline to make it, but "Béchamp's process," Perkin said in 1861, "is now employed for the preparation of aniline *by the ton*." What just a few years prior was "only to be met with in the laboratory", was now ubiquitous. For a couple of years around 1860, Perkin's purple dye was all the rage in ladies' fashion from London to Paris and beyond, making him wealthy.

But Perkin's wasn't the only one who saw aniline's potential. In Germany, the *Badische Anilin- und Soda-Fabrik* (Baden Aniline and Soda Factory) or BASF, now the world's largest chemical company, was established to find uses for Béchamp's cheap and promising chemical.

In 1869, Perkin found a method to commercially produce the red dye alizarin from aniline, but BASF had patented the same process one day before he did. Synthetic alizarin quickly disrupted the production of natural alizarin from the madder plant. Later, in the 1890s, BASF used aniline to make synthetic indigo and it quickly disrupted natural indigo production.

Perkin made purple dresses so popular that people deemed them the 'Mauve Measles'. In a late 1860s party game when asked his favorite color, communist Friedrich Engels, who himself was very wealthy from textile manufacturing, replied, "any one not Aniline."

To Dye For

Aniline's potential did not stop with dyes. Paracetamol (also called acetaminophen) is based on aniline, as was salvarsan, the first successful chemotherapy agent, and prontosil, the first antibacterial drug. Fentanyl can be made from aniline too.

Aniline is also used as an antioxidant in rubber processing and to make herbicides like Propanil, which is used to kill weeds that grow around rice. But most aniline today is used to make polyurethane.

Perkin had done more than just make a plague of purple petticoats. Prior to his discovery, the chemicals doctors and scientists worked with were extracted from nature. People would collect specimens we thought might have some sort of medicinal property or textile dyeing ability or other useful function, and then we would extract and purify the useful components the best we could.

But mauveine was not known from nature and to this day we have not found it in any plant, animal, or mineral. When Perkin synthesized the first batch, he was holding in his hands all of the mauveine known in the entire universe. And there is no good reason to believe with high certainty that any exists anywhere else. Mauveine contains twenty-six carbon atoms (plus some hydrogen and nitrogen). Ask any chemist how long it would take to synthesize every compound that contains twenty-six carbon atoms (plus as much hydrogen or nitrogen as they want) and they will probably puff out their cheeks and look off into the distance as if you just asked them how much time it would take to walk to Saturn. The correct answer is 'essentially infinite'.

What Perkin did by making the first chemical not known in nature was to open the door a crack into the possibility of making anything we could conceive. He did not just discover one molecule, he essentially invented all of them.

From synthetic chemistry, we were able to produce things that are found in nature, like indigo dye and the red dye alizarin, but also things that are not, like paracetamol, prontosil, Propanil, polyurethane and, of course, Perkin's purple. No longer were we limited by the chemicals nature had discovered, but in principle could make any ones we wanted.

However, it often takes a lot of time and hard work to find a way. It took until 1944, about ninety years after Perkin's experiments, for people to finally synthesize quinine, but since then we have developed other, better treatments for malaria. The chemical structure for mauveine itself – how each of the carbons, nitrogens, and hydrogens are connect to each other - was not finally identified until 1994.

By the mid-1900s, Friedrich Engels and Karl Marx's idea of Communism had spread to southeast Asia, where the Viet Cong, backed by China and the Soviet Union, were fighting the Republic of Viet Nam, backed by the US, South Korea, Australia, and other allies. Both sides experienced heavy casualties due not only to fighting, but also to tropical diseases like malaria. As a result, in 1967, Communist scientists began a program to screen traditional herbs to find compounds with anti-malarial effects.

In the early 1970s, Chinese chemist Tu Youyou, who later won the 2015 Nobel Prize in Medicine, extracted artemisinin from sweet wormwood. Artemisinin, or one of its derivatives, is now recommended as the frontline therapy for all cases of malaria. As of 2006, quinine is no longer recommended by the World Health Organization (WHO) as a first course of treatment.

However, artemisinin is no longer only extracted from plants, as a research program begun in the early 2000s produced its precursor compound using precision fermentation by genetically engineered brewer's yeast. As the final product costs roughly the same as the one extracted from plants, the biosynthetic artemisinin is slowly being adopted.

◆ ◆ ◆

"When looms weave by themselves, man's slavery will end."

— Aristotle

◆ ◆ ◆

Pants, Pants, Revolution

> *For millennia, weaving was a far more popular way to make large amounts of fabric than knitting. But industrial knitting machines are now rapidly displacing weaving, and consumers are casting aside their stiff, woven blue jeans for more-elastic, more form-fitting (but less-durable) yoga pants and sports leggings*

We have been weaving fabric for ten thousand years. Weaving started the Industrial Revolution and punch-card looms were instrumental to the creation of computers. But over about the past fifteen years, weaving looms have been getting disrupted in industrial applications by automated knitting machines. In *The Fabric of Civilization: How Textiles Made the World*, published in 2020, Virginia Postrel writes: "knitted fabrics occupy about 57% of the world market." But even just a couple of decades ago, woven fabrics were dominant.

Imports to the US of women's pants made from blue denim, a woven fabric, have hovered around two hundred million pairs per year since at least 2005. If

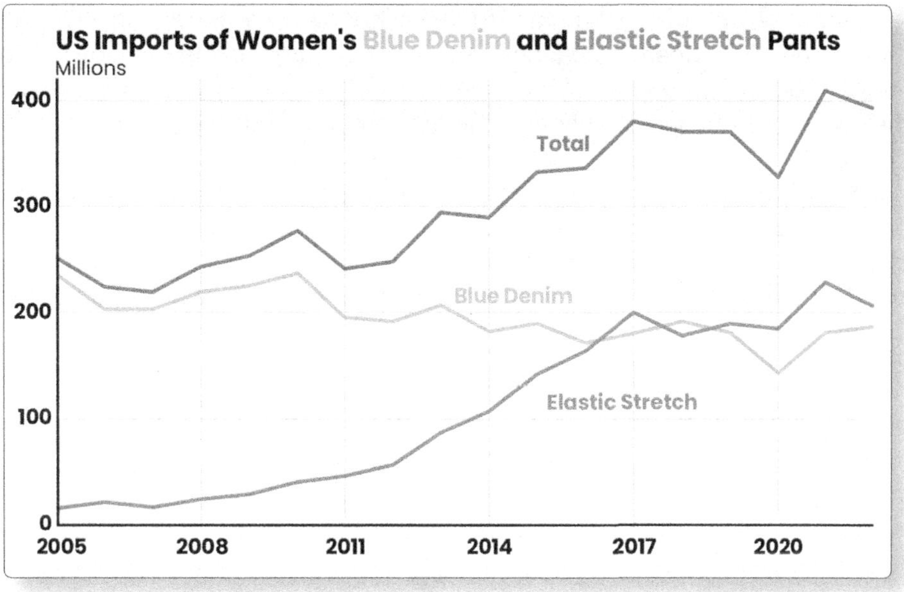

Figure 64 – Imports of Women's Blue Denim and Elastic Stretch Pants to the US, Millions

anything, the trend has been on a slightly downward slope for most of that time. In contrast, elastic stretch pants, which are knitted and made from either cotton or synthetic fibers, have grown dramatically from fewer than seventeen million pairs imported in 2007 to more than two hundred million ten years later. In 2007, close to a dozen pairs of women's blue denim pants were imported to the US for each pair of elastic stretch pants. By 2017, more women's elastic stretch pants were imported to the US than women's blue denim pants.

Postrel writes that "after more than ten thousand years of dominance, weaving no longer rules the textile world. Knitting has staged a coup… Knits now outsell woven textiles nearly two-to-one, and the ratio for garments is even more lopsided."

In the pandemic year of 2020, imports of women's blue denim pants fell more than 20% compared to 2019 and imports of elastic stretch pants were 30% higher than imports of women's blue denim pants.

These figures are only for imported women's pants – girls' pants are categorized separately. The numbers only include woven denim and elastic knits. They do not include wool pants nor corduroy nor pants sold as parts of suits or ensembles. But even so, the market for just these two categories of pants – essentially blue jeans and yoga leggings – changed dramatically, going from almost entirely jeans, to closely split in about ten years. This was not a one-for-one replacement. The total imports of blue denim pants and elastic stretch pants combined grew 75% in the decade from 2007 to 2017.

As with other disruptions we have studied, it was outsiders who led the change, in this case companies like Lululemon and American Outfitters, while incumbents like Levi's and Kontoor Brands (the owner of the Lee and Wrangler names) have suffered.

In response, companies like San Francisco-based Huue, headquartered in the former cow town where blue jeans were invented, are fighting back by helping reinvent denim production. While the synthesis of indigo dye, according to their website, is "highly polluting to the environment and ... requires dangerous chemicals – like benzene, formaldehyde, and sodamide", Huue's approach uses genetically modified microbes to brew indigo from sugar, much the way that beer- or wine-making microbes convert sugar to alcohol. According to the company, their method creates "microbes that mirror nature's process" for making indigo, potentially being more efficient and better for the environment. But if the blue denim market keeps dropping as it has been, their potential sales might, too.

MSG: From Plants to Petroleum to Precision Fermentation

Monosodium glutamate, or MSG, is a chemical naturally found in *kombu* seaweed that is used as a flavor enhancer in stews, gravies, and other savory food. The human tongue has taste receptors for glutamate, making savoriness one of the basic tastes, along with saltiness, sweetness, bitterness, and sourness.

MSG was first produced beginning in the early 1900s in Japan by soaking wheat gluten or de-oiled soybean flakes in hydrochloric acid, a time-consuming and laborious process that exposed both workers and equipment to heated, concentrated acid. But the 1957 discovery of a bacterium now called *Corynebacterium glutamicum* sparked the beginning of a revolution in how MSG was obtained.

According to *Van Nostrand's Scientific Encyclopedia*, production of MSG via fermentation was first practiced in 1961. "By 1972, the world production of glutamate was 180,000 tons of which 90% was made by fermentation". (Considine, 2008) So, in only about eleven years from first commercial production, and fifteen years from the discovery of the *C. glutamicum* bacterium, production of MSG went from being done entirely by using plants to almost entirely by fermentation.

"The impact of glutamic acid bacteria on monosodium glutamate cost was dramatic," the Encyclopedia reports, "in 1950, MSG was obtained from natural products (i.e., soybean) and cost $4/kg. When the first fermentation process started up in 1961, the price dropped to $2/kg. By 1970 the price settled to about $1/kg."

However, the switch from MSG extracted from plants to MSG made by precision fermentation was a little more complicated than that. Chiaki Sano of the Ajinomoto Company of Tokyo, Japan, the company that first brought MSG to commercial production using wheat gluten, reported in *The American Journal of Clinical Nutrition* that one of the unintended consequences of the post-war development of Japan's acrylic fiber industry was greater availability of acrylonitrile, a chemical that can be converted into MSG, "and acrylonitrile could be supplied at a cost lower than that of other potential starting materials." Like the bacterial fermentation method, MSG production by acrylonitrile began in 1961, but it ended in 1973, as the Arab Oil Embargo made raw materials too expensive and as bacterial fermentation took over the market.

But that was not the end of the story for Ajinomoto. In the mid-1990s, the company discovered that one of the by-products from MSG fermentation could be made into a non-electrically-conductive film. The typical way of making insulating layers for semiconductors at the time was to paint on an ink and let it dry, but that was slow. Ajinomoto Build-Up Film (or ABF), made by computer-industry outsiders, was faster and better, and quickly became the standard way to make insulators for computer chips.

A Bright Idea: Coal Gas Started Disruptions in Lighting, Steel, and Dyes

> *Artificial lighting was expensive for centuries until the invention of coal gaslighting brought the price down 60% in fifteen years in the early 1800s and kicked off two centuries of cheaper, better lighting disruptions, continuing with kerosene, natural gas, and electricity. Even in just the past decade we are seeing LED lights disrupting fluorescents, halogens, and incandescents. Coal gaslighting also produced coke, which enabled a steelmaking making revolution, and coal tar, which kick off the investigations into the chemistry of artificial dyes that started the pharmaceutical industry*

Sometime in late 1795 or early 1796, observers recall, forty-two-year-old engineer William Murdoch had set an iron chamber called a retort into a fireplace at the Neath Abbey Iron Works, near Swansea, Wales. The fire was burning wood, but inside the retort was coal, and the heat from the wood fire drove gas out of the coal and through an iron tube coming out of the retort. To the iron tube was affixed a rifle barrel and Murdoch took a burning stick and touched it to the barrel's mouth. With a pop, the gas coming from the barrel ignited, giving off a flame just as brightly as if a candle or oil lamp was burning. This was a demonstration Murdoch had done many times over the past few years, according to John Griffiths in *The Third Man: The Life and Times of William Murdoch.*

Murdoch was born in Lugar, about twenty-five miles directly south of Glasgow, Scotland, and in 1777 at the age of twenty-three, walked to Birmingham to seek employment from fellow Scotsman James Watt and his business partner Matthew Boulton, who were manufacturing steam engines for the booming mining business. Murdoch quickly proved himself an astute engineer and Boulton and Watt dispatched him a couple of years later to Redruth, Cornwall, in England's southwest tip, where their customers were pumping water from the tin mines.

Boulton and Watt's company made money from a portion of the fuel costs that they saved their customers, so the company had an incentive to make its steam engines as efficient as possible. Murdoch, who also used the spelling 'Murdock', improved the operations by inventing the sun-and-planet gear, which Watt patented in 1781, and a few years later discovered a new kind of cement made from iron filings that was useful in steam engines. Boulton wrote: "We want more Murdocks, for of all others he is the most active man and the best engine erector I ever saw".

In subsequent years he invented a machine for drilling wooden pipes and a pneumatic tube system for message delivery. Using a heated, almost-closed vessel – an arrangement that foretold his later work making coal gas – he produced compounds that were later useful in the invention of aniline dyes and made a tar for protecting ship bottoms that worked almost as well as copper plating. He even built the 'Murdoch Flyer', a steam-powered, three-wheeled model carriage, similar in concept to an automobile, that is the first recorded example in Great Britain of a machine that could move around under its own power.

In 1795, he found a means to extract a chemical agent from local codfish for clarifying beer which was so much cheaper than the imported Russian agent made from sturgeon that British customs authorities tried to have it banned due to how much money in lost duties it would cost the state. This would not be his last time developing a disruptive technology.

Just a couple of years prior, Murdoch had built a hand-held coal gas lamp, and by 1794 had installed a retort in his back yard and run a pipe into his house in Redruth, making it the first residential building to be lit with coal gas. Moving back to Birmingham in 1798, he partially lit the Boulton and Watt factory by gas, and then illuminated a cotton mill in Manchester. In the 1978 book *Theatre Lighting Before Electricity*, Frederick Penzel reports that "the Phillips and Lee factory spent £2,000 a year for candles as opposed to £600 a year for gas. Even the insurance companies approved of it."

An Explosion of Lighting

In modern terms, switching to gas was a 'no brainer'. In the coming years, Murdoch lit streetlamps along Pall Mall, London, and then Westminster Bridge. He did not patent his work and had no way to prevent copying outside Britain, so coal gas illumination spread quickly from Newport, Rhode Island, to Baltimore, Maryland, to St. Petersburg, Russia. Griffiths estimates that "the new streetlamps had eighteen times the light output of the ones they replaced."

And light was not the only product of gas lighting. Not all the coal fuel was 'gasified' in the retort – the leftover solid was coke, a high-carbon product with few impurities in it that kicked off a revolution in steelmaking in the mid-1800s, and which is useful for brewing lighter beers which would otherwise take on bad flavors if other fuels were used. Making coal gas also leaves behind a residue of coal tar, investigations of which led to the first artificial dyes and, by extension, virtually all synthetic chemistry, from pharmaceuticals to artificial fibers. Historians believe that brighter, cheaper lighting benefited manufacturing, helped drive down urban

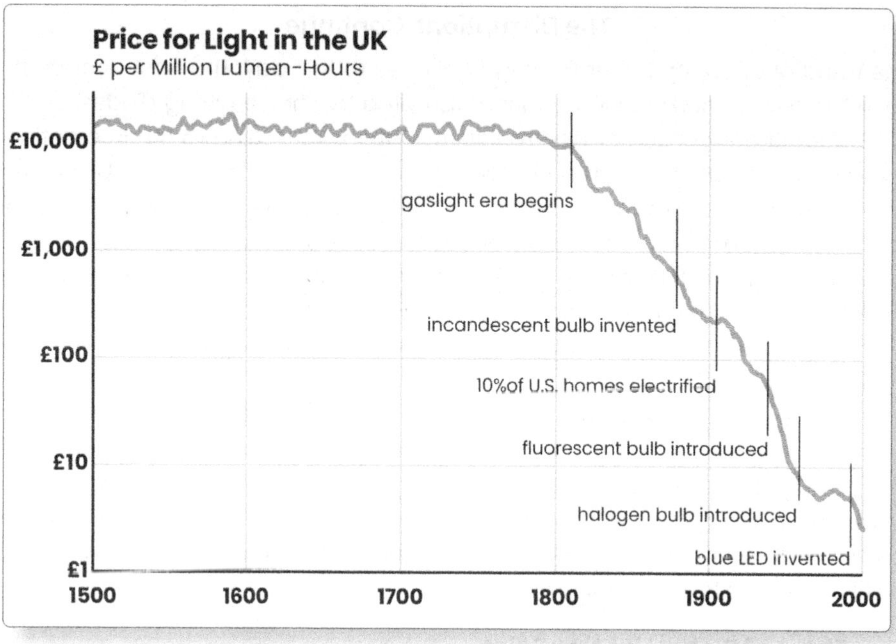

Figure 65 – Price for Light in the UK. *(Roser, 2018)*

crime rates, and helped propel literacy. Max Roser, the founder of Our World in Data, wrote, "if you don't have light, your day ends at sunset."

Coal gaslighting marked a turning point in the history of illumination, quite literally. Considering the cost of lighting in the United Kingdom over five centuries, from the year 1500 to the year 2000, the price per million lumen-hours was above £10,000 for the first three hundred of those years. (One million lumen-hours is about how much light a hundred-watt incandescent bulb produces in about one month of continuous use.) Coal gas dramatically changed that, with the price falling 60% in just the fifteen years starting in 1810.

Lighting saw numerous waves of disruption in the two centuries since then. Coal gas and whale oil were the most-popular sources of fuel in the early 1800s, while kerosene and natural gas became more popular in the latter part of the 1800s. The electric light bulb dominated the 1900s, going from less than 10% of the energy used for illumination in 1933 to 80% just seventeen years later, in 1950. (Fouquet, 2008) But even electric lighting saw its own waves of disruption, as incandescent light bulbs lost market share to fluorescents and halogens. Sales of fluorescent lamps started in 1938 and by 1951, just thirteen years later, more light was produced in the United States by fluorescent lamps than by incandescent ones.

Nevertheless, throughout all those disruptions, the price of lighting kept marching downward for two straight centuries. Over the course of the 1800s the cost of lighting fell by a factor of about forty and over the course of the 1900s, it fell by another factor of ninety. These were not one-for-one disruptions. The total amount of artificial illumination in the UK grew by a hundredfold the 1800s and then by another factor of one hundred during the 1900s.

The Disruptions Continue

The trend of cheaper, better lighting continues to this day. The past decade has seen the global market in illumination disrupted by light-emitting diodes, or LEDs, a product of the consumer electronics revolution. According to the International Energy Agency, an intergovernmental organization based in Paris, global lighting sales went from less than ten percent LED in 2014 to more than 50% just six years later in 2020, with every indication of eventually reaching nearly 100%. It is easy to understand why LED lighting is taking over the market so quickly. According to Hiroko Tabuchi in the *New York Times*, a five-watt LED "provides nearly the same amount of light" as a typical 60-watt incandescent bulb, which uses twelve times the electricity.

Changing the source of the light is also changing the metrics by which we measure it. With radically more-efficient bulbs, it no longer makes sense to compare bulbs by their power consumption (or 'wattage'). Instead, manufacturers have begun listing the intensity of light produced (in 'lumens') and the so-called 'color temperature' (measured in 'kelvin') a number that describes how reddish, whitish, or bluish the light appears.

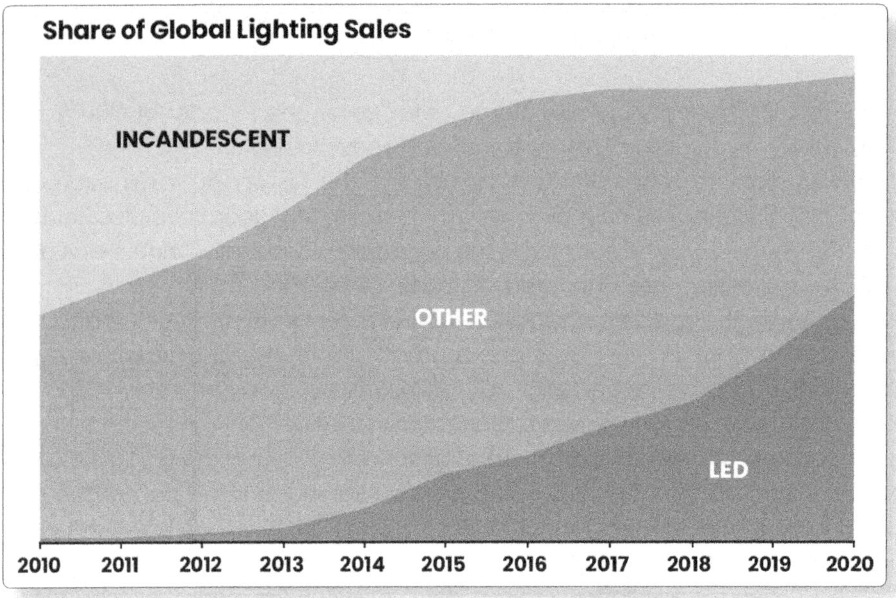

Figure 66 – Share of Global Lighting Sales. *(IEA, 2021)*

From a historical perspective, lighting is essentially a solved problem. Just as salt is now a small fraction of a fraction of the price it has historically been, as making phone calls and taking photos have become dramatically cheaper, technology has made our spaces brighter, and at lower cost. Dr. Adam Dorr, speaking in October 2021 at a TEDx event in Glasgow not far from William Murdoch's birthplace, said, "We live lives our ancestors 10,000 years ago couldn't begin to imagine and their most pressing problems are trivial to us. And the only difference between us and them is that we have better tools."

◆ ◆ ◆

"How did you go bankrupt?" Bill asked.
"Two ways," Mike said. "Gradually, then suddenly."
— Ernest Hemingway, The Sun Also Rises

◆ ◆ ◆

From Whale Oil to Kerosene

> *Dr. Abraham Gesner's discovery of how to produce kerosene disrupted the whaling industry over about a decade, dramatically decreasing the amount of whale hunting in the late 1800s. But it also later enabled the fossil-fuel powered ships that brought whaling back, much stronger, after World War II*

Eighteen hundred and fifty-six was a bittersweet year for Captain Tristram Pease Ripley. On April 27 his whale ship, the *Charles W. Morgan*, arrived in Massachusetts having taken 268 barrels of spermaceti oil (8442 gallons, or about 32,000 liters) from the heads of sperm whales, 1958 barrels of whale oil (61,677 gallons, or about 233,000 liters) made by rendering blubber, and 22,700 pounds of whalebone (about 10,000 kilograms). Bringing in a combined market value of over $76,000, it was the ship's most-lucrative voyage in the fifteen years since she had first touched water.

The lucky ship *Morgan* had set sail on September 20, 1853, the captain's thirty-second birthday, and been at sea for 950 days. Ripley, now thirty-four years old, had married Eliza Mayhew, eight years his junior, in June 1851 (the year 'Moby-Dick' was published), and they had built a home together before his departure overlooking the idyllic harbor in Edgartown, on the eastern end of the island of Martha's Vineyard.

In the 1953 book *Whaling Wives*, Emma Mayhew Whiting describes the captain as "a handsome man" having "dark hair worn long and brushed at the sides", and "suitably bearded". Eliza was pregnant when he had set sail in 1853 and their daughter, Sarah, was born after his departure.

Upon his return, Tristram and Eliza had not seen each other for two years, seven months, and one week. But now she came to him empty-handed. "While he was at sea," Whiting reports, "his first child, Sarah Mayhew, died at the age of ten months", and she was buried without the captain ever having laid eyes upon her. Eliza must have been pregnant again shortly after her fifth wedding anniversary as a son, Tristram Everett Ripley, was born to her the following year, in the spring of 1857.

But something else important had happened in the nearly three years when the *Morgan* was gone. Something that would change the whaling industry, and the rest of the world, forever.

Figure 67 – Captain Tristram Pease Ripley (left) and Dr. Abraham Pineo Gesner (right).
Illustrations by Alperen Moluluo

In June of 1854, Canadian medical doctor by training and geologist by vocation Dr. Abraham Gesner received US patents for a method of obtaining gasoline and kerosene from bitumen (essentially, asphalt). He actually had discovered kerosene much earlier, essentially as a hobby. "What first set Gesner upon his new lamp oil enterprise, we can only speculate," Kendall Beaton described in an article in *Business History Review*. "In 1818, he had visited the 'tar springs' on Barbados and on this, or another voyage, had called at Trinidad and made a close inspection of the famous 'Pitch Lake'."

The La Brea Pitch Lake, on the southwest part of Trinidad, was used by the Spanish as a source of caulking material for their carvel-built ships starting soon after Columbus' discovery of the island on his third voyage in 1498. In a book he wrote in 1860, Gesner says that "it was from this bitumen that the author first obtained kerosene."

Gesner gave the first public demonstration of the use of the new lamp fuel in Charlottesville, Prince Edward Island, in August 1846 and then spent several fruitless years being tied up in lawsuits and failing to secure financial backing to bring this new kind of oil to market. In 1853, nearly fifty-six years old, Gesner packed up his family and moved to New York. Patents were granted by the US government the following year, but Beaton reports that "by the time Dr. Gesner's patents were granted in June, 1854, the establishment of a works to make Kerosene, in accordance with the methods he had described, was already under way."

"From engineers' drawings of the plant which have survived," Beaton continues, "we can appreciate Dr. Gesner's very real abilities as a practical manufacturing chemist. His plant was laid out in orderly fashion and the individual pieces of equipment, shown in detail drawings, were well planned and well constructed, differing very little from similar pieces of refinery equipment being built as late as the time of the First World War."

Kerosene Delivered 10X the Light for the Same Money

"Kerosene", according to Gesner's measurements, "produced a light 13 times brighter than... Whale Oils" and "6 times as bright as sperm oil", Beaton wrote. "Kerosene was then [in 1856] priced at $1.00 a gallon" and in comparison, "the oil of the right whale, priced at a dollar... and sperm oil, $2.25 a gallon". That is, on a performance-to-price basis, kerosene was at least 10X better compared to either whale oil or sperm oil, according to Gesner.

When he arrived in September 1856, Captain Ripley did not get quite these prices for his products. For whale oil, which sold for a dollar a gallon, Ripley received $0.80, and for spermaceti oil, which sold for $2.25, he received $1.62. But even so, 1856 was a good year, financially, for the Captain as well as for many others involved in the whaling business.

The Whaling Business Peaked Just After Gesner's Kerosene Works got Underway

According to tables in *A History of the American Whale Fishery* by Walter Tower, in 1856 American whale ships brought in more than 2.5 million gallons of spermaceti oil, over 6.2 million gallons of whale oil, and 3.4 million pounds of whalebone, for a total industry revenue of $10.6 million dollars. "At its height," wrote Derek Thompson in the article 'The Spectacular Rise and Fall of U.S. Whaling: An Innovation Story' in *The Atlantic* magazine in 2012, "the whaling industry was the fifth largest sector of the economy."

In dollar terms, 1856 was second only to 1854, when the industry brought in $10.9 million. The following year, 1857, would be the last year that the whaling industry generated more than $10 million total. Of the five best years in whaling history, all of them came in the 1850s. But, as with much else in America, the 1860s would bring an unprecedented disruption to the whaling business.

By 1859, Gesner's New York facility, according to Beaton, had a production capacity of five thousand gallons of kerosene per day, working from coal-like starting materials. And at that point Gesner was far from the only one in the business. "By the fall of 1859", Beaton wrote, "there were in this country at least 33 'coal-oil' refineries actively engaged in the manufacture of hydrocarbon illuminating oils". These facilities combined, Beaton reports, had a production capacity of 22,750 gallons of kerosene per day, or roughly 8.3 million gallons per year, nearly as much as the entire whaling industry had produced in whale-derived oils, in volume terms, in 1856.

Some of the other components of crude oil, like gasoline, were considered waste products. Lawrence Burns wrote in the 2018 book *Autonomy: The Quest to Build the Driverless Car*, "before the Model T, gasoline was considered such a useless by-product that refineries had pumped it straight into rivers rather than bothering to sell it."

Though the industry he helped create became phenomenally successful, Gesner himself did not make much money from it. His financial backers in New York had incorporated the company with Gesner as a salaried employee, not an owner, and when they decided to replace him, he turned to writing books about geology and chemistry and to practicing medicine again.

Kerosene Ignited the Petroleum Industry

Kerosene production would only get easier over time. In western Pennsylvania, Samuel Kier had recently become the first American to refine crude petroleum oil into kerosene and Edwin Drake, in 1859, used a steam engine-powered drill to drive a cast iron pipe into the ground in Titusville, becoming the first American to successfully drill for oil. Lining the borehole with pipe to prevent the hole from collapsing is still the standard method used today when drilling for hydrocarbons. According to the *New York Times*, Drake's well produced twelve to twenty barrels per day. (Ripley's voyage, in comparison, took nearly one thousand days to bring home more than two thousand barrels of oil, or about two barrels per day. Thus, in both financial and volume terms, even the first crude oil well in the US was up to 10X more productive than a whale ship.)

John D. Rockefeller got his start in kerosene refining in the early 1860s, and the petroleum business made him the wealthiest American of all time with a peak net worth above $400 billion in today's dollars. Using a self-reinforcing process of reducing refining costs, borrowing heavily, buying less-efficient competitors, and consolidating operations, Rockefeller relentlessly expanded his business, officially incorporating Standard Oil in 1870 in Cleveland, Ohio. In 1872, in what is known as "the Cleveland Conquest", Rockefeller bought twenty-two of twenty-six local competitors over the period of just a few months. By the end of the 1870s, Standard Oil was refining over 90% of the oil in the U.S. (Segall, 2001)

Winning the War by a Couple of Inches

Just before World War II started, according to a 2018 blog post by the National Institute of Standards and Technology (NIST) called 'The Big Inch: Fueling America's WWII War Effort', "within the U.S., 95 percent of petroleum products were transported by ocean-going tankers, which carried the precious stuff from the Texas Gulf Coast to refiners and distributors in the industrialized Northeastern states" amounting to "300,000 barrels of oil daily by ship." (Martin, 2018)

Historian Michael Gannon wrote in his book *Operation Drumbeat: The Dramatic True Story of Germany's First U-Boat Attacks Along the American Coast in World War II*, "when the US entered the war, all that transport was threatened by German U-boats, who attacked American ships in US waters along the Atlantic coast." In 1942, the Nazis sank nearly four hundred Allied ships off the East Coast of the U.S. As the Texas Eastern Transmission Corporation (TETC) wrote in *The Big Inch and Little Big Inch Pipelines: The Most Amazing Government-Industry Cooperation Ever Achieved*, "the U-boat assault on merchant shipping in United States waters during 1942 was a greater strategic setback for the Allies than the surprise attack on Pearl Harbor."

The solution to the problem was obvious in principle: build a land-based pipeline from Texas to the East Coast. TETC wrote that by the 1880s, pipelines "had clearly proven their superiority for transporting oil over long distances", but for most of their history pipelines were built by rules of thumb, personal experience, and fabrication limited to eight inches in diameter. "Eight inches became the standard pipe size and remained so until the early 1930s," TETC wrote, "for it was the largest diameter that could function at the normal operating pressures of the times."

But replacing coastal shipping would require what TETC said was "the biggest, heaviest, and longest pipeline built up to that time" – two of them, actually. A 24-inch (60-centimeter) diameter pipe, dubbed the 'Big Inch', would carry crude oil while a smaller 20-inch (50-centimeter) line, the 'Little Big Inch', would carry refined petroleum products like gasoline, diesel, and kerosene. Twenty-four inches is three times the diameter of eight inches, meaning that the cross-sectional area of the Big Inch would be nine times the area of a typical one of the time. Building a pipeline like that could not be done solely from personal experience and 'rules of thumb'.

Fortunately, according to the NIST blog post, thirty-one-year-old engineer William Heltzel "beginning in 1926... introduced a more rational, science-based methodology to designing pipelines" through a series of published articles. "Upon Heltzel's death in 1959," the post continues, "a petroleum industry trade publication noted that 'No other development in the catalog of pipeline achievements has contributed so much to the industry as [Heltzel's] concept of rational pipeline design.'"

Through heroic efforts, the two 'Big Inch' pipelines were completed, significantly under budget, just one year after construction began in June 1942. Thanks to dynamite, steel, steam- and diesel-powered vehicles, it was an achievement that would have simply been impossible a century before. Without Heltzel's insights, it would not have been possible even a couple of decades before.

"The pipelines delivered more than 500,000 barrels of oil per day for the war effort," NIST wrote, "and all of it on a land route safe from Nazi submarines." So, this was not a one-for-one replacement of shipping capacity. The new pipelines delivered two-thirds more than the previous ship-based system.

Though the pipeline network was built with government funding and coordination, after the war was over the network was declared 'war surplus property' and sold to private companies, according to *The Big Inch and Little Big Inch Pipelines*, "for the full cost of its construction". The Inch Lines are still in use today.

But newer pipelines have been built since then, such as the 40-inch and 36-inch Colonial Pipelines, built in the 1960s, one of which primarily carries gasoline and the other jet fuel, heating oil, and diesel.

Petroleum was not a One-for-One Replacement for Whale Oil

The staggering rise of the petroleum industry was far from a one-for-one replacement for oil from whales. So much so that it would be difficult to even plot revenue from both industries on the same scale. US sales of crude oil in 1864, the year Dr. Abraham Gesner passed away and just five years after Drake succeeded in Pennsylvania, were more than $17 million, significantly higher than the best year ever of selling oils from whales.

If we add together total revenue from oils from whales with petroleum oil, the combined oil market went from 100% whale oils in 1859 to consistently over 80% petroleum oil just ten years later, in 1869.

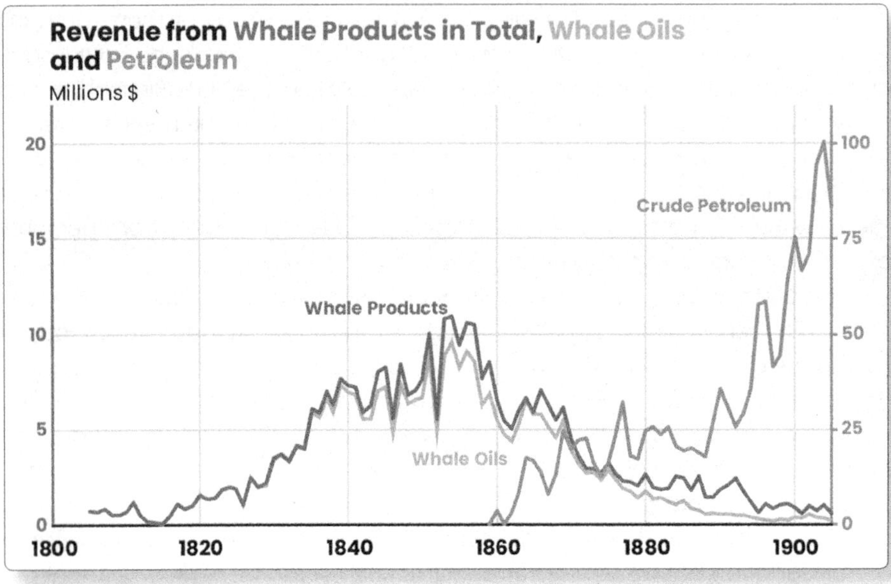

Figure 68 – Revenue from the Whaling Industry Compared to the Petroleum Industry

In the early 1880s, the petroleum business was revolutionized again when the brothers M.C. and C.E. Baker of South Dakota developed the first rotary drilling rig, which quickly replaced the then-standard percussion drilling technique to find and extract oil, helping to propel US crude oil production to $75 million in sales by the year 1900.

But the end of hunting whales for their oil was not the end of hunting whales. As whale oils found it more difficult to compete with petroleum oils, the whaling industry slowly declined and shifted its focus from liquid products to whalebone, a fingernail-like material found in the mouths of some kinds of whale and used in the manufacture of products like corsets and buggy whips. On Captain Ripley's 1853-1856 journey, the sale of whalebone accounted for 12% of the trip's revenue, a typical figure for the time. But between 1875-1885, the portion of industry revenue that came from whalebone rose to be regularly more than half of whaling industry revenues.

The industry clung on through the early 1900s. The last commercial voyage of the *Charles W. Morgan* was in 1921. The ship still exists to this day. Now converted to a museum and docked in Mystic, Connecticut, it ranks as America's oldest commercial ship still afloat. Only the USS *Constitution*, a military vessel, is older. For that matter, the Ripley's house still stands too and is now the Harborside Inn in Edgartown Harbor, Massachusetts. Captain Ripley passed in 1881 at the age of sixty; his wife Eliza in 1889, also aged sixty.

Diesel Caused a Re-birth of Whaling

Far from petroleum killing off whale hunting, the industry instead saw a dramatic resurgence in the 1900s, especially after World War II, when diesel-powered ships made whaling easier. According to a 1984 article in *Marine Fisheries Review* called 'The Sperm Whale: Physeter macrocephalus' by Merrill Gosho from the National Marine Mammal Laboratory in Seattle, Washington, the number of sperm whales hunted in the 1900s was more than three times the number killed in the 1800s, even though there was relatively little whaling in the last twenty years of the 1900s.

The decline of the whaling industry in the late 1800s and its ultimate near-termination in the late 1900s both had interesting effects on the prices of the whale products that were being produced. Whaling in the 1800s focused largely on oils, with spermaceti oil being the preferred product and whale oil made from rendered fat being of secondary importance. As the industry declined in terms of the annual volume of oils produced, the price of whale-derived oil rose dramatically. As a result, the industry's best years, financially, were actually ones of declining production.

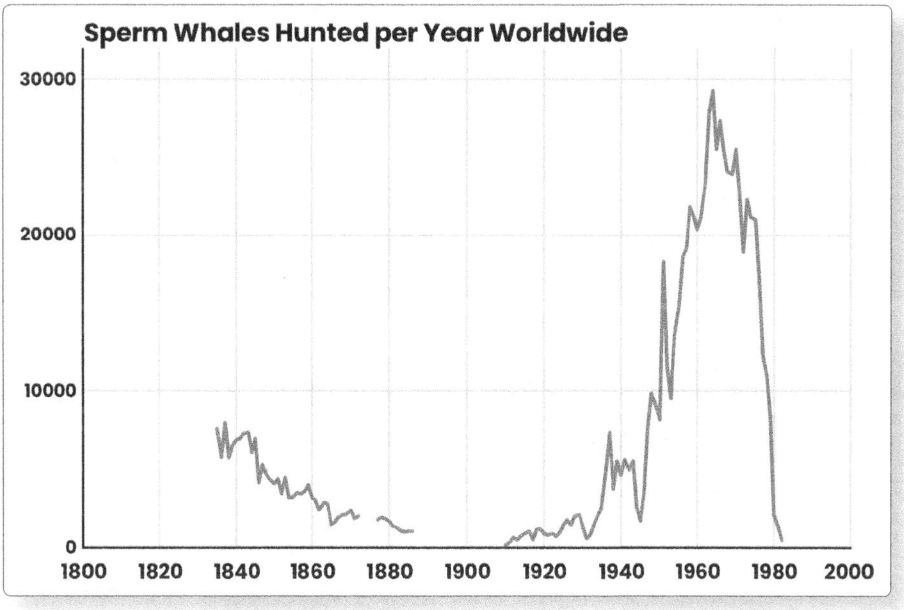

Figure 69 – Sperm Whales Hunted per Year Worldwide

In this way, the whaling industry was largely sheltered from its own decline for most of the 1850s. As production volumes decreased, prices rose. In fact, prices for both spermaceti oil and for blubber derived whale oil peaked in 1860s, as petroleum went from 0% to 80% of the value of the oil market.

A similar dynamic played out again in the late 1900s, as the market for whale meat collapsed. In the 1800s, with no way to store meat at sea, most whale meat was simply discarded. In the 1900s, meat was the primary product sought from whale hunting. From 1978 to 1982, according to an article by CNBC, the amount of whale meat available at major Tokyo markets fell nearly half. But the price per

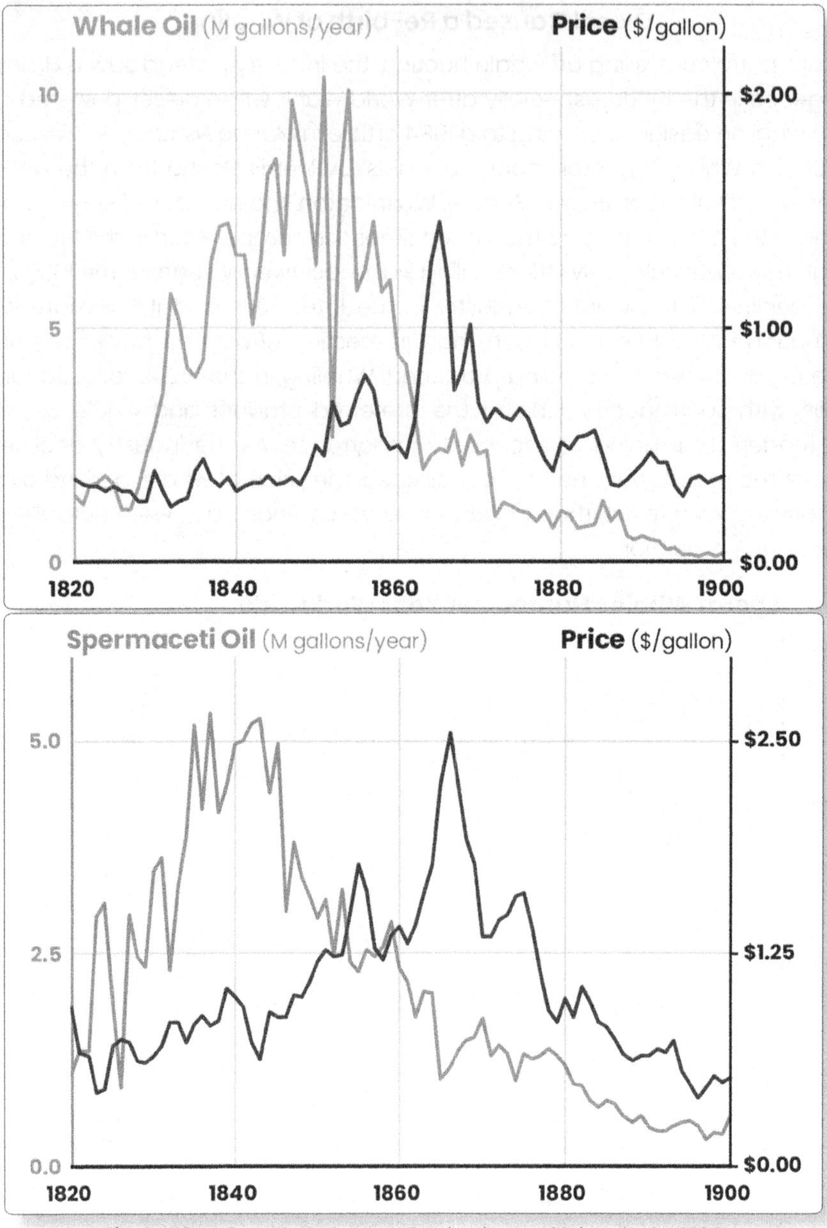

Figure 70a & b – Whale Product Production and Prices in the 1800s

pound rose 40% in that time too. So, the total revenue to the whale fishing industry only fell about 25%. As sales volumes collapsed over the subsequent decade, prices continued to rise.

The end of whale hunting caused problems for the automotive industry. In 1975 the *New York Times* reported that some of General Motors' cars "have had major transmission problems because of G.M.'s difficulties resulting from a law banning the killing of some whales. General Motors had used whale oil in its automatic transmission fluid. But the sperm whale was put on the endangered species list in

1971, forcing the company to change its transmission oil."

In a sense, though, whale hunting has continued to this day. Tom Nicholas in his 2019 book *VC: An American History* found that modern venture capitalists, people who give money to start-up companies in the hope of earning a large return for their money if the company becomes successful, on average have similar chances of being profitable today as whaling voyages were in the 1800s.

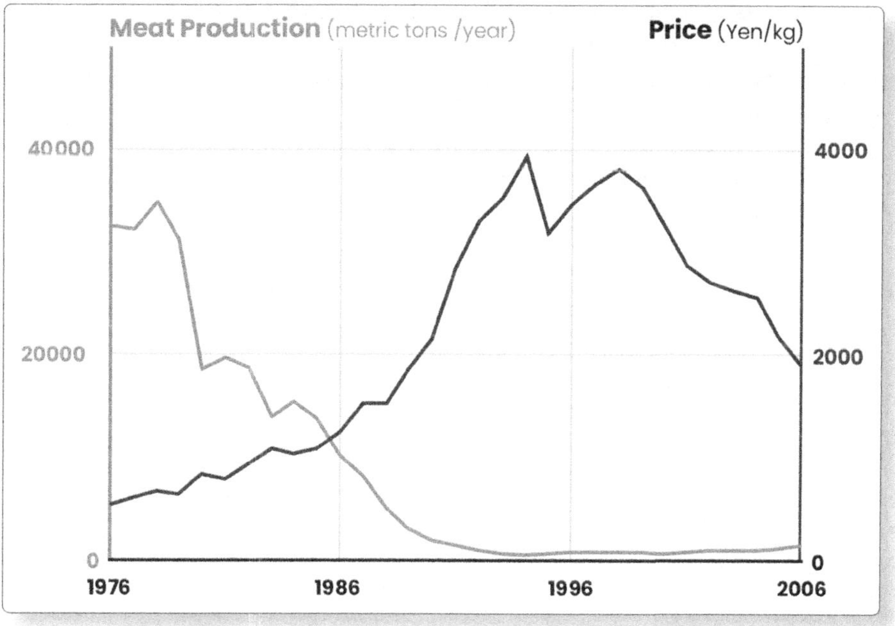

Figure 71 – Sales and Prices of Whale Meat in Tokyo. *(Fahey, 2015)*

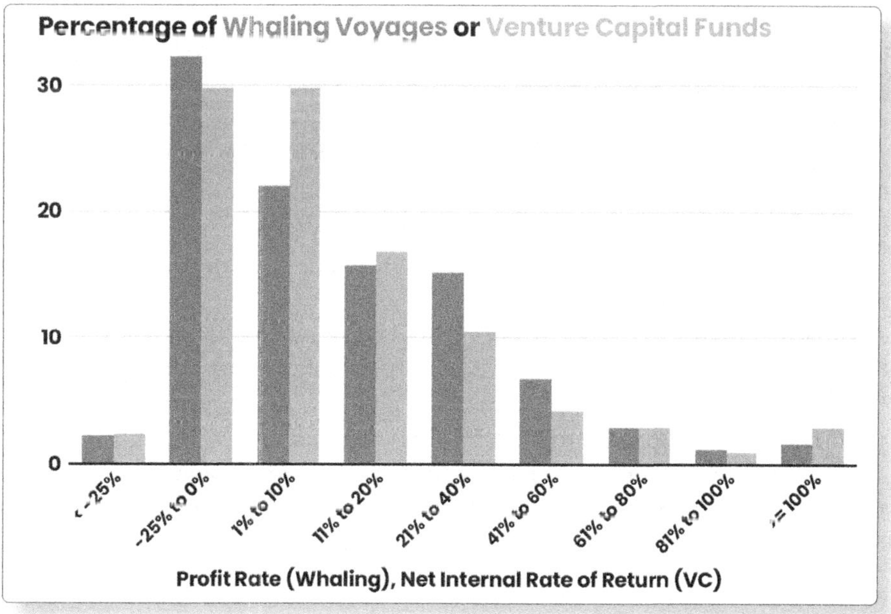

Figure 72 – Investment Returns from Whaling Voyages and from Venture Capital Funds. *(Nicholas, 2019)*

From Coal Gas to Natural Gas

The discovery of natural gas in the North Sea in the 1960s profoundly affected the economies of Britain, the Netherlands, and Norway. Though it meant plentiful energy and prosperity for many, natural gas also made old infrastructure obsolete, toppled old industries, and threatened to upend entire economies.

Natural gas meant that the pipes laid for town gas – which came from heating coal – all had to be removed and replaced with new pipes, due to the different composition of the new gas. In Britain, over forty million appliances were converted in the decade from 1967 to 1977.

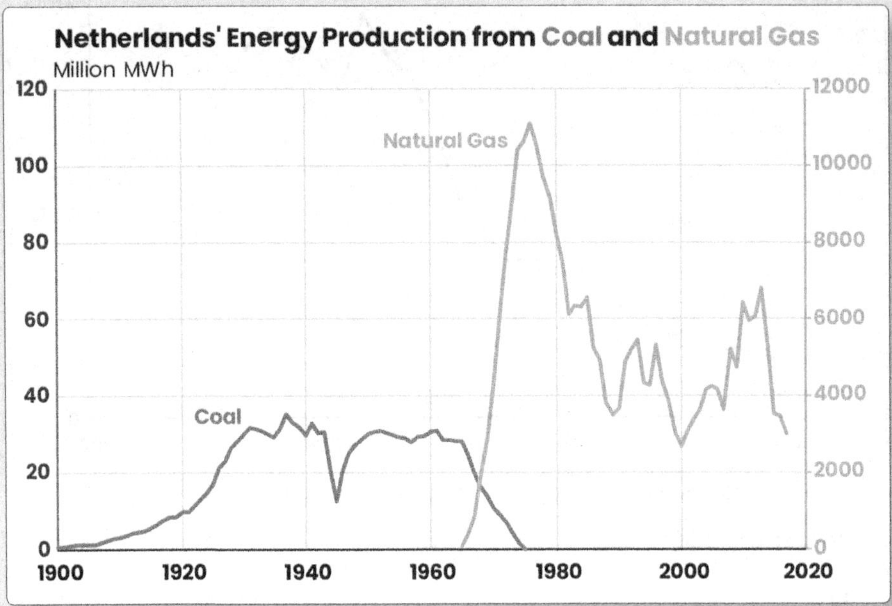

Figure 73 – Coal and Natural Gas Production in the Netherlands, measured in millions of megawatt-hours of energy the products would produce if burned.

In the Netherlands, coal mining collapsed completely from 1965 to 1975 as the natural gas industry quickly ramped up. And in Norway, oil and gas quickly grew to become the largest segment of the economy, now accounting for more than half of all exports.

Both the Dutch and Norwegian economies were affected by this new and very lucrative industry. The tendency of a very successful industry to attract talent and investment interest away from other parts of the economy was labeled 'Dutch Disease'. To combat this effect, the Norwegian government established a sovereign wealth fund (colloquially called the 'Oil Fund'), which went from the smallest sovereign wealth fund in the world to the largest over just fifteen years.

◆ ◆ ◆

"Anybody who ever lighted a fire without matches has probably gained some proper respect for 'low' or 'primitive' or 'simple' technologies; anybody who ever lighted a fire with matches should have the wits to respect that notable hi-tech invention."

— Ursla Le Guin

◆ ◆ ◆

Reinventing Fire: The Safety Match

> *Humans used fire for at least a half-million years without having a standard way to start one. But over just a decade in the 1860s, the red-tipped safety match reinvented one of our oldest technologies*

The first successful friction match was invented in 1826 by John Walker, an English chemist. But these matches contained white phosphorous, which is poisonous, making them both dangerous to manufacture and dangerous to use as flaming bits (not just sparks but actual tiny embers) would fall to the ground or into the ruffles of clothing during lighting. Additionally, they were deadly if accidentally (or intentionally) ingested. Marketed as 'lucifers', they could be lit by dragging them across any rough surface, which is why old cowboy movies sometimes show men lighting a match on their boot heel or on the stubble of their chin. As an incendiary munition the deliberate use of white phosphorous (sometimes called "Willy Pete") against civilian targets is forbidden by the Geneva Conventions. This is not the sort of stuff the average person today would want to keep around their home if they care about the safety of themselves and their family and, for that matter, the continued existence of their house.

The idea of using red phosphorous as a striking surface was patented in 1844 by the Swede Gustaf Erik Pasch. (Red phosphorous and white phosphorous are different allotropes of the same element. Just as diamond and graphite are both made entirely of carbon simply in different atomic arrangements, red and white phosphorous both consist entirely of phosphorous in different atomic arrangements.) British chemist Arthur Albright developed an industrial process for manufacturing red phosphorus at large scale, which he demonstrated at The Great Exhibition in London in 1851.

Brothers Johan Edvard and Carl Frans Lundstrom, who ran a match factory in Jönköping, Sweden, got a sample of Albright's red phosphorous at that fair, but apparently did not figure out what to do with it until just prior to the Paris Exhibition of 1855, when they began manufacturing match sticks tipped with potassium chlorate and sulphur, which could be lit on a red phosphorous striking surface.

According to the Jönköping Match Factory's website, "Jönköping's match factory came from around 1860 onwards to take a leading position in the world for many years in the production and sale of safety matches." The Wikipedia page for Swedish safety match mogul Ivar Kreuger says that the invention of the safety match "made matches the most important Swedish export" and "made Sweden the leading exporter of matches" including, by the early 1900s, having legal monopolies through Europe, Central and South America on the match trade.

From the consumers' perspective the 'safety match' represented a 10X disruption, as it had a much lower chance of causing a house fire, a painful burn, or even death, to anybody who used one. And like many disruptions, it had both an incubation period and a period of rapid adoption. It was eleven years from the invention of the red phosphorous striking surface by Pasch in 1844 to mass production of the safety match by the Lundström brothers in Jönköping, and from there about fifteen more years for Swedish safety matches to conquer the world.

Steel

> *Coal gas produces coke (unburned coal) as a by-product, opening the possibility of manufacturing steel in large quantities. Several generations of steelmaking techniques have disrupted each other in turn since the mid-1800s*

Steel has been prized for its usefulness in tools, weapons, and armor since antiquity. The earliest-known pieces, about 3800 years old, were recovered at Kaman-Kalehöyük, in what is now central Türkiye, and throughout the ages people have independently discovered various methods to turn iron into steel in Sri Lanka, Tanzania, and China, among other places. But the modern industrial technique for making steel in larger volumes at reasonable cost (and therefore steel's usefulness in construction or making engines, sewing machines, railroad ties, and ships) were developed in the middle of the 1800s.

In 1856, Henry Bessemer patented a device that blows hot air through molten iron (the 'Bessemer converter') as well as a variation that used oxygen instead of air (the 'basic oxygen furnace' or 'oxygen blast furnace'). Carl Wilhelm Siemens, the brother of the founder of the German industrial giant, developed the 'open hearth' furnace in the 1850s, which French engineer Pierre-Émile Martin used for making steel starting in 1865. According to David Landes in *The Unbound Prometheus* "together, the Bessemer, Siemens-Martin, and basic processes drove the real cost of crude steel down some 80 or 90 per cent between the early 1860's and mid-1890's".

"The combined production of Britain, France, Germany, and Belgium in 1861 — before the Bessemer process had taken hold — was approximately 125,000 tons", Landes reports, "In 1870 the total was perhaps 385,000 tons; in 1913 it amounted to 32,020,000 tons, a gain of 83 times (10.8 per cent per year) over the forty-three-year period." As a result of these new methods of production, the percentage of all iron products that were steel rose dramatically, going from less than 10% in 1870 to over 75% by 1910.

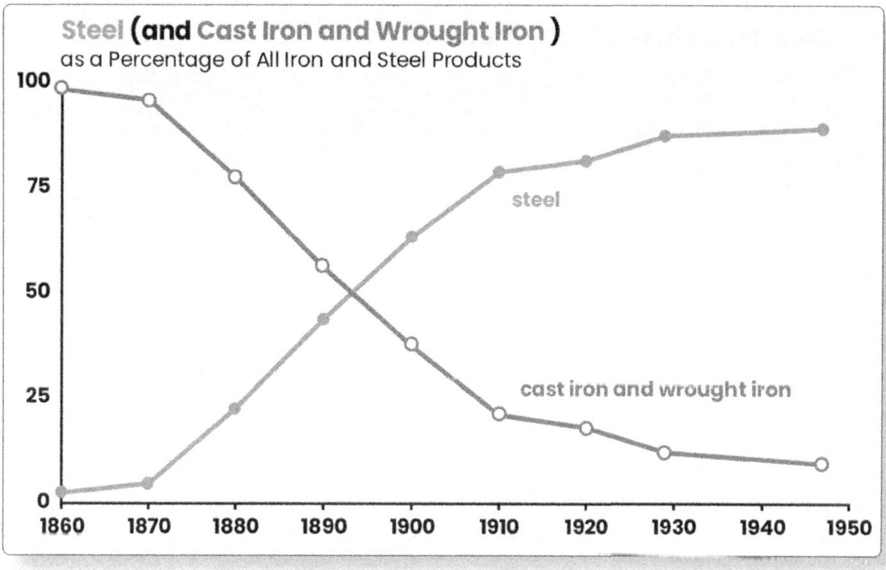

Figure 74 – Steel (and Cast Iron and Wrought Iron) as a Percentage of All Iron and Steel Products. *(Ayers, 1989)*

Cheap steel revolutionized and created too many industries to count. Steel-framed skyscrapers with steel elevators dramatically increased urban land values by letting us build upward. Steel trains, automobile frames, and steel ships revolutionized transport. And steel cans and kitchen gadgets remade how we store and prepare food.

The amount of steel produced in the US by the Bessemer process averaged about ten million tons each year in the first two decades of the 1900s, but because production from the Siemens process went up tenfold in that same time, from about four million tons per year to about forty million, the relative portion of steel which was produced by the Bessemer process dropped from about two-thirds in 1900 to about 20% twenty years later.

Thus, the Siemens–Martin process was a complement to the Bessemer process rather than a replacement for it. Although the Siemens-Martin open hearth process could handle larger batches than the Bessemer process, those batches also took longer time. Nevertheless, by the start of World War II, the Siemens' open-hearth process was responsible for over 90% of US steel production, with most of the remainder being due to the Bessemer process and a few percent produced from electric arc furnaces, which are typically employed for recycling end-of-life steel products and which were first used commercially at the end of the 1800s.

In the early 1900s, the slower production time of the Siemens' process was an advantage, as it gave workers time to take samples of the molten steel and analyze them as the steel was being produced, to maximize the quality of final product. But by the later part of the 1900s, with faster analytical techniques and better process control, those inherently long production times became a severe drawback.

After World War II, with the US economy and population booming and much of Europe in need of rebuilding, steelmaking was ripe for disruption. The Russian

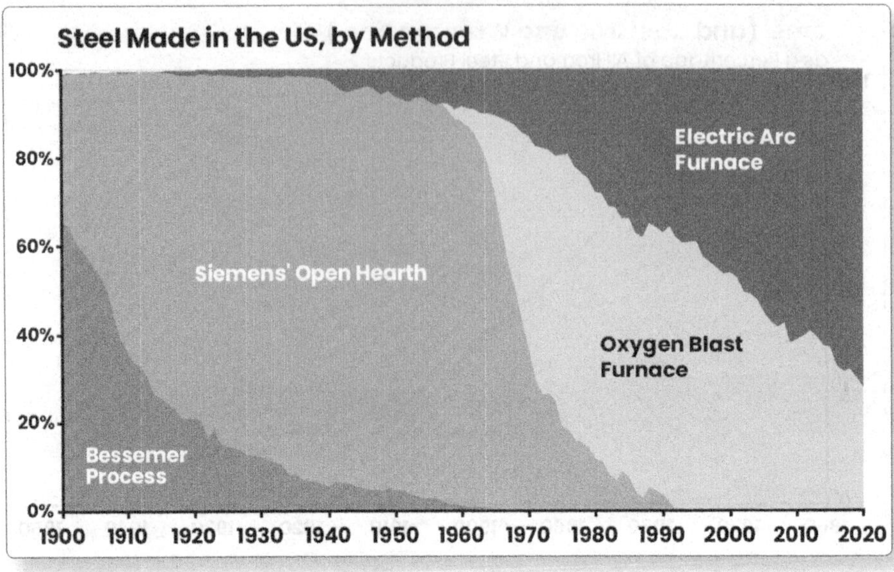

Figure 75 – Percentage of Steel Made in the United States by various methods

physicist Pyotr Kapitsa, who later won the Nobel Prize in Physics for his work, invented the turboexpander, a device that uses air to produce purified oxygen in industrial quantities.

Air is only about 20% oxygen, so Bessemer's air-based converter requires much more gas to be heated than is required using only oxygen. Air also introduces nitrogen, which makes steel brittle. Kapitsa's invention opened a whole new possibility for making steel, one that Siemens had envisioned in the 1800s, but could not realize as he did not have a cheap source of purified oxygen.

Outsiders Pioneered the New Approach

Engineer Robert Durrer, head of the Von Roll steel company in Austria, revolutionized steelmaking starting just after World War II by producing the first small batch of oxygen blast steel in 1948. Oxygen blast furnaces not only require far less heated gas (and therefore, less energy) than the air-based methods, they also require less time, taking about forty minutes to produce a batch of steel, compared to ten to twelve hours for an open-hearth furnace. This alone represents a disruption factor of about 15X.

US steelmakers were behind the curve when it came to adopting the new technology. It took ten years for the first oxygen blast steel to be produced commercially in the US, but then oxygen blast steel went from nothing in 1958 to over 50% of US steel production just thirteen years later, by 1971.

It was not a One-for-One Replacement

The replacement of open-hearth steel production with the oxygen blast process was essentially one-for-one in the US, but not globally. US production peaked in the early 1970s, just as oxygen blast furnaces began to dominate, but worldwide steel production rose from about 185 million tons annually in 1950, when oxygen blast was 0% of the market, to 350 million tons annually in 1960 (when it was still less than 5% of global production) and to about 630 million tons in 1970 (when oxygen blast steel had reached 40% of the market). It was more than 50% just a couple of years later. (Holappa, 2017)

As with many other incumbent technologies, open hearth furnaces saw a gradual decline followed by a sudden one. The total amount of steel produced in the US by the Siemens' open-hearth process peaked in 1955, just before the introduction of the oxygen blast furnace. Though total open-hearth steel production was still 90% of its 1955 peak ten years later in 1965, by 1975 it was only 20% of its peak level, and was only about 5% of its peak in 1985. The very last open-hearth steel was produced in the US in 1991.

The oxygen blast process was a disruption in a way that no other steelmaking technology had been. It was due to a convergence of technologies, namely the century-old oxygen blast concept combined with turboexpanders, the means for producing oxygen. The disruption was led by a player outside the 'big steel' industry, in particular a post-war European producer that had far more to gain from the new technology than to lose from it. It had long incubation and development times, followed by a rapid takeover of an established industry in about fifteen years. And the incumbent technology saw its business decline gradually, and then suddenly.

Clayton Christensen's classic book *The Innovator's Dilemma* covers the growth of electric arc 'minimills' in the 1970s and 1980s which disrupted integrated steel mills. As the availability of end-of-life steel has grown and as electricity prices decline, the electric arc process should continue to gain in popularity. The decline of steel-intensive industries like the oil and gas and maritime shipping in the coming decades will require us to rethink entire categories of products, as we have done many times before.

Let us look now at one example: cut iron nails replacing hand-wrought nails, and then being disrupted by steel wire nails.

Nails: From Hand-Wrought to Cut Sheet to Steel Wire

> *Nails were once made laboriously by hand by pounding a rod with a hammer, then by cutting tapered slivers from iron sheets, and now mostly from cutting round steel wire. Each style took over the market in essentially the same way, revolutionizing the construction business in the process*

The humble nail has undergone major transformations in the last couple of centuries, mostly in the way they were produced. At Şanlıurfa Archaeological Museum in south-central Türkiye, there are stone nails on display which are from the Neolithic Era. Yes, stone. We have been making nails since before we have made metal. For most of the time since we started using metal, more than five thousand years ago, nails were then made in the same fashion – a blacksmith or 'nailer' would pound the end of a thin iron rod to a sharp point, cut the rod to a good length, and then beat the other end flat to form the nail head. But between the late 1700s and the late 1800s, this fundamental invention was dramatically re-invented.

Hand-Wrought Nails Were Laborious to Make

Thomas Jefferson founded a nailery at Monticello in 1794, a few years before he was elected President, selling nails as a source of income and becoming "one of the first to establish a commercial nailery" in the independent United States, according to 'The Nail Making Industry in Early Virginia', the 1970 Master's degree thesis of John Tracy Keene at The College of William and Mary.

Jefferson's 'Farm Book', a collection of his hand-written agricultural records, lists nine of his enslaved boys who were between the ages of ten and eighteen at of the beginning of 1795 and who were likely involved in nailmaking. Their names (and ages) were: Joe (14), Wormeley (13), Burwell (11), Moses (15), Shepherd (12), Barnaby (11), Davy (11), Ben (10), and Jamey (11 or 12).

The Jefferson Foundation's website says that "up to fourteen young male slaves, aged ten to twenty-one, worked at the forges of the nailery". From 1794 to 1796, Jefferson calculated the efficiency of the nailers, "each day weighing their nail rod and the nails they produced." In 1795, "workers at Monticello are making from eight to ten thousand nails per day," the website says, or about 250,000 per month, which equates to fourteen people producing one nail each per minute for twelve hours per day, twenty-five days per month.

Nail-Cutting Machines were Developed by Outsiders

But at this time a new means of producing nails was being pioneered by Jacob Perkins and other inventors, credited with making machines that would cut wedge-shaped slices from a flat sheet of iron. Jeremiah Wilkinson of Cumberland, Rhode Island, according to Amos Loveday's 1983 book *The Rise and Decline of the American Cut Nail Industry*, "produced textile card tacks by shearing thin slivers of metal from the ends of a metal plate instead of cutting them from a rod". The same idea for making tacks would later be applied to larger brads and nails. This was no small enterprise. According to Loveday, in the thirty years "between 1790 and 1820, patents for nail making machines and improvements accounted for slightly more than 2% of all the patents granted by the US Patent Office", a remarkably high percentage for something that is now such an unremarkable product.

In nail production, the American colonies were outsiders. Dr. Fredrik Sjögren, in his 2019 Ph.D. thesis at the University of Birmingham (UK), reported that "in 1760, between 40 and 50 per cent of British metal exports to America consisted of nails". In fact, according to Loveday, "throughout most of the colonial period British mercantile policy discouraged manufacturing nail rod in the American colonies." However, Sjögren continues, "by 1800, the figure [the percentage of British metal exports to America that were nails] had fallen to 14 per cent."

Why were the first nail-cutting machines developed in the US in the late 1700s and not in Great Britain, which was more economically developed? First, of course, there was more wood in the Americas and so more wooden construction compared to England. As a result, the per-person demand for nails in the US was higher. Second, the US had a labor shortage compared to Great Britain, which favored the mechanization of labor-intensive activities. The combination of greater demand for nails and greater need for mechanization caused early America to become the global center of cut nail production.

Early cut nails were not as good as hand-wrought nails for most applications, as they were headless and needed to be headed by hand. They were also typically cut across the grain of the iron fibers, which made them easier to snap than later ones which were cut along the grain. But the incentive to switch from hand-forging nails to cutting nails from sheet of iron was clear, as cut nails could be produced at a substantially lower cost than hand forged ones.

Cutting Nails at least Doubled Productivity

How much cheaper were cut nails than hand-wrought ones? In early 1796, Jefferson procured a machine that could cut four-penny nails and Monticello's website reports in that year were "workers producing a ton of nails per month." The 'Farm Book' reports 250 four-penny nails to the pound, so one ton of nails would have numbered about 500,000.

This suggests that the nail-cutting machine at least doubled the output of the nailery, if the number of workers stayed the same. If the number of workers in the nailery decreased from 1795 to 1796, then the productivity per person increased even more than two-fold. So, early cut nails were likely around half the price of hand-wrought nails, and maybe even less. Jefferson's 'Farm Book', in a section called 'Rates of estimating builder's work', lists 450 four-penny nails costing 3 shillings and 9 pence, or about $16 in 2012 money, or 3.5 cents per nail. So, cut nails were likely half of this cost, or less than 2 cents (in 2012 money) per nail.

Cutting Nails Required Less Skilled Labor

Let us be clear here: even though the hand-forging of nails was conducted largely by uneducated enslaved boys, it was skilled work. Jefferson's records estimate that a boy with six months' experience could forge five hundred nails per day, while a boy with one year of experience could make eight hundred. An adult laborer, he estimated, could make one thousand. Thus, it took time and experience to get good at making nails.

But later cut nails and steel-wire nails were made entirely by machines, of which one person could monitor several at a time doing a little more than feeding the machine and waiting to clear any jams. An advertisement from Wallaceburg, Ontario, from 1889 said that "one boy can fully attend to four machines" each of which "will produce 600 pounds" of nails per day, or more than one ton per worker per day. That is, we went from about six hundred nails per day per worker to about six hundred *pounds* of nails per day *per machine*, in less than a century. In 2022 Daniel Sichel, a professor at Wellesley College in Massachusetts, reported that adult workers currently attend to up to seven nail machines simultaneously, each of which can make about five hundred to two thousand nails *per minute.* So, the number of nails produced per worker per minute is now several thousand times larger than it was in the days when Wormeley, Burwell, Barnaby and Jamey made them.

As we have seen in other examples in our 'pattern of progress', skilled labor was replaced by unskilled labor. This affected not only *how* nails were made, but also shifted *where* the work was done, from a cottage industry to a factory enterprise. According to Loveday, "by the 1840s, commercial nail manufacturing was largely a factory enterprise confined to no more than 150 units [in the United States] that accounted for over 85 percent of the nation's nail output."

In the Winter 2022 issue of the *Journal of Economic Perspectives*, Sichel estimated that prices in Philadelphia averaged about 1.5 cents per nail (in 2012 money), for

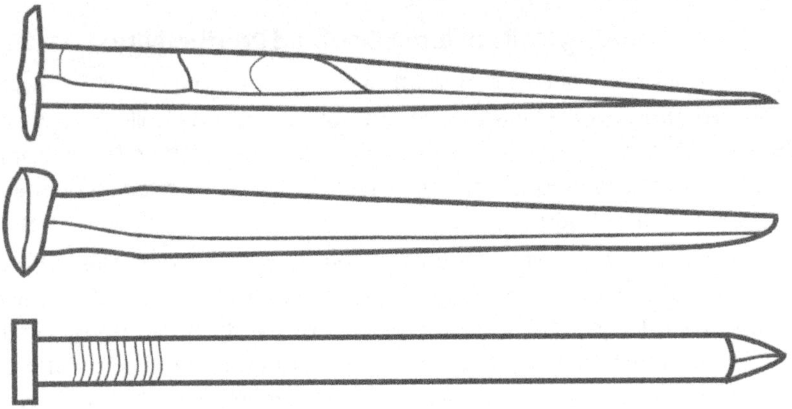

Figure 76 – A hand-forged iron nail (top) typical before the year 1800, a cut iron nail (middle) from the 1800s, and a steel wire nail (bottom) from about 1890 onward

assorted kinds of nails, from 1784 to 1813. And an article in the *Boston Masonic Mirror*, 'Invention of Cut Nails', from January 1832, a time when Sichel's calculations suggest cut nails cost about 1.1 cents each, says "we are now able to manufacture, at about one-third of the expense that wrought nails can be manufactured for; and nails which are superior to them for at least three-fourths of the purposes to which nails are applied, and for most of those purposes, they are full as good." All these values suggest that the earliest cut nails were about half the cost of hand-wrought nails and by the 1830s, cut nails were perhaps one-third the cost.

This not only changed the market for nails, but also our language about it. "By 1800," according to Loveday, "not only was the cut nail finding a market, but the term 'cut nail' entered the American language". As with many other innovations, our use of language changed faster than the market itself.

The Market Transformed over about Fifteen Years

According to archaeologist William Adams of Oregon State University, by "1810 roughly 70% of the nails made in the United States might have been made on a nail machine". In a period of only about fifteen years, from 1795 to 1810, machine-cut nails replaced most of hand-wrought nail manufacturing. Additionally, Adams estimates that shaping nail heads entirely by machine replaced the hand-shaping of nail heads in the period between 1825 and 1840.

This was not a one-for-one replacement of hand-wrought nails by cut nails. The overall production of nails dramatically increased. According to Loveday, in 1810, American nail manufacturers produced nearly sixteen million pounds of nails. Forty years later, they made over two hundred million pounds. The US population grew during that time, of course, but still per-capita production was nearly three-and-a- half times larger per person in 1850 compared to 1810.

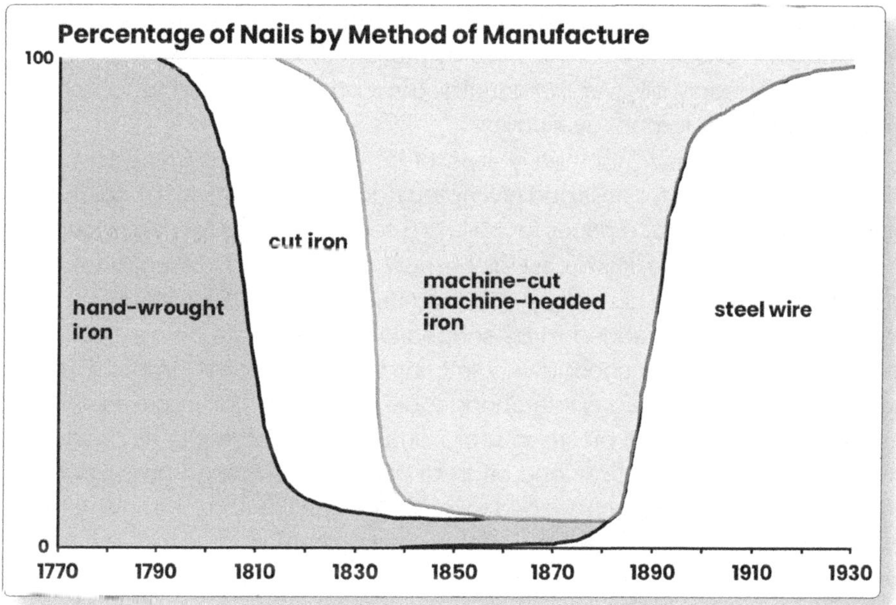

Figure 77 – Percentage of Nails by Method of Manufacture. *Adapted from (Adams, 2002)*

Cut Nails Revolutionized Construction, Replacing Beams with Dimensional Lumber

The rise of machine-cut nails had unexpected consequences for the construction industry. Mass production of nails in standard sizes prompted a revolution in building framing starting about 1830, with dimensional lumber consisting of boards of standardized sizes replacing the traditional post-and-beam construction that was designed to minimize nail usage.

Cut nails not only disrupted hand-wrought nails, they made us rethink the entire building process from one based on post-and-beam interlocking timbers held together by mortise-and-tenon joints to a new means of construction, the 'balloon frame', which relied on dimensional lumber pieces that are nailed together.

Loveday described an 1835 article in The *American Agriculturist* which claimed that the balloon frame was "just as good and far cheaper than ordinary frames" and another article from 1869 which said that the balloon frame could be constructed for "forty percent less money than the mortise and tenon frame". Brian Potter, Senior Fellow at the Institute for Progress think tank in Washington, D.C., and author of *Construction Physics*, which he describes as "a newsletter about the forces shaping the construction industry", wrote that "by omitting the heavy timber frame, the balloon frame used 1/3rd of the wood as a post and beam house, and could be erected much faster."

The balloon frame revolution was due not just to cheaper nails, but to a convergence of technologies that included steam-powered sawmills and steam-powered railroads. Steam-powered sawmills could produce far more cut lumber than the water-powered mills. Potter writes that "in the early 1800s, a water-powered mill could produce perhaps 1500 board feet of lumber per day. By the 1850s, steam powered mills were routinely producing nearly 30,000 feet per day, which increased to nearly 100,000 by the 1890s... Railroads allowed lumber to be shipped cheaply all over the country, allowing it to be used for construction regardless of local lumber availability."

The shift to this new nail-heavy style of construction, according to Loveday, "stimulated demand for nails and prompted additional growth in the nail industry", helping to further reduce prices for nails in a self-reinforcing feedback loop.

It is understandable from cost alone how the nail and dimensional lumber disrupted the mortise-and-tenon joint and thick timbers. The timbers for the old style of construction needed to be shaped when they were on the ground and then assembled in one specific way. In contrast, dimensional lumber pieces are standardized and therefore, interchangeable. The construction crew only needs to keep the dimensional pieces in a pile and take from the pile as needed. This meant significantly less time and effort to assemble building frames, but it also meant less skilled labor was needed to do so. "As significant as the savings in time was," Loveday wrote, "the replacement of the skilled mechanic with the unskilled workman in the construction industry was even more important."

The cut nail really changed every industry. Iron cut nails did not just cause us to rethink construction methods. As simple commodity pieces made from a uniform material with no moving parts, iron cut nails also gave rise to cost accounting, a standardized way to measure the efficiency of manufacturing.

The nail industry underwent another revolution in the late 1800s as the Bessemer process made steel much less expensive, so nails could be made from stiff steel wire, rather than iron. Adams says that there are "excellent production figures ... for the transition from machine-cut nails to [steel] wire nails in the United States" in the late 1800s.

Again, this disruption was driven by economics. Nails made from steel wire were about half the cost of cut nails. In 1886, less than ten percent of the nails produced in the United States (by weight) were made from steel wire. But by 1892, just six years later, more steel-wire nails were being produced than iron-cut nails, and by 1897, wire nails were 81 percent of all nails made in the US.

During this time, total nail production increased. US nail production in the year 1880 totalled five million 100-pound kegs, all of them cut nails, rising to 8.8 million kegs in 1886, 9.2 million in 1892, and 11.1 million in 1897. That is, the total market (by weight) for nails more than doubled in less than twenty years.

The effect of the rise of the steel wire nail on incumbent iron cut nail manufacturers such as those in Wheeling, West Virginia, once a major center of nail production in the US, was profound. Henry Scott, in his 1929 book *Iron and Steel in Wheeling* said that in 1886, "Wheeling's capacity was one-third of the whole nation's output." As with many other innovations, incumbent manufacturers took little interest in the new technology until it was too late. "The final chapter in the history of Wheeling's nail industry", Loveday writes, "is the story of a tenacious adherence to the cut nail that had put Wheeling on the map".

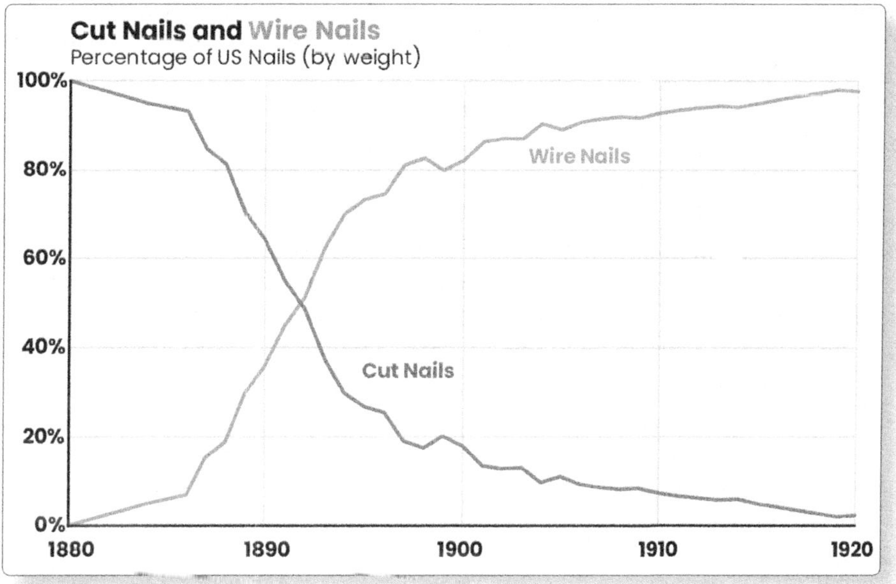

Figure 78 – Percentage of Nails (by weight) Produced in the US that are Cut Nails or Wire Nails. *(Adams, 2002) Table 3*

The steel wire nail did not come as a surprise to the iron cut nail manufacturers. In fact, according to Loveday, "all of the Wheeling manufacturers briefly looked at using steel instead of wrought iron as the basic material for nails". However, "there can be no question that, in the first few months after the wire nail threat became apparent, the Wheeling manufacturers displayed an aloof attitude... When the profits disappeared in 1887, 1888, and 1889, the manufacturers assumed that the problem was a matter of overproduction, as it had been in 1883 and 1884. Their initial response was to attempt to control production and thereby drive prices up." In the face of a new, rapidly growing alternative technology, the managers of the Wheeling factories did the worst thing possible: assuming the decline in demand for cut nails was a temporary result of a weak market, they reduced production of cut nails in order to maintain high and stable prices. As a result, steel wire nails would go from less than 10% of the market in 1886 to more than 80% just eleven years later. Loveday writes, "insofar as can be determined, no Wheeling manufacturer ever actually switched to wire nail production."

Henry Scott wrote, "the curious fact is that not one of these nail companies saw fit to adopt the logical substitute for their dying cut nail – the new wire nail.

They were established in the nail business, had been for years in close contact with every buyer of nails in the country, and enjoyed a nation-wide reputation for the unsurpassable quality of their product. How simple it would have been to take up the manufacture of the wire nail and continue unbroken their business relationships the nail trade! But the Wheeling nail makers fought a dogged losing fight; their profits shrank and finally disappeared, but they kept on making cut nails till there was but a trifling demand."

"With every attempt to limit production, the wire nail simply gained new customers," Loveday wrote. "That the cut nail manufacturers did not understand this is almost beyond comprehension in the light of hindsight; at the time, however, such a misreading of the situation was not only plausible but quite possible." Due to the lack of accurate sales information, according to Loveday, the idea that overproduction was "solely responsible for the price decline made an accurate assessment of events almost impossible."

Nails were made in essentially the same way for at least a millennium. "The same basic tools" that were found in Roman archaeological sites from the third century AD, according to Loveday, "could be found thirteen hundred years later in colonial America." However, "the idea of cutting nails from sheets of metal rather than bars jolted tradition enough to encourage a complete rethinking of the manufacturing process. Within twenty years after the cutter was first used, several mechanics and inventors could construct a totally automated production system – a remarkable change, particularly for an industry that had remained essentially unchanged for centuries."

We made nails mostly from the same material, iron, and in basically the same way for five thousand years. And then, cutting the nail with a machine and shaping the head by machine each revolutionized nailmaking, both over roughly fifteen-year periods just a couple of decades apart in the early 1800s. The modern wire steel nail revolutionized the nail market a third time, again over about fifteen years, in the late 1800s due to the falling cost of making steel. "Wrought nails continued to be made until late in the 19th century," Adams says, "in small quantities and particularly in remote areas. An insignificant quantity continues to be made up to the present time."

Aluminum

> *It was only over a couple of decades in the late 1800s that aluminum went from being more expensive than silver to almost disposable. In doing so, it turned soft drinks from something to buy at a soda counter to a home beverage, and enabled powered flight*

Aluminum production saw at least two large step decreases in cost in the 1800s: one in the 1850s when the first commercial production began and another in the late 1880s. Before the 1850s, aluminum was more expensive than gold. From the 1850s until the late 1880s, it was the price of silver. Now we use it to make Coca-Cola cans, electric car bodies, electrical grid components, and many other things. Aluminum became the first new metal to attain widespread use since the prehistoric discovery of iron. Today, more aluminum is produced than all other non-ferrous metals combined.

Figure 79 – Tip of the Washington Monument, Washington, D.C.

The image above shows the tip of the Washington Monument. It is a pyramid of aluminum about eight inches high and weighing about six-and-a-half pounds. At the time it was cast in 1884, this was the largest single piece of aluminum ever made. It was so valuable and unique that it was put on public display at Tiffany's jewellery store in New York City before being transported to the US capital for installation on the Monument.

New Science Decreased the Price of Aluminum

The element aluminum is abundant in the Earth's crust but locked up in mineral compounds. The existence of aluminum as an element was discovered by Danish scientist Hans Christian Ørsted in 1825 and the first commercial-scale production process was developed by French chemist Henri Étienne Sainte-Claire Deville in 1856, causing the price to drop by an order of magnitude, from $500 to $40 per pound, a more than ten-fold decrease over just a couple of years.

At the time the tip of the Monument was cast, aluminum cost about $1 per ounce, roughly the same as silver. By coincidence, a workman working on the Monument was paid $1 per day, so the aluminum in this piece cost about 100 days' income, or four to five months.

Just two years after the Washington Monument tip was cast, other ways of purifying aluminum were discovered, bringing the price down even further. In 1886, after five years of tinkering, the 23-year-old Charles Martin Hall developed an inexpensive electrocatalytic method for producing aluminum. At about the same time in Normandy, France, Paul Héroult, also twenty-three years old, developed a very similar process, which he had been working on since he was 15. (Interestingly, both men were born in the same year, 1863, and died in the same year, 1914, each just a few weeks after their 51st birthdays.) The Hall–Héroult process, which reduced the price of aluminum by a factor of about two hundred, is still used to produce aluminum to this day. Why was this process for making aluminum developed in 1886 and not 1876 or 1896? Because it relied on the use of large amounts of electricity and the combustion-engine-driven technologies for making electric power in large quantities were being developed at that time.

In 1855, when aluminum was more expensive than gold, the Monument's tip would have cost more than $3000, about what a person earned in twenty years. Today, six pounds of scrap aluminum sells for about $3, less than thirty minutes income at minimum wage.

Aluminum's Popularity Has Soared, Creating and Transforming Entire Industries

As the price of aluminum has fallen its popularity has grown exponentially. From essentially nothing in the year 1900 the industry reached twenty-five million tons per year by the year 2000 and double that fifteen years later.

Aluminum has played instrumental roles in numerous industries, including aerospace, as it is lighter weight than steel for the same strength. In 1888, Hall opened the first large-scale aluminum production plant in Pittsburgh. When the Wright brothers' needed a light-weight engine crank case for their airplane, they used aluminum from the Pittsburgh Reduction Company, the world's leading producer, later renamed Alcoa.

Social Perceptions

> *Technology can change quickly. Humanity went from using trebuchets to cannons in a couple of decades, from horses to cars in thirteen years, and from slide rules to pocket calculators in just a few years.*

But social perceptions can change quickly too: In 2004 the only US state to legally recognize marriage between homosexuals was Massachusetts. By 2015, everywhere in the US did.

New products can take over a market in just a few years, and so can new ideas. Smoking indoors went from normal to strange in a decade. Wearing fur went from opulent to shameful in a decade or two as well. These do not seem to be just passing fads, like clothing or music styles that drift out of fashion and then come back into style again – it is difficult to think how indoor smoking in the presence of people who do not wish to be exposed to second-hand smoke will ever make a comeback, even in a world with electronic vaporizers.

Those Lazy Germans

Perceptions of national characteristics can change quickly too. A 2016 report by the United Nations Economic Commission for Africa called 'Transformative Industrial Policy for Africa' describes how "until the mid-19th century, when German economic development started, British visitors to German states would often describe the Germans as being mentally slow, laid back, dishonest, and excessively emotional." The report gives the example of John Russell, "an early-19th century British traveller in Germany", who remarked: "The Germans are a 'plodding, easily contented people ... endowed neither with great acuteness of perception nor quickness of feeling ... It is long before [a German] can be brought to comprehend the bearings of what is new to him, and it is difficult to rouse him to ardour in its pursuit'." When travelling in Germany, Mary Shelley, the author of Frankenstein, complained that "the Germans never hurry."

But here's the funny thing: the Roman historian Tacitus in his text *Germania* also described Germans as 'idle' and 'indolent', remarked on their 'unpunctuality', and their tendency to simply take 'what can be got quickly by the loss of a little blood' rather than work to 'accumulate slowly by the sweat of his brow'. And that was in the year 97 AD.

So, for at least one thousand seven hundred years, other Europeans described Germans as 'slow', 'laid back' and 'lazy'. But over just a generation or so in the late 1800s, Germans suddenly became regarded as 'skilled', 'rational', and 'efficient'. Did Industrialization suddenly reshape all German culture, or just foreigners' perception of the Germans? Either way, according to the UN report, it happened to the Japanese and the Koreans too. Before Industrialization, they were often called negative terms. After, they were suddenly described in positive terms.

Abstract Concepts Spread in about Fifteen Years

Perceptions of abstract concepts can change just as quickly. Searches using Google's Ngram Viewer tool, examining books published in English, shows that "high-quality Japanese products", a phrase that rarely appeared in American books prior to the 1960s, suddenly became very popular starting in the late 1970s through the early 1980s and has slowly declined in popularity since then. (Interestingly, the phrase "high-quality Chinese products" has been used so seldomly in American books that it is not even possible to make a graph of its frequency. Perhaps this will change quickly in the coming decades.)

Likewise, phrases like "Women's Liberation", "the Pill", "smoking ban", and "gay marriage" all saw a similar ramps up in popularity over periods of about a decade, with the last example being interesting because it most likely was kicked off by a single 1989 article in *The New Republic* magazine called 'Here Comes the Groom: A (Conservative) Case for Gay Marriage', by author Andrew Sullivan.

More recently, since the turn of the 21st Century, we have seen sudden increases in the popularity of phrases like "recreational marijuana" (about fifteen years after the rise of "medical marijuana") and "the right to be forgotten", a new legal concept which maintains that since personal details are now easier than ever to find through internet searches, individuals have a right to have old, embarrassing information about themselves expunged from public view after a while.

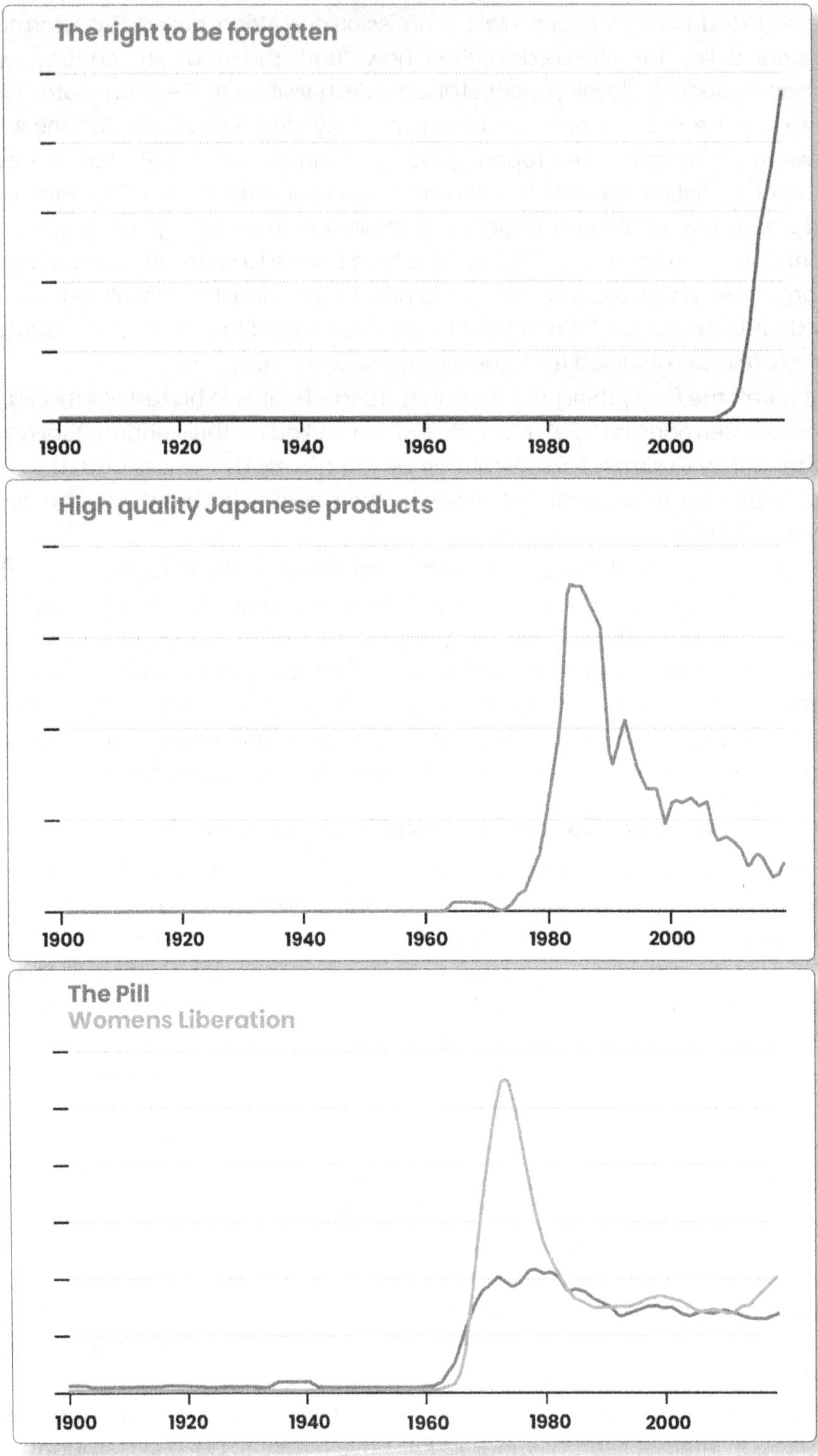

Figures 80a to f – Frequency of some phrases in American English-language books, according to Google Books.

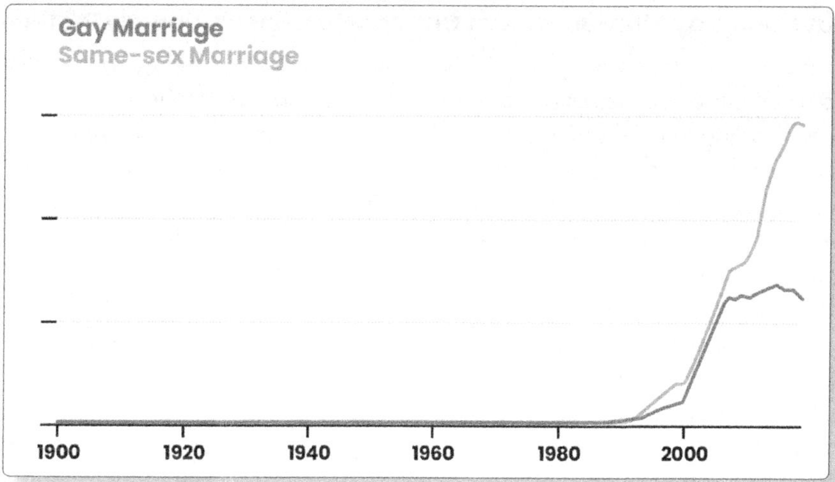

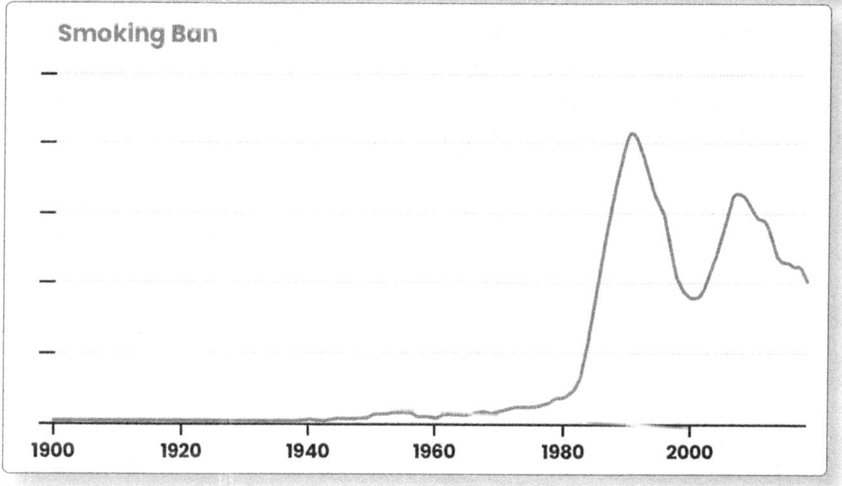

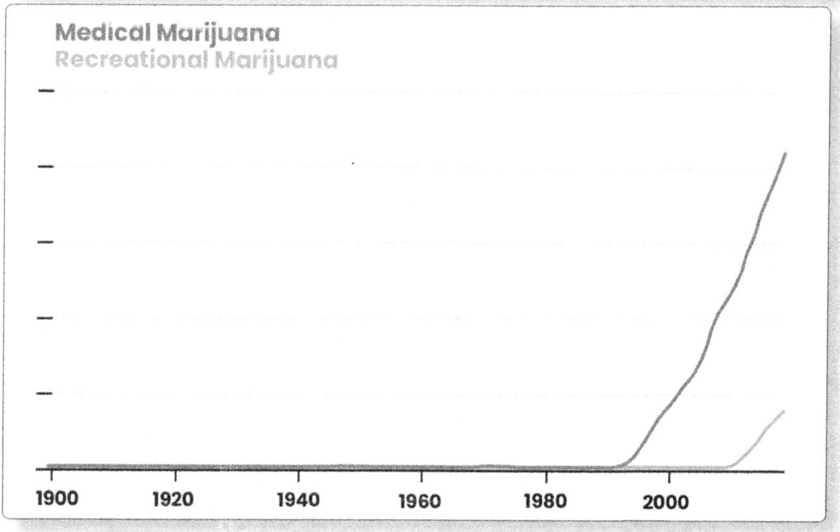

(Graphs are on different scales, so lines can only be compared on the same graph, not between different graphs.)

Gays and Gay Marriage: From Criminal to Constitutionally Protected

It is not just usage of phrases in books that can change quickly, but real-world impacts can, too. The use of phrases like "gay marriage" and "same-sex marriage" took off starting in the early 1990s and then, about a decade later, real legislative action began on recognizing marriages between homosexuals (not only in individual US States, but also in various European countries). In the US, it took roughly twelve years of individual States successively recognizing homosexual marriage before a Supreme Court ruling made it universal throughout the US.

The gay rights movement in the US has progressed by a series of S-shaped curves, each taking about thirteen years. Homosexuality was decriminalized in about half of US states (by population) primarily in the 1970s. After a decade-long pause largely corresponding to the Reagan/Bush era, a second wave of decriminalization started in the early 1990s, ending with the US Supreme Court ruling in *Lawrence v. Texas* in 2003 that saw homosexuality decriminalized throughout the country.

A period of rapid recognition of same-sex marriage began the following year, first in Massachusetts, although various places allowed 'domestic partnerships', 'civil unions', and other marriage-like legal arrangements before then. Same-sex marriage was recognized everywhere in the US with the Supreme Court ruling of *Obergefell v. Hodges* in 2015.

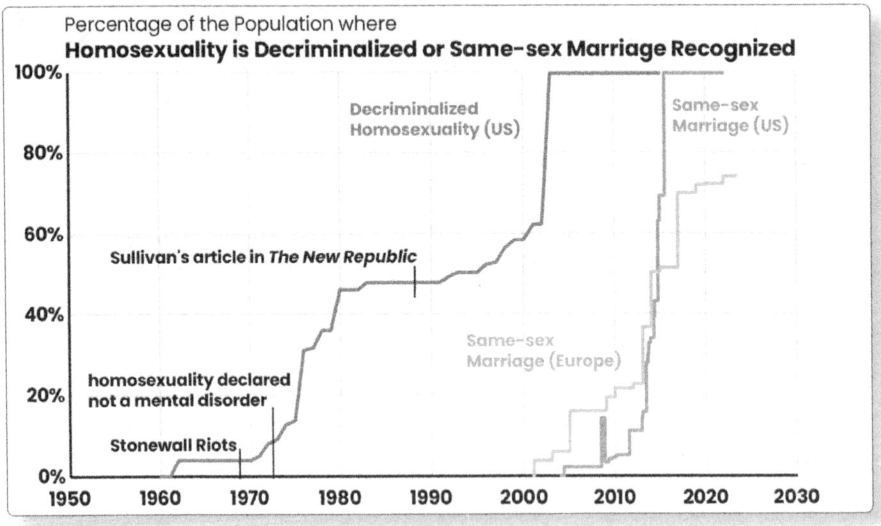

Figure 81 – Percentage of the Population where Homosexuality is Decriminalized and where Same-sex Marriage Recognized. *Data from Wikipedia*

In recent years we have been seeing individual States legalize recreational marijuana possession and use, implying that it too might become universal in the US in the coming years.

Just like we saw with words like "automobile" and "television", where adoption of the terminology and the increase in the frequency of the use of these words preceded the market adoption of the actual devices themselves, the adoption of phrases like "gay marriage" and "recreational marijuana" has preceded the legal adoption of those concepts.

Chapter 4:

The Modern World

◆ ◆ ◆

"You can't learn in school what the world is going to do next year."

Henry Ford

◆ ◆ ◆

Electric Motors Helped End Child Labor

> *Electric power caused a revolution in lighting and then in mechanical work. As steam-powered mechanical drive was replaced by electric motors, the source of the power to produce the electricity was disrupted, spawning a new business model, the electric utility company. Soon, electric power might undergo another disruption, as solar panels and batteries enable some manufacturers to bring power generation in-house once again*

As with William Murdoch's coal gas, the earliest application of electricity was lighting, both in public use (streetlights) and industrial settings (factories). In the late 1870s, both Thomas Edison in the US and Joseph Swan in England developed incandescent light bulbs using a mercury-based vacuum pump invented by German-born chemist Hermann Sprengel in 1865, about thirteen years prior. By 1881, Edison had built a direct current (DC) generator to power lighting for some stores in Niagara Falls, New York, while in the same year the world's first electric street lighting was turned on in the town of Godalming, England, southwest of London. The following year, 1882, the Edison Illuminating Company began serving eighty-five customers in New York City.

Electricity was probably first used for driving machinery in 1883, but what really kicked off the electrification of manufacturing was Nikola Tesla's 1888 invention of the alternating current (AC) induction motor. Prior to this point, factories relied on mechanical drive trains: a rotating shaft running down the center of the building turned pulleys and belts which transferred power to workstations. But over just a few decades in the late 1800s and early 1900s, a series of revolutions changed how the central shaft was turned, replaced the central shaft with many small motors situated at each workstation, and then replaced on-site power generation with centralized electric utilities. Now solar panels and battery-based energy storage are poised to disrupt utilities, returning us once again to local power production.

Some people saw the electric revolution coming. Werner Siemens, the founder of the German industrial giant that bears his name, said in 1878 that "the electric motor will in course of time produce a complete revolution in our conditions of work". (Klemm, 1959) According to Gerald Berk in his 2009 book *Louis D. Brandeis and the Making of Regulated Competition, 1900-1932,* "newspaper printers were the first U.S. manufacturers to electrify, making the shift in the 1890s" since electrification "improved plant layout, reduced space requirements, reduced power costs, increased control over individual machines, and made it possible to install better lighting".

In an article published in December 1967 called 'The Introduction of Electric Power in American Manufacturing', Dr. Richard B. DuBoff, a professor emeritus of economic history at Bryn Mawr College in Philadelphia, wrote that "dismantling steam engines and purchasing electric power a firm might cut power costs by 70 to 83 per cent", when considering all costs related to fuel, labor, and capital. So, in cost-saving terms, the switch from direct mechanical power to electric power was therefore as big of a no-brainer as the switch to coal gas had been about a century prior.

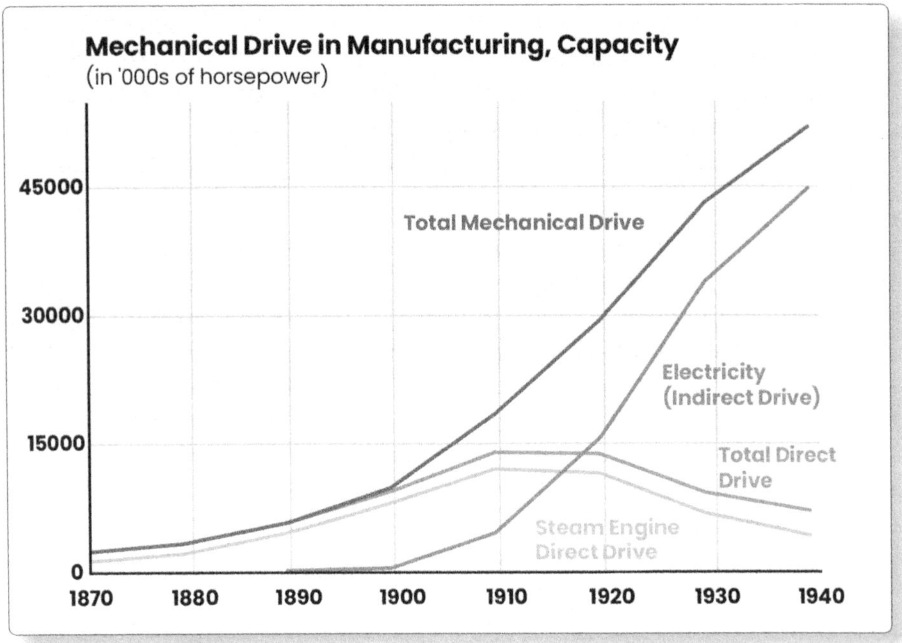

Figure 82 – Mechanical Drive in Manufacturing. *(Devine, 1982)*

As a result, the adoption of electric power in the manufacturing industry was explosive. In 1899, there was about 475,000 horsepower of electric-driven machinery in the United States, just under five percent of the ten million horsepower of total mechanical drive at the time. More than eighty percent of all mechanical drive in the US was steam powered. By 1919, just two decades later, electric-driven machinery outpowered all direct-drive machinery combined. Ten years after that, eighty percent of the mechanical drive in manufacturing was electric-powered.

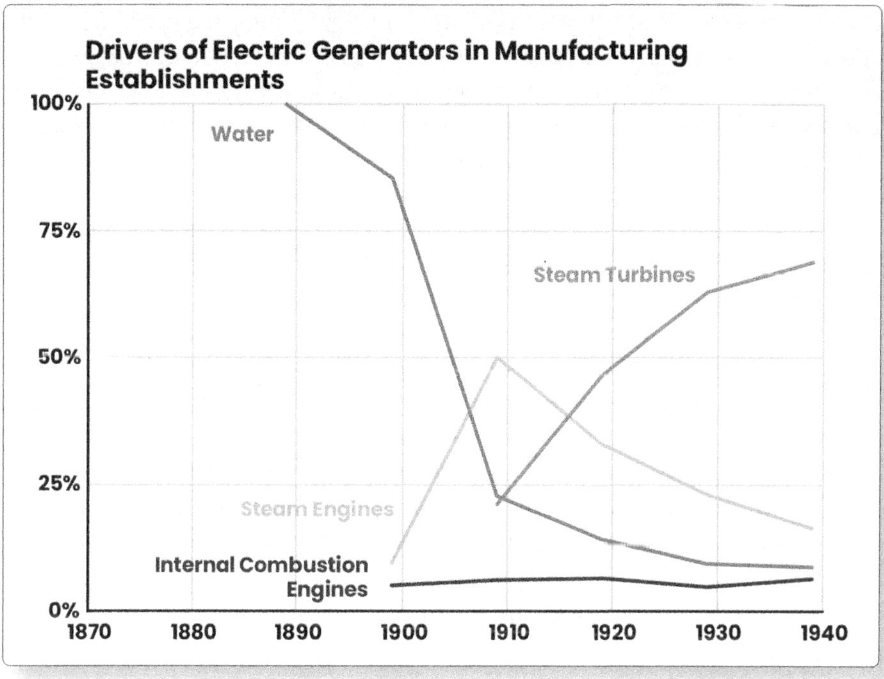

Figure 83 – Drivers of Electric Generators in Manufacturing Establishments. *(Devine, 1982)*

This was not a one-for-one replacement. The amount of horsepower capacity in 1929 was more than four times the amount in 1899. As DuBoff wrote, "the rise of electric power is … the story of the demise of the steam engine."

As manufacturers switched from direct mechanical drive (delivered primarily

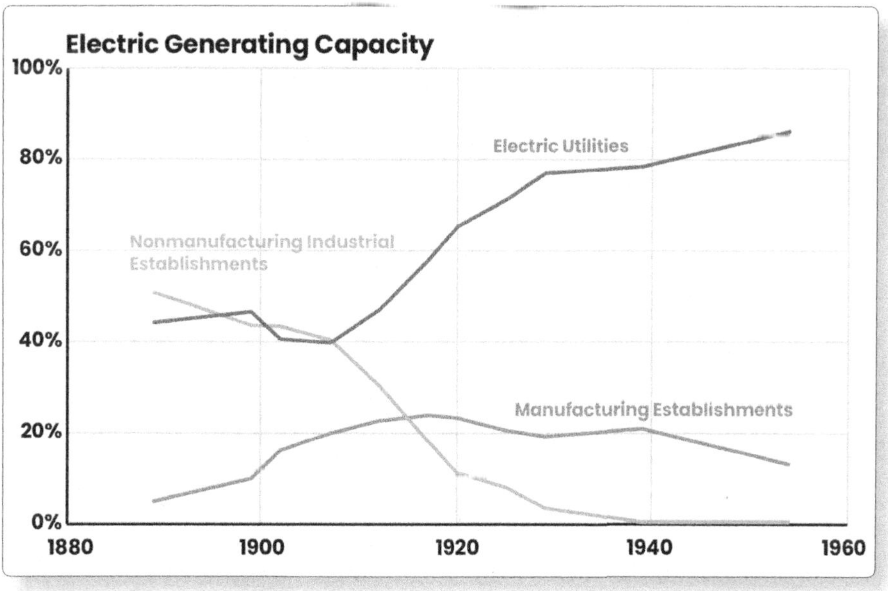

Figure 84 – Electric Generating Capacity owned by Non-manufacturing Industrial Establishments, Manufacturing Establishments, and Electric Utilities. *(Devine, 1982)*

by steam engines) to electricity, the source of power for the electric generators underwent disruptions as well. At the end of the 1800s, most electricity was generated by waterpower. By 1909, half of electricity in manufacturing establishments came from steam engines. And over the course of the following decade, electricity generation became dominated by steam turbines.

The switch from waterpower to steam engines and then to steam turbines brought with it a business model disruption over the course of the decades of the 1910s and 20s, as non-manufacturing industrial establishments ('railroads, mines, farms, and services') largely stopped producing their own power and simply purchased "power as a service" from utility electricity providers.

Textile mills were also among the early adopters of electric power. In 1900, the Bibb Manufacturing Company bought a hydropower dam site on the Chattahoochee River near Columbus, Georgia, on the border with Alabama. At eight thousand horsepower, this facility alone contained more than 1.5% of the total industrial electric power production capacity in the United States at the time. A factory complex was built there that eventually would become the largest cotton mill in the country.

The Machines that Beat Children

Photographer Lewis Wickes Hine spent sixteen years between 1908 and 1924 working for the National Child Labor Committee (NCLC), documenting working and living conditions of children in the United States, often under threat of violence by adult supervisors. His work later proved instrumental to the development of the country's first child labor laws. One of the first places he visited was 'Bibb Mill No. 1' in Macon, near the geographic center of Georgia, where he took a photo of these

Figure 85 – 'Some adolescents in Bibb Mill No. 1, Macon, GA. Jan. 19, 1909'.
Photo by Lewis Hine for the National Child Labor Committee

girls. The last remaining building in this complex was gutted by fire in March 2022 and later demolished.

Perhaps more than any other factor, electrification disrupted child labor in American factories. According to the US Census survey of 1900, in that year there were 1.75 million child workers in the US. More than one-in-five children ages 10-14 were gainfully employed, constituting about 6% of the US workforce. By 1930, the

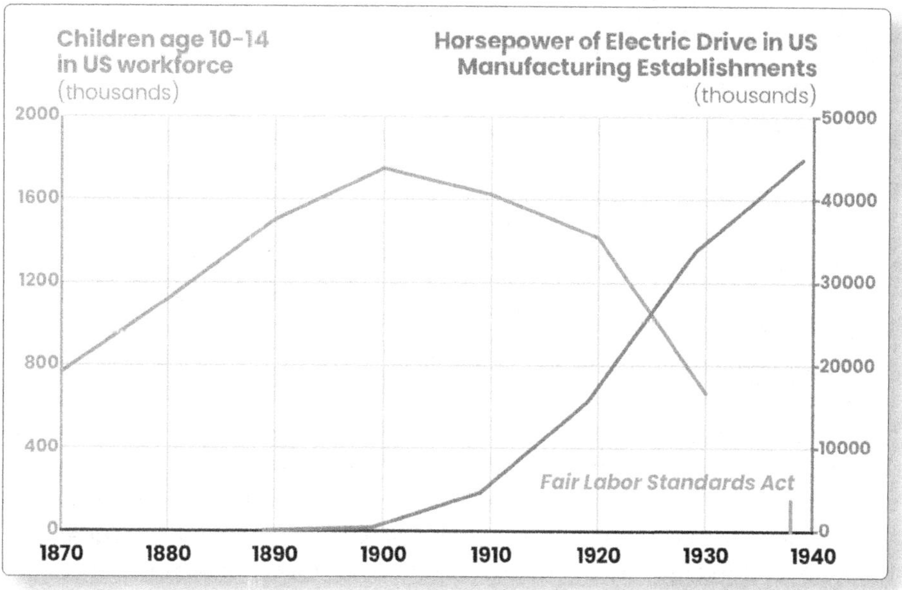

Figure 86 – Children aged 10-14 in US Workforce, compared to the amount of Horsepower of Electric Drive in US Manufacturing Establishments. *(US Census, 1975) and (Devine, 1983)*

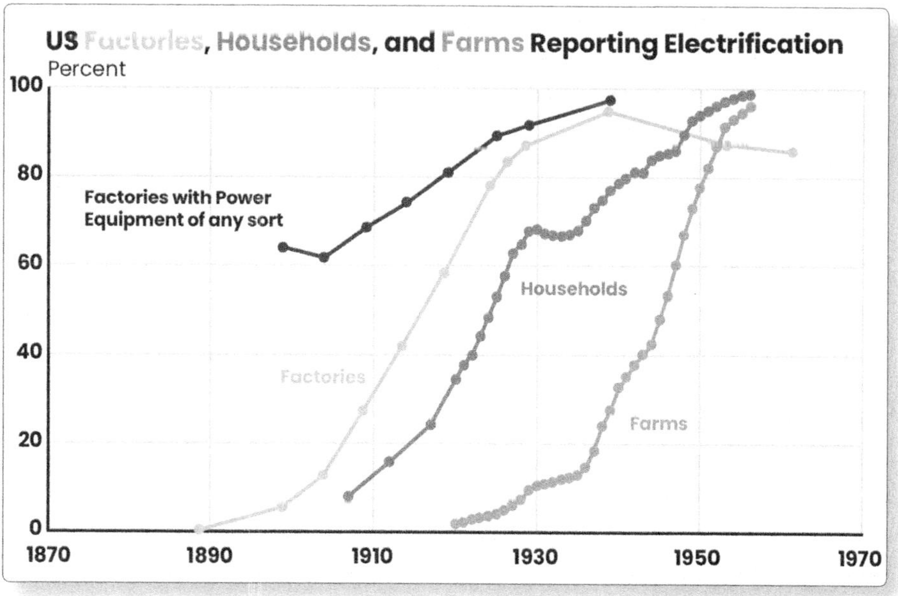

Figure 87 – Percentage of Factories, Households, and Farms Reporting Electrification in the US. Factory and Household data *(Ayers, 1989), farm data (Kline, 2000) and total factory data (Devine, 1983)*

total US workforce was 70% larger, but the number of child workers had dropped to 667,000. If the child work force had stayed at about 6% of the US total, then there should have been about 2.9 million employed children. Instead, in 1930, about 2.2 million children were "missing" from US factories and there were thirty-four million horsepower of electric motors installed in their place.

In time, factories, households, and farms throughout the US became electrified, with each category adopting electric power over the course of a couple decades each.

Now we are on the cusp of another disruption in powering manufacturing facilities: solar panels, wind turbines, and batteries working together to disrupt grid-supplied electric power.

Food for Babies and Animals

Electrification might have helped children in other ways besides eliminating the use of their labor. The primary costs of producing canned foods include preparation by cutting, chopping, and grinding it, and the preservation through heating. Electric motors and electric heating dramatically reduced the difficulty and costs of all of these.

Tin cans are made largely from steel (with a thin coating of tin). By 1910, steel constituted 80% all iron and steel products in the US. By 1930, it was nearly 90%. Glass jars require enormous amounts of energy to produce and sanitize as well.

The convergence of electrification and widely available steel resulted in a revolution in food preservation. Harold and Anna Clapp created the first canned baby food, Clapp's Vegetable Soup, in 1922.

According to Amy Bentley in her book *Inventing Baby Food,* "canned goods, especially canned produce, though commonly available in the late 1800s, were too expensive for most. By the 1920s, however, manufacturers produced canned goods in sufficient quantity to allow Americans to consume fruits and vegetables year-round."

Also in 1922, the Chappel brothers of Rockford, Illinois, introduced Ken-L Ration, the first canned dog food in the United States. The rapidly rising popularity of the automobile caused the price of horses to collapse about 60% in the ten years from 1910, the year that the US horse population peaked according to the US Department of Agriculture, to 1920. Ken-L Ration's primary ingredients were horse meat, ground horse bones, and oats (horses' food). Oats production in the US peaked just after World War II and fell dramatically as production of corn and soybeans rose to feed chickens, pigs, and cattle.

Bentley says it is interesting "how quickly and dramatically infant food and feeding practices changed in the United States with the rise of industrialization. In the space of just over a half century, from the late nineteenth century to the mid-twentieth, infants in the United States went from consuming nearly exclusively breast milk, whether provided by mother or by wet nurse, until late in the infant's first year, to bottle feeding and beginning solids at six weeks postpartum."

◆ ◆ ◆

"We have been so desensitized by a hundred and fifty years of ceaselessly expanding technical prowess that we think nothing less complex and showy than a computer or a jet bomber deserves to be called "technology" at all. As if linen were the same thing as flax - as if paper, ink, wheels, knives, clocks, chairs, aspirin pills, were natural objects, born with us like our teeth and fingers - as if steel saucepans with copper bottoms and fleece vests spun from recycled glass grew on trees, and we just picked them when they were ripe...."

— Ursla Le Guin

◆ ◆ ◆

More Bang for the Buck

> *Nuclear bombs allowed a single airplane to deliver the destructive force that previously took one hundred. In just fifteen years, the US built its entire nuclear arsenal. The Soviets were not far behind. These weapons forced a re-think of the structure of the major militaries of the world. But they also required the creation of an entirely new industry unlike any the world had seen before*

When the US National Security Council met in January 1952, the geopolitical world map was being redrawn nearly every day. The Korean War was in full force and the Cold War with the Soviet Union was growing ever deeper. The Soviet's blockade of Berlin in 1948-1949 had split the city in half. Pakistan became an independent country in 1947, China a Communist People's Republic in 1949, and India an independent republic in 1950. In August 1949, the Soviets detonated their first nuclear device, a plutonium-based bomb equivalent to 22,000 tons of TNT, nearly identical in size and design to the U.S.'s 'Fat Man' bomb dropped on Nagasaki. However, the US would not test the first, far-more-powerful hydrogen bomb, which delivered the equivalent of more than ten million tons of TNT (about five hundred times the energy delivered by the 'Little Boy' bomb on Hiroshima), until the end of 1952.

With President Harry Truman in attendance, Secretary of Defense Robert A. Lovett, according to the book *Atomic Audit: The Costs and Consequences of U.S. Nuclear Weapons Since 1940*, "pointed out how expensive high explosive was compared with fissionable material 'from the standpoint of energy release for dollar expended... something on the order of twenty to one in favor of nuclear detonations.'" (Schwartz, 1998) The first self-sustaining artificial nuclear reaction was demonstrated at the University of Chicago by physicist Enrico Fermi in December 1942. Ten years later, it was being debated if the American military would be structured to accommodate nuclear weapons as the central pillar.

Getting 'more bang for the buck' became a prime argument for nuclear weapons under Eisenhower, who was elected president in late 1952 and who began serving early the next year. Truman and Eisenhower knew probably better than anyone the potential for how disruptive nuclear technology could be.

On the night of March 9 and early hours of March 10, 1945, according to Yuki Tanaka's 2009 book *Bombing Civilians: A Twentieth-Century History*, about three hundred B-29 bombers conducted Operation Meetinghouse, a devastating firebomb attack on eastern Tokyo. In one essay in Tanaka's book, "A Forgotten Holocaust: U.S. Bombing Strategy, the Destruction of Japanese Cities, and the American Way of War from the Pacific War to Iraq", Mark Selden wrote: "probably more persons lost their lives by fire at Tokyo in a 6-hour period than at any time in the history of man ... No previous or subsequent conventional bombing raid came close to generating the toll in death and destruction of the great Tokyo raid of March 9-10".

It is difficult to compare the destructive power of the bombs dropped during Operation Meetinghouse to other kinds of bombs, since firebombs are incendiary devices designed to start fires, not explosives which create shock waves. However, in terms of deaths and property destruction, the Tokyo bombings of the spring of 1945 were comparable to the Hiroshima attack a few months later, leaving up to 100,000 dead and one million homeless, with large areas obliterated.

However, the number of bombs and aircraft involved in the atomic attacks on Hiroshima and Nagasaki was considerably less than the number involved in the firebombing of Tokyo.

One Plane Could do What Previously Took an Armada

The Hiroshima mission carried a single bomb, Little Boy, containing one hundred forty pounds of highly enriched uranium in a 10,000-pound (5 ton) package. It was carried by the Enola Gay accompanied by two other B-29s, for a total of three aircraft. The bomb dropped on Nagasaki, Fat Man, contained fourteen pounds of plutonium in a 10,000-pound (5 ton) device, delivering the equivalent blast of 22,000 tons of TNT. That mission used five B-29s. That is, in terms of aircraft used and bomb mass dropped, the atomic bombings were about a factor of 100X smaller than the Tokyo mission. But they delivered more destructive power.

If the Operation Meetinghouse planes had instead carried conventional high explosives, then the roughly 1600 tons they dropped on Tokyo could have delivered the equivalent of up to 2200 tons of TNT. This is about 10% as powerful as the bomb dropped on Nagasaki. That is, even if all the bombs dropped in the March attack on Tokyo had instead contained high explosives, the Nagasaki atomic blast alone still would have been about ten times more powerful than all the Tokyo bombs combined.

The bombs were so powerful that new terms needed to be invented to describe how powerful they were. B-29 bombers could only carry about five tons of payload each, but by the early 1950s, nuclear bombs had the destructive power of tens of thousands of tons of TNT, hundreds of thousands of tons of TNT, or even more. Engineers devised the term 'megaton' (literally 'a million tons') to describe how much power nuclear warheads contained, and chillingly, the term 'megadeath' (literally 'a million deaths') to describe how many people each one might kill.

In a September 1951 speech, Senator Brien McMahon of Connecticut argued that nuclear weapons were not just better in terms of explosive power delivered, he believed they were also more cost-effective. "Dollar for dollar," he said, "atomic deterring power is actually hundreds of times cheaper than TNT... Money spent upon the atomic bomb could pulverize a dozen enemy war plants at no more expense than destroying a single plant with TNT, to say nothing of the fact that one plane can deliver one A-bomb as against the huge armadas needed to deliver an equivalent cargo of block-busters".

Even Military Programs Can Follow an S-shaped Curve

Throughout the 1950s and into the 1960s the US military conducted a program to greatly increase the size of the US nuclear arsenal. Though in 1952 the US had fewer than one thousand warheads, by 1967, fifteen years later, its stockpile was 31,225, the highest it would ever be. More than 11,000 of the net change in warheads came in just two years, 1959 and 1960. As late as 1955, the number of warheads in US possession was less than 10% of its peak value but by 1962 it had reached 80% of the peak. That is, most of the change in the number of warheads happened in just seven years. This is surprisingly like what we have seen with the adoption of many other technologies in the consumer and commercial marketplaces, even though nuclear weapons are different from consumer products in basically every way imaginable.

The power of individual weapons increased dramatically during this time as well. While the bomb dropped on Nagasaki in 1945 was the equivalent of 22,000 tons of TNT, one of the ones tested at Bikini atoll at the end of 1952 yielded the equivalent of more than ten million tons of TNT, nearly five hundred times more powerful.

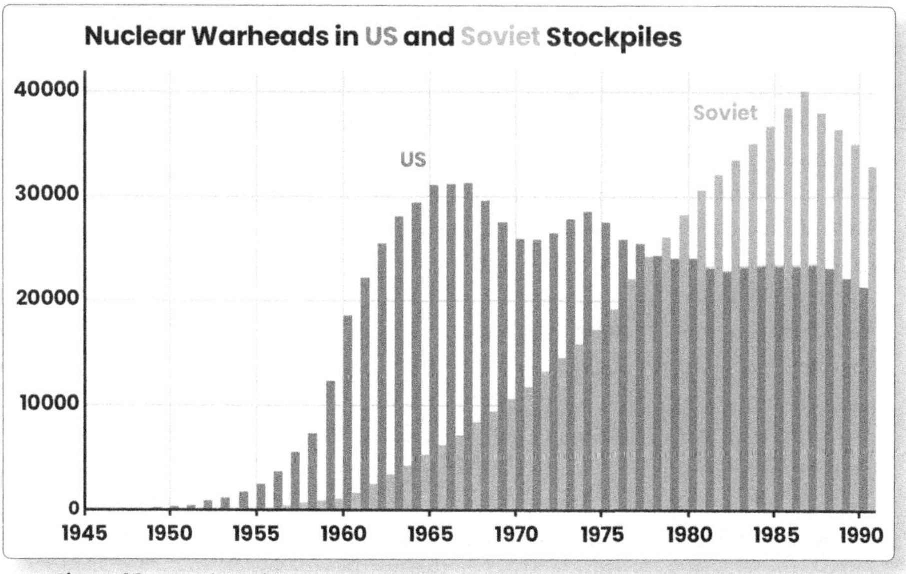

Figure 88 – Number of Nuclear Warheads in US and Soviet Stockpiles. *(Kristensen, 2018)*

Less than the Cost of a Single Tank

In his September 1951 speech Senator McMahon also said: "If we mass-produce this weapon, as we can, I solemnly say to the Senate that the cost of a single atomic bomb will become less than the cost of a single tank." The Washington-D.C.-based Nuclear Threat Initiative estimates that from 1940 to 1996, "the total incurred costs of the U.S. nuclear weapons program exceed $5.8 trillion. Of the $5.8 trillion, just seven percent ($409 billion) was spent on developing, testing, and building the actual bombs and warheads." (Schwartz, 2008) In its history, the US nuclear weapons program is estimated to have produced 70,000 warheads. That means an average of about $5.8 million per warhead, which in fact is less than the acquisition cost of a single M1-Abrams tank.

In fact, the marginal cost of warheads is likely to have been significantly lower than that average cost of $5.8 million. If nuclear warheads follow Wright's Law, as many other technologies do, then the cost of building one additional warhead is probably less than $3 million and reached less than $5.8 million (in constant dollars) somewhere around 1960, less than ten years after McMahon's speech. Examples like the cost of nuclear warheads and the cost of the F-35 fighter jets (as seen in Chapter 1) show that Wright's Law is not confined to free-market commercial products, but can also be seen in military hardware development as well.

In his speech, Senator McMahon also recognized the civilian use of this technology to provide electric power. "Even if this [nuclear] material is not exploded," he said, "it can be useful for our peacetime economy." In 1958, sixteen years after the inauguration of Fermi's pile, the first civilian reactor for generating commercial electric power in the US went into operation in Shippingport, Pennsylvania.

The French Revolution in Electricity

Some countries took civilian nuclear power to heart. France, for example, reconstructed its electric grid starting in the late 1970s, going from less than 10% nuclear-powered as late as 1977 to more than 75% fifteen years later.

The French, of course, were pioneers of early nuclear science. Henri Becquerel discovered radioactivity in the late 1800s after leaving pieces of radioactive ore on a photographic plate, winning a Nobel Prize for his work. Marie Curie, who spent most of her life in Paris, won two Nobel Prizes (the first person to do so) and was also the wife of a Nobel Prize winner, the mother of a Nobel Prize winner, and the mother-in-law to another Nobel Prize winner, with all their work related to radioactivity.

But other countries took the lead in turning France's scientific discoveries about radiation into weapons and civilian electric power. France did not detonate its first nuclear device until 1960, fifteen years after the US, and their first nuclear power plant opened in 1962, about five years after the US and UK had each started their own.

By the end of 1977, fifteen years after France's first nuclear power plant started operation, nuclear power still accounted for less than 10% of France's electricity, behind West Germany, the US, the UK, Switzerland, Sweden, Belgium, and even

Bulgaria. But just six years later, by 1983, France was number one in the world in the percentage of electricity derived from nuclear power. Since 1986 it has generated at least 70% of its power each year from its nuclear plants. The bulk of France's transformation from laggard to leader took only about a decade.

According to an assessment conducted at the International Institute for Applied Systems Analysis (IIASA), France's civilian nuclear power build-out "is legitimately considered as the most successful scaling-up of a complex and capital-intensive energy technology system in the recent history of industrialized countries." (Grubler, 2010)

Using data made public in the year 2000, IIASA found that France's nuclear program perversely exhibited "negative learning" – that is, the more nuclear capacity that was built throughout the 1970s, 80s, and 90s, the more expensive building additional reactors became, an inversion of Wright's Law. As each new nuclear project was undertaken, IIASA argues, regulators required more safety features, more locally made components, and new generations of reactors that effectively reset the learning curve to a new, higher starting point.

And this was true not just for France. The study found that "all countries with

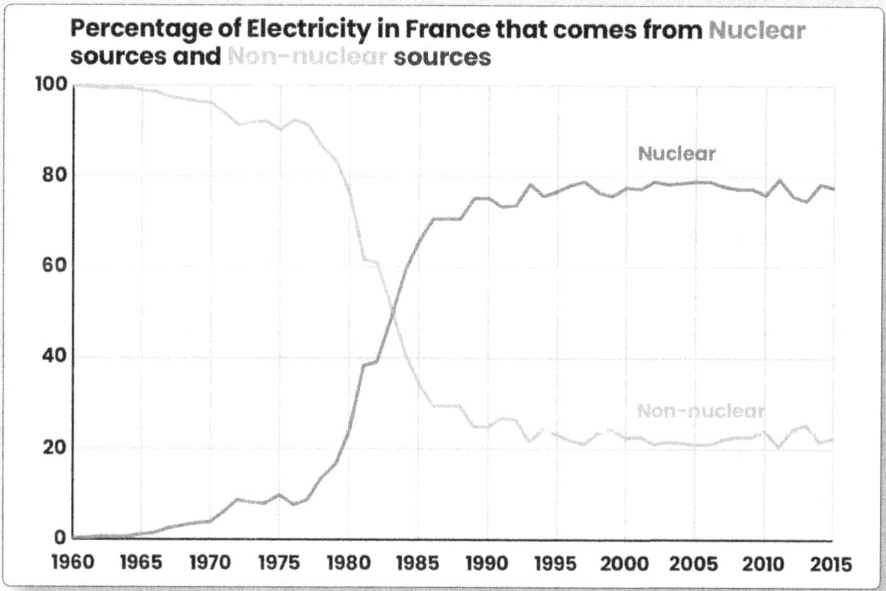

Figure 89 – Percentage of Electricity in France from nuclear sources. *Data from Our World in Data*

significant programs invariably exhibit negative learning" in nuclear power costs.

But history has shown that nuclear power can be quickly phased out. Japan nearly eliminated nuclear power quickly after the Fukushima disaster in 2011. In 2010, nuclear power accounted for more than 25% of Japanese electricity. Just two years later, it was less than 2%.

Due to dramatically changing economics driven by self-reinforcing feedback loops, we are now amid a new energy disruption. Except this time, nuclear power is the incumbent.

◆ ◆ ◆

"Things do not happen. Things are made to happen."
— John F Kennedy

◆ ◆ ◆

Dropping Silk Stockings Created a Revolution in Dental Care

> *Just as DuPont was bringing nylon to market, the start of World War II brought a sudden end to American access to Asian silk (a natural fiber whose production was itself disrupted in the early 1900s by mechanization and by new strains of silkworms). But nylon did not just replace silk, it also revolutionized dental care and launched the synthetic fiber and synthetic fabrics industries*

In February 1928, gifted chemist Wallace Carothers began working for DuPont after being recruited away from teaching at Harvard University. The 32-year-old initially was reluctant to leave his academic position as he feared his history of mental depression (which he called "neurotic spells of diminished capacity") might make him not well-suited for a corporate environment. But DuPont had big plans for a new department to do fundamental scientific research and wanted Carothers to oversee their organic chemistry efforts.

Within two years, Carothers' team had produced the world's first synthetic rubber, neoprene, and during his career also developed the first synthetic musk, Astrotone. (MIT, 2020) But the invention for which Carothers is best known came in 1935 with the discovery of nylon, the world's first synthetic fiber. As DuPont's initial press release said, nylon was made of nothing more than "coal, air and water". In 1935, not even yet forty years of age, Carothers became the first organic chemist elected to the National Academy of Sciences.

The invention of nylon couldn't have come at a better time, as US tensions with Japan rose throughout the 1930s. By the later part of the decade, the U.S. was importing four-fifths of the world's silk and Japan accounted for 90 percent of US supply. About 75 to 80 percent of US silk went into making women's stockings. (Cutlip, 2015)

The Silk Revolutions

As with the European and American textile industries, east Asian silk production underwent multiple revolutions in its underlying technologies. This was not a niche business. According to Debin Ma of Hitotsubashi University in Tokyo, "between 1850 and 1930, raw silk was the leading export item for Japan and China".

In the late 1800s, the hand-reeling of silk was disruptively displaced by machine-reeling technology and then, in the early 1900s, scientifically bred silkworm varieties quickly replaced more-traditional varieties of silkworm.

Ma estimates that just over 10% of Chinese silk production was machine-reeled in the 1880s. Just twenty years later, in the period 1900 to 1910, it was more than 60%. The transition to machine-reeling in Japan happened even sooner, being over 40% in the 1880s, and more than 90% from 1900 to 1910.

The mechanization of Japanese silk reeling is even more remarkable given that the country was largely closed to the outside world prior to 1868. In one of the world's most-dramatic cultural and geopolitical turns, Japan went from being intensely isolationist prior to the arrival of US Commodore Matthew Perry's fleet in 1853, to the newly enthroned Emperor Meiji declaring just fifteen years later that 'knowledge shall be sought throughout the world.' According to Virginia Postrel, author of *Fabric of Civilization: How Textiles Made the World,* "in 1872, the government sponsored the first reeling factory, importing French machinery... By the mid-1890s, hand reeling accounted for less than half of the country's raw silk production."

The rise of machine-reeling put pressure on silkworm breeders to both increase silk filament production and to make silk quality more uniform. Using Western inventions like the thermometer and hygrometer (humidity measurer) and the scientific method, in the 1910s, "Japanese scientists combined the country's long tradition of silkworm breeding with Mendelian genetics to develop a highly productive silkworm hybrid. Its cocoons were so superior, especially for machine reeling, that the new silkworm type swept the country."

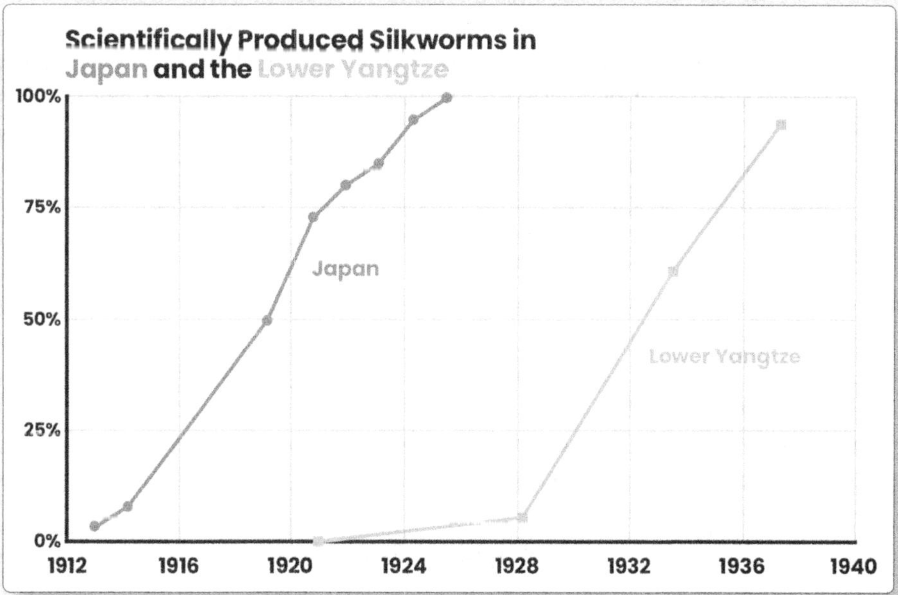

Figure 90 – Percentage of Silkworms in Japan and in the Lower Yangtze, China, silk industry that are scientifically produced varieties. *(Ma, 2004)*

According to Ma, scientifically produced hybrid silkworm varieties, which accounted for less than 10% of Japanese silk production in 1914, were more than 80% just eight years later. The Lower Yangtze region of China saw a similarly dramatic change in silkworm varieties, albeit about thirteen years after Japan's transformation, going from less than 10% hybrid in 1928 to more than 90% in 1937.

These changes did not result in a one-for-one replacement of Asian silk production. In fact, in the period of 1920 to 1930, silk production was more than six times larger than it was prior to the machine-reeling and scientific silkworm revolutions.

This new world of silk production created both winners and losers, with the winners being those who took advantage of the new technologies. Toyota, the Japanese conglomerate, was founded in 1926 as a manufacturer of automatic looms. Only later, in the 1930s, did they branch out into making their first cars and, in the 1940s, into steel.

Japan's lead in both the mechanization of silk reeling and in the adoption of hybrid worm varieties meant a shift in the center of Asian silk production away from China. According to Ma, "in 1873 China exported three times as much raw silk as Japan, but by 1905, Japanese raw silk exports exceeded the Chinese, and in 1930, Japanese raw silk exports tripled those of China."

But the era of mechanized and scientific silk production would not last long. DuPont first brought Carother's nylon fiber to market in 1938 and the attack on Pearl Harbor, just a few years later, would bring Asian exports of silk to a sudden halt.

Carothers would not live to see his best-known product come to market. In the early part of 1937, his sister died shortly before he celebrated his first wedding anniversary with his pregnant wife. While the winter lifted, his spirits did not. In April, just two days after his 41st birthday, the depression that had plagued him throughout his life finally defeated Wallace Carothers. He checked into a hotel in Philadelphia and drank a mixture of lemon juice and potassium cyanide.

DuPont's nylon stockings first went on sale in May 1940 and were a huge hit, even though they were slightly more expensive than silk. In Seba's categorization, nylons were a 'disruption from above'. DuPont's press release said nylon was "as strong as steel, as fine as the spider's web, yet more elastic than any of the common natural fibers and possessing a beautiful luster".

Susannah Handley wrote in *Nylon: The Story of a Fashion Revolution* that "nylon became a household word in less than a year and in all the history of textiles, no other product has enjoyed the immediate, overwhelming public acceptance of DuPont nylon." In just two years, DuPont earned 30% of the women's hosiery market.

The Oral Hygiene Revolution

The first product ever to use nylon was not women's hosiery, but toothbrush bristles. The toothbrush itself was probably invented in China during the Tang

Dynasty, about 1300 years ago. Before that, people around the world used a variety of ways to clean their teeth, including sticks whose ends were shredded so that they could work debris loose and mouth washes made with vinegar, salt, olive oil, alcohol, and a variety of other ingredients. For hundreds of years, brush handles were made from bone or ivory, with bristles made from the coarse hairs from a boar's neck.

In February 1938, the Weco Products Company of Chicago began selling its new "Dr. West's Miracle-Tuft", the first toothbrush to use synthetic DuPont nylon bristles. "Until now, all good toothbrushes were made with animal bristles," noted a 1938 Weco advertisement in Life magazine. Johnson & Johnson introduced their own nylon-bristle toothbrush the following year, in 1939. Today, virtually all toothbrushes have nylon bristles, even ones whose handles are made from bamboo or other natural materials.

Artificial polymers in general, and nylon in particular, sparked a revolution in personal oral hygiene that played out over the period of about a decade around World War II, when US soldiers were first ordered to brush their teeth. The newly developed nylon bristles made this possible. During the war, Dr. Charles Bass, who served as dean of the Tulane University School of Medicine from 1922 to 1940, invented waxed nylon dental floss which replaced silk floss just as nylon hosiery had quickly replaced silk stockings.

Celluloid and Bakelite plastics had replaced bone and ivory handles in the early 1900s, but in 1951 DuPont, the inventors of nylon, invented Lucite acrylic resin,

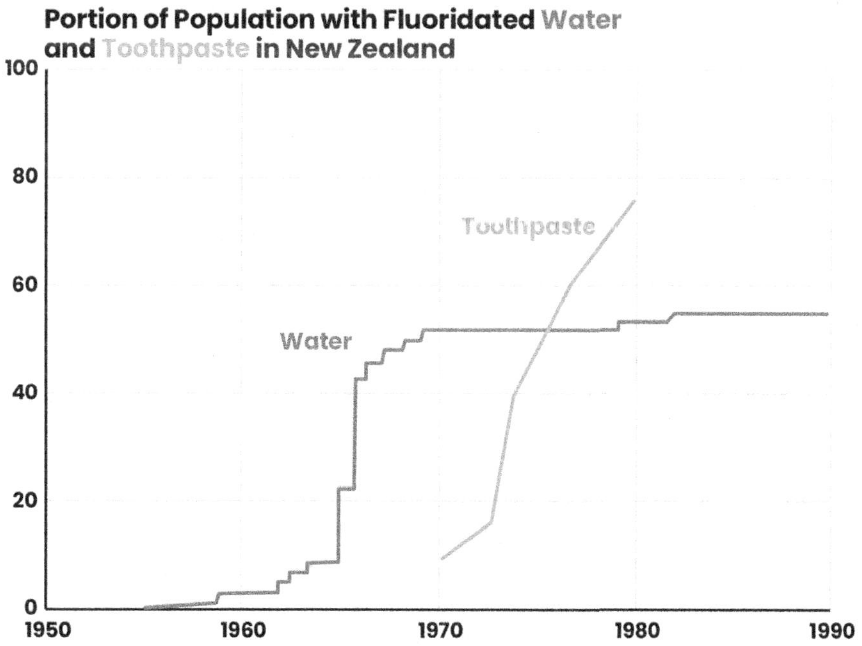

Figure 91 – Portion of Population with Fluoridated Water and Toothpaste in New Zealand *(Colquhoun, 1997)*

and clear plastic toothbrush handles of different colors appeared. Starting in the early 1940s, with regular toothbrushing on the rise, Proctor & Gamble began a program to develop a fluoride-containing toothpaste, which was launched in 1955 as the Crest brand and which remains the best-selling toothpaste brand in the US, effectively splitting the market with Colgate.

Prior to World War II, people used celluloid or Bakelite-handled toothbrushes with boar's hair bristles, toothpaste or tooth powders that were non-fluoridated and might contain chalk or charcoal or other abrasive ingredients. But over about a fifteen-year period the invention of nylon drove a personal oral hygiene revolution based on acrylic-handle nylon-bristle toothbrushes, waxed nylon dental floss, and toothpaste.

But the revolution did not end there. Fluoridation of both drinking water and toothpaste also played out, where it happened, over the course of a decade or so each. The example of New Zealand is shown in the graph, where drinking water was largely fluoridated in the 1960s and toothpaste in the 1970s.

Silk's Last Laugh?

Soon after the introduction of nylon, the sales of all women's stockings would come crashing to a halt as both the supply of Japanese silk abruptly ended and as DuPont diverted all nylon production to the manufacture of parachutes, rope, aircraft fuel tanks, flak jackets, hammocks, mosquito netting, and other products. Nylon became known as "the fiber that won the war." For four years American women had to 'make do' they best they could, using substitutes like leg make-up, a cosmetic painted onto the legs to make it look like the user was wearing stockings.

When the war ended, there was such a rush to buy nylon stockings again that fights broke out over the limited supply, the so-called 'Nylon Riots'. The sales of silk stockings never recovered. Over the course of about five years, most of the world's demand for silk had simply disappeared and nylons went from 0% to essentially 100% of the US women's hosiery market. Europe, too, went crazy for nylon. "In 1950, West German retailers sold just 900,000 pairs of ladies' nylon stockings", according to Tony Judt in *Postwar. A History of Europe Since 1945.* "Four years later, in 1953, they moved 58 million pairs."

But conquering 100% of the hosiery market was just the start of the synthetic fiber industry. "There's a whole series of synthetic materials that indeed came from the base idea that chemists can design and develop a series of materials that had certain properties, and the ability to do it from the most basic molecules," said Matt Hermes, associate professor in the bioengineering department at Clemson University. (Cutlip, 2015)

Today, synthetic materials account for more than 70% of world fiber production with nylon, acrylic, polyester, polyethylene, and polypropylene combined representing about 98 percent of synthetic fiber production by volume. But the scientific production of silk might get the last laugh. Several companies, including the San Francisco Bay Area's Bolt Threads, are using genetically modified microbes to produce silk proteins. Virginia Postrel, the author of *Fabric of Civilization: How Textiles Made the World,* says that "if the endeavor succeeds, silk won't need to come from insects. It will be brewed in giant fermentation vats like beer. And the process has implications beyond silk. Wool, too, is a protein polymer. So is cashmere. So are countless other fibers we can barely imagine." Japanese company Spiber has already put on the market several products containing lab-grown silk proteins, including a parka made in collaboration with The North Face, a sweater by Goldwin Sports, and a hoodie with Pangaia.

However, genetically modified microbes can also make nylon itself. An April 2023 by CNBC reports that popular athletic clothing manufacturer Lululemon has begun selling short-sleeved shirts made from nylon produced in fermentation vats. The shirts, according to the article, cost the same as ones from conventionally sourced, fossil-fuel-based nylon.

Cultured Pearls Disrupted Natural Pearls, Helping the Rise of the Oil Industry

Natural pearls are rare, and ones that are symmetrical in shape and uniform in color are even more so. But that rarity and beauty have made them obvious objects of desire for millennia.

For at least seven thousand years, people have collected pearls and since before the time of Christ we have intentionally fished for them. The world's oldest natural pearl was discovered among flint arrowheads and beads made of shell and stone at a Neolithic settlement site on an island off the coast of the United Arab Emirates in 2019. Radiocarbon dating showed it to be 7600 to 7800 years old.

Dr. Robert Carter of University College London in Qatar, in a 2005 article in the *Journal of the Economic and Social History of the Orient,* reports that "a small string of pearls" was discovered in Uruk, in what is now Iraq, dated from 3200-3000 BCE (the time that the wheel was revolutionizing Mesopotamian life), that pearling was mentioned a thousand years later in the epic poem of Gilgamesh, and that "400-500 pearls" in a three-strand necklace were found in an Achaemenid sarcophagus. In the Gulf between Arabia and Iran, "pearling was certainly an established industry by the time of Alexander".

It was not just the Middle East that supplied pearls, as a wide variety of mollusks in saltwater and freshwater bodies around the world produce them, from the Sea of Japan and the South China Sea, to islands of the South Pacific and islands off the northern coast of Venezuela. Pearls have been revered for millennia from China and India to ancient Greece and Rome as obvious symbols of purity, perfection, opulence, and femininity.

But it was in Europe after the rise of Christianity that pearls became the world's most valuable gems. As the persecution of Christians ended under Emperor Constantine in the 400s, "pearls and pearl shells were symbols of resurrection among early Christians," according to Robin Donkin in the 1998 book *Beyond Price: Pearls and Pearl-fishing.*

Throughout the Middle Ages and Renaissance, pearls were not just valued as gems but also for their supposed medicinal value. Powered and dissolved in lemon juice or vinegar, pearl concoctions supposedly provided relief for heart palpitations, stomach disorders, and eye afflictions and, according to Donkin, would help "relieve melancholia or excessive apprehension". Pearl was also recommended "to stimulate the flow of milk in women".

Europe developed such an insatiable demand that people made artificial pearls from glass beads or from compressing powered glass, fish bones, eggshells, egg whites, or a myriad of other substances. Leonardo da Vinci provided a recipe for making large pearls that involved dissolving small seed pearls or mother of pearl in lemon juice, drying the powder, adding egg white to make a paste, shaping the paste into a ball, and polishing until lustrous and pearl-like.

In Columbus's third voyage to the New World, an enormous supply of pearls was discovered in shallow-water oyster beds off the northern coast of Venezuela. According to Stephen Bloom in *Tears of Mermaids: The Secret Story of Pearls*, Nueva Cadiz, the first Spanish settlement in South America, on the island of Cubagua and inhabited only from 1509-1536 "served as the sole New World processing center for hundreds of thousands of exceptional pearls, spirited back to the Spanish crown and then to the rest of pearl-hungry Europe... Pearls exceeded by ten times the value of all other goods combined that the Spanish brought back from the New World. For a short quarter of a century, the greatest number of pearls the world had ever known passed through this now forgotten island."

The Old World oyster beds survived this short-lived threat in the early 1500s, but they would not survive a major technological innovation led by industry outsiders in the early 1900s. William Saville-Kent, a British marine biologist living in Australia in the late 1800s, figured out how to culture pearls. His breakthroughs were commercialized in Japan, most successfully by Kōkichi Mikimoto. By 1893, Mikimoto was able to make hemispherical pearls by seeding oyster with a small amount of pearl shell. These were first introduced to the market at a marine products exposition in Norway in 1897. But it took another twelve years to create high-quality spherical pearls that were indistinguishable from natural ones. Commercially viable harvests were only first obtained in the 1920s. Pearls, formerly the domain of royalty, suddenly became affordable to the masses.

The new technology enabled Japan's cultured pearl industry to expand quickly. Though the entirety of Japan produced fewer than one million pearls in 1930, by the time of Japan's invasion of China in 1937, they were producing more than ten million annually. The following war-years were not kind to either the Japanese cultured pearl industry or its Arabian natural pearl counterpart. But while the Japanese industry rebounded after World War II, the Middle Eastern business did not. Today, more than 99% of all pearls sold worldwide are cultured.

According to Carter, "the development of cultured pearl farming in Japan

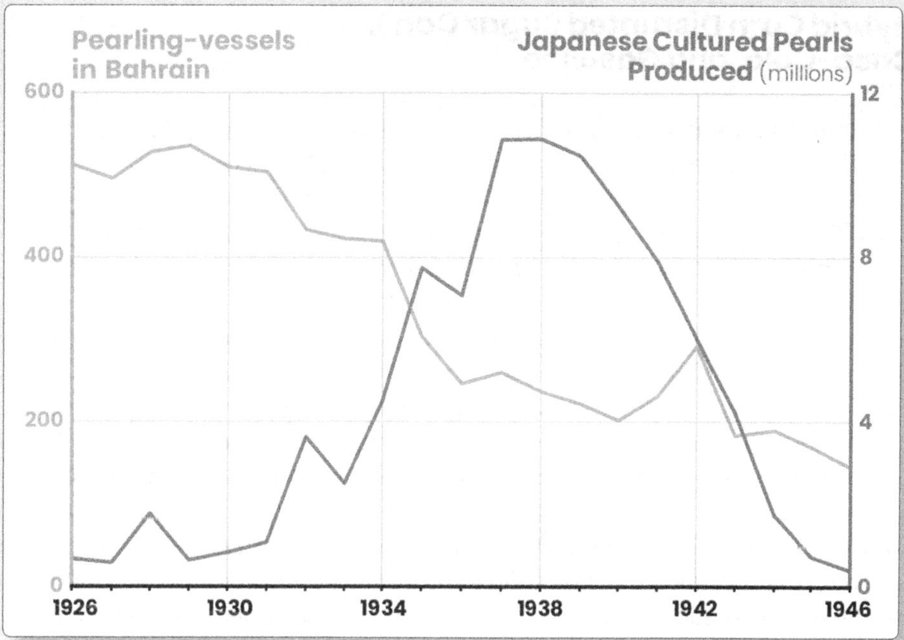

Figure 92 – Production of Japanese Cultured Pearls (millions) and number of Pearling-vessels in Bahrain

precipitated a catastrophic collapse in the pearl industry" in Arabia as "the economy of every single emirate on the Arabian side of the Gulf was dependent on the pearl industry." Carter claims that "around half or more of the population of Ajman, Dubai, Abu Dhabi, and Qatar were involved in the pearl trade. If one assumes an approximate 50/50 ratio of men and women, it implies that the entire male population of these places was employed in pearling." By fortuitous timing, the nascent oil industry in the Middle East was likely one of the key beneficiaries of the sudden increase in the availability of male labor.

In the Middle East, the natural pearl industry, according to Carter, "lingered, much-reduced, throughout the 20th century, its final demise being marked by the official closure of Kuwait's pearl-oyster market in the year 2000, which brought to an end over 7000 years of pearling in the region."

But the Japanese pearls that disrupted this trade were themselves largely disrupted in the late 1990s and early 2000s by freshwater pearls from China. According to Stephen Bloom, writing in 2011, "a strand of medium-sized, near-perfect Chinese freshwater pearls can be bought wholesale today for under $150. Such reverse sticker shock is freaking out just about every other national producer of pearls. To make matters worse, to most consumers, such a strand is virtually identical to strands that sell for five and ten times as much (and sometimes more)."

Hybrid Corn Disrupted Sugar Cane, Coca-Cola, and Gasoline

> *In the 1930s, a single hybrid strain took over the corn (maize) industry. High-fructose corn syrup replaced cane sugar in Coca-Cola and Pepsi in just four years. Genetically-modified (GM) corn disrupted non-GM in the 1990s. Then, corn-derived alcohol changed the gasoline market*

How did yellow corn (maize) become what historian Chris A. Kniesly calls "the biggest success story in agricultural history"? In his TED talk 'How Corn Conquered the World' he says, "corn currently accounts for more than one-tenth of our global crop production [and] over 99% of cultivated corn is the exact same type: Yellow Dent #2. This means that humans grow more Yellow Dent #2 than any other plant on the planet."

But it was not always that way. Modern maize was derived in Mesoamerica from teosinte, a virtually inedible grass with hard seeds. Over many centuries the native people of the Americas bred it into both a staple food as well as a source of fiber. Isotopic evidence from human skeletal samples from two rock shelters in Belize suggest that "individuals dating before ~4700 calendar years before present show no clear evidence for the consumption of maize. Evidence for substantial maize

Figure 93 – Multicolored 'flint corn'. *Photo: Markus Winkler*

consumption (~30% of total diet) appears in some individuals between 4700 and 4000 calendar years before present. Isotopic evidence after 4000 calendar years before present indicates that maize became a persistently used staple grain comparable in dietary significance to later maize agriculturalists in the region (>70% of total diet)." That is, corn went from zero percent of the diets of the humans examined to 70% in a period of about seven hundred years. (Kennett, 2020) The oldest remains were from people who ate wild tropical plants, fruits, nuts, and meat from animals. In other words, the shift in diet from essentially no corn to 70% corn represented not just a change in food preferences, but also likely a complete revolution in living style, from forest hunter-gatherers to settled farmers.

The Full Spectrum of Corn is Replaced by a Single Strain

When Europeans arrived in the New World and until the late 1800s, a wide variety of corn remained available with differing colors and characteristics. But over the course of the 1930s, a bewildering array of distinct corn varieties was replaced by a single strain. (This is essentially the same story that happened in Japanese silk in the 1910s, where the rapid rise of a hybrid silkworm resulted in greater uniformity of the final product.) The rise of this hybrid strain, produced from two distinct lines of corn, is one of the examples Everett Rogers covers in *Diffusion of Innovations*, his classic 1962 book on the adoption of new technologies. The first hybrid corn was sold commercially in Connecticut in 1922. Recognizing what new breeds might mean for his home state of Iowa, Henry A. Wallace stated in 1925, "corn means so much more to us in the corn belt than it does to the New England farmer that I am hopeful our farmers will be easy to educate to the use of the new seed corn which I believe will be all the rage ten years hence." (Brown, 1983) Wallace, who founded the Hi-Bred Corn Company in 1926, was right.

However, he experienced some initial setbacks. In 1927 he cultivated forty acres of seed corn and then, in 1928, fifty acres. But while expanding operations in the autumn of 1930, his old corn processing plant was destroyed by fire, resulting in the loss of some parent corn. Wallace did not let this stop him. "In 1931, despite the Depression and low farm prices, Wallace's young seed company continued its expansion. Seed acreage was increased from 170 [acres in 1931] to 220 acres [in 1932]." "The farm economy was entering a serious depression in the early thirties," Brown continued, "corn prices were dropping and farmers were hesitant to pay relatively high prices for seed to produce a crop of decreasing value."

According to Rogers "in 1928, hybrid seed was made available to Iowa farmers [for the first time]. The hybrid seed was adopted rapidly. By 1941, about thirteen years after its first release, the innovation was adopted by almost 100 percent of Iowa farmers." Even more remarkable was that the adoption by any individual farmer was often not rapid. "The innovation-decision process involved considerable deliberation, even in the case of an innovation with spectacular results. The average respondent [farmer] took three or four years after planting his first hybrid seed, usually on a small trial plot, before deciding to plant 100% of his corn acreage in hybrid varieties."

According to Rogers, individual decisionmakers go through five steps on their way to choosing to use a new technology: Awareness, Interest, Evaluation, Trial, and finally Adoption. "The farmer-to-farmer exchange of their personal experiences with hybrid seed was at the heart of diffusion." The experiences of the earliest adopters of the hybrid seed were driven by a convergence of factors, like the availability of the new hybrid corn strains as well as by the mechanization of labor as internal combustion engine farm vehicles were developed. The new strains were compelling by themselves. According to Rogers, "hybrid corn yielded about 20 percent more per acre than the open-pollinated varieties that it replaced. It was also more drought-resistant and better suited to harvesting with mechanical corn-pickers."

A Convergence of Factors: Mechanical Harvesters, Synthetic Fertilizer, and Fickle Customers

Mechanical harvesting was at least a 10X disruption. As Kniesly says, "following the technological developments of World War II, mechanized harvesters became widely available. This meant a batch of corn that previously took a full day to harvest by hand could now be collected in just 5 minutes." With synthetic fertilizer "farmers could plant dense fields of corn year after year, without the need to rotate their crops and restore nitrogen to the soil." It is no wonder that, according to Kniesly, "trade boards in rail hubs like Chicago encouraged corn farmers to breed one standardized crop," hybrid corn.

Corn is the most plentiful crop in the US. Even in 1930, before the bulk of the yellow corn revolution, the number of bushels of oats that the US produced was only about two-thirds as many as corn, wheat was less than half as many as corn, and rye

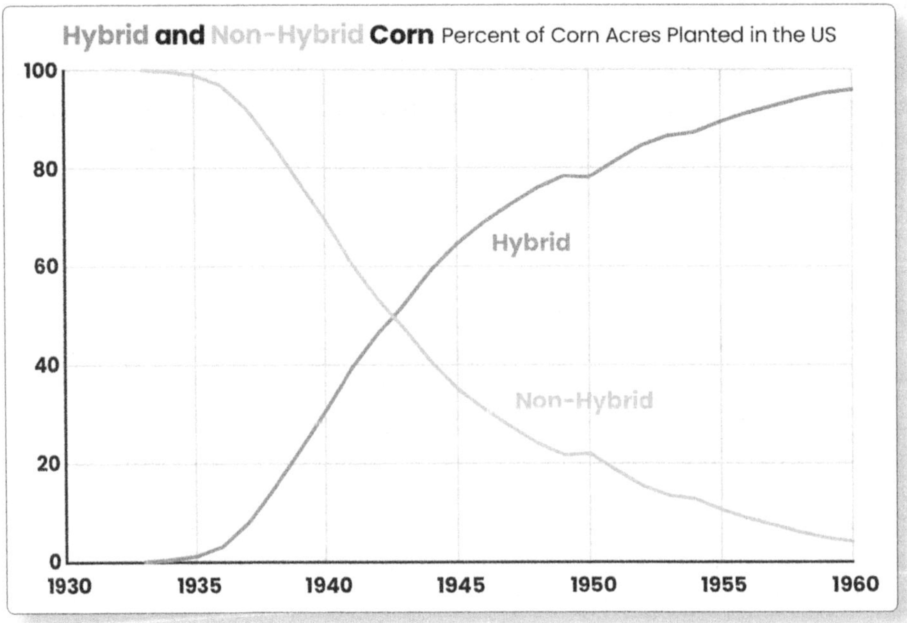

Figure 94 – Percent of Acres Planted with Hybrid and Non-Hybrid Corn in the US

was only about 2% as many. (USDA, 1962) We've been growing corn in the Americas for thousands of years, and then over about fifteen years in the 1930s and 1940s we completely changed how we grew our most important food item, replacing all the different varieties of corn with just one kind that could be fertilized with synthetic fertilizers, was drought-resistant, and could easily be harvested by machines.

But it did not stop there. In his Ph.D. dissertation work, Zvi Griliches, a Holocaust survivor who moved to the US after World War II, investigated the adoption of corn in various parts of the United States. Grilliches showed that hybrid corn had been adopted by at least half of Iowa corn farmers by 1938, half of Wisconsin by 1940, half of Kentucky by 1944, half of Texas by 1948, and half of Alabama by 1953. In each case, the state-level adoption curve was S-shaped, with the bulk of the change taking place over a period of about a decade. Across the US, according to Department of Agriculture data, hybrid corn went from 10% of corn acres planted in 1938 to 80% of corn acres planted in 1951, just thirteen years later.

This was not a one-for-one substitution of old corn varieties for the new one. The total number of acres planted by the late 1950s was 20% lower than in the early 1930s, but because hybrid corn produced more per acre, the yield-per-acre doubled, and the total harvest (in bushels) almost doubled in that time. While the yield of corn in the US remained between about twenty to thirty bushels per acre prior to the advent of hybrid corn, it has risen steadily since then and now tops one hundred sixty bushels per acre, due to a convergence of technologies: not only hybrid corn, but also diesel-powered equipment, and cheap, plentiful artificial fertilizer made from natural gas.

The US grows so much corn today that food for people is not its most-common use. Over 40% of all corn grown in the US is used as livestock feed and an additional

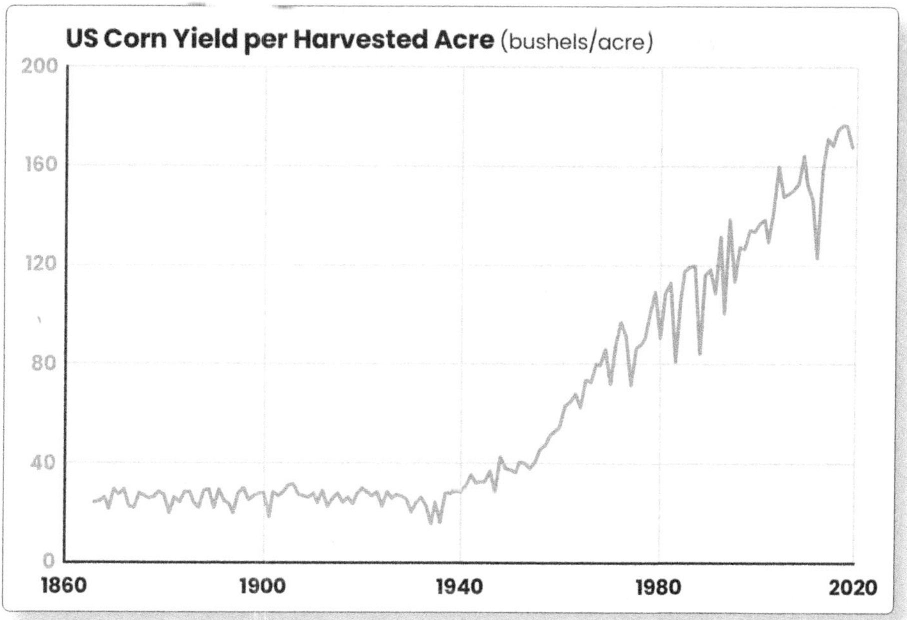

Figure 95 – Yield per Acre of Corn in the US

30% is used to make ethanol as a fuel additive, 15% is exported and only about 10% is used for direct human consumption or to make food products like corn oil and high-fructose corn syrup.

Subsequent Disruptions of Soft Drinks, Gasoline and Hybrid Corn Itself

Introduced to the market in 1978, HFCS 55, a sweetener containing 55% high-fructose corn syrup, replaced cane sugar in both Pepsi Cola and Coca-Cola over just four years from 1980 to 1984. With this one switch, "U.S. sugar consumption decreased by more than 500,000 tons a year", according to historian James Bovard, "driving sugar prices so high that it wrecked the market for sugar".

The corn business has seen additional disruptions after the hybrid revolution, particularly because of advancements in genetic science. Today, more than 90% of the corn grown in the US is genetically modified to make it tolerant to herbicides, resistant to certain insects, or to have multiple added traits. As with hybrid seeds themselves, these revolutions played out over a period of only about a dozen years. According to Alexandre Magnier in his article 'Product Life Cycles and Innovation in the US Seed Corn Industry', "while the market share of conventional hybrids stood at 95% in 1997, the share decreased to 10% in 2009."

Other key grains have seen similar disruptions. Genetically modified soybeans, which are largely fed to livestock, went from about 5% of the US market in 1995 to about 95% from 2008, yet another revolution in just thirteen years. As a result of these disruptions in conventional and modified grains, the US has enjoyed among the world's highest availabilities of meat per person, primarily chicken, pork, and beef, which have caused disruptions of their own.

Production of ethanol as a fuel additive is now one of the largest uses of corn, a

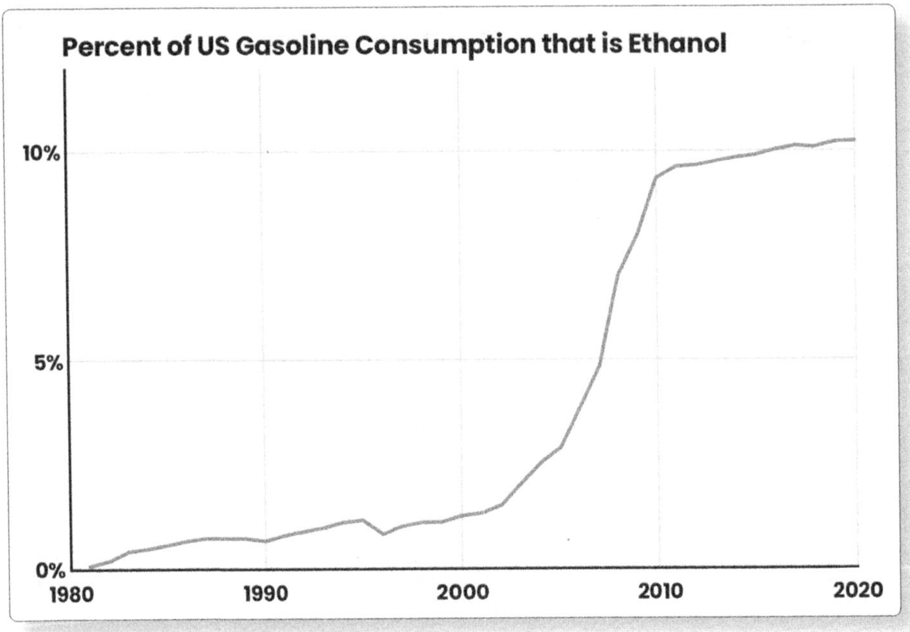

Figure 96 – Percentage of US Gasoline Consumption that is Ethanol. *Source: US EIA*

change that played out over about ten years. As late as 2002, less than 2% of the gasoline consumed in the US consisted of ethanol, but ten years later it was nearly 10%, the maximum quantity that most engines can tolerate.

Hamburgers Turned Restaurants into Real Estate Businesses

> *Henry Ford's assembly line ideas, applied to the hamburger business, turned McDonald's into the largest restaurant chain in the world, and a major owner of real estate. Autonomous delivery vehicles might make their locations obsolete*

In 1940, brothers Richard (Dick) and Maurice (Mac) McDonald opened McDonald's Bar-B-Q drive-in restaurant in San Bernardino, California, east of Los Angeles. Like many drive-up restaurants of the time, McDonald's was a hang-out for teenage boys who were less interested in spending money on food than they were in getting the attention of the young women servers.

After World War II, according to Richard, the world "was kind of getting into an age of jet propulsion and this [drive-in restaurant the McDonald brothers owned] was really a horse-and-buggy operation. So, we knew we had to do something to speed things up."

The McDonalds reformulated the menu in 1948, cutting the number of items offered from twenty-five to just their best-selling nine and focusing on food that could be served quickly: burgers, fries, and drinks. Everything about the new McDonalds' restaurant reflected the Massively Transformative Purpose of eliminating the least-enjoyable part of the dining experience, the waiting time between ordering and receiving your food.

A Convergence of Ideas Caused a 10X Service Improvement

The brothers took good ideas from other popular restaurants, like A&W and White Castle, devised new ideas of their own, and mixed them together into a convergence of concepts, none of which individually unlocked their goal, but all of which together amounted to a revolution. The waiting staff did not visit you. Instead, you went to a window counter to order and receive your food. The kitchen behind the counter was designed like Henry Ford's assembly line, with cooks manning stations and each doing a limited number of tasks. The burgers were pre-cooked and then kept warm under heat lamps. Flat-plate griddles are easier to clean than grills, so the grills were gone. Tim Harford, author of *The Undercover Economist*, wrote: "The brothers worked with a local craftsman to invent a new kind of spatula, a new dispenser that squirted the same amount of ketchup and mustard every time, and a rotating platform to speed up the process of assembling of burger, bun and condiments." Even the drinks were accelerated. Ray Kroc, a milkshake machine salesman, sold them five-spindle machines, which could make five milkshakes at a time, rather than the simpler machines most restaurants were using.

What the McDonald brothers devised was the "Speedee Service System" that is the basis of the modern fast-food industry. Harford wrote, "What Henry Ford had done for cars, the McDonald brothers did for hamburgers and French fries: they broke down processes into simple, repetitive tasks." With these changes, the McDonald brothers cut serving time down from twenty minutes to thirty seconds. In fact, Ray Kroc was known to offer refunds to customers who needed to wait longer than five minutes. "The customers' standard expectation seems to be three minutes [of waiting time]," said Marianela Aran, vice president in charge of research and development at Burger King, in a New York Times article in 1997. "The level of satisfaction drops off at four minutes and goes way down at five minutes." (Feder, 1997) By delivering orders in half a minute, the McDonald's were serving their customers 10X faster than that level. And by delivering quickly, they could deliver cheaply.

This New System Led to the Rise of Empires

McDonald's was on the leading edge of a wave of fast-food establishments: In-and-Out Burger opened their first location in 1948, Kentucky Fried Chicken in 1952, Burger King in 1954, and Pizza Hut in 1958. The McDonald brothers began franchising their system in 1953. Ray Kroc, the milkshake machine man, started his own location in Illinois in 1954, fourteen years after the McDonald brothers opened their drive-in restaurant. The brothers wished to have only a small number of restaurants, but Kroc had bigger plans, so he eventually bought them out in 1961.

Under Kroc, the pace of innovation did not slow down. Quite the contrary. 1962 saw the first McDonald's restaurant with indoor seating. In 1960, Gilbert Lamb invented the 'water gun knife', a French fry cutter that used a fire hose to shoot potatoes at an array of sharp steel blades, and now his company, Lamb Weston, is a major supplier of McDonald's famous shoestring fries. The McDonald's Corporation's Ralph Weimer filed a patent application in 1964 for a waste receptacle with a swinging door in the lid, in 1965 for the bagging scoop, a specially-designed device which scooped fries and made it easy to deposit them into a paper sleeve, and in 1968 for a vertical bun toaster that used a conveyor belt to toast buns, depositing them onto a tray at the bottom when they were done.

McDonald's system had unintended consequences. Reducing the menu offering not only sped up food delivery after the order was placed, it sped up ordering too by taking away the 'cognitive burden' of deciding what to have. Just as Ford supposedly said, "any customer can have a car painted any color that he wants, so long as it is black", the McDonald's brothers offered diners any meal they wanted, so long as it was a burger, fries, and a Coke.

Using an assembly line not only sped up the rate of delivery of food, it also reduced the variability of product quality. Each hamburger was as close to being identical to the one before it as they could possibly be. Ray Kroc said, "People don't want the best burger in the world, they want one which tastes just like the one they had last time." In a presentation called 'The psychology of digital marketing', Rory Sutherland, vice chairman at the advertising company Ogilvy UK, elaborated on this concept: "McDonald's is the most successful restaurant in the world because it's very, very good at not being terrible. You can criticize it as much as you like, but the fact is you're never disappointed. Because the food is always identical, you're never disappointed, you're never ripped off."

Perhaps the key factor that made McDonald's "the most successful restaurant the world" was the realization by Harry J. Sonneborn, the company's first president and chief executive, that the most valuable part of any fast-food restaurant is the land on which it sits, since it is the "location, location, location" (as the old real estate adage goes) that determines the number of people who could visit the restaurant each day.

A Business Model Innovation Unlocked the Burger Chain's Potential

Purchasing plots of land and then leasing that land to franchisees was a business model innovation that provided "greater capital for expansion," as Sonneborn's character in the 2016 biopic film *The Founder* says to actor Michael Keaton playing Ray Kroc, "which in turn fuels further land acquisition, which in turn fuels further expansion. And so on and so on." A self-perpetuating feedback loop of explosive growth. "You're not in the burger business", Sonneborn says, "You're in the real-estate business." Potential franchisees from all around America and all around the world would line up at Kroc's door each holding a fat stack of cash, metaphorically speaking, for the opportunity to start their own McDonald's restaurant.

According to the Wikipedia page about Sonneborn, "the 'Sonneborn model' persists to this day within the corporation, and might have been the most important financial decision in the company's history. McDonald's present-day real estate holdings represent $37.7Bn on its balance sheet, about 99 percent of the company's assets and about 35 percent of its global revenue."

Fast food became an American export. According to Tom Standage in *A Brief History of Motion*, "the top fifteen fast-food chains in the world, ranked by number of branches, are all American. Surprisingly, McDonald's was a relatively late adopter of the drive-through model, opening its first drive-through in Arizona in 1975. But drive-through sales soon accounted for 70 percent of its sales in America — an indication of the enduring association between cars and fast food."

Now, in a post-pandemic world where smart phones are common and self-driving delivery vehicles might soon be too, McDonald's business is undergoing a rethink. McDonald's is developing drive-through-only restaurants to minimize person-to-person interaction and take-out-only restaurants for larger urban areas. A future with self-driving delivery vehicles might be a future with no McDonald's restaurants at all or, at least, one where their locations are no longer very important.

◆ ◆ ◆

"When I was seven the Wright Brothers suddenly flew and my memory is vivid enough at seven to remember for about a year the engineering societies were trying to prove it was a hoax, because it was absolutely impossible for man to do that."

— Buckminster Fuller

◆ ◆ ◆

Shipping Containers Moved Factories to Asia and Helped End Famines

> *Cheap steel from the post-War oxygen blast furnace revolution enabled containerized shipping, making distant manufacturing locations cost-competitive with established ones. And it also has nearly put an end to famines*

On April 26, 1956, at Port Newark, New Jersey, forty-two-year-old entrepreneur Malcom McLean loaded fifty-eight steel shipping containers onboard a modified tanker ship, the *Ideal-X*, and then that ship set sail for Texas. The previous year, thirty-five-year-old engineer Keith Tantlinger had watched at Bethlehem Steel in Baltimore as his customer, McLean, and executives from McLean's company jumped up and down on top of two metal shipping containers Tantlinger had built, to make certain they were secure enough to handle the journey.

Ships had been loaded and unloaded the same way for thousands of years, literally since the time of the Phoenicians, with crews of men carrying boxes and bags onboard or off. But McLean remade the entire shipping industry over just a couple of decades, with consequences for the global economy that are still playing out today. Memorializing McLean shortly after his death in 2001, then-Secretary of Transportation Norman Mineta said, "Malcom revolutionized the maritime industry in the 20th century. His idea for modernizing the loading and unloading of ships... has resulted in much safer and less expensive transport of goods, faster delivery, and better service."

An Outsider Saw an Easier Road to Market

At the time of that first voyage, McLean was not a maritime shipping industry expert. In fact, he was, in the words of Marc Levinson in *The Box: How the Shipping Container Made the World Smaller and the World Economy Bigger*, "an outsider who had no experience with ships". After graduating high school in 1931, McLean bought a used truck and prospered from the tobacco boom of that era, hauling tobacco from his native North Carolina to destinations in the northeast. He was aggressive in growing his business helped by, according to Levinson, his "obsessive focus on cutting costs". By 1940, he had thirty trucks and by 1954 McLean Trucking "had become one of the largest trucking companies in America".

But McLean's business had always been hampered by the inherent inefficiency of loading and unloading goods and by the fact that the US Interstate Highway System, constructed largely over a fifteen-year period from the late 1950s to the early 1970s, had not been built yet. Standardized 'intermodal' containers which could be loaded onto trucks, rail cars, or ships might be the solution. In 1955, McLean sold off his trucking business. According to Levinson, "the country's best-known trucking magnate walked away from the business he had built in order to build a new one, based on some untested ideas about shipping."

McLean was not the first person to try shipping goods in containers, far from it. According to Levinson, "different container systems came into use on railways... during the 1920s in response to a new competitive threat - the truck." The Central Georgia Railroad tried moving loaded railroad cars by ship and the Missouri Pacific Railroad promoted 'speedboxes,' aluminum containers on wheels. A 1955 census of shipping containers in non-Communist Europe found that "almost all European containers were made of wood, and many had no tops; the user piled the goods inside and covered the load with canvas". And they were small. Levinson reports that the majority of them were smaller than 106 cubic feet (3 cubic meters), "less than the volume of a box 5 feet on a side."

But the containers Tantlinger built were different. McLean's first containers were thirty-three feet long. "These boxes were at least seven times the size of any containers then in common use," Levinson wrote. They were made largely of steel. And, unlike loaded railroad cars or wooden boxes with canvas tops, they could be stacked on top of each other. Tantlinger acquired seventy-nine patents in his lifetime, including ones for the Twistlock latches that connected containers together, and for the spreader bar system that allowed a crane to grasp all four corners of the container to lift it.

There was much more to the container revolution than just building corrugated metal boxes. McLean devised what in Levinson's words was "a concept of containerization quite different from anything that had come before". In addition to specially outfitted container ships, McLean also used dock-side cranes for lifting the containers into and out of the ships and ports designed to move the containers quickly on to or off from truck trailers or rail wagons. Special refrigerated containers had built-in diesel-powered cooling units that allowed the contents to be maintained at a steady temperature.

The result was a system that was much more efficient than what had ever been done before. Levinson quotes McLean Industries telling shareholders after two years of operation that "a containership can be loaded and unloaded in almost one-sixth of the time required for a conventional cargo ship and with about one-third of the labor". But loading was only one step in the shipping process. "Analysts for the Port of New York Authority calculated that sending [Ballantine Beer] to Miami on board a traditional coastal ship... would cost four dollars a ton" while "the container alternative... was estimated to cost twenty-five cents a ton," a sixteen-fold difference.

The result, understandably, was a rapid and profound reorganization of the global shipping industry. "Container technology developed far more quickly and affected transportation industries far more significantly than even its most ardent proponents had imagined," Levinson wrote. "The Propeller Club, the association of top shipping company executives, devoted a full day of its 1958 convention to containers. No one could doubt that conventional shipping would soon be in trouble." That same year the Port Authority of New York and New Jersey began the massive development effort to make Port Elizabeth, located just south of Port Newark where the Ideal-X was loaded and "designed from the start as a port for containers". In January 1959, the world's first purpose-built container crane went into operation in Alameda, on the east side of San Francisco Bay. The International Organization for Standardization (abbreviated 'ISO') issued its first specifications for shipping containers in 1961 and, according to Levinson, "by 1965, half a dozen ship lines announced plans to launch container services to Europe." The next fifteen years, from 1966 to 1981, saw 80% of the maritime countries of the world adopt at least one containerized port, according to Dr. Gisela Rua, an economist with the Board of Governors of the Federal Reserve System.

It was Not a One-for-One Replacement

The revolution McLean kicked off was clearly not a one-for-one replacement of loose freight with containerized shipments, as the amount of global trade today, both in absolute terms and as a percentage of global economic activity, is higher than ever. Today, about 90% of global international trade goes onboard a ship at some point in its journey and maritime shipping costs are so low that they are often treated by economists as negligible compared to the value of the goods transported.

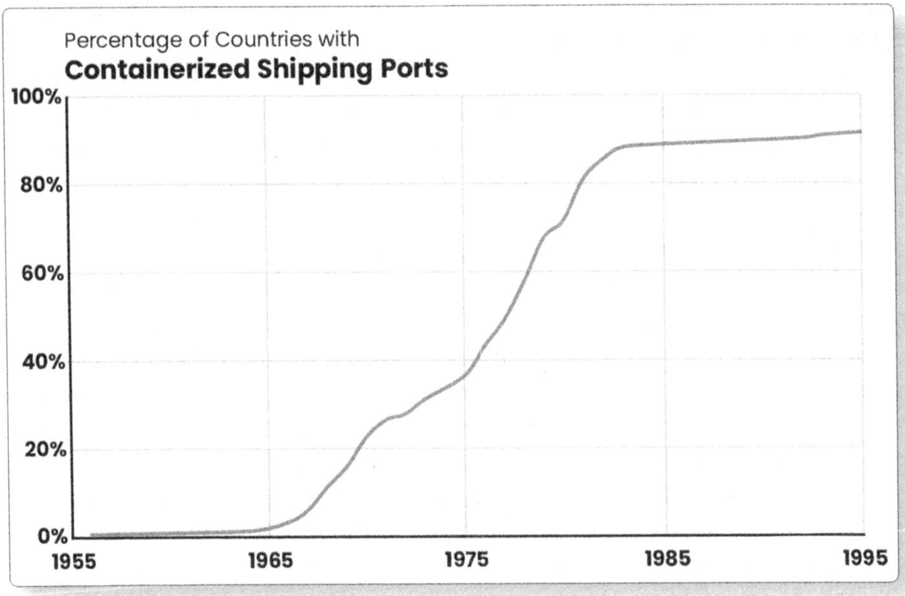

Figure 97 – Percentage of Countries with Containerized Shipping Ports. *(Rua, 2014)*

A Convergence of Factors: Trucks and Steel

Why did Malcom McLean's containerization revolution happen when it did, and not a few decades sooner or later? Shipping containerization in both the US and around the world was enabled by a convergence of factors. First among them, of course, was an unparalleled post-War era of global peace and sustained economic growth.

A key component of the containerization revolution was the growth in number of motorized heavy vehicles, something that lagged the growth in cars. Carrying containers onboard a ship was only one segment in a logistics chain that often included the containers traveling as well by rail and by truck. So, the containerization disruption could not happen until the trucking industry was well-established, and that took until about the end of World War II. In fact, McLean himself played an important role in the development of truck capabilities. As Levinson reports, "at a time when most trucks had gasoline engines, McLean Trucking was the first major company to install diesel engines in its tractors". And, as we mentioned above, the rise of containerization both benefited from and coincided with the development of the Interstate Highway System.

Another key component of the containerization revolution was the availability of cheap steel, not only to make the trucks, trains, and ships that would carry containers, but also the cranes at ports to load and unload the ships, and of course, the containers themselves. During the decades of the 1950s and 1960s, there was a revolution in steelmaking as the Siemens process was disrupted by the oxygen blast furnace. Professor Lauri Holappa of Aalto University in Finland estimates that essentially no global steel production was by oxygen blast furnace (OBF) in 1950 and still only about 5% of global steel production was from OBF in 1960. The Siemens process supplied 70% of global steel production at that time. But by 1970, the oxygen blast furnace surpassed the Siemens process and was unequivocally dominant by the early 1970s. This newer, cheaper, better means of producing steel literally provided the raw material for the containerization revolution.

Winners and Losers under the New System

Immediate losers from the opening of Port Elizabeth container port included the docks in Brooklyn and Manhattan, on the other side of New York Harbor. Levinson reports that "by the middle of the 1970s, the New York docks were mostly a memory. Lighters carried a grand total of 129,000 tons of freight to waiting ships in 1974 - less than one-tenth of the load moved in 1970, one-fiftieth as much as in 1960," a 98% drop in fourteen years.

What was once an advantage, Manhattan's dense development, now became a liability. Levinson says, "containerization eliminated one of the key reasons for operating a factory in New York City: ease of shipment. A New York City location had long offered transport-cost advantages for factories serving foreign or distant domestic markets, as local plants could get their goods loaded on ships with much less handling than could factories inland. The container turned the economics of location on its head."

Manufacturing moved outside of New York City, and in fact helped to propel the economic development of much of east and south Asia. Levinson says, "cities that had been centers of maritime commerce for centuries, such as New York and Liverpool, saw their waterfronts decline with startling speed, unsuited to the container trade". But "even as it helped destroy the old economy, the container helped build a new one. Sleepy harbors such as Busan and Seattle moved into the front ranks of the world's ports, and massive new ports were built in places like Felixstowe, in England, and Tanjung Pelepas, in Malaysia, where none had been before."

Containerization "began as a means of shaving a few dollars off the cost of sending Malcom McLean's trucks between New York and North Carolina", Levinson notes, but McLean's 'obsessive focus on cutting costs' overturned entire economies in the process.

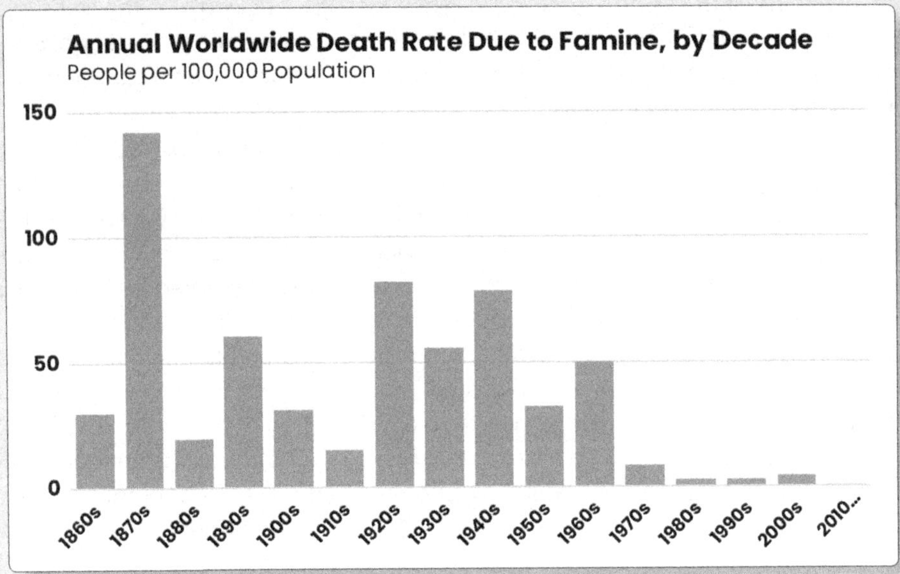

Figure 98 – Annual Worldwide Death Rate due to Famine. *(Hasell, 2017)*

The Disruption of Famine

In the blockbuster 1968 book *The Population Bomb*, Paul R. Ehrlich wrote, "in the 1970s the world will undergo famines – hundreds of millions of people are going to starve to death." If you only had the historical data available to Ehrlich at the time, just a few years after the end of China's 'Great Leap Forward' that killed perhaps tens of millions of people in the deadliest famine in human history, his prediction might have been a reasonable one to make. In the century prior to Ehrlich's writing, no decade saw fewer than about fifteen people die per 100,000 due to famine. In the 1870s, the rate was nearly ten times higher than that as catastrophic droughts hit South America, Africa, Persia, India, and other parts of Asia. But the 1970s did not become a decade of famines and starvation. Quite the opposite.

Per capita deaths from famine worldwide fell more than 90% over the single decade of the 1970s. Even as the world population was growing dramatically, from about two billion in the 1930s to three billion in the 1960s to five billion by the 1990s, only about half as many people died from famine from 1970 to 2016 as died during just the 1960s.

The website *Our World In Data* suggests that increased food supply, improved health, reduced poverty, the spread of democracy, and the demographic transition (fewer children per woman) all played key roles. But the fact that the drop happened so quickly, over the span of less than ten years, indicates that another key factor might have been a disruptive technological innovation like containerized shipping, which spread rapidly in that time period. Lower cost and higher reliability of transportation might have had a disruptive effect on the death rate from famines.

In his 1996 book *The Lexus and the Olive Tree*, Thomas Friedman proposed his famous 'Golden Arches Theory of Conflict Prevention', where he half-jokingly claimed that 'no two countries that both had McDonald's had fought a war against each other since each got its McDonald's.' His reasoning was that "countries with middle classes large enough to sustain a McDonald's have reached a level of prosperity and global integration that makes warmongering risky and unpalatable to its people." His theory is not true: in 2008, Russia invaded Georgia, even though both had McDonald's restaurants at the time. Then, it invaded Ukraine. But, as a general idea, it is intriguing.

Likewise, a not-entirely-accurate theory as to why famines have largely disappeared is: No country has experienced a famine after they adopted containerized port infrastructure. We can compare the list of famines compiled by Our World In Data to the year each country established its first containerized port, as compiled by Dr. Gisela Rua. For example, the USSR's last famine was in 1946 in Ukraine and Belorussia. The USSR adopted containerization in 1971. The USSR has experienced no famines since then. Ethiopia experienced famines in 1957 (Tigray), 1966 (Wallo), 1972 (Wallo and Tigray), and 1983. Ethiopia adopted containerization in 1983. Ethiopia has experienced no famines since then.

Only three reported famines have happened after adoption of containerization: Nigeria had one in 1968, one year after containerization. India had a famine in Maharashtra in 1972, one year after Rua says containerization officially began. (However, in the 2008 book *Logistics Management*, Reji Ismail claims that "India began to seriously adopt containerization only in 1978.") And Congo had a famine in 1998 after adopting containerization in 1979. (In Congo's case, there was a war in the late 1990s and early 2000s that was fought in the eastern part of the country, far from the coast.) All four famines in the first two decades of the 21st Century have been in non-containerized countries.

It is understandable why Ehrlich, writing before the cusp of the most dramatic decline in famine deaths in human history, was so pessimistic, but his pessimism proved to be wrong. Famine, like many infectious diseases, seems to largely be a thing of the past.

According to Edward Humes in *Door to Door. The Magnificent, Maddening, Mysterious World of Transportation*, it was not just ocean freight that became containerized. "Containerization also came to air freight, and competition — in the air and on the ocean — drove down the cost of flying goods, particularly small ones, by an order of magnitude."

◆ ◆ ◆

"I think someday we'll have commercial flights between New York and London."

— Charles Lindbergh

◆ ◆ ◆

The Jet Engine Hurt Gander but Helped Florida

> *Powerful jet engines required airplanes to be re-designed. But because jets could fly faster, higher, and more-efficiently, they enabled airlines to be profitable by carrying only passengers, not just packages and postal mail, thus re-writing the global air route map, to the benefit of some places and detriment of others*

In 1901, as Rivka Galchen wrote in The New Yorker, "after one particularly disheartening summer at Kitty Hawk, Wilbur confided to Orville his feeling that 'not in a thousand years will man fly.' Two years later, they flew their plane".

The Wright brothers' first flight in their heavier-than-air vehicle was due to convergence of factors, including a four-stroke internal combustion engine (invented in 1876) made from aluminum, a material which could be obtained cheaply due to the Hall-Héroult process that was discovered thirteen years before the brothers began their powered-flight experiments. The brothers also benefitted from the rise of bicycles, which provided parts for their plane as well as the income to finance their airplane hobby. David McCullough, in his 2015 biography The Wright Brothers, said they even built a bicycle that ran over sand, to make moving themselves and their equipment easier.

According to *Van Nostrand's Scientific Encyclopedia,* airplanes improved dramatically in the decade after the Wrights' first flight in December 1903. Between 1908 and 1914 the airplane altitude record went from under 1000 feet to over 25,000 feet, the duration record from a couple of hours to more than 24 hours, the distance record from under 100 miles to over 1000 miles, and the speed record from about 30 miles per hour to about 120 m.p.h. (Considine, 2008)

That year, 1914, also saw the start of the world's first scheduled passenger airline service, between St. Petersburg and Tampa, Florida, a trip which according to ARK Invest analyst Sam Korus, cost $400 (in 1914 prices, or more than $10,000 today) and lasted twenty-three minutes.

According to Arthur C. Clarke in his 1973 book *Profiles of the Future: An Inquiry into the Limits of the Possible*, astronomer Willian H. Pickering, speaking after the invention of the airplane, dismissed it as a toy for the rich:

> "The popular mind often pictures gigantic flying machines speeding across the Atlantic and carrying innumerable passengers in a way analogous to our modern steamships.... It seems safe to say that such ideas must be wholly visionary, and even if a machine could get across with one or two passengers the expense would be prohibitive to any but the capitalist who could own his own yacht."

"By the time he died in 1938 at the ripe age of eighty," Clark reports, "Professor Pickering had seen airplanes traveling at 400 m.p.h." The first transatlantic commercial flight took off the following year, in 1939, and cost $375. That is, it cost less in 1939 to fly across the ocean than it had cost twenty-five years before to fly across Old Tampa Bay. While Columbus' original voyage had taken seventy days to cross the Atlantic, by the 500th anniversary of his trip in 1992, it was possible to cover that same distance in about seven hours, more than 150X faster, and still at prices of about $375 each way.

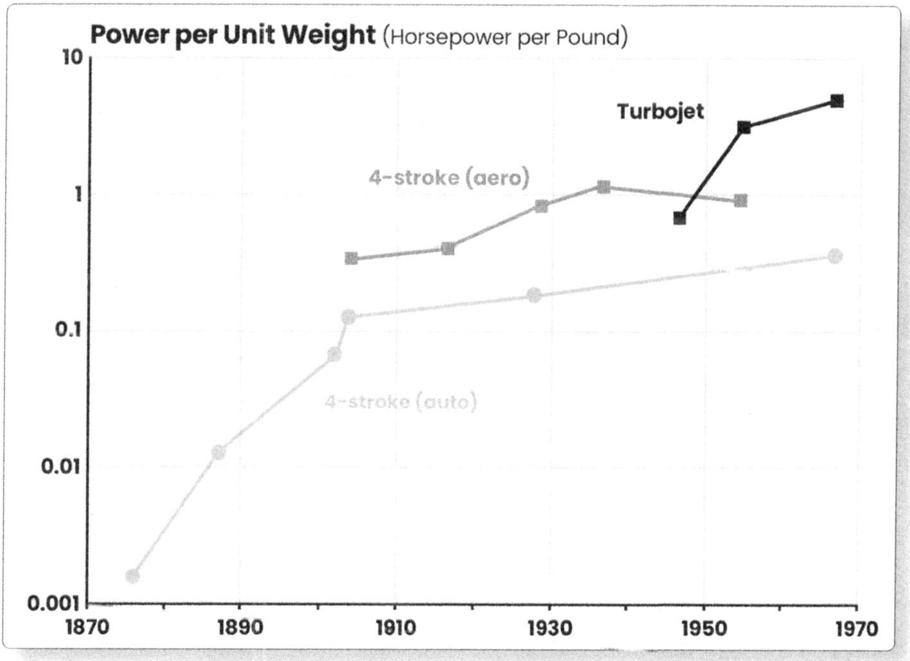

Figure 99 – Engine Power per Unit Weight. *(Ayers, 1989)*

What made this level of performance possible was a new type of engine, the turbojet, patented by British Royal Air Force cadet Frank Whittle in 1930. The Depression meant that Whittle's design received little attention until World War II. But by the end of the War, both the Axis and Allied Powers had brought their first jet-powered aircrafts into service.

The piston-engine, propeller-driven aircraft popular in the early 1940s were not able to make direct flights across the Atlantic Ocean, which is why airports from Gander, Newfoundland, to Bangor, Maine, to Windsor Locks, Connecticut, were requisitioned and upgraded to be able to handle military aircraft just before North American involvement in the war began.

By the mid-1950s, turbojet engines had grown powerful enough to spark a revolution, not only in aircraft propulsion and design, but in virtually every aspect of aviation. According to an analysis by the International Institute for Applied Systems Analysis (IIASA) in Vienna, Austria, by 1954 the turbojet engine was three to four times as powerful (per unit weight) as the four-stroke piston-driven engines that powered propeller aircraft, and they got even more powerful after that.

For a period, it seemed that a hybrid style of engine, the turboprop, might become the dominant design over the piston, propeller engine. But by the mid-1960s, the turbojet was the clear winner. A chapter called 'The Rate of Innovation in the Commercial Aircraft Jet Engine Market' in the 1976 book Technological Substitution: Forecasting Techniques and Applications reports that the market share for 'first-generation turbofans [turbojets] was about 15% in 1959 and rose to over 80% just three years later in 1962. Likewise, 'second-generation turbofans', released about ten years later, were just over 10% of the commercial aircraft jet engine market in 1969, over 75% in 1972, and over 80% in 1973.

Unintended Consequences of More-Powerful Engines

Changing the engine from propeller-based to jets did not just change the way the plane moved. It also changed how quickly the plane could fly, which meant it changed the height at which planes fly. A plane that flies higher encounters less air resistance, so it can be wider and longer, to hold more people, and can be more fuel efficient, flying them at a lower cost per person.

The British firm de Havilland test flew the Comet, the first jet airliner, in 1949 and the craft entered commercial service in 1952. It was more spacious than propeller-driven aircraft and offered a quieter, 'vibration-free' ride, but was plagued by numerous technical problems brought about by the more extreme flying conditions, which caused a couple of high-profile crashes that doomed the plane commercially. For example, the plane's windows were not rounded, causing stress at the corners during high-altitude flight. But the plane did have some lasting effects on commercial aircraft design. "Since the introduction of the first commercial airliner, the ill-fated British de Havilland Comet, in 1952 the body and wings of commercial aircraft have been made largely of aluminum," reported Ian Baker in his 2018 book *Fifty Materials That Make the World*.

It was not until thirteen years after the war that Boeing introduced the 707, the first commercially successful airliner that could be profitable carrying only passengers (and their luggage) and therefore did not financially need to have a contract to carry air mail. The jet engine changed the business model of flying from one focused on cargo and airmail to one focused on people. This changed who got to fly - from only businessmen and celebrities to a much wider audience, including

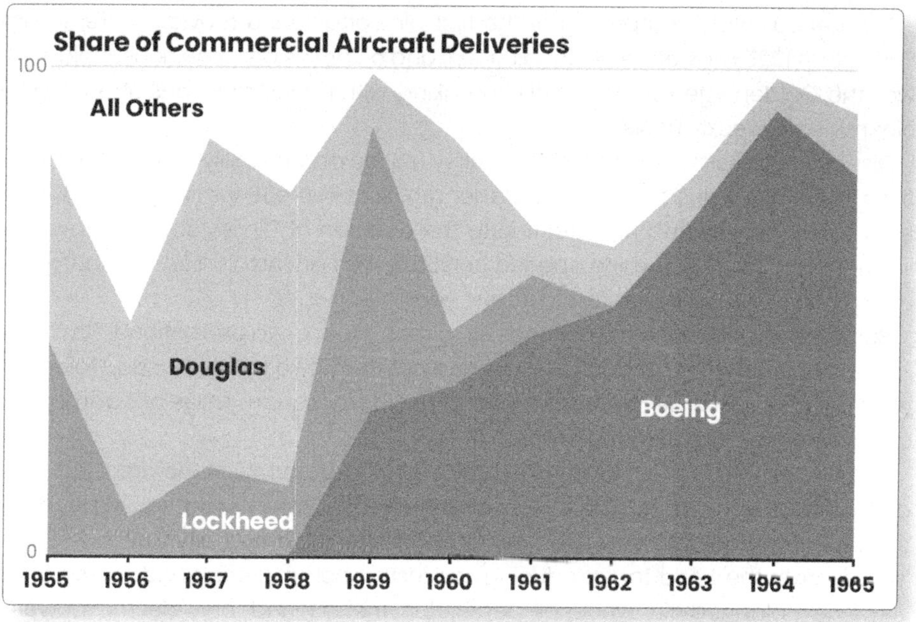

Figure 100 – Share of Commercial Aircraft Deliveries. *(Nelson, 1982)*

families. And it changed the places to which planes flew. Before the jet liner, planes would fly between major population centers (because that's where the mail and businessmen were going). After the 707, airlines could fly profitably from a major city to newly realized 'tourist destinations'.

Though Boeing's name is now synonymous with passenger airliners, prior to the 707, Boeing was a commercial industry outsider. In the ten years prior to the introduction of the 707, Boeing had not sold any commercial aircraft at all, leaving the market to Douglas, Lockheed, and a few other major players like Convair and Martin. But in the first year of deliveries of the 707, Boeing took over 30% of the commercial aircraft market, a figure that rose to more than 90% five years later in 1964.

The effect that these new long distance jetliners had on traditional transit hubs was devastating. What containerized shipping of goods did to New York's commercial waterfront, the jetliner did to its transoceanic passenger ship terminals and to transit stop airports. Gander, which had undergone a major renovation in 1959, saw its passenger traffic plummet as flights went straight to their final destinations, as did other stopover locations like Shannon, Ireland, or Prestwick, UK.

The Florida Project

Not everyone lost out on this new technology. Jet flight gave rise to the recreational cruise industry as old transatlantic liners were repurposed for pleasure cruises after jets made them obsolete for serious passenger transport. But it was in a small plane that Walt Disney took a flight over central Florida in 1959. Walt Disney had built his fortune and legacy from technological disruptions, including motion film and Technicolor, of course, but also the air conditioner. Now he was looking to expand his empire.

The period after the opening of the first air-conditioned movie theater in New York City in 1925 but before air conditioners and color TVs became household items was the 'Golden Age' of Hollywood filmmaking, with most Americans attending the cinema on a weekly basis.

Disney's success was built first on Mickey Mouse and the 'Silly Symphonies' short films in the 1930s and then on feature-length films in the 1940s and beyond. In the early 1950s, he devoted his attention to the creation of Disneyland, a theme park in Anaheim, California which opened in 1955. It was an instant success, attracting more than three million visitors in its first year.

But Disney's ambitions did not end with a theme park. In a promotional film made in the months before his death, Disney laid out the vision for his 'Florida Project', an enormous futuristic development spread over forty square miles of swamp land, about twice the size of Manhattan.

The plans called for a new theme park covering about 1% of the area. But, in his own words, the 'heart of everything we'll be doing in Disneyworld' would be EPCOT, the 'Experimental Prototype Community of Tomorrow', a new city that would be a 'showcase to the world' for 'introducing and testing new materials and new systems'.

The new planned city would be pedestrian-friendly and "only electric-powered vehicles will travel above the streets of EPCOT's central city", relying on monorails and short-distance 'peoplemover' trains for mass transit. Cars and trucks would be relegated underground to a series of tunnels. At its center was originally planned a hotel, but this was later replaced with the now-iconic Spaceship Earth, a geodesic sphere designed by engineer Buckminster Fuller.

One thing that made Disney's vision possible was part of what contributed to his success in filmmaking: air conditioning. Florida's population grew approximately 40% every decade from the 1870s through the end of the 1950s. But with the advent of household air conditioning, the population nearly doubled during the 1960s.

The rise of air conditioning changed land values. In the early 1960s, Disney started his 'Florida Project', realizing that cheap land in a Florida swamp plus declining prices of family airfares (thanks to the 707) plus air conditioning mean that hordes of northerners would want to come South in the winter, just for a family holiday. He died before Disney World finally opened, but it is now the number one tourist resort in the world, with more than fifty million visitors a year.

Don't count propeller aircraft out yet. A new generation of small, electric-powered 'drone' aircraft and 'air taxis' are being developed both for goods deliveries and to transport people short distances. And their designs use propellers.

The Kodak Moment: When the Disruptor Got Disrupted

> *Kodak got disrupted by digital cameras, a technology they invented. But from dry plates to roll film, paper-base to cellulose, the razor-and-blades strategy, black-and-white to color, the introduction of safety film and many other innovations, the Eastman Kodak company knew how to cause disruptions too*

Ask any businessperson to name a company which got disrupted by new technology, which was faced with a fundamentally new way of doing business and which badly bungled their response and one of the most popular answers might be Kodak getting nearly wiped out by digital photography in the first decade of the 2000s. According to an article in *The Economist*, in the 1970s "Kodak commanded 90% of film sales and 85% of camera sales in the U.S." But due to the rise in popularity of digital cameras at the end of the 1990s, revenue from color film sales and processing, Eastman Kodak's core business, peaked in 2001 (when digital camera sales were still only about half of film camera sales) and then their core business experienced a rapid decline in the ten years that followed. In 2012, Kodak declared bankruptcy. At the end of the 1970s, according to Kaitlyn Tiffany in *The Atlantic* magazine, "Kodak employed more than 50,000 people in Rochester... By the time Kodak filed for bankruptcy, in 2012, it employed just over 5,000 people".

The company's memorable expression 'A Kodak Moment', which meant a time in their customers' lives worth being preserved on film, was re-interpreted after Kodak's bankruptcy to mean 'the instant an incumbent realizes their business is doomed'.

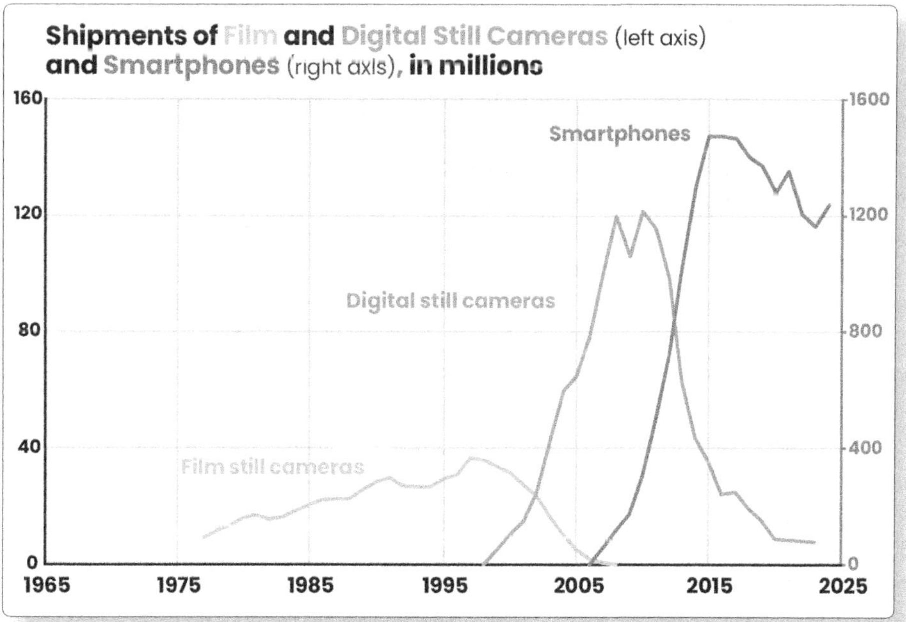

Figure 101 – Shipments of Film and Digital Still Cameras and Smartphones, by year.
Data from (CIPA, 2021)

Photography has not died out, of course. On the contrary, taking photos is far more popular than ever. What got disrupted was simply the use of film. According to statistics from the Camera and Imaging Products Association (CIPA), the sales of film still cameras peaked in 1997. Media and technology analyst Benedict Evans claims that in 1999, "the peak of the film-camera industry" (in his words), consumers took around eighty billion photos. In late 2022, Apple revealed that three trillion photos were taken with iPhones over the previous year, or roughly three thousand photos per phone. And there are more non-Apple phones in the world than Apple phones. Reasonably speaking, there are probably about one hundred times as many photos taken per year now compared to the late 1990s.

A Flurry of Photos

This was not a one-for-one replacement. In each of the years from 2007 through 2011, digital camera sales were more than one hundred million units per year, more than three times the peak annual sales of film still cameras just eleven years before. The sales of film cameras had fallen so low that CIPA simply stopped compiling and reporting data about them after January 2008.

It is not as though Kodak was unaware of digital imaging. In fact, they had invented it. A Kodak engineer, Steve Sasson, made the first hand-held, self-contained digital camera in 1975. In 1986, Kodak's research labs developed the first mega-pixel digital camera. And former Kodak executive Vince Barabba recounted in his book *The Decision Loom: A Design for Interactive Decision-Making in Organizations* how in 1981 he conducted an analysis of the "core technologies and likely adoption curves" behind digital photography, concluding that the physical film business was under little threat at the time and into the foreseeable future. According to an article in Forbes magazine, "History proved the study's conclusions to be remarkably accurate, both in the short and long term." (Mui, 2012)

So, Kodak saw the potential of digital technology, and they tried to take advantage of it too. In fact, by 2005 Kodak was the #1 company in the U.S. in digital camera sales. In an article in *Harvard Business Review*, Scott D. Anthony, co-author of *Eat, Sleep, Innovate*, claims, "where they failed was in realizing that online photo sharing was the new business, not just a way to expand the printing business". A great example of this is the 1996 introduction, just before digital cameras took the world by storm, of Kodak's Advantix Preview, a film camera that used digital technology to allow users to preview their photos. Kodak simply saw digital technology as a way to sell more film, photo paper, and processing.

Ironically, stand-alone digital cameras, which had seen about thirteen years of dramatic growth, increasing from essentially zero in 1998 to more than one hundred million units per year by the time of Kodak's bankruptcy, were themselves largely disrupted by mobile phone digital cameras. Within four years of Kodak's bankruptcy, stand-alone digital camera sales had fallen by more than 75% to fewer than twenty-five million units per year, a number that would fall by more than half again before the end of the decade. Digital still camera sales in the year 2020 were down nearly 93% from their peak just ten years prior.

Kodak's History as a Disruptor

But digital technology was not the first disruption Kodak faced. In fact, Eastman's company was based on disruption. Twenty-three-year-old George Eastman had taken up photography in 1877 and within a couple of years invented and patented a machine to coat the glass plates photographers used. In April 1880 he formed The Eastman Dry Plate Company to bring his product to market. To be clear, Eastman did not invent photography. Far from it. The first process to make an exposure to light become a permanent image was invented by the Frenchman Louis Daguerre fifteen years before Eastman was even born. That is, Eastman did not invent photography, he *re-invented* it to turn it from something complicated that only experts could do, to one that nearly anyone could do.

In 1884, he patented the first film roll using a coating on a paper base, and in 1888, gave up his profitable dry plate efforts to begin selling the Kodak camera to turn photography from a technically difficult hobby to something that would be popular among a wider audience, using the slogan "You press the button, we do the rest".

A couple of years before the first Kodak had been sold, Reverend Hannibal Goodwin in 1886 filed for a patent on producing photo film using celluloid plastic as the base, but Eastman began manufacturing the Kodak camera with celluloid-base film before Goodwin's patent was awarded. (A twenty-year legal battle ensued, which ended with the Eastman Kodak company finally settling with Goodwin's estate).

Glass plates were heavy and fragile, and Eastman's original paper-based film was not very good quality, but an advertisement from the Celluloid Manufacturing Company (now called Celluloid Corporation) said of celluloid, "the rapidity with which this material... has been received by the Photographic world, will cease to be wondered at when it is known that 100 'Celluloid' negatives can be packed into the space of an inch, that their weight is only about one-fifteenth that of glass".

When the owner had exposed all one hundred negatives, the camera could be returned to Eastman Kodak for processing, and for the camera to be reloaded with more film. In business terminology, this is called the "shaver and blades" strategy: sell a device at low cost, but then charge a premium for a consumable component, like the film or the razor blade. (In fact, Gillette did not begin to manufacture razors with disposable double-edge blades until 1904, sixteen years after Kodak camera was introduced, so maybe the "shaver and blades" strategy in the business literature should instead be called the "Kodak camera and film" strategy.)

One of the unintended consequences of the use of transparent celluloid instead of glass plates or opaque paper backing was that it enabled an entirely new kind of photography, motion pictures. What is believed to be the oldest film in existence, 'The Roundhay Garden Scene', shot in Leeds, England, in October of 1888 and only about two seconds long, was recorded on Eastman Kodak film. For his contributions to still photography and to motion picture photography, George Eastman is the only person to have two stars on the Hollywood Walk of Fame in Los Angeles, California.

By 1930, Eastman was in intense pain caused by a spinal disorder. He had already witnessed his mother's decline and death from a similar ailment. Wishing to avoid a similar fate, on March 14, 1932, George Eastman sat down at his home in Rochester, New York, wrote a note which read, "To my friends: my work is done. Why wait?", and then shot himself in the heart.

But the company he founded continued. In 1935, they introduced Kodachrome color film and in 1948 launched cellulose triacetate film, also called "safety film", which was far less flammable than the cellulose nitrate film that had been the standard for sixty years. Within two years they stopped producing rolls of the old film and by 1951 had stopped with the motion picture film as well, disrupting their own business over a period of just three years.

It is fair to say that the Eastman Kodak company eventually got disrupted by digital photography. But from dry plates to roll film, paper-base to cellulose, the razor-and-blades strategy, black-and-white to color, the introduction of safety film and many other innovations, the Eastman Kodak company knew how to cause disruptions too.

The Printed Page: Two Thousand Years of Disruption

> *Gutenberg's Bible was a convergence of technologies many centuries in the making, primarily paper, the codex, ink, moveable type, and the winepress.*

The invention of paper is typically attributed to Cai Lun, a Chinese court official, in 105 CE when he probably was in his early fifties. When Emperor An assumed power in the year 121, Cai Lun was investigated for his role in the death of An's grandmother, Empress Jingyin, who had committed suicide with poison thirty-nine years prior in 82 CE after being accused of witchcraft. Expecting to be sentenced to death, Cai Lun bathed and dressed in court clothes, and then himself drank poison. Paper had existed for centuries, but Cai Lun is credited with devising a popular formulation that used bamboo, hemp, old rags and bark from mulberry trees, the leaves of which are the sole food for silkworms. Papermaking spread slowly over centuries from China to central Asia, the Middle East, and then Europe, where high-quality papers were being made in Gutenberg's time from linen and other fibers.

By the mid-1400s, the codex, the idea of binding pages together in a protective cover, rather than simply rolling parchment into a tube, had been the preferred form of storing written text in Europe for a thousand years. The conversion of the Roman population to Christianity saw the decline of the scroll and the rise of the codex book, but the reasons for this are debated among historians. Perhaps it was because Christianity (like Islam and Judaism, and unlike the pagan faiths that came before them) has a sacred text that impels adherents to make the finest copies possible and to treat these documents with great respect.

Ink as well was an ancient invention. Various recipes called for lamp soot, iron salts, tannins from nuts, pine tree resin, bark boiled in water, or wine. The Chinese made ink in the form of solid sticks, some of which would be shaved off and mixed into water. But Gutenberg's idea to apply ink to metal cast letters meant that he could not use these water-based recipes, as the ink would not adhere to the metal type. Printing with a press, rather than writing by hand with a quill, required a rethink of ink.

Richard Schwab and co-workers at the University of California at Davis shot pages of the Gutenberg Bible with a low-energy beam of protons that was so narrow that they could target a region on a page as small as the punctuation mark at the end of a sentence. The proton beam caused the atoms in (or on) the page to emit X-rays and from those X-rays the scientists could determine which elements were in the paper and the inks that Gutenberg used.

Surprisingly, the scientists found that Gutenberg's black inks contained copper, iron, manganese, and lead and the red inks were "rich in mercury", while the paper in the Bibles did not have any of these. Some copper oxides are black and "lead oxides turn a dark purple", perhaps contributing to the ink's "famous glossy black color." In all, Schwab and co-workers found that Gutenberg's oil-based inks were unlike writing ink and instead were so much more like the paint that oil painters used that he "may have drawn his recipe quite directly from the oil painters" of the time.

With potential customers for a Bible, paper and ink, Gutenberg also needed many pieces of metal type, each of which held a single letter or character, arranged in the form of a page, and a means to press them onto the paper.

According to *The Verge: Reformation, Renaissance, and Forty Years that Shook the World*, by Patrick Wyman, printing the original run of one hundred eighty copies of the Bible "required the casting and setting of 100,000 individual pieces of type" and six printing presses. Additionally, forty of Gutenberg's one hundred eighty copies were not set on paper, but rather on vellum, which were preserved animal skins. According to Wyman, "the skin of about 170 calves would have been required for each copy of the Bible printed on vellum", for a total of nearly seven thousand calves.

Wyman says that the actual printing run "entailed twelve printers working with six compositors... for 330 working days". In comparison, "a scribe might need three years to produce a manuscript Bible". If we assume scribes worked 310 days per year (six days a week for fifty-two weeks), then a manuscript Bible took 930 person-days of labor. In comparison, Gutenberg and his eighteen men took 330 working days (or 6270 person-days) to make one hundred eighty printed Bibles, or an average of about thirty-five person-days per Bible.

So, not considering the time and effort to make the presses or moveable type (which were re-useable), to make the paper or vellum and inks, or the effort of setting up a print shop or scriptorium, the printing press was at least a twentyfold improvement over scribes in the rate of production. And the quality of the product was better as well. In *The Gutenberg Bible of 1454*, author Stephan Füssel says that "this first printed work is still frequently upheld as being the most typographically accomplished ever to be produced... the printed lettering itself was of an astonishing clarity, and correctly set, and to such a degree that one could easily read the text without spectacles."

In a personal communication Tony Seba said, "I see the printed book as a "Big Bang" disruption. The Gutenberg bibles of 1454 were cheaper (by ~10x) and better (higher quality) than manuscripts... Printing technology improved in cost and capability over the next 450 years and opened up incredible new possibilities, but the manuscript was finished [as a means of text production] on day one."

The numbers back him up. In a 2009 article in *The Journal of Economic History*, Eltjo Buringh and Jan Luiten Van Zanden of Utrecht University in the Netherlands estimated that from the beginning of the Sixth Century through the end of the Fifteen Century (that is, between the years 500 and 1500), there were nearly 10.9 million manuscripts produced in Western Europe. In contrast, during the first decades of the printing press (from Gutenberg's first works through 1500) there were more than 12.5 million books printed, more than the number of manuscripts produced since at least the year 500, and likely more than the number of manuscripts ever produced in Western Europe. The authors did not even attempt to estimate the number of written manuscripts produced after the year 1500, as it was so low.

The rate of book production only increased. The number of books printed from 1501 to 1550 was more than six times larger than the number printed before 1500. As Seba said, the manuscript was finished on day one. "In the year 1550 alone," the authors report, "some 3 million books were produced in Western Europe," more than one-quarter of the number of manuscripts ever produced.

The ripple effects of printing were staggering. "Not only did printing revolutionize the book publishing trade," Wyman says in *The Verge*, "but it kick-started a massive overhaul in communications, transforming everything from intellectualism to religious belief in the relative blink of an eye. Indeed, none of the enormous shifts of the sixteenth century — from the upheavals of the Reformation to global exploration to the approaching Scientific Revolution — could have happened without the emergence of the printing press."

The conversion of northern Europe to Protestantism suddenly made large numbers of old manuscripts written in Latin worthless. When automobiles disrupted horses for transport hundreds of years later, people began selling excess animals for human consumption and for pet food. Likewise, in the late 1400s, many manuscript documents were cut up and used for other purposes, like to reinforce book bindings.

Professor Joseph Henrich of Harvard University argues that literacy not only changed our cultures, it also changed our brains. In his 2020 book *The WEIRDest People in the World: How the West Became Psychologically Peculiar and Particularly Prosperous*, "learning to read forms specialized brain networks that influence our psychology across several different domains, including memory, visual processing, and facial recognition. Literacy changes people's biology and psychology".

Printing saw many more disruptions over the centuries. An episode of *The History Guy* on YouTube describes the series of disruptions that constitute the 'Paperback Revolution', a shift from expensive hardcover books to 'dime novels', 'pocket books', and 'paperbacks', from roughly 1860 to 1960 as demand for cheap reading material grew due to the rise of rail transport and as printing technology improved due to cheaper steel, cheaper paper, and eventually, electronics. One publisher's advertisement from 1860 read: "Books for the millions! A dollar book for a dime. 128 pages complete. Only 10 cents." When The Beatles recorded their song 'Paperback Writer' in 1966, the term was younger than the members of the band.

Figure 102a & b – Popularity of 'dime novel' in the 1860s and 'paperback' about a century later.
Graphs from Google Books Ngram Viewer.

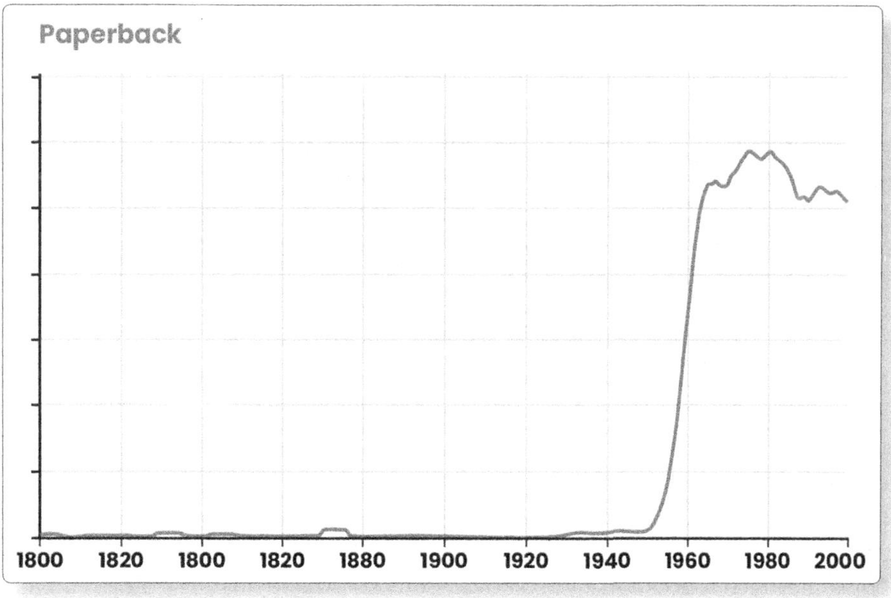

These were not one-for-one disruptions. They instead increased the overall demand for books. They sparked an explosion in entirely new genres of writing with mass appeal, like crime novels, "blood and thunder" action novels, science fiction, and other sorts of 'pulp fiction'.

Cheaper books also changed where books were sold, from designated bookstores to train stations and drug stores. By 1960, paperback books outsold hardcovers for the first time ever.

Glossy Magazines and Catalogs

> *Over about five years in the late 1950s and early 1960s, magazine covers went from largely black-and-white to largely color, as better printing technology driven by the offset press enabled cheaper and better color images*

The cover of *National Geographic* magazine is so iconic, with its bright yellow border framing a colorful, striking photo, that it is easy to forget that they originally did not have photos at all but instead consisted only of a plain black-and-white table of contents. The magazine was founded in 1888, but the first color photo, an image of the US flag that occupied only a portion of the page, did not appear on the cover until July 1959, just over sixty years ago. (Put another way, there have been more years when the covers of *National Geographic* magazine did not contain photos than years where they have contained photos.)

This new format, with a partial photo and plain text table of contents continued until May 1962, when the first full-photo cover appeared. Since then, every *National Geographic* cover has been a full-color photo (or hologram). That is, it took just three years, from 1959 to 1962, for *National Geographic* to go from no color photos on their cover to only color photos.

The first issue of *Playboy* magazine, published in December 1953, contained only black and white photos, plus one full-color single-page photo of Marilyn Monroe, the 'Sweetheart of the Month'. Even the cover was black and white. Covers that partly contained photos were published in the following years. And the first three-page fold-out centerfold was March 1956. That means that years before *National Geographic* had even a partial color cover, *Playboy* had a full-page color photo cover and a three-page centerfold.

From No Photos to Entirely Photos in Five Years

It's not just *National Geographic* and *Playboy* that changed like this. So did *Cosmopolitan*. So did cookbooks, which used to just be textbooks of written recipes, but which have become basically coffee table books of only full-color photos. Swedish furniture and housewares retailer IKEA is known for their colorful catalogs. In 1958, the cover consisted entirely of a multi-colored drawing. From 1959 to 1961, it contained a partial photograph. And from 1962 it was a full-page photograph. That is, it took IKEA about five years to go from no photos on their catalog cover, to partial photos, to entirely photos.

What happened in the late 1950s that enabled magazines and catalogs to start printing full-page, full-color photos all at about the same time? Cheaper, better color film (and access to airplanes with jet engines) meant that photographers

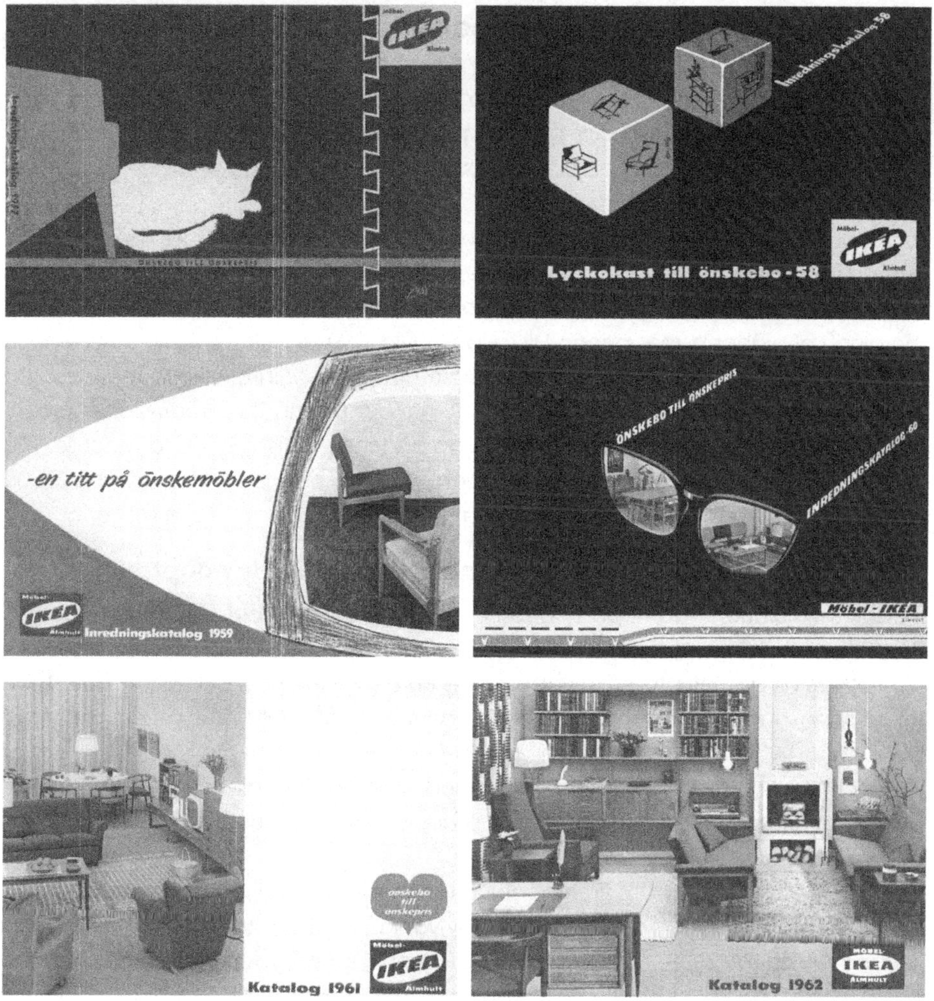

Figure 103 – IKEA catalog covers from top left 1957 to bottom right 1962. *(IKEA Museum, 2020)*

could shoot more photos in more places. A less-expensive way of printing that used new electronic technology based on the transistor (the first generation of which had only become available at the beginning of the decade, and which was rapidly getting cheaper). And high-speed offset presses that mounted printing plates automatically in only a few minutes, reducing preparation time, and thus printing cost.

Older style printing plates required a raised ink-bearing layer to be added to the plate, but the newer plates instead used a more-durable layer that was selectively removed, making the plates able to print more copies at higher speed. Ultimately though, these photolithography techniques used by the printing industry contributed greatly to the development of computer technology, and thus to the disruption of printing. This was not a one-for-one replacement; the amount of data stored and transmitted digitally is now many orders-of-magnitude larger than what is put onto the printed page.

Newspapers: Wood-Pulp Paper Disrupted Linen Rag

> *Until the late 1800s, paper was made largely from linen. But powered machinery for grinding trees into pulp completely revamped the newspaper industry in a short time*

For about a thousand years, people used rags to make paper. Gutenberg himself printed most of his Bibles on paper made from linen. But by the middle of the 1800s, with literacy rates rapidly rising, newspapers in the US and Europe were feeling the pinch of higher paper prices. Historian David C. Smith wrote that, "as the price of rag paper soared, the search for a substitute product intensified. [In] 1854... the price [of rag paper] rose two and a half cents per pound that year alone. In 1855 the price rose again by a half-cent a pound and the publishing world reacted. The London *Times* found itself in such a position because of paper prices, that a £1,000 reward was offered for the discovery of a new and readily available material."

Outsiders Remade Paper, which was Quickly Adopted

In fact, the solution had already been invented a decade before. In 1844, twenty-three-year-old Canadian Charles Fenerty, an inventor who was also well-known in his lifetime as a poet, after five years of experimentation had perfected a method for making paper from ground and bleached wood pulp, but never patented his idea. In 1845, working independently, twenty-nine-year-old German weaver Friedrich Keller patented a similar method, but quickly ran out of money and eventually let his patent lapse. It took a couple of decades to make their ideas commercially feasible.

"In 1867," Smith wrote, "the first machine for grinding wood into pulp was installed in the United States at Manayunk, Pennsylvania... By the early 1870's newspapers around the country were beginning the shift from paper made of rags, with its high cost, to paper made of wood pulp at half the cost." Smith notes, "some newspapers started early: the *New Yorker Staats-Zeitung*," a German-language newspaper, "was experimenting as early as January, 1868... The *New York Times* started using wood-pulp stock on August 23, 1873, and by 1874 had changed over completely... One can say that by 1882 nearly every paper in the United States with a large circulation had changed over." That is, it took only about thirteen years from the first newspapers trying wood-pulp paper until 'nearly every paper' had switched.

Wood Pulp Transformed the Newspaper Industry

The impact of wood-pulp paper on the economics of the newspaper industry was profound, not only in the cost of the paper itself, but also in labor and other factors. According to a 2002 article by Ted Smythe, professor emeritus at California State University-Fullerton, production costs "plummeted and the time spent in preparing news for publication had decreased until it was cheaper, faster, and easier to print 100,000 copies than it had been to print 10,000. The decreased cost of white paper as a result of the new [wood pulp] technology was reflected in lower prices per copy as well as an abundance of newsprint, both of which permitted extraordinary circulations to develop."

Smith reported that "the circulation of American newspapers increased 2.4 times" in the twenty years from 1860 to 1880, far above the rate of population or economic growth, and that "between 1880 and 1890 the amount of paper used by American newspapers" rose from over one hundred million pounds per year to nearly seven hundred million pounds, a sevenfold increase in just ten years. The price per copy of newspapers plummeted. Smith writes that "the New York *Herald* which had sold for 4 cents dropped to 3 in 1876, and then 2 cents in 1883. The *World,* which was a 5 cent paper in 1866, sold for 2 cents in 1882, and this for a paper which had doubled its size to 8 pages in the same year."

It was not just the price and size of newspapers that changed in this time, the entire industry required a rethink. With a wider readership and more single-copy purchases, the style and layout of newspapers were changed to add attention-grabbing large-font headlines, pictures, and illustrations.

The content changed as well. Smythe writes, "the shift in content was reflected in a drop in political news. The percentage of space devoted to politics in a group of quality newspapers declined from 50 percent during antebellum years to 19 percent in 1897, while space devoted to crime and courts, accidents, society, and leisure activities (including sports) increased three-fold or more."

The publishing frequency changed, with many daily papers in large urban areas offering both a morning and evening edition, rural weekly papers shifting to a daily format, and most urban weeklies simply being discontinued. Even the structure of the newly enlarged newsroom staffs changed, as managers began organizing and overseeing personnel, something that, according to Smith, "had not been done systematically before the 1880s."

Wood-pulp paper was not the only disruption that affected the newspaper business in the late 1800s. The rise of steel starting in the 1870s revolutionized printing machinery. Gas lighting made it easier both to read and to make newspapers. And cheaper transport brought urban newspapers to suburban readers and suburban commuters into large cities.

Now, of course, printed newspapers are being disrupted by online news and social media. From late 2018 to late 2022, the circulation of *USA Today,* the largest national newspaper in the US, dropped more than 93%. And in 2023, the world's oldest national newspaper, Austria's *Wiener Zeitung,* printed its final edition after three hundred and twenty years in business. But it is fair to say that the work of two paper industry outsiders, a Canadian poet and a German weaver, forever reinvented paper in the mid-1800s with effects that continue to this day.

Unintended Consequences: Bigger Dictionaries and Retail Catalogs

It was not just the newspaper industry that was affected by cheap paper: the dictionary was too. Webster's original dictionary from 1828 contained roughly 70,000 entries and in 1850 'The Imperial Dictionary', published in Glasgow, Scotland, and based on Webster's dictionary, added British terms, expanding the size to 100,000 entries. Merriam's 1860 dictionary was about 114,000 entries, Webster's 1890 edition contained about 175,000 entries, with a supplement published in 1900 that expanded it to 200,000 entries. The 1909 edition had more than 400,000 entries and by 1993, just before the advent of the internet to the public, Webster's contained nearly 500,000 entries.

That is, between 1900 and 1909, Webster's dictionary *doubled* in size. Were that many new words really added to the American English language in that time? Or was it simply this: wood-pulp made paper cheaper and as paper got cheaper, publishers expanded the number of dictionary entries. But compiling new editions of the dictionary takes many years, so the wood-pulp paper revolution of the late 1800s took some years before it showed up in the form of a greatly expanded dictionary.

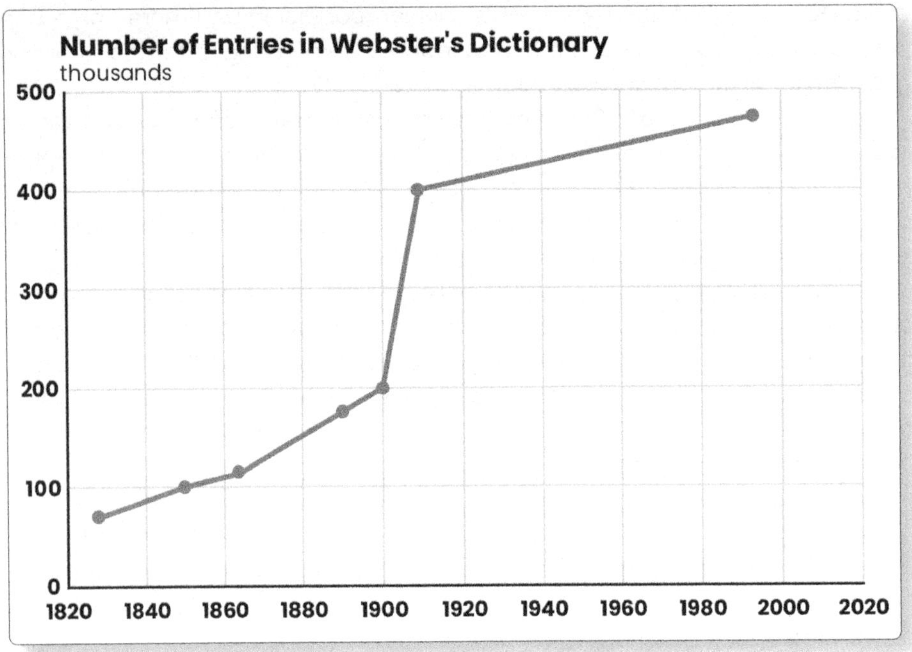

Figure 104 – Number of Entries Webster's Dictionary

Cheap paper changed more than newspapers and the dictionary. In 1886, twenty-two-year-old Richard Sears, working as an agent at a railroad station in Minnesota, began selling pocket watches to fellow railroad workers. His company printed its first advertising catalog in 1893, offering only watches, but within a few years had expanded to a wide range of items, all of which could be shipped to customers in remote locations by railroad. Sears' business eventually became the largest retailer in the US.

Disposable Pens Reinvented Writing, and Led to the Birth Control Pill

> *Plastics turned writing pens from steel-bodied, silver-plated luxury items to something almost disposable. The man who brought the best-selling ballpoint pen to America then invested his money in a chemical company that invented the birth control pill*

By all accounts the crowds were buzzing with excitement one day in late October 1945 at Gimbels department store, located in Manhattan near the Empire State Building, to see the shiny new "Reynolds Rocket". With a silver-plated steel body and refillable chamber, an estimated *five thousand* people showed up on the day it debuted for the chance to lay down twelve dollars and fifty cents (over $220 in 2025 dollars) to get their hands on one. It was like an Apple product launch day in the days before Apple.

American businessman Milton Reynolds had first seen the idea on a trip to Argentina. In the 1930s, fountain pens were essentially 100% of the writing pen market. A fountain pen that held a refillable reservoir of ink was patented in 1884 and the first ballpoint pen in 1888. But both kinds were prone to leaking.

Brothers László and Georg Bíró realized that it was the ink itself that needed a rethink, just as Gutenberg had done. At the Budapest International Fair in 1931, they presented a ballpoint pen that did not leak. Unlike previous ballpoints, the Bíró model used newspaper ink, which was more viscous than the pen inks of the time, but which also dried quickly, so it did not smudge after being applied to paper.

The Ballpoint Becomes a New Luxury

Reynolds copied the brothers' idea, modifying it just enough to avoid patent issues, and began selling his luxury, refillable ballpoint pens in the US. In an America ready to splurge after years of scrimping during the war, the Rocket became the first commercially successful ballpoint pen and made Reynolds wealthy. According to *Time* magazine, Gimbels alone sold more than $5 million worth in the first six months. (Rothman, 2015)

Reynolds sold his pen company, but had numerous other interests. He set an around-the-world airplane flying record, wrote a book about his travels in South America, and became one of the first investors in Syntex, a Mexican company that developed the oral contraceptive pill that revolutionized birth control starting in the 1960s.

But if you had to pinpoint the peak of the glory days of the luxury, silver-plated pen, that day at Gimbels in late October 1945 when the Reynolds Rocket debuted is a good choice because, also in the year 1945, the Bíró brothers sold the rights for their idea to Italian-born Frenchman Marcel Bich. And Bich had radically different ideas for what a ballpoint pen could be.

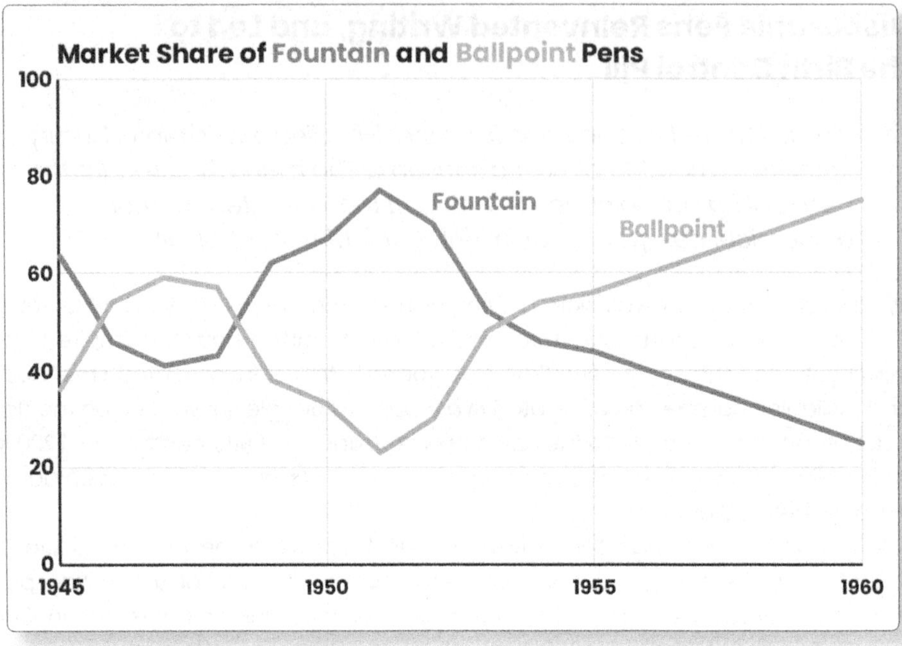

Figure 105 – Market Share of Fountain and Ballpoint Pens. *(Schnaars, 2002)*

A Competitor Emerges from a Convergence of Technologies

In a land devastated by war, Marcel Bich knew there would be no crowds eagerly waiting with hands full of money to buy luxury items of any kind. So he began developing a ballpoint pen that was low cost due to a convergence of technologies that were only newly possible. He made the pen body and ink reservoir from transparent plastics instead of silver, and used metals in the writing tip that workers at the time were extremely familiar with, including brass and stainless steel.

In 1950, Bich introduced the 'Bic Cristal' pen in Europe for a price of about eighteen US cents – a tiny fraction of the cost of Reynolds' Rocket – and by 1953 was selling forty million of them per year. According to Steven Schnaars in *Managing Imitation Strategies: How Later Entrants Seize Markets from Pioneers*, less than ten years after the Cristal debuted "by the late 1950s Bic held an astonishing 70 percent of the European ballpoint market."

This is part of the 'pattern of disruption' we have noted repeatedly: an industry outsider delivers a new product to the market that, in the case of the Bic Cristal, delivers equal or better performance at lower cost, and then that new product dominates the market within a decade or two.

Schnaars says plastic disposable Cristal pens from an industry outsider like Bich were "an idea that horrified the industry leaders" but that "low-priced pens forever changed the pen industry. They reconfigured the product from an expensive, almost jewelry-like, purchase, to an incidental disposable. Bic was more than successful. In fact, its entry relegated the industry giants – Parker, Sheaffer and especially Waterman – to the now much smaller high end of the market."

The Bic Cristal had sold one hundred billion copies by late 2006, more than a dozen for every person currently on Earth. This was not a 'one-for-one' disruption – the number of plastic disposable pens per person today far exceeds the highest number of fountain pens or luxury ballpoints at any point in history.

The Disposable Ballpoint Remade Writing

And this revolution forced a complete rethink of the act of writing. According to David Sax, the author of *The Revenge of Analog*, "the ballpoint pen was the equivalent of today's smartphone. Before then, writing was a stationary act that had to be done in a certain environment, on a certain kind of desk, with all these other things to hand that allowed you to write." The fountain pen required its own writing space, a desk with all the *accoutrements* of writing, not just the pen itself but also an inkwell and, because the pen was leak-prone, a desk blotter. But after the revolution, anyone could carry a pen in their pocket, write whenever they needed to, and not be upset if their pen got lost or stolen.

Like Reynolds, Bich was also an adventurer. An avid sailor, he was inducted into the America's Cup Hall of Fame in 1998, a few years after his death. His BIC company remains the world's leader in cheaply mass-producing mixed material consumer products, with disposable cigarette lighters, disposable razors, and of course disposable pens making up the bulk of their business.

◆ ◆ ◆

"The wireless music box has no imaginable commercial value. Who would pay for a message sent to nobody in particular?"
— David Sarnoff, founder of NBC

◆ ◆ ◆

Radio, Television and Movies

> *Sending signals wirelessly was just a parlor trick until Marconi's company built powerful diesel-powered electric generators and invented the vacuum tube, which gave rise to televisions and electronic computers*

Twenty-year-old Guglielmo Marconi first demonstrated a functional radio transmitter and receiver to his mother in December 1894, but it was little more than a curiosity, using a button on one side of a room to create a spark that generated the radio waves that wirelessly rang a bell on the other side of the room. By the summer of 1895, he had transmitted a signal a half mile and, using a higher power transmitter, extended the range to two miles later that year. To continue developing his ideas for radio, he relocated from Italy to England in 1896 and, through a series of improvements to his equipment, extended the range to four miles by 1897 and then to across the English Channel in 1899.

A Convergence Unlocked the Radio Spectrum: Combustion Engines and Vacuum Tubes

Doing more than that would require a higher-power transmitter than what Marconi could build himself. He hired English physicist and engineer John Ambrose Fleming to make a spark transmitter run by a 25-kilowatt combustion-powered alternator and used it to send the first radio transmission across the Atlantic from Cornwall, England, in December 1901.

But it would take a major technological advancement to truly unbridle the power of radio waves. In 1904, Fleming invented the vacuum tube, a device which allows electrons to flow only in one direction. Just as copper tools gave rise to the potter's wheel, the wagon wheel, the domesticated ox and the plow, the vacuum tube enabled commercial radio, television (which was broadcast by radio waves), Wi-Fi (which also operates on radio frequencies), radar, magnetic resonance imaging (MRI, which also uses radio waves), long-distance telephone transmissions, and early computers (which used Fleming's vacuum tubes).

The rise of radio was due to a convergence of technologies: the scientific discovery of radio waves, Fleming's vacuum tube, and the combustion-engine-based spark transmitter to power it all. Scottish physicist James Clerk Maxwell had predicted the existence of radio waves in 1867, but it took twenty years, until 1887, for the German physicist Heinrich Hertz to figure out how to produce them. When asked about the applications of his discovery, Hertz replied "nothing, I guess" and "it's of no use whatsoever." He could not have been more wrong. Serbian electrical engineer Nikola Tesla gave the first public demonstration of radio in St. Louis in 1893. The following year, Marconi showed his mother what Tesla had done. And by 1901, Marconi had figured out how to transmit and receive a radio signal across an ocean.

Radio was Quickly Adopted into Peoples' Homes, Despite the Depression

It took about fifteen years after that for the first commercial stations to begin 'broadcasting' radio signals (a term that comes from the agricultural practice of spreading seeds) from transmitters to multitudes of receivers. Station KDKA, the first commercially licensed radio station in America, began transmitting from Pittsburgh, Pennsylvania, in October 1919 and PCGG in The Hague, Netherlands, in December of that year. The popularity of radio skyrocketed after that. According to *Our World in Data*, in 1925 only 10% of US households had a radio. But by 1932, thirteen years after the first commercial radio transmissions, this number had risen to 60%, and it would reach over 80% in 1939, just fourteen years after radios reached 10% of US households, despite the Depression.

But just when there was a radio receiver in nearly every home, the brilliant electrical engineer and Columbia University professor Edwin Armstrong reinvented the game. Armstrong had attained the rank of major during his time in the Signal Corps during the First World War. For his work on radio, the French government awarded him the Legion of Honor in 1919. And in 1927, at the age of thirty-seven, he had become the first-ever recipient of the Medal of Honor by IEEE (the Institute of Electrical and

Electronics Engineers) for his work on self-reinforcing feedback circuits.

FM Radio Remakes the Game

In June 1936, Armstrong performed a demonstration at the headquarters of the US Federal Communications Commission (FCC), an agency that had been created only two years prior. First, he played a jazz record over a standard radio and then he switched to his new system. A reporter in the audience wrote, you "would have believed the jazz band was in the same room. There were no extraneous sounds... Several engineers said after the demonstration that they consider Dr. Armstrong's invention one of the most important radio developments since the first earphone crystal sets were introduced."

What Edwin Armstrong had shown off was frequency modulation (or 'FM') broadcasting, using a new kind of radio receiver circuit, the one on which modern analog radio receivers still work. The first commercial FM broadcast began in January 1941 and in January 1949, thirteen years after Dr. Armstrong's demonstration, the last of the old-style transmitters were shut down, making many old household radio receivers obsolete. This did little to slow the adoption of radio, which by then was in 93% of American households.

But fierce legal battles plagued Armstrong for the rest of his life, bringing both him and his wife Esther to the brink of financial and mental ruin. By the early 1950s, he claimed he was spending ninety percent of his working time on litigation against RCA, which owned the rights to some of his patents, and against twenty FM radio set manufacturers, who he claimed were infringing on some of them. In early 1954, after striking his wife with a fireplace poker so severely that she fled their New York City home to go live with her sister, Edwin wrote her a note of apology and, according to an article in *New Scientist* magazine, "just walked out of his thirteenth story apartment window to instant death." (Hope, 1979) Esther settled with RCA but continued the separate fights with the manufacturers and in 1967, thirteen years after her husband's death, "the most crucial case – against Motorola Inc – was decided in Armstrong's favor." The widowed Mrs. Armstrong collected $10 million for royalties owed just for the period of 1948 to 1950.

Radio Enabled Television

The boom in radio broadcasting enabled a different kind of wireless transmission, one that was even more disruptive than radio: television.

Though there were a variety of technologies developed in the late 1920s and early 1930s that contributed to television, the first all-electronic sets were sold to the public starting in 1938 by DuMont Laboratories. Naturally, World War II delayed the adoption of television by the public but when the war ended, the post-war boom, well, boomed. "By 1950, 9 percent of American households owned a television set, a proportion that had increased to 64.5 percent only five years later," wrote Robert J. Gordon in *The Rise and Fall of American Growth*. "This increase ... is the fastest diffusion of any appliance or device in history".

A 1960 episode of the comedy *Dennis the Menace* had the main character, played by then nine-year-old Jay North, lobbying his parents to buy him his own TV set. That is how quickly televisions were taken up by American homes: in the year of North's birth, few American homes had television sets. But before he was ten, it was conceivable that a young boy might have his own.

Just as early home computer manufacturers in the 1980s wrote the software that ran on their machines, early manufacturers of television sets, like DuMont and RCA, needed to produce content as well, or else there would have been nothing to watch. DuMont ran its own network for fourteen years, from the early 1940s to the mid-1950s. RCA founded NBC as a radio network in 1926 and then NBC got into television broadcasting in 1939 and later spun off ABC as a separate broadcaster in 1943.

The earliest television broadcasts were public events like the Olympic Games (first sent in 1936 from Nazi Germany), news broadcasts, opera, Shakespeare, and radio and stage plays. The late 1940s saw a flourishing of variety shows like *The Ed Sullivan Show*, Milton Berle's *Texaco Star Theatre*, and Jackie Gleason's *Cavalcade of Stars*, each of which essentially employed the vaudeville format simply shot with a camera. Some television shows were largely re-creations of previous radio programs, like Father Fulton Sheen's *Life is Worth Living*, believed to have been the most widely viewed religious series in the history of television, drawing as many as ten million viewers each week. Prior to television, Sheen had hosted *The Catholic Hour*, a radio program which ran for twenty years from 1930 until 1950. (So, he started in radio just as it was quickly becoming popular in American homes, and then switched to television just as it was becoming popular.) Comedian Groucho Marx produced two versions of his hit show *You Bet Your Life*, one for radio and one for television.

Just as a car is not just a faster horse, television is not just radio with images. Changing the medium changed the business model. Lucille Ball and Desi Arnaz, the producers and stars of one of the early 1950s biggest hit shows, *I Love Lucy*, are credited with employing a convergence of ideas concerning the modern situation comedy (sit-com), including using multiple cameras on a stage that contained adjacent rooms laid out to look like the interior of a house or apartment. They pre-recorded each episode on high-quality film using studio cameras from which the broadcast was made, rather than the common methods of the time, which included simply broadcasting live with cameras that could not record or pointing a film camera at a screen that was displaying a live broadcast and then re-broadcasting from that film. (By the early 1950s, all film used by TV stations was the non-flammable, safety film the Eastman Kodak company had introduced over just a few years starting in the late 1940s.)

But perhaps Lucy and Desi's biggest innovation was a business model disruption in that they retained the legal rights to their content so that they made money from merchandising, and they made money each time a station chose to 're-run' their program. They made so much money off retaining ownership rights that they were able to finance a host of other shows, including *The Andy Griffith Show*, *My Three Sons*, and *Star Trek*. News shows, political talk shows, and even game quiz shows typically do not fare well in re-runs, as their content often quickly becomes outdated. But a good sit-com or 'space opera', designed to have no continuity between episodes, is timeless. That is why episodes of *I Love Lucy* and many similar shows are still in syndication today, decades after they were made.

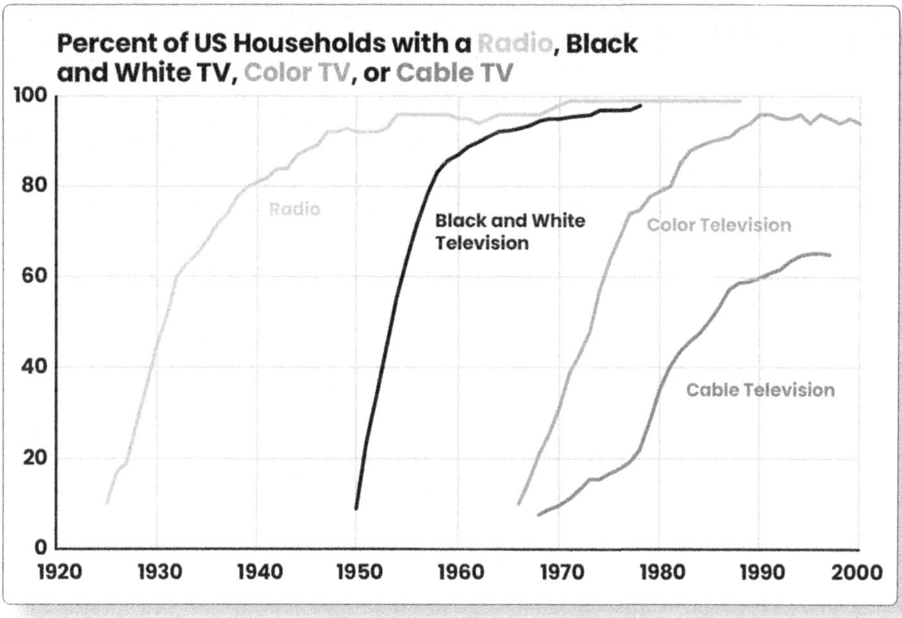

Figure 106 – Percent of US Households with a Radio, Black and White TV, Color TV, and Cable TV.
Data from Our World in Data

The television revolution required a rethink of radio. According to Margaret Blanchard's *History of the Mass Media in the United States: An Encyclopedia*, by the late 1940s "radio network talent, programs, and program production dollars were migrating steadily to the new medium" of television. "Radio was forced by television's competition to develop new and unique approaches to programming." Prior to television, radio stations used general formats, offering soap operas, documentaries, news, comedy, crime, religious shows, and other kinds of programs depending on the time of day. After television, radio stations began specializing in certain genres of music or taking on a news or talk format.

The disruptions didn't end there. The adoption of color TV was only slightly slower than of the original black-and-white television sets and happened about fifteen years later. According to the website *Our World in Data,* the percentage of US households with color TV was 10% in 1966 and 80% in 1981, going from 10% to 80% in just fifteen years total. Cable TV then greatly expanded the number of stations and brought radio's specialized formats to television, with cable channels tending to specialize in a single subject just as radio stations did.

As transistors and electronic components became popular, and easier and cheaper international shipping meant that the production of television sets was increasingly located in Asia, incumbent television manufacturers like RCA were replaced by Japanese competitors.

As Edward Humes wrote in *Door to Door: The Magnificent, Maddening, Mysterious*

World of Transportation, "significant parts of early TVs were mechanical as well. The first remote controls for TVs activated a motor that turned the mechanical tuner to another station with a series of impressive and loud ka-chunks. The economics of making digital technology, however, were different, requiring large investments in highly specialized fabrication equipment to forge processor chips, circuit boards, disk drives, and other complex components. It no longer made sense for a big electronics or technology company to invest in making the whole widget; the market favored factories focused on churning out large quantities of one major component to supply multiple and competing tech companies. The RCA do-it-all in-house style became a liability even as its technology became obsolete."

◆ ◆ ◆

"Books will soon be obsolete in public schools. Scholars will be instructed through the eye. It is possible to teach every branch of human knowledge with the motion picture. Our school system will be completely changed in ten years."

— Thomas Edison, 1913

◆ ◆ ◆

Videotape Disrupted the Kinescope

> *One of the first children to be saved by insulin grew up to invent the magnetic, re-writable videotape, which disrupted the film-based television camera for recording TV shows and became the basis for the home VCR*

Prior to videotape, TV stations on the East Coast in the US would broadcast live to East Coast audiences, and West Coast stations would use a 'kinescope' film camera to record that broadcast from a television screen. The West Coast stations would then have just a few hours to develop the film, using harsh chemicals, and to dry it so that the footage could be broadcast to West Coast audiences.

The process was labor-intensive, unpleasant, and expensive, and the resulting image quality was not very good. Recording just thirty minutes for re-broadcast could cost $4,000. But as television boomed so did the demand for film recording. By the mid-1950s the television industry in the US was using more film than all Hollywood studios combined.

Some better means was needed to record and replay television broadcasts and many large and small companies investigated ways to do that, including RCA, DuMont, General Electric, Alan Shoup Labs in Chicago, Bing Crosby Enterprises (BCE), Bell Television, and the BBC in the UK, among others. But the problem was solved by a team at Ampex, a small San Francisco company founded during World War II to work with radar, which later got into the field of audio tape recording.

The team was led by Charles Pauson Ginsburg, who was born in 1920 and as a child was one of the world's earliest insulin-takers. (It is fair to say that if insulin treatments had not been developed at the University of Toronto in 1922, the use of magnetic tape for video might not exist either.)

The fundamental problem the industry was facing is that a, say, ten-second recording of video simply contains much more information than a ten-second recording of only sound. Pulling a magnetic video tape across a reading device would require the tape to move at speeds that were simply impractical. Ginsburg's solution was to use a rotating reader with four heads and move the tape slowly across it – the quadruplex video tape that became industry standard for decades.

Ampex were not television industry insiders – far from it. According to Stewart Wolpin in 'The Race to Video', a Fall 1994 article in *Invention & Technology*, "here was a small group of newcomers running rings around [RCA's] enormous research complex... To RCA's engineers the research was simply a job" but Ginsburg's team members (including Ray Dolby, who later invented the Dolby NR noise-reduction system) "were passionate visionaries as well as engineers."

In preparation for their first public display, Ampex showed their technology to a select group including William Lodge, the vice president of engineering of the broadcast network CBS, and Jack Mullin, chief engineer of Bing Crosby Enterprises (BCE), who said afterwards, "It's all over for us. It was a beautiful picture, better than ours."

In April 1956, according to Wolpin, "more than two hundred managers of CBS affiliate stations from around the country were jammed into the Normandy Room of the Chicago Hilton". When Ampex projected a recorded video taken just a short time before of one of the speakers who was still on stage, "pandemonium engulfed the room" as engineers and businesspeople rejoiced at the thought of never having to pay for or develop film again. "Some in the audience just applauded, some stood on their chairs to get a better look, but most rushed toward the curtained area to examine the new electronic marvel... In four days Ampex took $5 million worth of orders for the new machines."

What Ampex had shown off was the quadruplex video tape, the same kind that would also be used decades later in home video cassette recorders (VCRs). "The introduction of magnetic tape recording sparked a revolution in the broadcasting industry," Wolpin said. Magnetic video tape quickly took over the professional market for television recording. The last 'kinescope' film-based recording of a television broadcast was made in the US in 1969, thirteen years after Ampex's public demonstration.

TV Technology Remade the Living Room and is Reshaping the World

> *In 1984, comedian and television talk show host David Letterman joked that not watching television was the equivalent of 'throwing your life away'. But technologies developed for TV, including the streaming video services that have disrupted videotape, liquid-crystal display (LCD) screens, and the rare-earth elements that are key to the powerful magnets at the hearts of electric car motors and wind turbines show that Letterman might not have been far off when he called television a 'miracle'*

It would take the transistor revolution and continued reductions in the size and cost of consumer electronics for video tape to be ready for home use.

Video cassette recorders (VCRs) were introduced to the US market in 1977. By 1985, 14% of US households had one and by 1995, they were in 79% of US households. That is, it took around thirteen years for VCRs to go from about 10% of American homes to about 80%. By 2005, 88% of US households had a VCR, but that was near the peak, as they were down to 58% by 2013. The last known manufacturer, Funai Electric Corp., quit making VCRs in July 2016.

As VCR ownership increased, so too did the fortunes of Blockbuster, which described itself in their annual 10-K filing with the US Securities and Exchange Commission for the year 2000 as "the world's leading retailer of rentable home videocassettes, DVDs and video games". By the end of 2004, the year of their

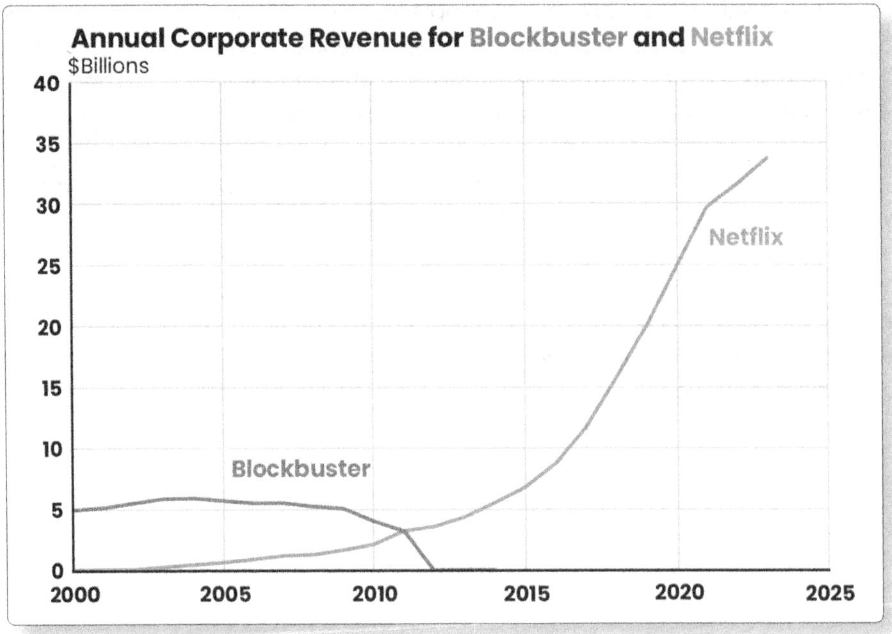

Figure 107 – Annual Revenue for Blockbuster and Netflix

peak revenue, the company had, in their own words, "approximately 9,100 stores in the United States, its territories and 24 other countries". In 2010, six years later, the company filed for bankruptcy and their last corporate-owned store was shut down in 2014, just ten years after their best year ever.

What brought about the rapid end of both the VCR and video rental, in a word, was Netflix. The ability for consumers to rent videos first by DVDs sent through the mail and later over the internet turned Blockbuster's 9,000+ locations from assets into liabilities. In their annual filing for the year 2000, Blockbuster estimated that "about 70% of the U.S. population lives within a ten-minute drive of one of our stores." And the number of stores in the US would go up another 20% in the five years after that.

But with streaming video, consumers no longer needed to drive to a video rental store. Blockbuster's many locations quickly went from being a competitive advantage to expensive liabilities. In 2004, Netflix's revenue was less than 10% of Blockbuster's. In 2011, they were equal. This was not a one-for-one substitution. In 2015, Netflix's revenue was higher than Blockbuster's best year ever and then it grew fourfold more by 2020.

But Netflix is not immune to disruption. Netflix had fewer than 150 million subscribers in late 2019, when Disney launched Disney+. By 2022, Netflix had grown to 222 million subscribers, while Disney+ reached 130 million, more than half of Netflix's level in a little over two years. Who knows what the future will bring?

As consumers were switching from physical video to digital, they were also changing the devices they used to watch those videos. According to Tony Seba in his 2014 book *Clean Disruption,* "the percentage of YouTube traffic that comes

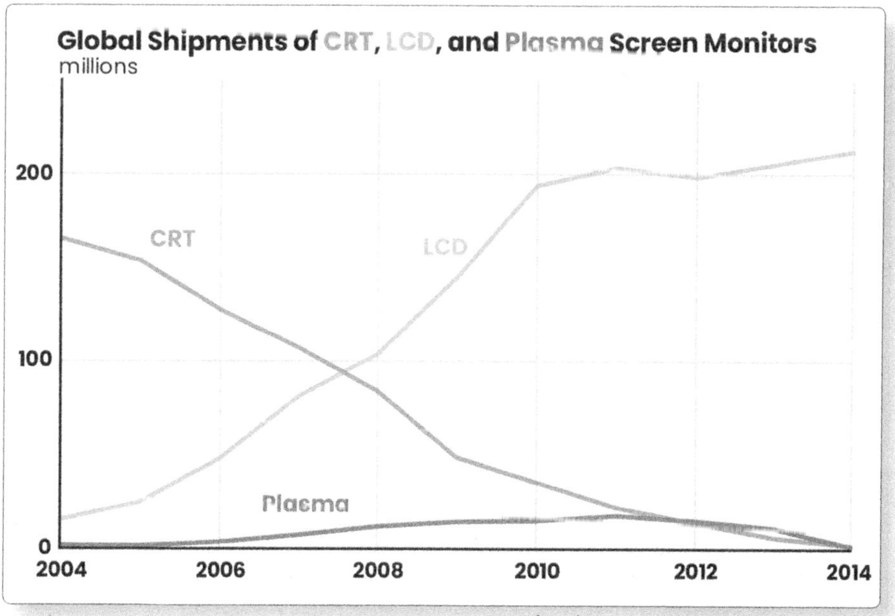

Figure 108 – Global Shipments of Cathode Ray Tube (CRT), Liquid Crystal Display (LCD) and Plasma screen monitors. *(Clausen, 2017)*

from mobile Internet went from 6 percent in 2011 to 25 percent in 2012 to 40 percent in 2013."

And, as our living rooms have taken the place of movie cinemas, they have needed to be renovated. In yet another thirteen-year-long disruption, from 2000 to 2013, televisions that use cathode ray tube (CRT) went from more than 99% market share to less than 5%. Over just a couple of decades, our TVs got larger, thinner, brighter, and far cheaper. *Our World in Data* estimates that the price-to-performance ratio of television sets in the US fell 99% from 1990 to 2020.

But the improvements in television technology and economics have had effects far beyond the small screen.

In October 1984, comedian David Letterman placed telephone calls during his popular late-night television show to Nancy DeSalvo, a local librarian in Connecticut who was promoting an event called 'TV Turn-Off', in which she encouraged the citizens of her town to not watch television, including Letterman's show, for the entire month of January 1985.

DeSalvo had originally championed the first 'TV Turn-Off' in January 1984, in which an estimated one thousand people participated. "One thousand human beings, many of them children," Letterman remarked, "took one month of their lives and threw it away" by avoiding what he called "the miracle of television".

Letterman was joking, but his comments about television being a miracle might not have been too far from the truth. The use of rare earth elements found its first great success in the mid-1960s in the form of color television. But in subsequent decades, these elements have found many other applications. Now, screens for electronic devices and powerful magnets for electric motors and wind turbines all owe their existence effectively to color TV. But as China has become a dominant

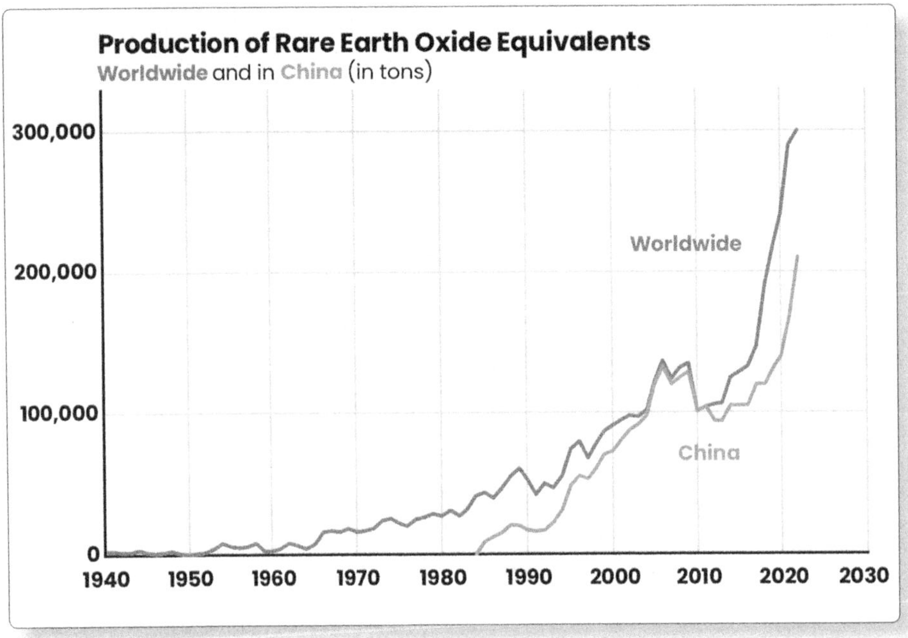

Figure 109 – Production of Rare Earth Oxide Equivalents, in tons

force in manufacturing, it has also taken over most of the global production of these critical resources. Though they produced essentially no rare earth elements in the mid-1980s, by the early 2000s, just a couple of decades later, essentially all the world's production of these materials was in China (though this has now started to moderate as other countries have realized how important these elements are to our future).

Digital Moviemaking Disrupted the Film Industry

> *Digital cameras and digital image post-processing software allowed less-experienced directors (particularly younger people, women, and people of color) to enter the movie business, smaller and riskier projects to be tried, and scenes to be filmed that were not possible previously. They also took a lot of power out of the hands of cinematographers and put it into the hands of digital effects directors. And they quickly took over Hollywood filmmaking*

As digital technology has improved, the way that movies are made has also been completely transformed in the past couple of decades. A blog post by Stephen Follows, who describes himself as a "data researcher in the film industry", shows that in the year 2000 none of the top 100 highest-grossing films were shot digitally. Even up to the year 2005, digitally made films were less than 10% of the top 100. In 2012, digital films were for the first time about half of the top 100 highest-grossing films of the year. And by 2015, they represented nearly 90%. That is, moviemaking went from less than 10% digitally produced to almost 90% digitally produced in just ten years.

In the middle of this disruption, in 2012, Keanu Reeves produced and narrated the documentary *Side by Side,* which describes some of the key effects that digital photography had on moviemaking. The movie is essentially an extended walkthrough of the principles of disruption that were laid out in Chapter 1.

First, Reeves noted that digital format is cheaper to shoot than physical film, so it resulted in more movies being made (not all of which were of great storytelling quality) and in established directors trying smaller, riskier projects. As a result of the experimental nature of the new medium, digital equipment was initially adopted outside Hollywood in places like Denmark, which has a much smaller movie industry. Easier filmmaking meant fewer people going to film schools, more women directors, and more young directors. People started telling the stories they wanted to, rather than taking time to first work their way up through the industry.

These directors were enticed by the wider 'possibility space' of recording stories. Smaller, lighter-weight cameras meant that directors could explore new camera movements, like flying through a crowded room or following people on a busy street, as was done in *Slumdog Millionaire.* Scenes that were impossible to make, or too expensive to make, suddenly became possible. In one scene in the horror-thriller *28 Days Later*, the main character was shown walking alone through London, but recording this was done without blocking large areas from traffic (which would have been expensive). Instead, the crew set up ten digital cameras on a short segment of street and blocked a street for only a couple of minutes, during which they recorded the actor walking. The director then had ten different angles of the same few seconds of action, which he later edited together to make the recording look like one, longer scene.

Essentially, all the dynamics of moviemaking changed because of this technology. With film, if the director wanted to film a scene where the foreground was very clear and the background was blurred, the cinematographer (also called a 'Director of Photography', or 'DP'), the chief person in charge of the technical aspects of filming, would decide which lighting, cameras, and lenses would get that desired effect. They would film the scene one day, develop the shots overnight, and review everything the next day. Now, with digital images, the director can see the shot on the digital screen during production and make any adjustments immediately – with much faster feedback and with much less reliance on the DP.

As a result, much more of the decision-making was shifted to the post-processing after recording. Digital editing allows directors to do things that could not be done with film. If the director wanted a patch of trees to be a different color than the trees were in reality, and the imagery was shot on physical film, only the red, green, or blue coloration or the brightness of the whole shot could be adjusted during film development. But when shot digitally, colors of small regions could be adjusted, items that were not in the original shot added, things could be taken away, etc. Digital post-production editors took on many of the duties previously done by the director of photography.

With the low cost and high flexibility of digital moviemaking, metrics which were formerly important, like the ratio of 'time shot to time used' in the movie, became meaningless. With digital cameras, producers (and financers) simply stopped caring about how much of the shooting went into the final product and how much was left out.

It is not just the mainstream film industry that was affected by digital technology. When adult movies were shot on film, people called "fluffers" were employed to ensure that the male talent was ready to perform as soon as the cameras started rolling. But Listverse's *Ultimate Book of Bizarre Lists: Fascinating Facts,* written by Jamie Frater and published in 2010, says that "digital cameras and drugs like Viagra have removed much of the need for fluffers." Since the cost of using digital cameras is low, so is the need for people who can get male talent aroused.

Digital filmmaking is also eliminating another kind of actor: animals. Computer-generated digital images of human faces still often look odd and a little creepy. The human brain is hardwired to recognize faces and viewers are sensitive to even the slightest movement or appearance that seems unnatural or unusual. But animals do not look or move like people, so animating them digitally is easier than animating humans and easier in general than working with actual animals.

The rise of movies shot digitally created a push to convert theaters to digital projectors. In 1999, there were four digital cinema projectors in the US. In 2002, according to Side by Side, there were one hundred fifty. By 2012, thirteen years after the first projectors, the country was "50% of the way to having all theaters be digital". Ultimately, this resulted in more consistency and uniformity of the viewing experience for customers. When showing physical film, some theaters (especially those in large cities) simply had better equipment than others, but with digital projection, the image quality is similar in all cinemas.

As a result, all major camera producers stopped development of film cameras. They simply no longer make them.

Music

> *The means by which we make and consume music have changed dramatically since Edison's invention of the phonograph. But these new technologies have not just affected how we listen to music, but also which kinds of music get made, and how much the industry got paid*

In the past century or so, music has been re-shaped by technology more radically and more frequently than probably any other human activity. The wax-cylinder phonograph, electric amplifiers, headphones, AM radio, FM radio, magnetic tape, the vinyl long-play (LP) record and turntable, television, the 8-track cassette, the cassette tape, the boombox, the Walkman, the compact disc (CD), MP3s, and streaming music services, among others, have all been disruptive technologies that have ensured that no decade since Edison's 1877 invention of the phonograph has been similar in style of music or means of consuming it to any other. (And those are just technologies for listening to music. Means for producing it – including microphones, electric pick-ups, the piezoelectric sensor, multitrack recording, synthesizers, the transistor, the foil electret, digital audio editing and pitch-correction software, again among many others – have each caused revolutions behind the scenes in how music is conceived, made, and recorded.)

In the January 2016 issue of *Smithsonian* magazine, Clive Thompson describes in detail how the phonograph, which Marshall McLuhan in his book chapter 'The Medium is the Message' said "owes its origin to the electrical telegraph and the telephone", changed how music was produced and consumed. Edison's invention is simple in principle: a cone receives sound, causing a needle to vibrate as it cuts a groove into a wax cylinder. Later, turning the cylinder causes the needle to vibrate in the same way as when the groove was cut, reproducing the sound out of the cone.

But, as Thompson describes, this simple means for recording and re-playing sound had profound implications for the world of music. Prior to the phonograph, musicians made their livings from live performances and the sales of sheet music. The phonograph, in contrast, favored musicians who could record music well and whose recordings sold well. So, the phonograph changed which genres of music were popular – toward blues, jazz, and "hillbilly" music, which favored short songs that would fit onto a phonograph cylinder, and away from opera and orchestral music, which often had very long, complex, or nuanced pieces. "The three-minute pop song is basically an invention of the phonograph," Thompson quotes Mark Katz, a professor of music at the University of North Carolina at Chapel Hill and author of *Capturing Sound: How Technology Has Changed Music*, as saying.

The phonograph affected which instruments were used, as early recording was difficult with lighter-sounding instruments like dulcimers and violins. It affected which people got to make music, favoring men's deep voices over women's lighter ones, and favoring artists who could perform an entire song without making any mistakes over ones who might be more musically gifted at improvization, but also more error prone. And it affected where, when, and how people listened to music. For the first time ever, people could passively listen to music at home, 'on demand', and alone.

Each new technology made music easier to access, increased its fidelity, reduced its cost, or opened new possibilities:

- **Radio** allowed access to new music, including performances that were happening live.
- **Vinyl records** held more recordings and were more durable than wax or shellac discs.
- **Cassettes,** according to Matthew David of Leaving Records, were "the most accessible, easiest, cheapest way for anyone to record a piece of audio". (Ulaby, 2011) In forty years, Sony sold more than four hundred million units of their Walkman cassette player.
- **Compact discs** (CDs) promised every play would be as high quality as the first, with no rewinding, driving a tripling of recorded music revenue in the US in the period from 1984 to 1999, when CDs accounted for nearly 90% of sales (in financial terms).
- **Sales of personal cell phones** drove the sales of ringtones, growing from essentially nothing in the year 2000 to a peak of nearly $900 million per year in 2007, before the smartphone drove ringtone sales back to zero over the next seven or eight years, ending a fifteen-year industry.

But in the past couple of decades, music has been transformed by digital technology as much as it was by the phonograph starting in the late 1800s. Downloading digital music files and then streaming services turned the industry from more than 90% of sales coming in the form of physical media (like vinyl, cassettes, and CDs) to more than 90% in digital form (MP3s and streaming subscriptions) over the period of only about fifteen years.

As with the phonograph itself, digital recording and distribution did not just change how music was delivered, but also which kinds of music get made. With the next song only a swipe away for the listener, musicians find themselves forced

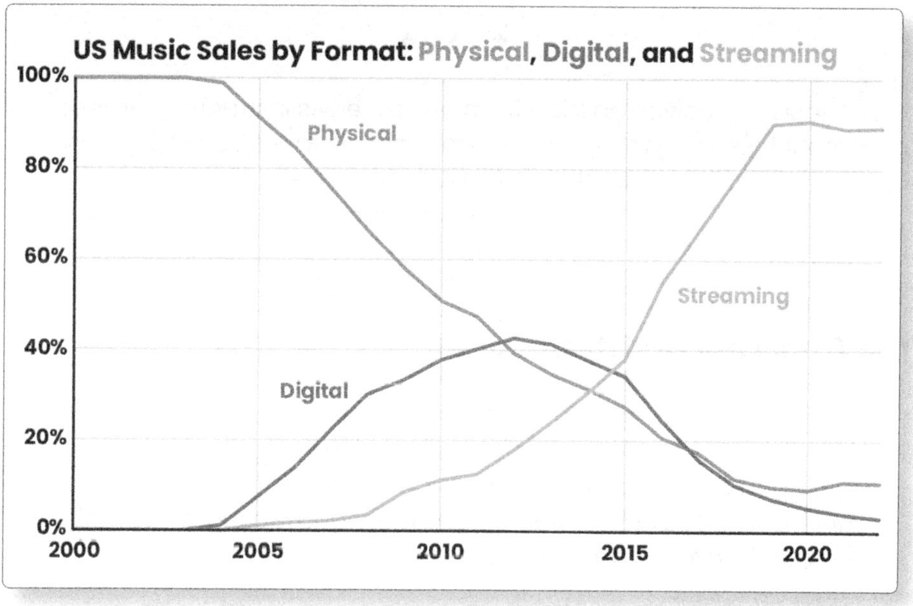

Figure 110 – Percent of Music Sales by Streaming vs. Physical Media. *Data from RIAA*

to 'write to the algorithm' – that is, to tailor songs so that they will do well on the biggest music platforms. This has favored trying to craft hit singles rather than trying to sell full albums, writing shorter songs that get straight to the main riff without much build-up, and putting the chorus at the start of a song.

The rise of song streaming has crushed album sales. From 2001 to 2021, there were forty-two music albums that sold more than ten million copies (so called 'diamond-selling records'). Thirty-nine of those were from the years 2001 to 2010. Only three have been from 2011-2021, all released in the first half of that decade, a 90% drop over ten years. No album released since 2016 has turned diamond.

Auto-Tune, invented in 1997, was developed by the oil and gas industry to use sound reflections from explosives to find hydrocarbon deposits. But within just a couple of years, it found use in the music industry to help correct the voice of singers who were slightly off-key. Cher's "Believe", released in late 1998, became the first major hit to use the technology, going well beyond minor adjustments to instead give the singer's voice a distinctive, robotic sound. The song became a smash hit, eventually winning *Billboard Magazine*'s title of #1 'Top Hot 100' song of 1999. Understandably, the software quickly took over the industry. Simon Reynolds of the music news website *Pitchfork*, wrote that some speculate that Auto-Tune now "features in 99 percent of today's pop music."

From 2000 to 2015, the industry saw years of declining revenue as digital downloads and streaming services were disrupting physical formats. But the effects were temporary. In the years after 2015, industry revenues have grown dramatically (largely due to the rise of paid subscription services) until the point that in 2021, they reached the highest level (in non-inflation-adjusted terms) ever, and they have continued to grow since then.

◆ ◆ ◆

"He who receives an idea from me, receives instruction himself without lessening mine; as he who lights his taper at mine, receives light without darkening me."
— **Thomas Jefferson**

◆ ◆ ◆

The Transistor and its Children

> *In terms of disruptive technologies, the transistor, an electric switch with no moving parts, is perhaps the largest ever. It spawned solar photovoltaic cells and solid-state data storage and revolutionized how we do computing, affecting almost all human activities by extension*

Sitting at his desk at Bell Labs in Murray Hill, New Jersey, in December 1947, forty-five-year-old physicist Walter Brattain glued a strip of gold foil over the tip of a plastic triangle, took out a razor, and carefully sliced through the gold, creating two pieces of foil separated by a very thin gap. Using a spring, he held the tip of his device against an ingot of germanium, an element which is chemically like silicon.

Brattain applied a relatively large voltage to the block of germanium and a much smaller voltage to one piece of gold. Nothing happened. But when he slightly increased voltage on the gold foil, current suddenly began flowing out from the germanium and through the second gold foil. Turning the voltage down on the first piece of gold foil did not reduce the current slightly; it shut it off completely. Turning the voltage up again restored the current.

What Brattain had made that day, with significant help from his co-workers John Bardeen and William Shockley, was the point-contact transistor, essentially a switch with no moving parts. And it was not only a switch but also an amplifier which could take a small signal and turn it into a much larger current. In terms of disruptions, 'the little plastic triangle that changed everything' was perhaps the largest ever. Like the vacuum tubes that enabled radio, radar, TV and early computers, transistors could act as switches and amplifiers, but it took about one million transistors to consume as much power as a single vacuum tube. Computation, communication, and even data storage all got dramatically cheaper, easier, faster, and better as a result.

Brattain, Bardeen, and Shockley were awarded the Nobel Prize in Physics in 1956, but the world of semiconductor electronics was just getting started. (Bardeen won the Nobel Prize in Physics again in 1972 for his work on superconductors, making him the only person to be awarded the Nobel Prize in Physics twice.) In 1951, the point-contact transistor was supplanted by the bipolar junction transistor. Bell Labs then fabricated the first silicon transistor in 1954. In 1959, Mohamed Atalla and Dawon Kahng at Bell Labs invented the metal-oxide-semiconductor field-

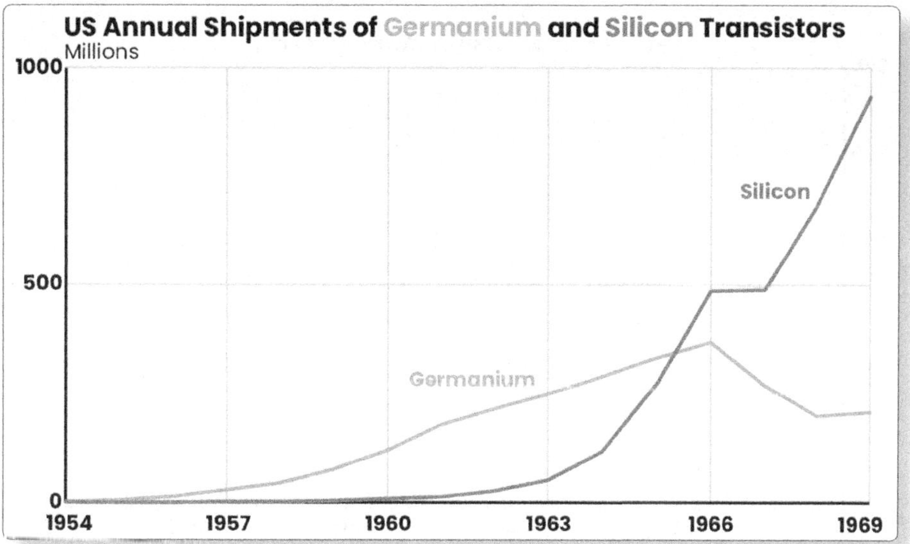

Figure 111 – Annual Shipments of Germanium and Silicon Transistors from US Factories.
Data from (Brock, 2003)

effect transistor, or 'MOSFET'. In 1961, Robert Noyce was awarded a patent for the integrated circuit, which could pack many transistors onto a small chip. Since Atalla, Kahng, and Noyce's inventions, almost every transistor produced has been a silicon MOSFET in an integrated circuit.

In 1954 the first transistor radio, the Regency TR-1, which sold for just under $50, used four Texas Instruments germanium transistors. At the time, the first silicon transistors cost nearly $24 each, making them uneconomical for a $50 radio. But over the subsequent years the price of silicon transistors fell much faster than germanium ones. (Transistors did not just change how we listened to the radio, they also changed what we listened to. The rise of solid-state transistors meant the rise of 'effects' – electric circuits which take an instrument's sound and add distortion, echo, reverb, tremolo, fuzz, sustain, or other modifications that gave rise to rock-and-roll music.)

As their price dropped throughout the late 1950s and early 1960s, the popularity of silicon transistors eclipsed germanium ones. In 1963, shipments of germanium transistors from US factories outnumbered silicon ones by a ratio of about 5-to-1. By 1969, shipments of silicon transistors outnumbered germanium ones by about 5-to-1. This was not simply a one-for-one replacement of one kind of transistor for another; in these few years, annual shipments of transistors increased fourfold.

Interestingly, a silicon transistor was more expensive than a germanium one during this entire time, being almost ten times more expensive in 1957 ($17.81 for a silicon transistor vs. $1.85 for a germanium transistor) and dropping being the same price as each other ($0.37) twelve years later in 1969. However, silicon transistors were also better performing and more stable at higher temperatures, making the silicon transistor a 'superior' product (higher price, but higher performing) compared to the germanium transistor. The replacement of germanium transistors by better-performing, more-expensive silicon transistors was a 'disruption from above'.

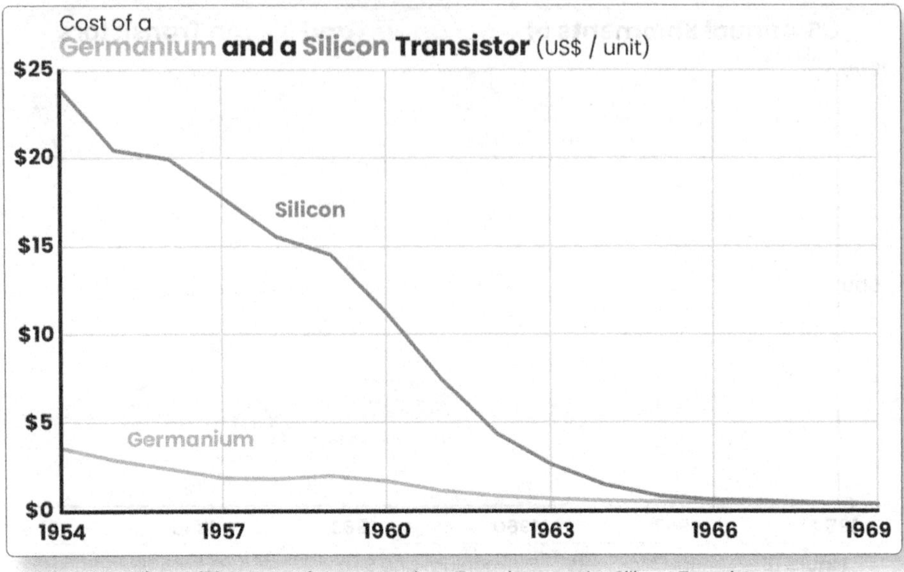

Figure 112 – Cost of a Germanium Transistor and a Silicon Transistor.
Data from (Brock, 2003)

But silicon transistors were about to get much cheaper. According to Ernest Braun in *Revolution in Miniature: The History and Impact of Semiconductor Electronics,* the year 1969 was the peak in shipments of discrete transistors, as afterward the integrated circuit largely replaced discrete components. In 1963, ten percent of all electronic components were in the form of integrated circuits. By 1967 it was nearly 80%, by 1970 over 90%, and by 1971 nearly 95%. From 10% to 95% in under a decade.

In 1969, the peak year of shipments of discrete transistors, silicon devices cost $0.37 each according to Gerald Brock in his book *The Second Information Revolution.* When the Intel 4004 microprocessor chip was introduced in July 1971 for a price of $60 and containing more than two thousand transistors, the price per transistor had fallen to less than $0.03 each, a drop of more than 10X in just two years.

The microchip revolution was due to a convergence of technologies, including photolithography, which took techniques directly from the photography and printing industries to lay out circuit designs in silicon. But the semiconductor base was not pure silicon. To work properly, part of the silicon base was doped with impurities like boron or aluminum, and another part of the base was doped with phosphorus.

In his 1998 book *Crystal Fire: The Invention of the Transistor and the Birth of the Information Age*, Michael Riordan describes how one of the key discoveries about how to make silicon semiconductive was made at Bell Labs back in the late 1930s by metallurgist Henry Theuerer. While cutting silicon with a diamond wheel, Theuerer noticed an odor that reminded him of "the smell you used to have on these acetylene lamps that you had on automobiles before [they] had electric lights." But the smell was not acetylene. Theuerer knew it was due to phosphorus. There were impurities in the silicon that were making it act as a semiconductor. It

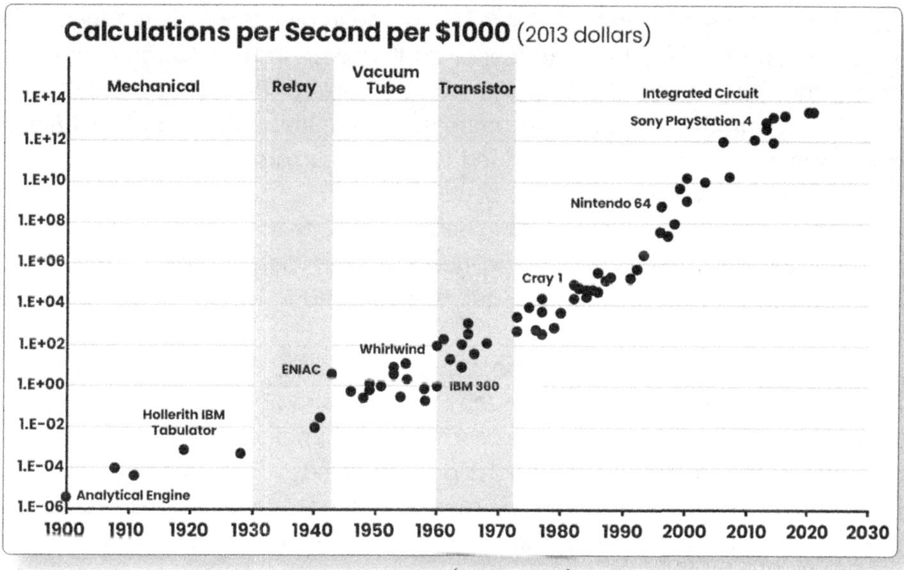

Figure 113 – Calculations per Second per $1000 (2013 dollars), for various computing machines.
Source: (Dorr, 2016) plus additional data from Dr. Adam Dorr

was phosphorus that made silicon act like a switch and had Theuerer not been exposed to early automobiles as a younger man, he might not have made the key discovery that enabled today's silicon computing revolution.

As remarkable as silicon semiconductor-based computers have been, what is perhaps more remarkable is that they in fact are little more than a continuation of a trend in dramatically cheaper and faster numerical computation that has been going on for more than a century, using a variety of mechanisms. From mechanical computers that used gears and levers, to ones based on telephone relays, to vacuum tubes, to discrete transistors, to integrated circuits, the ability to do computation has improved over many orders of magnitude.

And what might have kicked it all off was something you might not expect: the United States Constitution. By law, the US Federal government is required to conduct a census every ten years. But as the population increased, tabulation of the results of the tenth census, conducted in 1880, was so daunting that it took almost a decade to complete. It did not take genius to see the growing problem here. But it would take one to see a solution.

In the February 1989 issue of *Scientific American* magazine, author Keith Reid-Green describes a conversation overheard during the tabulation of the 1880 results where young Census Office employee Herman Hollerith described his idea for a mechanical tabulating machine that would work "something on the principle of the Jacquard loom", combined with the new technology of electricity. Hollerith patented his invention in 1884 when he was twenty-four years old.

"By the end of the 1890 census", Hollerith had developed and refined his electric tabulating machine, and invented a keyboard-based 'pantograph punch' for making the cards the tabulating machine took in. "As a result", Reid-Green writes, "efficiency at the Census Office skyrocketed."

By the end of the 1900 census, Hollerith had added an automatic feeder system to process cards in batches. The feeder itself was an impressive advancement. While in the 1890 census "the average clerk was able to feed almost 8,000 cards per day through the tabulator", in processing the 1900 results, "a record 84,000 cards were processed in one day, and the average rate was at least six times faster than what the hand tabulator could achieve."

Hollerith founded a company that in 1911 merged with several others to become the company now called International Business Machines, or 'IBM'. But the idea of using machines to do calculations quickly flourished in the form of a variety of similar companies.

The now slightly sexist-sounding book titled *A Better Day's Work at a Less Cost of Time, Work, and Worry to the Man at the Desk*, published by the Burroughs Adding Machine Company in 1910, describes some of the benefits of owning one of their mechanical tabulators. The Burroughs document says their "machine lists and adds a column of figures three to six times as fast" as a person can do it and that, when it comes to multiplication, their machine is "more than three times as fast as it is possible to multiply mentally".

Better, faster, cheaper computing has resulted in a self-reinforcing feedback loop, as one of the primary uses of faster and better computing technology has been to design and manufacture even-faster and even-better computing technology.

As computers have gotten faster and cheaper, they have revolutionized industry, government, and business. Vaclav Smil, a professor at the University of Manitoba, Canada, wrote in his 2006 book Transforming the 20th Century: Technical Innovations and Their Consequences, that when it came to the design of aircraft,

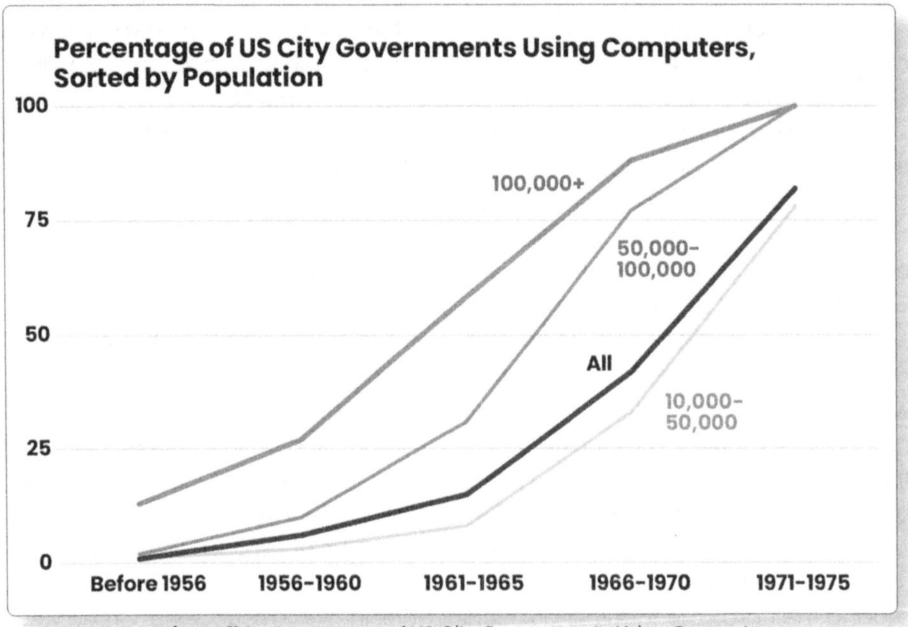

Figure 114 – Percentage of US City Governments Using Computers.
Data from (Perry J. , 1979)

companies like Boeing switched their process from one that used paper-based technical drawings to one based on computer systems as "even the earliest CAD [Computer-Aided Design] systems often halved the time spent on a particular task, and later advances reduced labor requirements to between 10% and 20% of the original need."

In the 1980 book *Technological Innovation in American Local Governments: The Case of Computing*, James Perry reported the results of a 1975 survey of US cities and counties which found that computer use in US city governments went from 6% to 82% in fifteen years. In US counties over 100,000 people, it went from 11% to 78% in the same time.

According to the US Census, by 1984, just thirteen years after the release of the Intel 4004 chip, the first integrated circuit, about one-in-four of all adults who had jobs used a computer at work. (Kominski, 1988)

◆ ◆ ◆

"There is no reason anyone would want a computer in their home."

— Ken Olsen, founder of Digital Equipment Corp., 1977

◆ ◆ ◆

The Foil-Electret Microphone

In 1962, thirty-one-year-old scientist James West, whose mother was one of NASA's human computers profiled in the hit book and movie *Hidden Figures*, filed a patent for a sound sensor made from an electret, a statically charged foil of Teflon (which itself was a disruptive innovation, invented the same year and place as nylon). You might have made a similar foil yourself if you have ever rubbed a balloon against your head to charge it with static electricity. When sound hits the foil, causing it to vibrate, the motion can be detected by the small current it induces in a nearby transistor. (The kind of transistor in the foil electret microphone is the kind that Bardeen, Brattain, and Shockley were trying to produce when they instead invented the point-contact transistor in 1947. The first field-effect transistor, the kind used in West's microphones, had to wait a few more years until 1953.)

The commercial foil-electret microphone debuted in 1968. An interview with National Public Radio described his microphones as "inexpensive but sensitive" compared to the then common carbon microphones. It is not surprising then that West's microphone design rapidly became popular. A July 1975 article in *DB Magazine* said, "today, the manufacture of electret microphones represents a significant part of the audio industry." (Lane, 1975) And the *Acoustics Source Book*, published in 1988, twenty years after the first foil-electret microphone, said "condenser microphones, particularly of the electret-biased type, are now the most widely used microphones." (Parker, 1988)

In just twenty years, West's microphones went from their first commercial application to being commonplace. But they did not just replace carbon microphones on a one-for-one basis. West's microphones opened a new 'possibility space' of devices, including mobile phones, laptops, blood pressure monitors, hearing aids, and baby monitors. Since the device also works in reverse – applying a sound generates a current, but applying a current generates a sound – the foil-electret is also useful as a speaker, not only in mobile phones and laptops, but also in headphones. If you have a phone in your pocket, headphones on your head, or a laptop in front of you, you are probably in arm's reach of several foil-electret devices right now.

West's invention is not only sensitive to audible sounds, but also to infrasound, those sounds whose frequencies are too low to be heard by people. So, they were used to detect the sound waves generated in the atmosphere by radiation from far-way pulsar stars and they were also able to hear from six hundred miles away the launches of the Saturn V rocket made possible in part by West's mother and the other 'Hidden Figures' mathematicians.

◆ ◆ ◆

"We didn't do anything wrong, but somehow, we lost."

— Stephen Elop, CEO of Nokia, 2013

◆ ◆ ◆

Cell Phones

> *Few technologies have changed daily life as dramatically, for most people and industries, as the cell phone.*

The first call ever from a handheld mobile phone was made in April 1973, when Martin Cooper, an engineer for Motorola, called a rival company to let them know he was talking to them from a mobile phone. In 1984, Motorola released the first commercially available cell phone, the DynaTAC 8000X, made famous by the movie 'Wall Street', starring Michael Douglas.

In addition to the high price tags, batteries were the Achilles' Heel of cell phones. For example, the DynaTAC 8000X initially cost nearly $4000 (nearly $12,500 in 2025 dollars) and offered only thirty minutes of talking time for a ten-hour charge on its nickel-cadmium batteries.

But that was about to change. Lithium batteries were proposed by British chemist M. Stanley Whittingham while working for Exxon. The rechargeable lithium-ion battery was patented in 1977 and the first practical lithium-ion battery, invented by John B. Goodenough and Akira Yoshino, was brought to the market by Sony in 1991. Over the course of the next fifteen years, lithium-ion batteries took over the

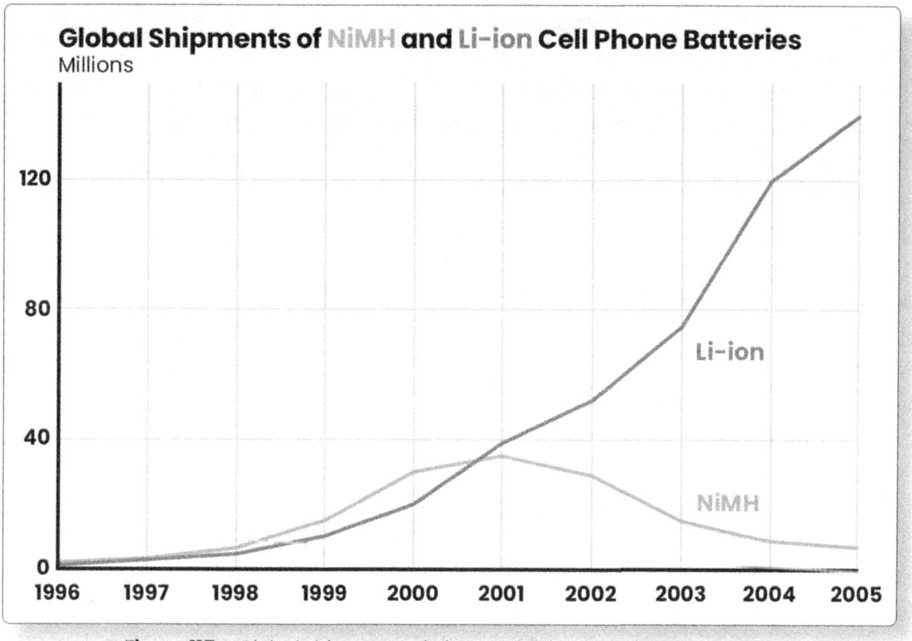

Figure 115 – Global shipments of NiMH and Li-ion cell phone batteries.
Data from (Wilburn, 2008)

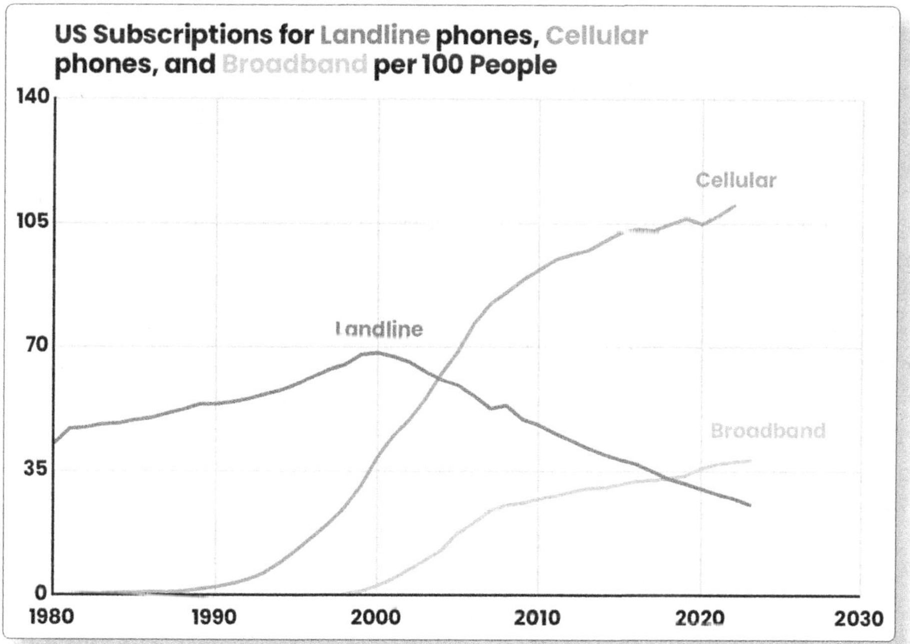

Figure 116 – US Subscriptions for Landline Phones, Cellular Phones, and Broadband per 100 People.
Data from (Ritchie, 2017)

cell phone market and, as a result, cell phones proliferated. In the year 2000, global shipments of phones with nickel metal hydride batteries exceeded shipments of ones with lithium-ion batteries by 50%. Just two years later, in 2002, shipments of phones with nickel metal hydride batteries were 50% of shipments of ones with lithium-ion batteries. Two years later, the market was more than 90% lithium-ion batteries.

According to *Our World in Data*, in 1994, ten percent of US households had a mobile phone. By 2009, fifteen years later, it was 85%. The Pew Research Center found similar numbers when asking about the percentage of US adults with phones, as opposed to households. According to their surveys, as of early 2021, 97% of US adults have a cellphone and 85% have a smartphone.

When we look at the number of subscriptions to landline service, cellular service, and broadband service, the story is a bit more complicated, but still one of rapid disruption. Access to landline phones in the US peaked in 2002 at 95% of households. Fifteen years later, fewer than half of homes still had one. As with other disruptions we have seen, this was not a one-for-one replacement of an old technology with a new one. At the peak in the year 2000, there were about sixty-eight landline service subscriptions per one hundred people in the US. A couple of years later, the number of cellphone subscriptions exceeded the number of landline subscriptions. As of 2014, there were more mobile phone subscriptions in the US than people. Today it seems that people mostly have landline subscriptions as part of their broadband internet package, but rarely use them.

In a January 2020 presentation at the North Carolina Department of Transportation Summit, Tony Seba wondered why the smartphone revolution happened when it did. "Both the iPhone and the Android came out the same year 2007. Not '05 or '09. What was the magic of 2007?", he asked. "2007 is when all the technologies that made the smartphone possible converged. So, all the techno-computing power, broadband, data storage, touch screen, all of these technologies that made a $600 smartphone possible happened in 2007. So any of thirty, forty companies could have actually built a smart phone. Nokia could have done it. Motorola could have done it. Remember them? Blackberry could have done it. Remember them? So why didn't they do it?"

Why they didn't, according to Seba, is that these companies were all cell phone incumbents, so making a smartphone was left to outsiders. "Disruptions happen from the outside," Seba continued. "Neither Google nor Apple had ever built a phone before and yet they're the ones who own 90+ percent of the smartphone market. Disruptions happen from the outside. It's very rare the company that can disrupt itself."

Because disruptions start when they are first technologically possible, innovative technologies are often introduced at the same time by multiple companies. As Seba said, the iPhone and the Android phone were both introduced in 2007. But the Model T and General Motors were also launched in the same year, 1908. The Boeing 707 and the DC-8 jet airplanes were both introduced in 1958, the last year that the propeller-driven DC-6, DC-7, Lockheed Constellation, and Lockheed Starliner were sold. Betamax and VHS video cassettes were introduced within a year of each other in 1975 and 1976.

◆ ◆ ◆

"When wireless is perfectly applied the whole earth will be converted into a huge brain, which in fact it is, all things being particles of a real and rhythmic whole. We shall be able to communicate with one another instantly, irrespective of distance. Not only this, but through television and telephony we shall see and hear one another as perfectly as though we were face to face, despite intervening distances of thousands of miles; and the instruments through which we shall be able to do this will be amazingly simple compared with our present telephone. A man will be able to carry one in his vest pocket."

— Nikola Tesla, 1926

◆ ◆ ◆

Smartphones are Biting into Chewing Gum

> *Unintended consequences and ripple effects often appear in areas that do not seem immediately related to a new technology. Cell phones disrupted more than just the telephone market; they affected many other aspects of life as well.*

The consequences of the widespread adoption of new technologies are often unforeseen, even unforeseeable. It was clear to many from the beginning that cell phones would be useful to traveling salespeople and other businesspeople on the go and that they would upend the telecommunications industry. But few foresaw the effects that they would have on taxis, mafia organizations, banking, and even the sales of chewing gum.

A 2017 article on the *Vox* news website said that over the ten years after the launch of the iPhone in 2007, chewing gum sales in the US dropped about 15%. They reason that gum is often an impulse purchase made when waiting in line at a store, but now "we're so consumed with our phones that we're not reaching for a pack of gum". If you were a chewing gum manufacturer in 2007, how could you have possibly anticipated this effect? (Molla, 2017)

The taxi business has felt an even greater impact from smartphone apps. Tony Seba, speaking in 2020, said: "Uber and Lyft are business model innovations. They were made possible by the smartphone and the cloud. So, it's no coincidence that Uber started in 2009, two years after the smartphone came out. See? One thing makes the other possible. Now, in seven years Uber had more bookings than the whole taxi industry in America. Talk about a disruption. Seven years from a little apartment to more bookings than the whole taxi industry in America."

Computer programmer Todd W. Schneider maintains a website that tracks data about taxis and ridehailing services in New York City. As late as mid-2015, ridehailing services like Uber, Lyft, Juno, and Via accounted for under 20% of the half-million taxi rides given each day in New York City. By mid-2019, just four years later, they accounted for three-quarters. Since the COVID pandemic started, ridehailing services have made up at least 80% of trips every month.

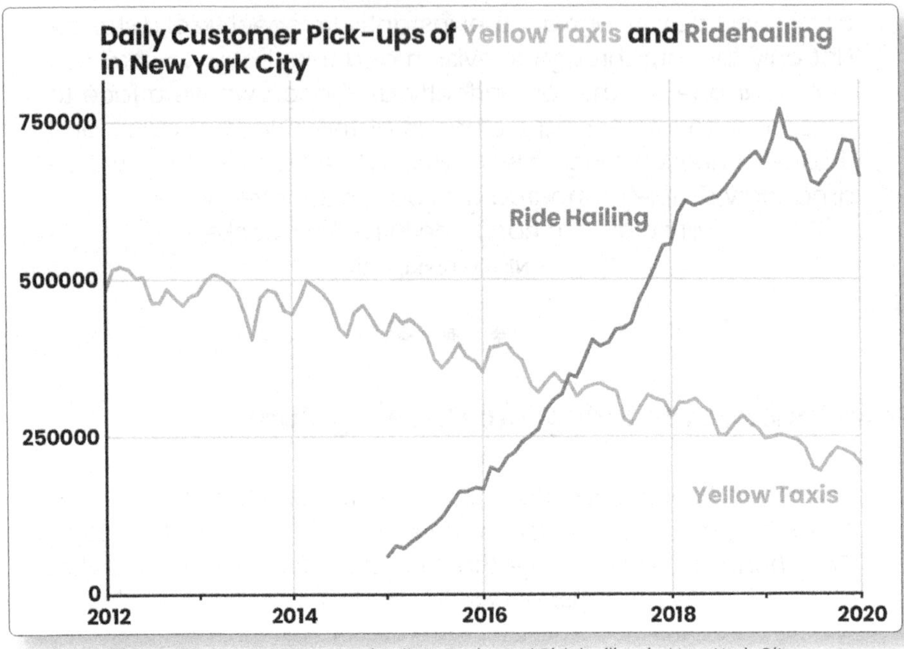

Figure 117 – Daily Pick-Ups of Yellow Taxis and Ridehailing in New York City.
Data from toddwschneider.com

Much of this change happened over just two years. In January 2016, cars on ridehailing apps made half as many trips as New York's Yellow taxis. In January 2017, cars on ridehailing apps made slightly more trips than Yellow taxis. And in January 2018, they made almost twice as many trips as Yellow taxis.

From a financial perspective, ridehailing services (and the COVID pandemic) have been nothing short of devastating for New York City's taxi drivers. The total revenue (or 'farebox') for Yellow taxis, averaged over $6 million per day as recently as June 2015 and were over $4 million per day for the last time in late 2019. The total farebox, of course, plunged during COVID, but even as of early 2025 still only averages less than $3 million per day, for a total drop of about 50% in the decade from early 2015 to early 2025. Ridehailing revenue, in comparison, reached its pre-COVID level again in early 2022 and has only grown since then, now averaging more than $22 million per day.

This has resulted in mass unemployment for many taxi drivers. The number of drivers per day working Yellow taxis dropped 60% from early 2015 to early 2025 (and was even lower during the peak of the COVID pandemic). In New York City, taxis must be authorized by the city in the form of a metal 'medallion' that is fixed to the hood of the vehicle. Not long ago, these medallions could cost hundreds of thousands of dollars each, for which drivers would take loans to purchase them, and then be obligated to pay back the loan at a rate of thousands of dollars per month. Even though industry revenue has plummeted, drivers still owe the debt for their medallions, which are now selling for a fraction of their price just a few years ago. According to an Al Jazeera report from October 2021, "the average owner-driver owes $550,000 on medallions that are worth just over $100,000 apiece."

Richard Chow, a New York City taxi driver who immigrated to the US from Taiwan in 1987, described how ridehailing has pushed some drivers beyond the brink. In 2021, some New York City taxi drivers, Chow among them, launched a hunger strike intended to get politicians to step in and help. For Chow's brother Kenny, also a New York City taxi driver, it was too late. Burdened by high debt and a business rapidly losing ground to ridehailing apps, Kenny Chow took his own life in 2018. (Osgood, 2021)

Mobile phones have become so integral to modern life, even criminals are choosing to go straight rather than give them up. In Japan, new laws in the past decade have, among other things, prohibited ownership of a cell phone to members of the Yakuza mafia organization. According to data from Japan's National Police Agency reported at *Nippon.com* total membership in the Yakuza, including both full members and associate members, declined about 70% in the ten years from 2010 to 2020, in an S-shaped pattern.

But for some people, cell phones have opened new opportunities and made possible things that were previously difficult, like banking. In *Doughnut Economics*, economist Kate Raworth described how Kenya's M-PESA mobile money service launched in 2007. "Within six years," she wrote, "three quarters of all Kenyan adults had used the service, including 70% of those in rural areas, and - astonishingly - over 40% of Kenya's GDP was passing through M-PESA." Since then, digital payments have exploded around the world, bringing banking services to people who formerly had none. In India, according to Banque de France, "54.2% of the population owned a smartphone in 2020, up from 4.7% in 2011." (Revelin, 2022) As a result, India's digital payments system, Unified Payments Interface (UPI), has averaged about 80% growth per year since its launch in early 2016. In Brazil, the Pix service "had the fastest adoption curve amongst all surveyed real-time payment systems in the world," according to the World Economic Forum (WEF). (Vélez, 2022) Digital payments have grown so rapidly that the WEF concludes "it's time for a rethink on financial inclusion". Perhaps traditional banks are on the cusp of being the next in line to be disrupted.

The Pocket Calculator Disrupted the Slide Rule

The slide rule was used by mathematicians, scientists, and engineers for over two hundred fifty years until it was suddenly disrupted by the electronic calculator over just a few years in the early 1970s. The first electronic desktop scientific calculators were introduced in 1965, but the first handheld unit that could fit into a shirt pocket was the HP-35, launched by Hewlett-Packard in 1972 at a cost of $395. In an interview in 1981, Apple founder Steve Jobs said "the HP-35, the first scientific handheld calculator, was introduced by Hewlett-Packard in 1972. Well, we all know where that's gone in just nine years. But another interesting sidelight is in 1978 the largest manufacturer of slide rules in the world stopped making them entirely because the market had gone away. Six years."

The speed with which electronic calculators took over the slide rule market surprised even insiders. A February 1975 article in New Scientist magazine called 'The electronic slide-rule comes of age' said "Even those closest to the pocket calculator business have been overwhelmed by the pace of developments and the nosedive in prices... It is now possible to buy pocket-sized scientific calculators capable of performing a range of trig, log and power functions for a third the price charged two years ago. Prices will fall further, but not so dramatically." In fact, prices fell even more dramatically the following year when, in 1976, Texas Instruments started selling the TI-30 for under $25.

The International Slide Rule Museum, a website that calls itself "the world's largest free digital repository of all things concerning slide rules" says, "in ISRM's opinion, the official date of the death of the slide rule occurred on June 13, 1976 when Texas Instruments introduced the single chip TI-30 scientific slide rule calculator for $24.95 USD, which was below the cost of a comparable slide rule. There was no cost benefit to new students, technicians and engineers to buy a slide rule anymore, as everyone could afford the TI-30. Coincidently, barely one month later on July 11, 1976, Kueffel & Esser, the oldest and largest slide rule manufacturer in the United States, produced their last slide rule. In the next month of August, 1976, Pickett Industries followed suit".

As with many other inventions, the electronic scientific calculator was not a one-for-one replacement for the slide rule. The HP-35 sold over 300,000 units by the time it was discontinued in 1975, just three years after its introduction, and Wikipedia says the "TI-30 sold an estimated 15 million units during its lifespan from 1976–1983." Throughout the 1980s and 1990s, electronic pocket calculators became the norm in high school and even middle school mathematics classes. While only engineering and science students had them twenty years prior, now most students had one. They got cheap enough that banks gave them away as token gifts for opening savings accounts. But today the handheld pocket calculator itself has largely disappeared as calculator apps have now become standard tools on desktop computers and smart phones, including those made by Apple.

Advertising

> *In 1935 nearly seven out of ten US homes had a radio, but almost 90% of US advertising spending was in newspapers and magazines.*

As the way we listened to and watched news and entertainment changed, so did the ways that advertisers tried to reach us.

According to media and technology analyst Benedict Evans, the world of broadcast advertising (that is, money spent on radio and TV) changed dramatically in the early 1950s as we rolled those brand-new RCA televisions into our homes and tuned into Lucy or *What's My Line?* for the first time.

In 1950, out of every $1000 in total US economic activity (GDP), television advertising accounted for about $0.56. By 1955, it was more like $1.98. In contrast, advertising on the radio accounted for $1.68 out of every $1000 in US economic activity in 1950 and by 1955 it was $0.54. (Evans B. , 2020) That is, spending on advertising on radio and TV essentially switched places over just five years, going from about 75% radio to 78% television. (These calculations only include money spent on broadcast advertising – that is, radio and television advertising. Combined spending on print advertising, which is found in newspapers and magazines, was about three times larger in the early 1950s than spending on broadcast advertising.)

The next fifty years were essentially ones of continuous long-term growth for the spending of advertising dollars (and therefore the media that depended on them) as the wider economy grew. Ad spend varied, of course, but accounted for between 1.0% to 1.5% of US gross domestic product (GDP) in the entire latter half of the 1900s.

Then came the Internet. Newspaper classifieds and magazine ads could not compete with Craigslist and Google search ads. Spending on both newspaper and magazine ads both peaked in the year 2007 and then magazine ad spending fell more than 50% over the next thirteen years. Newspaper advertising is down 75% over the same period. (Molla, 2018) Employment at newspapers followed the decline in revenue. According to the U.S. Bureau of Labor Statistics, the number of employees of US newspapers by the early 2020s was down by about 80% compared to its all-time peak in the early 1990s. Local newspapers were once at an advantage, with their expensive printing facilities posing a barrier to entry, or 'moat' in the parlance of famed investor Warren Buffett. But in the face of the Internet, these same facilities became a disadvantage.

It's not just print advertising that has been affected. Television advertising spending peaked in 2014 and was down 27% in just the six years to 2020 as spending on search ads, social media, online video, and e-commerce have all grown, almost exponentially. Only radio has weathered the digital storm, remaining in the range of about $30 to $35 billion each year since the year 2000. Within just digital advertising, *eMarketer.com* divides its tabulations to money spent on desktop/laptops compared to phones/tablets. According to them, in 2011 advertising on mobile devices was 5% of all digital advertising spend. By 2016, mobile accounted for 65% of digital advertising dollars.

In August 2013, *eMarketer.com* estimated that advertising on mobile devices would make up half of digital advertising dollars by 2017, four years later. Ten months later, in June 2014, they estimated that mobile spend would be half of digital advertising by 2016, two years later. And in September 2015, they said that advertising on mobile devices already constituted more than 50% of digital advertising that year. That is how quickly the market changed!

Digital advertising went from about 2.3% of all advertising spending at the turn of the century to about half of all advertising spend over just fifteen years. And within digital advertising, the market went from 5% spent on ads for mobile devices to 65% on ads for mobile devices over just five years.

How Couples Met

Another kind of advertising of sorts, online dating, has revolutionized how couples meet.

Surveys by Professor Michael J. Rosenfeld of Stanford University of over 3500 American adults found that the percentage of homosexual couples who met online went from zero in the early 1990s to more than one-half just fifteen years later. For heterosexual couples, the shift was slower, reaching half of couples about fifteen years after homosexuals.

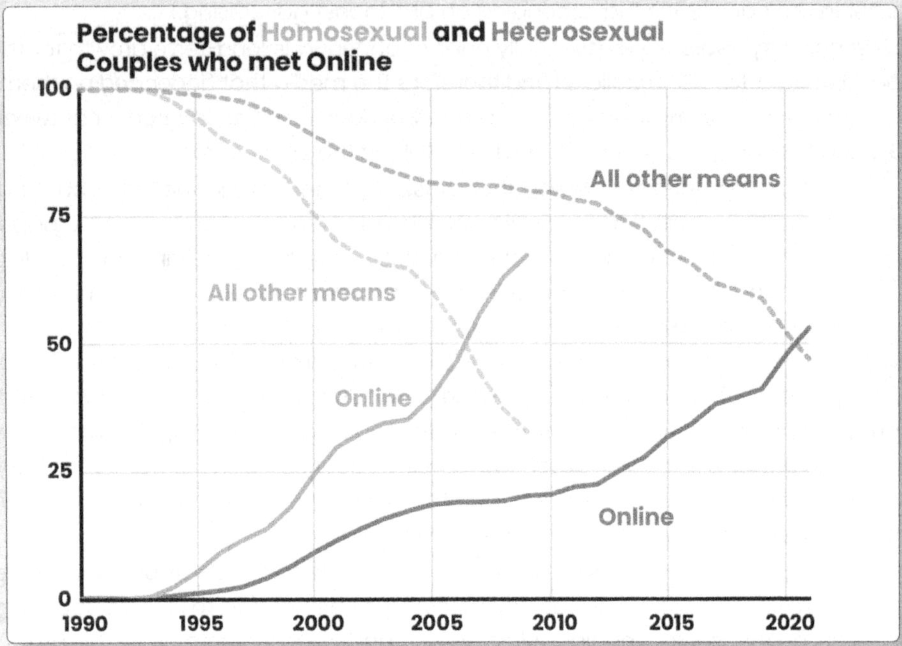

Figure 118 – Percentage of Homosexual and Heterosexual Couples who met Online.
Data from (Rosenfeld, 2019) and later updates

Now that news is mostly delivered electronically, rather than on newsprint paper, analysis by David Rozado and co-workers at the Te Pūkenga–New Zealand Institute of Skills and Technology of millions of headlines from 2000 to 2019 that appeared in dozens of major news outlets popular in the US suggests that headlines have increasingly used angrier and more fear-mongering terminology.

Newswriters are trained, either by their employers or by the bitter reality of online news economics, to write headlines and opening paragraphs that are likely to attract our attention, get us to click on links and ultimately, to see the ads on the content page. They tailor their writing for search engine optimization, using key words in their headlines and opening text, intentionally evoking emotions in their headlines rather than keeping a neutral tone, and including attention-grabbing 'power words', among other things.

The purpose is not just to get the reader's attention, but also to get them to share content with their friends, perpetuating the self-reinforcing feedback loop that drives some content to 'go viral'. In his book *Contagious,* Jonah Berger argued that strong emotions drive people to share content. He claims that "anger and anxiety lead people to share because, like awe, they are high-arousal emotions. They kindle the fire, activate people, and drive them to take action."

Sliced Bread

> *The steel bread slicing machine reinvented bread. Bread recipes need to be reformulated so that the sliced bread did not dry out. Packaging was made air-tight. Slicing was easier if the bread was square in cross-section, so American preferences quickly became for loaves that were fluffy, white, perfectly baked, flavorless, and low in nutrition. But this also meant that children could make lunch for themselves, freeing up Moms. And it almost caused the Allies to lose World War II*

Otto Frederick Rohwedder of Davenport, Iowa, was a jeweler by training, but an inventor and engineer at heart. Born in 1880, in 1916, at the age of thirty-six, he sold his chain of jewelry stores to finance his real passion, the development of a machine for bakeries that would mechanically slice and package a loaf of bread. Like James Albert Bonsack, the inventor of the cigarette rolling machine, Rohwedder lost his prototype to a fire. But by 1927 he had perfected his design and delivered his first bread-slicing machine. Like Bonsack, Rohwedder's machine was a business-to-business disruption: marketed to manufacturers rather than to households. (In some European countries, bread slicing machines to this day are available in supermarkets, who let the customer slice their bread in the store.) The world's first loaf of pre-sliced bread, automatically wrapped in waxed paper, was sold the following year at a bakery in Chillicothe, Missouri on July 7, 1928, Rohwedder's 48th birthday.

His timing was fortunate. The pop-up electric toaster had been patented by Charles Strite in 1921 and, in 1925, the Waters-Genter Company introduced the Model 1-A-1 Toastmaster, the world's first pop-up toaster that could brown bread on both sides simultaneously and eject the final product automatically. The Toastmaster was advertised as making "perfect toast every time - without watching, without turning, without burning". It was a huge commercial success. Of course, there were toasters before Strite's, including simple electric ones and, before those, rudimentary iron frames that would hold bread slices in front of the fire of an open hearth, but those required constant attention and most designs only toasted one side at a time.

In 1905, the 28-year-old engineer Albert Marsh had formulated a mix of nickel and chromium that could withstand repeated heating cycles, becoming the basis of the electric toaster heating element. The first electric toaster, the D-12, was invented by Frank Shailor of General Electric and first put on sale in 1909. But the device was marketed to restaurants as in that year only about 10% of American households had electric power. By the time the first pre-sliced loaf was sold in 1928, that number had risen to 64%.

Sliced bread was an instant hit. Rohwedder's son told a Chillicothe newspaper in a 2003 interview that the product was so popular that the Chillicothe bakery where the first loaf was sold increased its bread sales by twenty-fold in just two weeks. In 1930, Continental Baking Company introduced pre-sliced 'Wonder Bread', which has been one of the best-selling brands of bread in the US since its introduction, in part by having a recipe formulated so that it would not dry out when pre-sliced.

Pre-slicing required many aspects of bread to be rethought. Americans' preferences for the form factor of bread changed, so that rectangular loaves were more common instead of baguettes or rolls. And the electric toaster favored lighter-colored grains instead of black bread or pumpernickel. As Aaron Bobrow-Strain wrote in *White Bread: A Social History of the Store-Bought Loaf*: "slicing put bread's structure on display as never before. Crumb irregularities and unevenness that were once acceptable, or easily blamed on customers' deficient skills with bread knives, were now immediately apparent to anyone opening a package of bread. This alarmed bakers. Large uneven holes, so esteemed by artisan bread lovers today, had no place in the modernist aesthetic. Each one was an unacceptable reminder of bread's natural life, a tiny realm of imperfection unconquered by science. The perfect tapered loaf, the ideal slice thickness - all that came to nothing if bread's face looked worm-eaten. Thus, scientific bakers threw themselves into developing new dough-mixing equipment, loaf-shaping technology and, most importantly, chemical dough conditioners to ensure that every slice revealed the exact same architecture of tiny even cells. Consumers, for their part, admired the new look of bread, and accepted uniformity as a mark of quality. Indeed, only one aspect of the high modernist reengineering of bread's appearance stirred major controversy during the first decades of the twentieth century: flour whitening."

In 1873, Edmund LaCroix, a French-American living in Minnesota, was hired by Cadwallader Washburn (later a founder of the General Mills company) to develop a flour mill that efficiently separates wheat's germ and bran from the white endosperm, producing refined white flour. White wheat flour can be lightened to an even brighter shade of white through natural oxidation, but storing large amounts of flour is expensive for manufacturers. In the first few years of the 1900s, German chemists had found ways to make ammonia from nitrogen in the air and water and nitric acid from ammonia and oxygen. Chemical bleaching using nitrogen dioxide sped up the oxidation process, making white wheat flour cheaper than ever.

Bread has been a staple of the western diet for thousands of years. Archaeologists suggest that we have been baking it for at least 16,000 years, longer even than we have had domesticated grains. But over the course of about fifteen years, from just before World War I to the introduction of Wonder Bread in 1930, bread was re-invented by a convergence of technologies: Rohwedder's bread slicing machine, powered by an electric motor, Shailor's restaurant toaster and Strite's electric, pop-up household toaster, containing Marsh's heating element, and chemical bleaching agents that made white wheat flour cheap and widely available.

By 1933, five years after the introduction of Rohwedder's bread slicer, US sales of pre-sliced bread exceeded sales of unsliced bread for the first time. After thousands of years of baking bread at home and slicing it by hand, people could go to a bakery and buy perfect-looking, pre-sliced white bread. As with any technology, there were unintended consequences: sliced bread was so convenient that bread sales skyrocketed, along with the sales of spreads like jams and peanut butter. A 1901 recipe by Julia Davis Chandler in the *Boston Cooking School Magazine* described a new sandwich made from "three very thin" slices of bread, one containing peanut butter and one jelly, pressed together with a plain slice in the middle. The slicing of bread, thus, also changed who prepared the food. Making a peanut butter and jelly sandwich or a sandwich with pre-sliced meat and cheese on it went from a task that could only be completed by an adult with a large, sharp knife to something any child could do.

As such, the introduction of sliced bread was not a one-for-one disruption. In fact, total bread consumption rose because of pre-slicing to the point that, when World War II began, the US government was concerned that many potential US soldiers were chronically malnourished. In January 1943, the US government temporarily banned the sale of sliced bread in part to reduce consumption of bread and toppings, to reduce the possibility of wartime rationing.

From Medicines to Meat

Insulin Stopped Diabetes from Being a Death Sentence

> *Finding that animal insulin could treat diabetes turned the disease from a death sentence to an inconvenience in just a few years in the 1920s. Insulin production was revolutionized by genetically modified microbes in the early 1980s. And then the industry was revolutionized again by human-analog insulin about fifteen years later. Now, the technology is being used to make the hemeglobin that makes beef-free hamburgers taste like beef, threatening to disrupt the meat industry.*

Diabetes affects nearly 9% of the world's adult population, killing more than four million people per year and resulting in health care costs that amount to, in the US alone, roughly $1000 per year for every citizen. There are different forms of the disease, but it is characterized by the inability of the body to regulate its level of blood sugar (glucose), requiring some people to inject insulin.

In January 1922, thirty-year-old Frederick Banting and his 22-year-old assistant Charles Best, working in the laboratory of John Macleod at the University of Toronto, performed the first clinical treatment of diabetes in a human using insulin they had extracted from dog pancreases. The results were nothing short of miraculous.

While the diagnosis of diabetes had previously been a death sentence, it was now suddenly a treatable condition. Mortality for children under age ten with Type 1 diabetes, according to the Joslin Clinic in Boston, Massachusetts, went from 386 per 1000 person-years in the period 1914-1922 to 61 in the next four years, "a sixfold reduction immediately after bovine insulin became available". By 1940, the mortality rate was less than 1% of the pre-insulin level. (Cowie, August 2018)

Banting and Macleod shared the Nobel Prize for their work in December 1923, a few weeks after Banting's 32nd birthday, making him the youngest Nobel laureate ever in the field of Medicine or Physiology. (In 1972 the Nobel Foundation formally acknowledged that omitting Best was a mistake.)

For the next sixty years, insulin extracted primarily from pig pancreases was the standard treatment for diabetes. However, pig insulin is not identical to human insulin, so using pig insulin in people sometimes causes severe allergic reactions as a side effect. Additionally, it takes the pancreases from about 50,000 animals to make a kilogram (2.2 pounds) of insulin. According to Dr. Catherine Tubb, director of research at Synthesis Capital in London, this is only "enough to treat fewer than a thousand diabetic patients for just one year".

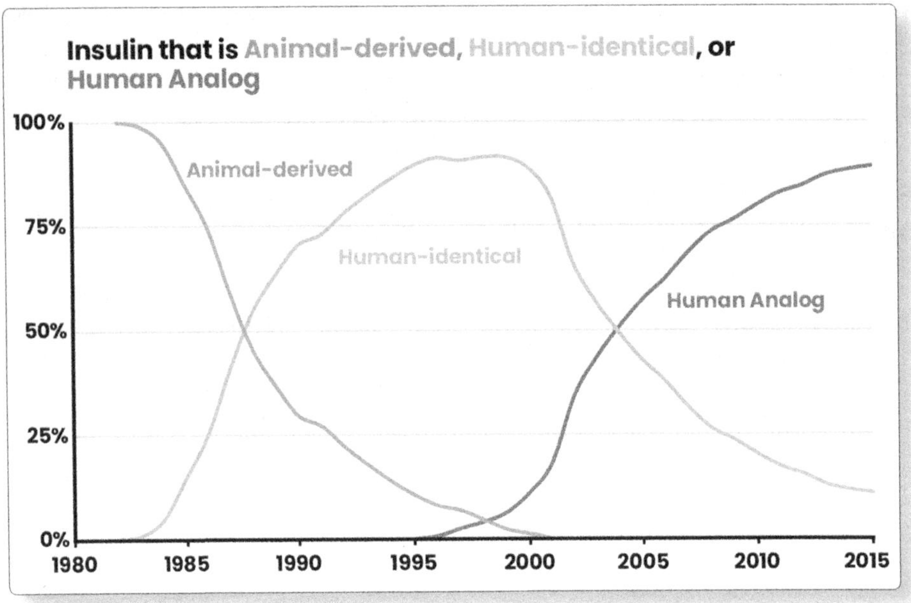

Figure 119 – Percentage of Insulin that is Animal-derived, Human identical, and Human Analog

In 1978 human insulin was produced for the first time using genetically modified bacteria by San Francisco-based Genentech, a company which was founded only two years prior. Genentech's human insulin was approved for pharmaceutical use in 1982 and within thirteen years accounted for about 90% of the insulin market.

In 1996, the US Food and Drug Administration (FDA) approved an alternative form of insulin in which scientists had modified the amino acid sequence of human insulin into 'analog' forms that act more rapidly and that act longer than natural human insulin. However, like animal-derived insulins, these modified insulins are not identical to human insulin, so they can induce allergic reactions in some patients and are therefore not perfect substitutes for human-identical insulin. Mixing analog insulin with other forms of insulin was not approved in the US until late 1999. So, even though analog insulin was slower to gain traction in its first few years than human-identical insulin, analog insulin mixtures now dominate the market, rising to about 90% of insulin within fifteen years of government approval.

Penicillin and Vaccines Ended Many Diseases, and Put a Chicken in Every Pot

Penicillin and other antibiotics helped reign in many dreaded diseases in the early 1900s. The car dramatically increased the portion of women who went to hospitals to give birth, rather than giving birth at home. And vaccines not only improved human lifespans, but made industrial farming easier, decreasing the cost of meat.

Scottish scientist Alexander Fleming was forty-seven years old and already well-known and highly respected for his work when he went into his lab in late 1928, having left some plates of bacteria growing while on holiday. He found that some of the plates, contaminated by penicillium mold, had regions where bacteria did not grow. "When I woke up just after dawn on September 28, 1928, I certainly didn't plan to revolutionize all medicine," Fleming later said, "but I suppose that was exactly what I did."

Antibacterial treatments had been used for millennia. Since antiquity people have known that 'magical clays' from the Greek island of Lemnos, for example, were beneficial body scrubs and that moldy bread or certain kinds of plants sometimes help wounds to heal, although they did not understand why, as bacteria themselves were not discovered until van Leeuwenhoek first saw them with his microscope in 1676 (and we did not know for nearly another two hundred years after that that they cause disease). In the late 1800s the German physician Paul Ehrlich noted that certain chemical dyes stained some bacteria but not others. (You might recall that synthetic dyes had only recently begun to be manufactured from coal extracts.) By systematically testing dye-like chemicals that might be absorbed into the living cells, Ehrlich discovered that arsphenamine helped fight syphilis in infected rabbits. But even Ehrlich himself only called this 'chemotherapy' (literally 'chemical treatment') and not antibiotic therapy. Marketed under the name Salvarsan, arsphenamine was the most-frequently prescribed drug until Alexander Fleming's discovery came to market.

Throughout the late 1800s, knowledge from textile dyeing was instrumental in giving birth to the pharmaceutical industry. As David Mowery wrote in the 1999 book *Sources of Industrial Leadership*, "the birth of the modern pharmaceutical industry can be traced to the mid-19th century with the emergence of the synthetic dye industry in Germany and Switzerland. At that time, Switzerland and Germany were the leading centers of the synthetic dye industry (due in part to the strength of German universities in organic chemistry, and in part to the fact that Basel was close to the leading silk and textile regions of Germany and France). During the 1880s, the medicinal effects (such as antiception) of dyestuffs and other organic chemicals were discovered. It was thus initially Swiss and German chemical companies such as Ciba, Sandoz, Bayer, and Hoechst, leveraging their technical competencies in organic chemistry and dyestuffs, who began to manufacture drugs (usually based on synthetic dyes) later in the 19th century. For example, salicylic acid (aspirin) was first produced in 1883 by the German company Bayer."

Fleming's discovery of the mold that made penicillin, the world's first-known true antibiotic, for which he shared the Nobel Prize in Physiology or Medicine in 1945, eventually revolutionized the pharmaceutical industry. Penicillin proved to be breathtakingly effective in treating not only syphilis, but also pneumonia, tuberculosis, diphtheria, tetanus, and a variety of other bacterial diseases. But Mowery wrote that "throughout the 1930s, it was produced only in laboratory-scale quantities and was used almost exclusively for experimental purposes."

(Interestingly, the chemical structure of penicillin was confirmed by British chemist Dorothy Crowfoot Hodgkin, who also discovered the structure of both insulin and vitamin B12. It was for the vitamin that she won the Nobel Prize in Chemistry in 1964, becoming only the third woman to have done so, after Marie Curie and her daughter Irène.)

Pure penicillin was first isolated only in 1940, twelve years after its discovery. Producing it in large quantities proved problematic. The first attempt at treating a patient in the UK ended with the patient's death simply because the doctors ran out of the drug. In 1942, the first-ever successful use of penicillin was in America with thirty-three-year-old Anne Miller, who was treated for streptococcus septicemia due to a miscarriage, but doing so required half of the US's total supply available at the time.

With the war in full force and wounded soldiers literally piling up every day, some of Fleming's collaborators moved to a US Department of Agriculture laboratory in Peoria, Illinois, to find ways to scale up penicillin production. Part of the solution came from bacteriologist Mary Hunt, who scoured local grocery stores for molding fruit. She found a cantaloupe infected with *Penicillium chrysogenum*, a strain of the mold which, according to science writer Steven Johnson, author of *Extra Life: A Short History of Living Longer*, "produced two hundred times the amount of penicillin" as the strain Fleming had found.

"The next step sounds like it came right out of science fiction," Johnson continues. "The *chrysogenum* was zapped with X-rays to cause mutation. This mutated mould now produced a thousand times the amount of penicillin." All the penicillin in the world today effectively comes from that one moldy cantaloupe that Mary Hunt found.

A Convergence of Technologies Enabled Large-Scale Penicillin Production

With a highly productive strain in hand in 1943, fifteen years after penicillin's discovery, both the British and US governments began intense programs to mass-produce the drug. The problem was solved through a convergence of technologies. Microbiologist Andrew J. Moyer had already discovered that the penicillium mold would produce penicillin when cultured in corn steep liquor, a by-product of making purified cornstarch. Mowery recounts that "Pfizer, which had production experience in fermentation, developed a deep-tank fermentation process for producing large quantities of penicillin." And west of Philadelphia, Granville Raymond Rettew, a chemist who studied mushrooms, found that using a dairy cream separator to continuously extract the penicillin from the fermentation tank greatly boosted production. His lab, located in a renovated automobile mechanic shop, quickly became the producer of most of the world's penicillin supply. Johnson writes, "By D-Day in 1944 there was enough penicillin to treat every soldier in need. By 1945 a million people had been treated with penicillin compared to fewer than 1000 in 1943."

The impact of penicillin on long-standing diseases like syphilis was dramatic. You might recall from Chapter 2, in the section describing the cannon disrupting the trebuchet, author Clifford Rogers claimed that "when the French marched into Italy in 1494, their artillery could do in a few hours what in Italy used to take days." But the French might have brought more than just cannons to Italy on that trip. Shortly after the arrival of French troops, there are the first European written records of an outbreak of syphilis around Naples, although scientists still do not know if the disease was brought to Europe by sailors who had visited the New World, or if there was a mutation in the bacterium around that time that suddenly made it more virulent.

The disease spread quickly throughout Europe and beyond, with every afflicted people seemingly naming it after some nearby group who they did not like. The Italians, English, and Germans called it the "French disease". The Ottomans named it the "Christian disease" and the Dutch "Spanish pocks". In India, Hindus and Muslims named syphilis after each other. It ravaged Europe for nearly five hundred years, being transmitted not only sexually but also congenitally, from mother to child. According to an assessment of London in the 1770s, the cumulative incidence then of having ever been treated for syphilis by the age of thirty-five was over twenty percent of men. (Szreter, 2020)

The End of Syphilis was the Start of the Sexual Revolution

People were plagued by syphilis for nearly five hundred years and then it disappeared as quickly as it had appeared. In about a fifteen- or twenty-year period after the widespread availability of penicillin, syphilis (and many other severe infectious diseases) largely vanished. (Prior to antibiotics, one of the primary treatments for syphilis was to intentionally infect a patient with malaria, which causes a fever that hinders syphilis. Malarial fever itself can be deadly, but is less likely to be fatal than syphilis.) Between 1940 and 1960, the reported mortality rate from syphilis fell about 90% in the US. In 1968, a revision of the World Health Organization's International Classification of Diseases (ICD) system which required diagnosis based on the presence of a specific bacterium, rather than a collection of symptoms, revealed that the disease had nearly been eradicated in the US. Of course, this decline was not entirely due to penicillin, as the decline began around 1940, a couple of years before large quantities of the drug were available. Nevertheless, penicillin played a key role in creating a new era in which the burden of infectious diseases was much lower than it historically had been.

The immediate post-War era was one of the most remarkable times of advancement in public health in world history. As Dr. Julian Davies wrote in the *Canadian Journal of Infectious Diseases and Medical Microbiology*, "The period from 1950 to 1960 was truly the golden age of antibiotic discovery, as one-half of the drugs commonly used today were discovered in this period." Fleming's drug unleashed a worldwide hunt for other antibiotics and in this time people found streptomycin (discovered in 1943), cephalosporin (1945), chloramphenicol (1947), and the tetracyclines (1948).

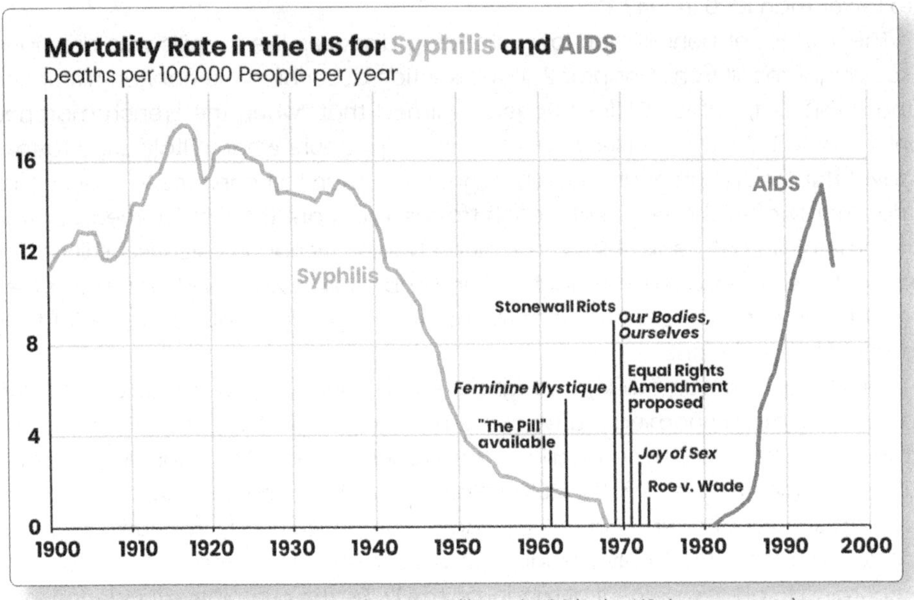

Figure 120 – Mortality Rate from Syphilis and AIDS in the US. *(Armstrong, 1999)*

Science made great progress against more than just bacterial diseases in this time too, as numerous vaccines to prevent viral diseases were also discovered, including ones for polio and smallpox. The last natural outbreak of smallpox in the United States occurred in 1949. Cases of polio fell more than 90% in the five years after Jonas Salk discovered his famous vaccine in 1955 and were nearly zero just a few years after that. Some other diseases that now see annual cases less than 1% of pre-War levels include diphtheria, pertussis (whooping cough), rubella, and tetanus. (Vanderslott, 2013)

The children born just after World War II – the Baby Boom generation – were not only more numerous than any American generation before them, they also got the healthiest start of probably any group of people in history up to that point. Too young to fight in or remember the war in Korea, they grew up in a brief window of unparalleled peace and tranquility, the first time ever when there was, as the 1928 slogan associated with Herbert Hoover went, "a chicken in every pot". And one Montana farm boy perhaps did more than any other man to make that slogan literally a reality.

Dr. Maurice Hilleman led the development of more than forty vaccines during his career, most of them after joining Merck & Co. in 1957. Of the fourteen vaccines routinely recommended in current American vaccination schedules, Hilleman and his team developed eight of them, including ones for measles, mumps, hepatitis A, hepatitis B, chickenpox, meningitis, streptococcus pneumonia and haemophilus influenza. In the US, cases of measles fell more than 90% in just four years after Hilleman's vaccine was introduced in 1963. Cases of chickenpox, which numbered more than 150,000 in the year before the varicella vaccine was introduced, fell more than 85% over six years, and then fell by half again over the following decade. According to an article by Alan Dove in the April 2005 issue of *Nature Medicine*, even today "his vaccines save nearly 8 million lives each year."

The effect of Hilleman's discoveries on the economy have been staggering. According to Alicia Atwood in the *American Economic Journal: Economic Policy*, if measles was as present in modern society as it was prior to the vaccine, then the US gross domestic product (GDP) would be 0.5% smaller than it currently is. That might sound like a small number, but actually few innovations have had a greater effect than that. And this increase in national income, according to her, came "not from an increase in hours worked but rather from greater productivity."

Hilleman's advancements were due to a convergence of discoveries. In the early 1930s, Ernest Goodpasture developed a technique for producing viral material by cutting a small hole in the shell of a fertilized chicken egg and injecting fluids from an infected person. And from Max Theiler, Hilleman learned that viruses could be weakened by growing them in a slurry of animal cells. The Wikipedia page for Hilleman says that he "credited much of his success to his work with chickens as a boy".

As much as the poultry business affected Hilleman's work, his work also affected the poultry business. In his 2007 book *Vaccinated: One Man's Quest to Defeat the World's Deadliest Diseases*, Dr. Paul Offit recounts how chickens used to be susceptible to 'Marek's disease', which causes them to be "thin and weak" or have "rock-hard tumors under their skin". The condition was highly contagious, being spread from a diseased bird to other chickens even by them only contacting the diseased bird's dander — the lightest, fluffiest feathers that chickens shed. Animals infected by this mysterious disease had to be culled and their bodies simply discarded, unfit for consumption.

"By the early 1960s, researchers found that Marek's disease was caused by herpesvirus", Offit wrote, and was a form of cancer. Researcher Ben Burmeister in East Lansing, Michigan, had isolated herpesvirus in turkeys and Hilleman used it to make a vaccine against herpesvirus for chickens. According to Offit, "In 1974... Hilleman's vaccine, the first to prevent cancer in any species, revolutionized the poultry business. Excess production caused the price of chickens to drop from $2 per broiler to forty cents and of eggs from fifty cents per dozen to five cents. Soon everyone could afford chickens, and for a while, Merck, one of the more conservative pharmaceutical companies in the United States, was the biggest chicken and egg producer in the world."

A 5X change in the price of chicken meat and a 10X change in the price of eggs had profound ramifications for the US meat industry, not just the poultry business. Vaccinating against this virus dramatically increased production of eggs and chicken meat as breeders were for the first time given the freedom to explore the 'possibility space' of animals that grew quicker and larger and that laid more eggs, without any regard to their susceptibility to Marek's disease.

As the difficulty and cost of commercially raising chickens decreased, and new breeds were produced that increased egg and meat production, the size of chickens grew dramatically. This illustration shows two typical commercial broiler chickens, raised solely for their meat, one in 1957 (at 905 grams, or two pounds) and the other in 2005 (at 4202 grams, or 9¼ pounds) when both were fifty-six days old. Commercial birds today can be more than four times the size they were fifty years prior at the same age. However, commercial broilers today are often slaughtered when they are about forty-two days old, so they rarely live long enough to fully mature.

Figure 121 – Commercial broiler chickens illustrated. One in 1957 weighing 905g (left) and in 2005 weighing 4202g (right), when each was 56 days old.

The vaccine against Marek's disease in the mid-1970s helped propel chicken from being the least-popular of the three largest sources of meat in the US – the other two being beef and pig – to becoming the most-popular over about fourteen years. In 1976, US chicken meat production topped five million tons for the first time and chicken accounted for 22.5% of all non-seafood meat produced. By 1990, chicken production had more than doubled and chicken had become the most-popular meat in the US. Today, chicken production has more than doubled again and chicken meat accounts for nearly half of all meat produced in the US. According to Dr. Carys Bennett at the University of Leicester in the UK, "broiler chickens, now unable to survive without human intervention, have a combined mass exceeding that of all other birds on Earth."

Hilleman's work was not the only factor that helped make chicken and eggs ubiquitous. In the late 1920s, Japanese farmers discovered a way to determine

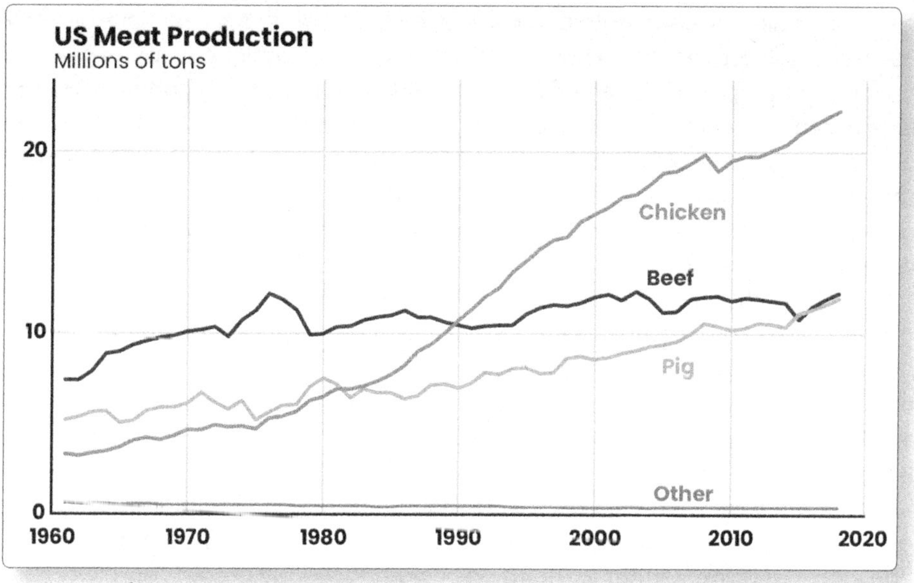

Figure 122 – US Meat Production, in Millions of tons annually. *(Ritchie, 2017)*

the sex of baby chicks by visual inspection, thus allowing egg farms to not need to raise baby males. Visual sexing has largely been replaced as new breeds of chicken have been developed where female chicks have longer feathers than males, allowing so-called feather sexing. And the practice of culling baby males of the most productive egg-laying breeds continues to this day, with the chicks simply being shredded or gassed to death as soon as possible.

Judging that we might already be pushing the physical limits of how much and how quickly we can extract products from chickens, companies are attempting to produce eggs and chicken meat without any chickens at all. Eat Just, a company based in California, has served at a restaurant in Singapore since late 2020 actual chicken meat that was grown in a bioreactor rather than inside a chicken. Other companies are pushing this disruption even further. The EVERY Company, also based in California, makes actual egg proteins by brewing them using modified microbes in much the same way that microbes ferment sugar into alcohol when brewing beer. Their egg proteins are being used in fruit smoothies sold by Pressed Juicery. The Israeli company Steakholder Foods is using an approach like Eat Just's technique of raising entire cells, but for the purpose of making chicken fat instead of lean meat. Mission Barns, based in Berkeley, is doing the same with pork fat. Their meatballs and bacon are likely to be the first cultivated meat products sold in the US. Through approaches like these, the poultry industry might see its offerings "unbundled", with different competitors offering new products that compete only with single items that chickens now offer, egg white proteins, chicken meat, chicken fat, or products even better than what chickens can produce.

Entrepreneurs like these are motivated by the recognition that modern animal agriculture is resource intensive, damages the environment, and some would say is immoral, but also that the industrial conditions under which farm animals are now raised directly contribute to the rise and spread of global infectious diseases. Zoonotic diseases, those that have jumped from animals to people, have repeatedly become major public health crises. Ebola, HIV, bird flu, swine flu, anthrax, dengue fever, leprosy, Creutzfeldt-Jacob disease, Rift Valley fever, COVID-19, and numerous others are all believed to have jumped from animals to humans. Eliminating or greatly scaling-down industrial animal-based agriculture as a source of protein and fat could reduce the rate at which new diseases arise, allowing us to better address the ones that are already concerns.

Now, conventional farmers who raise chickens for meat ('broilers') are faced with a new conundrum, as the most-common breeds, developed solely to produce meat as quickly and efficiently as possible, have left their birds susceptible to new maladies like 'woody breast', which gives chicken meat a gristly texture, and 'white striping', a malady characterized by excess strips of fat in what should otherwise be lean tissue.

Both conditions were rare only a decade or so ago and remain rare in slow-growing, heritage breeds. But now, according to a Bloomberg report, "77% conventional chickens that were 48 days old had woody breast... and two-thirds had white striping" (Pollard, 2021) Less than ten years ago, white striping disease was found in "an average of just 5%" of chicken breasts, according to a 2021 investigative report by The Humane League. This leaves breeders in the position of needing to reverse these recent developments, perhaps negatively impacting chicken meat economics, at precisely the time that new technologies are providing alternative, competing products.

Bring on the Barbecue Sauce

Mark Schatzker, in his 2015 book The Dorito Effect, argues that growing animals to be as massive as possible as quickly as possible has sacrificed both nutrition and flavor. He quotes the 1961 edition of chef Julia Child's best-selling cookbook Mastering the Art of French Cooking as saying that chicken "should be so good in itself that it is an absolute delight to eat as a perfectly plain, buttery roast, sauté or grill."

But over just a couple of decades, industrial agriculture made chicken meat so bland that its lack of flavor has become a running joke. In the 1999 dystopian movie 'The Matrix', one character who has realized that everything we consider to be reality is an elaborate computer simulation and who has escaped from that world, remarks to some other escapees on how the makers of the simulated world gave chicken meat no flavor. "Maybe they couldn't figure out what to make chicken taste like," he says, "which is why chicken tastes like everything."

In that empty space where flavor used to be, an endless variety of pre-mixed spices, dry rubs and hot sauces have popped up (including their own new metrics, like the 'Scoville unit', a measure of spiciness) as well as new gadgets and techniques to grill, smoke, deep fry, sous vide or otherwise impart flavor and the possibility of a joyful eating experience into what has become an otherwise unremarkable product.

The Dairy Cream Separator Got Disrupted

The mass production of penicillin in the mid-1940s depended on Granville Rettew's idea to use dairy industry cream separators, but the dairy cream separator itself became obsolete during the early 1950s.

In the book *Technological Substitution: Forecasting Techniques and Applications*, edited by Harold Linstone, author Dr. Devendra Sahal wrote that between 1950 and 1955 "the technology of dairy farming began to show numerous signs of change. The advent of milking parlor technology introduced a series of new or improved devices such as pipeline milkers, liquid manure handling and bulk milk tanks. By 1955, the 'diffusion' of the new dairy farming technology had proceeded so that the cream separator sales declined from 100,000 units to slightly more than 10,000 units... In 1960, the U.S. Census Bureau stopped collecting data on cream separators."

Over about ten years from 1950 to 1960, cream separators went from being standard dairy technology to being so obsolete that the US government no longer even collected data about them.

This raises a potentially interesting issue: if the strain of penicillin that Mary Hunt found had been discovered much later, perhaps there would have been no one around to think that a dairy cream separator might be a good way to produce the penicillin at scale, or no dairy cream separators readily available. Perhaps it could have been too late to find a way to produce the antibiotic at scale. Maybe some other technology would have been available, but maybe penicillin would have simply languished forever, and millions of people would have been far worse off as a result.

◆ ◆ ◆

"Whenever I get discouraged with the current situation I look back over the last decade or two and I cannot help but be thrilled by the vast progress that has been made. Then, the future appears very bright."

— **Dr. Alan Guttmacher,** *Babies by Choice or by Chance,* **1960**

◆ ◆ ◆

The Birth Control Pill Doubled Sales of Ballpoint Pens

> *Milton Reynolds, who brought the ballpoint pen to America, invested the money he made in a chemical company that later invented the birth control pill, kicking off the Sexual Revolution and allowing young women to pursue higher educations.*

By the start of the 1960s, Americans were living in a world that was unimaginable just fifteen years before and almost every aspect of life was undergoing a rethink.

Suburban living in every Levittown drove a dramatic expansion of the oil and automobile industries. Every home needed electric appliances in the kitchen, bathroom, and washroom, and a TV in every living room. Plastics meant colorful clothing from artificial fibers, colorful interiors, and stylish furniture. Antibiotics and vaccines virtually eliminated some of the biggest deadly and crippling diseases. Computers, nuclear power, nuclear weapons, jet airplanes, and the Space Race were changing the world in ways we could barely begin to imagine.

In 1960, there were just under three people in the US for each car. Sometime in the 1970s, it fell below two people per car. The sentence that came after "a chicken in every pot" in that famous 1928 campaign ad supporting Herbert Hoover read, in its entirety: "and a car in every backyard, to boot." For the first time in American history, it could fairly be said that both of those things were true.

It was also undoubtedly the best time ever (up to that point) to become a mother. Statistics compiled by *Our World in Data* show that a woman in 1945 had about a 70% lower chance of dying from pregnancy-related causes than a woman fifteen years before. In 1930 the maternal mortality rate in the US was 673 deaths per 100,000 live births. Put another way, out of every 150 children born, one mother would die. By 1945 maternal mortality had fallen to 207 and by 1960 it was under 38 deaths per 100,000 live births. That is, by 1960 the rate of maternal mortality was about 95% lower than it had been in 1930, and from 1960 to 1975 it fell about 65% again. And 1930 was not even an especially high year for the US - other years at the very beginning of the 1900s and around World War I were even higher.

Surprisingly, this dramatic transformation not only happened in the US but also in many other wealthy, industrialized countries at the *same time*. And it even continued to make progress during World War II, in the UK, which was an Allied country, in Japan, which was an Axis country, in Sweden, which was neutral, and even in the Netherlands, which was occupied by the Nazis. Maternal mortality was a spectre that haunted mothers-to-be for literally millions of years. And then, over the course of a couple of decades in the industrialized world, it largely disappeared. Even the average over the entire globe has now fallen to about 200 deaths per 100,000 live births, the level that many industrialized countries were at in the late 1940s, and headed downward still.

Obviously, a convergence of factors contributed to this dramatic decline, including antiseptics in the early 1900s and antibiotics after their widespread availability started after World War II. According to Richard and Dorothy Wertz in *Lying-in: A History of Childbirth in America*, in 1930 about 25% of births in the US were conducted in a hospital. By 1945, just fifteen years later, that figure was about 75 to 80%. Another important factor that contributed to the increase in hospital births was the rise of the automobile. "In rural areas", the Wertz's write, "where midwives and general practitioners had long attended home births, the automobile's increasing availability enabled women to travel considerable distances to hospitals even after labor began."

Other factors were important in the improvement in maternal mortality. "Starting in 1936," the Wertz's write, "and continuing until 1955, rates of mortality dropped

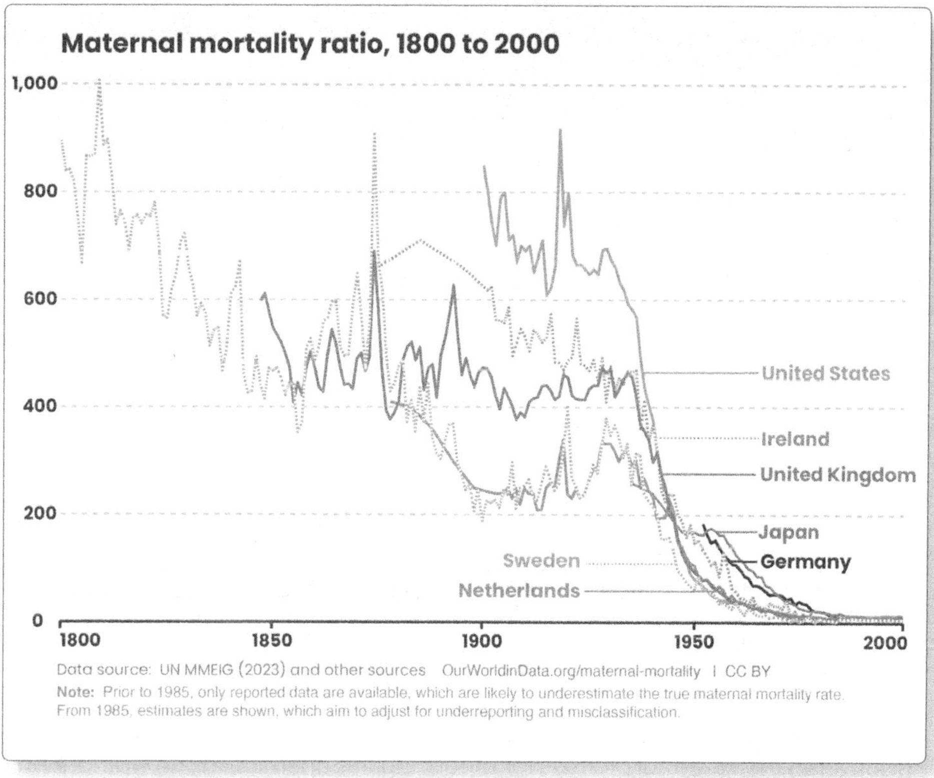

Figure 123 – Maternal Mortality from 1800 to 2000, for several countries.

precipitously... A number of medical discoveries and techniques improved the safety of birth. Hospital bloodbanks were established along with blood typing and transfusion techniques between 1936 and 1945, these removed much of the danger of postpartum hemorrhage." Enabled by the spread of electrification (and thus, the refrigeration necessary to store blood), the first hospital blood-bank in the United States opened in 1937, in Spain in 1936, and in the UK in 1938. Among other factors, they list "the discovery of oxytocic drugs to induce or speed up labor", "X-ray pelvimetry", "conduction anesthesias", "fetal-heart-monitoring machines", and the replacement of high and mid-forceps with Caesarean sections when necessary. Even the gradual spread of handwashing by doctors had an effect (something that happened shockingly slowly).

The Slow Spread of Handwashing

In the 1840s, Hungarian physician Ignaz Semmelweis had noticed that mothers at the Vienna General Hospital in Austria where he worked were twice as likely to die when a birth was attended by a male doctor as when it was attended by a female midwife. Semmelweis believed that the tendency of doctors to perform autopsies without washing their hands afterward might be the culprit.

An article in *National Geographic* magazine recounts, "in 1847, Semmelweis implemented mandatory handwashing among the students and doctors"

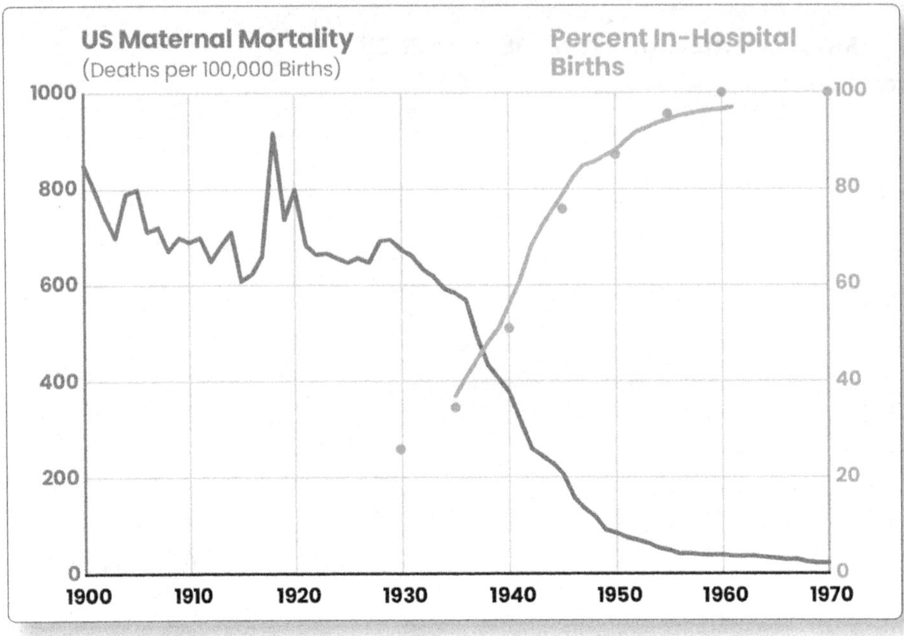

Figure 124 – US Maternal Mortality and Percent of Births in-hospital.
Maternal Mortality data from Our World in Data. Percent In-Hospital Births, line from (Devitt, 1977), dots from (Wertz, 1977)

using chlorinated lime, not just soap. (Remember, it was Charles Tennant who discovered the industrial process for producing this chemical as he was looking for artificial means to bleach white fabrics.) "The mortality rate in the physician-run maternity ward plummeted." (Strochlic, 2020)

Semmelweis's assertion, even backed up by scientific data, that doctors themselves were spreading disease generated vigorous resistance from the medical establishment. "The ensuing years were difficult for Semmelweis," the article continues. "He left Vienna and went to Pest in Hungary where he also worked in a maternity ward. He instituted his handwashing practice there and, as in Vienna, drastically reduced the rate of maternal mortality. Yet his success at saving lives earned no acceptance of his ideas." Semmelweis's own health deteriorated, and he died in a mental institution, aged 47.

"Surgeons began regularly scrubbing up in the 1870s, but the importance of everyday handwashing did not become universal until more than a century later. It wasn't until the 1980s that hand hygiene was officially incorporated into American health care with the first national hand hygiene guidelines."

We have already seen how dramatic improvements in public health in the 1930s, 40s, and 50s had taken away much of the mortality risk associated with childbirth and how penicillin and other antibiotics made sexually transmitted diseases like syphilis far less concern in the 1960s than they had been in previous decades as well, helping open the door for a Sexual Revolution that dramatically transformed American attitudes toward equality, women's liberation, gender roles, homosexuality and equal rights.

But one chemical discovery probably played a larger role than any other in bringing about that revolution, one discovery so profound that it is commonly referred to simply as "the Pill". And it all started with a bus ride.

Boarding a Bus

Russell Marker was a man who was good at finding answers and, in late 1941, he found in the book of a botanist friend an answer he had been seeking for a long time. *Dioscorea mexicana*, or the Mexican yam, is found throughout Mexico and down to Panama and can grow to several hundred pounds in size. Marker needed to get his hands on as much of it as he could. Unfortunately, this was just after the US had been attacked at Pearl Harbor and American authorities advised strongly against unnecessary travel outside the country. Thirty-nine-year-old Marker, unable to speak Spanish and with only some borrowed cash in his pocket, packed a bag and boarded a bus alone bound for Veracruz, Mexico, anyway.

Thirteen years prior, in 1928, scientists at the University of Rochester (New York) discovered that after ovulation, when the egg departs for the womb, it leaves behind a tiny sac that produces progesterone, a chemical that inhibits the release of additional eggs. In 1937, scientists at the University of Pennsylvania found they could prevent ovulation (and thus, pregnancy) in rabbits by injecting them with this chemical. But according to Bernard Asbell in *The Pill: A Biography of the Drug that Changed the World,* progesterone "was so expensive that even research scientists couldn't afford it. The minute amounts extracted from animals was bought up to improve fertility in world-class racehorses."

But Marker was a gifted chemist. At about the time of the discoveries in Rochester, he developed the concept of the 'octane rating', which is still used today for grading motor fuels. By the time of the ovulation experiments in Pennsylvania, he was a professor at Pennsylvania State when he had the idea to try to synthesize progesterone chemically, rather than extract it from animals.

A family of chemicals called sapogenins were known to be in the roots of many plants, and they are chemically like progesterone. Marker's idea was to harvest those chemicals from plant roots and then convert the chemicals to progesterone. But which plant? Many promising plants have roots so small that it would take a great deal of effort to harvest enough of them. Marker needed a plant whose roots were enormous, and when he found that plant, the Mexican yam, in the book of a friend, he set off for Mexico to find it in the flesh.

In Mexico, he made three kilograms of progesterone ("worth about $240,000" according to Asbell, or about $80 per gram) and in 1944 established the Syntex company with two colleagues and financial backing from Milton Reynolds who made his fortune by introducing the first commercially successful ballpoint pen to the US. As Marker and his colleagues made more of the chemical, it got cheaper to make even more.

Cheap Progesterone Opened Many Doors

In Asbell's book, Marker recounts that the price of progesterone dropped "from eighty dollars to ten dollars a gram and a few months later to five dollars a gram".

In May 1946, Marker turned over the process for making progesterone to a European company. "I remained for three years as a consultant," Marker recalled. "By that time [1949] the price of progesterone had been reduced to about two dollars a gram." A 40X cost decrease in about five years.

Abundant progesterone also meant abundant cortisone, a treatment for rheumatoid arthritis and for keloids, as cortisone can be made from progesterone.

In 1949, twenty-six-year-old chemist Carl Djerassi joined Syntex and quickly found other ways to synthesize cortisone, as well as ways to synthesize other hormones like estrone and estradiol. Asbell writes that, by the early 1950s, "Syntex became the major supplier of synthetic hormones to drug houses in Europe and America."

Not Just Cheaper, But Better

In 1950, Djerassi set himself the goal to find a progesterone-like chemical that was better than natural progesterone, which must be given by injection. He wanted one that could be taken orally. Djerassi knew of a form of testosterone, a masculine hormone, that had been produced to be taken orally described just a few years before in a research paper that was published shortly after Syntex had been founded. In 1951, he used a process like the one described in the paper, but with Marker's progesterone instead of testosterone, to yield norethisterone, which was tested successfully in rabbits, guinea pigs, monkeys, and then three women volunteers. According to Asbell, "Djerassi's new substance, when taken by mouth, was eight times more powerful than natural progesterone."

It took about a decade of careful medical testing but norethisterone, and another closely related chemical that is converted in the stomach into norethisterone, became key ingredients in the first generation of combined oral contraceptive pills made by several companies. The Searle company brought its birth control formulation to market before Syntex, the inventor of artificial norethisterone, in one of the great intellectual property coups of all time. The final step in the production of norethisterone is to treat another chemical with acid. So, the Searle company simply produced that penultimate chemical that gets converted by acid into norethisterone. When that chemical is swallowed, it enters the stomach and stomach acid converts it into norethisterone. The Searle company's formulation, sold under the brand name Enovid, was approved for contraceptive use in the US starting in 1960, before Syntex was able to bring its product to market. Syntex ultimately produced two formulations, one that was approved in 1963 and one in 1964, thirteen years after Djerassi first synthesized norethisterone. Over the next few years, other major manufacturers like Upjohn, Merck, and Wyeth made similar formulations available, and women suddenly had an abundance of oral contraceptive options to choose from. Asbell writes "the Pill has been swallowed as a daily routine by more humans than perhaps any other prescribed medication in the world." Their makers did not initially realize it, but cheap and effective birth control was not only about to launch a medical revolution, but sexual and social ones as well.

Djerassi's work made him enormously wealthy. He left for Stanford University to become a professor, bought a mansion in San Francisco and a 1200-acre ranch in the Bay Area which he called SMIP Ranch (for "Syntex Made It Possible"), amassed a collection of expensive artworks, and became a lifelong benefactor to the arts community.

Marker insists, according to Asbell, that "prevention of pregnancy as a goal never crossed his mind". He was simply trying to make progesterone, for science and for medical use. But methods to cheaply synthesize some hormones changed not only birth control, but much more beyond.

The Pill Stopped Unintended Consequences, and Caused Them

Between penicillin, hospital births, and the Pill, by the early 1960s for the first time ever, women had control of some of the most-important aspects of reproduction, being healthy, staying alive, and only getting pregnant when desired. (Contraceptives were not available to married women in all states until the Supreme Court ruling of Griswold v. Connecticut in 1965 and were not available to unmarried women in all states until Eisenstadt v. Baird in 1972.) The Pill not only reduced the number of unplanned pregnancies, it also reduced the number of unplanned weddings. And it likely played a key role in the (unsuccessful) push for an Equal Rights Amendment and the (temporarily successful) Roe v. Wade Supreme Court case for the legalization of abortion.

According to Claudia Goldin, professor of economics at Harvard University and winner of the 2023 Sverige Riksbanks Nobel Prize in Economics, the Pill profoundly affected the ability of women like her to get a higher education, especially in professional fields like law, business, and medicine. Goldin's research shows that the percentage of medicine, law, dentistry, and MBA students in the US who are women went from under 10% in the early part of the 1970s, just before the Eisenstadt v. Baird ruling, to over one-third about fifteen years later.

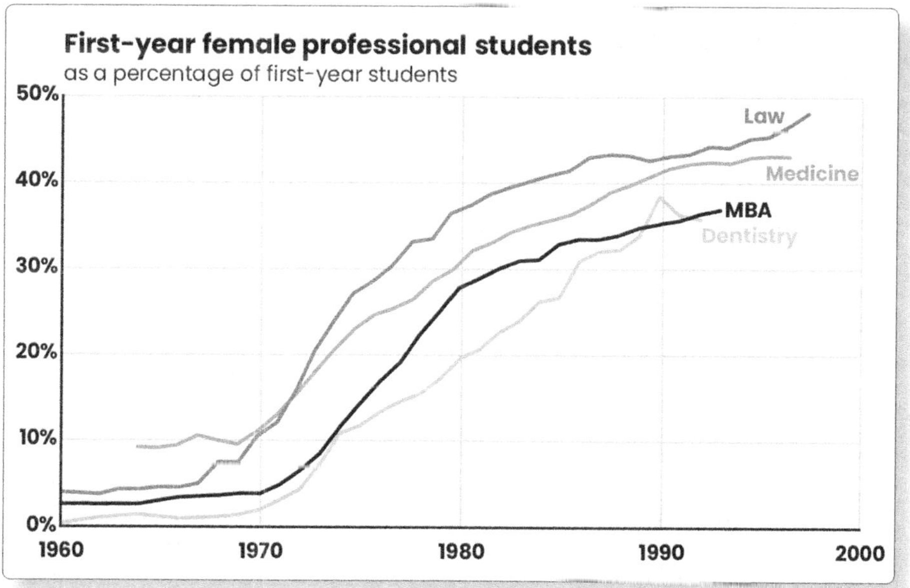

Figure 125 – First-year female professional students as a percentage of first-year students *(Goldin, 2002)*

This was not a one-for-one replacement: the percentage of first-year students who were female did not increase because women replaced males, it increased because the total number of students increased. According to the American Association of Medical Colleges (the AAMC), the number of first-year students in US medical schools more than doubled from 1960 to 1980, as did the number of new graduates. But this was not just about graduate degree programs. Since the 1980s, women have represented a majority of students in undergraduate programs in the US. And since the year 2020, women with college degrees have been a majority of the degreed US workforce.

In a roundabout sense, Milton Reynolds biggest contribution to the world of writing instruments might not have been his introduction of the ballpoint pen to the US, but rather his use of the money he made from pens to fund the birth control pill, thus effectively doubling the number of people who needed his product.

However, increasing the number of graduate students over the course of a decade or so, faster than the number of faculty, resulted in a significant increase in the amount of time required to get a PhD degree. In biology, chemistry, engineering, physics, and many other fields, the average number of years required to get a PhD degree increased significantly throughout the 1970s and 1980s.

The introduction of the birth control pill also significantly changed Americans' attitude toward family size preference. A 2013 report by the Gallup polling organization, compiling results from as far back as 1936, found that the portion of Americans reporting an ideal family size of four or more children fell from about 40% to about 10% in the fifteen years after the oral contraceptive pill became popular. Likewise, the portion of Americans reporting a desire for two children or fewer rose from about 20% to about 60% over the same time. (Newport, 2013)

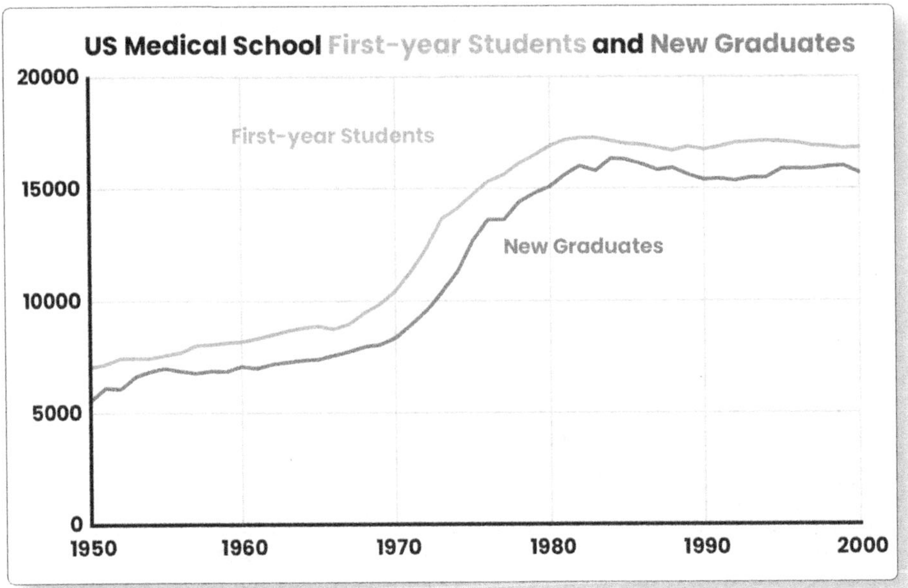

Figure 126 – Number of First-Year Students and New Graduates from US Medical Schools.
(AAMC, 2016)

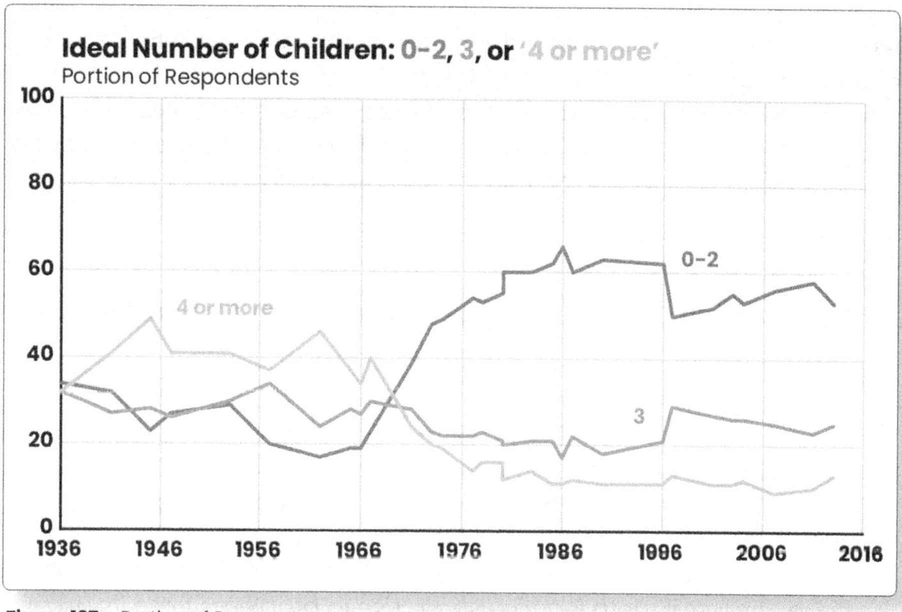

Figure 127 – Portion of Respondents Saying the Ideal Number of Children is 0-2, 3, or '4 or more'. *(Newport, 2013)*

The At-Home Pregnancy Test

We have now had at-home pregnancy tests for more than fifty years. Before the 1960s, the standard way to test for pregnancy was to inject toads with urine. At the beginning of the 1930s, researchers at the University of Cape Town, South Africa, discovered that urine from a pregnant woman injected into a female *Xenopus* toad would cause the toad to ovulate. According to Wikipedia, "this test was used throughout the world from the 1930s to 1960s, with *Xenopus* toads being exported in great numbers."

According to an article in *The Atlantic* by acclaimed science writer Ed Yong, the use of frogs was itself a disruption. The standard method before that was to inject urine into mice or rabbits, which were then "dissected to check the size of their ovaries. This meant that testing for pregnancy was laborious, expensive, and bloody work". Using frogs made the test cheaper, simpler, and faster, and it did not kill the test subject.

The *Journal of British Studies* says, "animal injections were made obsolete when Organon launched Pregnosticon, a cheap, mass-produced test kit, in 1962." A few years later "in 1965, commercial laboratories began serving women directly... The laboratories initially charged £2 for a test and (controversially) returned the result not to a doctor but to the woman herself." (Olszynko-Gryn, 2020) This was not only a technological disruption, but a business model disruption too, cutting medical doctors out of the act of diagnosing pregnancy.

In 1971, Margaret Crane launched the 'Predictor' at-home pregnancy test. According to Olszynko-Gryn, "Predictor was not predicated on any new technical innovation. Rather, it was a decade-old medical technology, Pregnosticon, redesigned for domestic use and thus full of disruptive potential... Predictor retailed for £1.75, undercutting most lab tests."

Today, the technology behind at-home tests is so cheap that IKEA has even used them in a magazine ad gimmick to sell baby cribs, by making a paper advertisement that revealed pregnancy when exposed to urine.

Synthetic Hard Drugs Disrupted Ones from Natural Sources

> *In 2006, a new US law made it much more difficult to obtain large amounts of pseudoephedrine, a chemical that is used in making remedies for the common cold as well as in making methamphetamines. This ban resulted in Mexican drug cartels synthesizing 'crystal meth' from legal chemicals which are cheap and easy to obtain in large volumes. The tenfold decrease in the price of crystal meth led to a sharp rise in problems from its use, including overdose deaths. Thus, a government policy designed to alleviate a problem made it far worse... And then came fentanyl.*

Examples of disruption are not limited to lawful activities. We also find them among criminal endeavors, like making synthetic methamphetamines and synthetic opioids, both of which radically changed the illicit drug market in the first couple of decades of the 21st Century.

In 2006 the U.S. Congress passed the Combat Methamphetamine Epidemic Act (CMEA) to make it difficult to acquire pseudoephedrine, the primary ingredient in Sudafed nasal decongestant, which can be used to produce methamphetamine.

By any reasonable assessment today, the CMEA did not solve the U.S.'s meth problem; if anything, it made it far worse as it drove drug producers to switch away from using pseudoephedrine to using wholly synthetic avenues of production.

The ephedra plant, native to Mongolia, eastern Russia, and northeastern China, contains several active molecules which act as stimulants and anti-asthmatic agents, which combined constitute 0.02 to 3.4% of the plant. A Japanese researcher in the early 1900s first converted these natural ingredients into crystal methamphetamine. According to journalist Sam Quinones in an October 2021 article in *The Atlantic*, "in the early 1980s, the ephedrine method for making meth was rediscovered by the American criminal world."

But "there was another way to make methamphetamine. Its essential chemical was a clear liquid called phenyl-2-propanone — P2P. Many combinations of chemicals could be used to make P2P. Most of these chemicals were legal, cheap, and toxic".

After the passage of CMEA made it difficult to acquire ephedrine-type chemicals in the US, it was outsiders, primarily Mexican cartels, who seized on the P2P synthetic route to make crystal meth in enormous labs. As Quinones explains, "drugs made in a lab were not subject to weather or soil or season, only to chemical availability: With this new method and full access to the world's chemical markets through Mexican shipping ports, traffickers could ramp up production of P2P meth in quantities that were, effectively, limitless."

These large-scale operations were much more efficient than ones based on extracting a precursor chemical from a natural plant, making the resulting product much cheaper, about ten times cheaper. Quinones says, "the price of a pound of meth fell to nearly $1,000 for the first time on U.S. streets by the late 2010s — a 90 percent drop from a decade earlier in many areas."

Over just a few years, the market went from essentially 100% meth made from ephedrine compounds to 100% meth synthesized from P2P, according to measurements made on seized drugs by Steven Toske of the U.S. Drug Enforcement Agency and published in the scientific journal *Forensic Chemistry.*

It is not just the methamphetamine supply that has seen a dramatic transformation in the past couple of decades, but also opioids. In the US, fully synthetic chemicals like fentanyl, made mostly in China, have disrupted Mexican heroin, which is grown and extracted from the poppy plant. It's no surprise why — fentanyl is ten to fifty times more potent than heroin. (Vardanyan, 2014) Because it is more potent, the same amount of fentanyl can be turned into many more doses than heroin, earning dealers much more revenue, or making smuggling much easier by making profitable loads smaller.

Science reporter Dan Vergano wrote, "a kilo of fentanyl can be split up many more times than a kilo of heroin, to make more money... a kilogram of heroin that costs about $4,000 dollars might be made into enough doses to garner the dealer

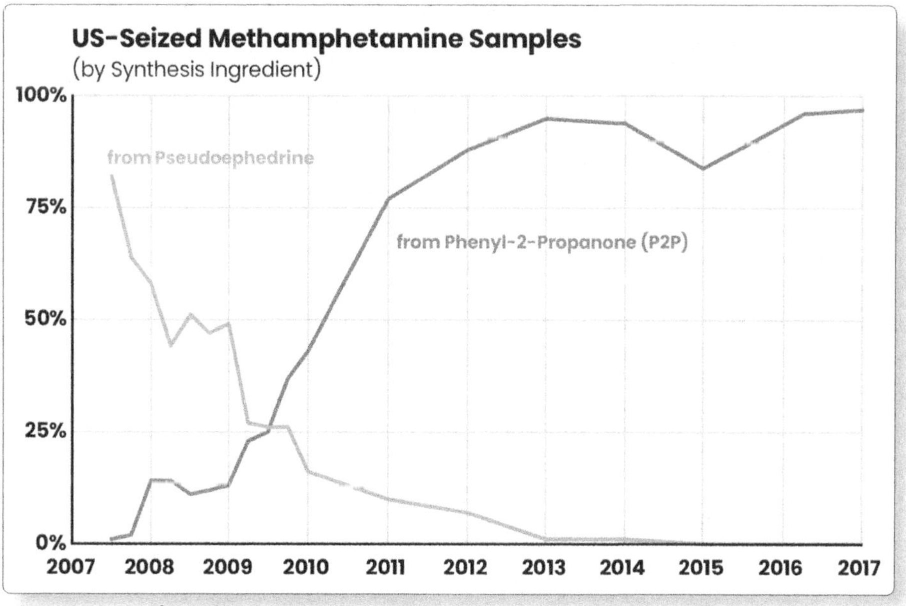

Figure 128 – Percentage of US Methamphetamines made from Pseudoephedrine and made from P2P. *(Toske, 2019)*

around $60,000. A kilogram of fentanyl might also cost a dealer $4,000 wholesale, according to the DEA, but it can be split into 20 times more doses, garnering $1.2 million or more."

(Heroin itself was disruptive. In August 1897, twenty-nine-year-old German chemist Felix Hoffmann synthesized diacetyl morphine, later trademarked as 'Heroin', from morphine. Eleven days earlier in the month, Hoffmann had synthesized acetylsalicylic acid, or Aspirin, for the first time. Both quickly became smash hits, in their own ways.)

These disruptions by synthetic chemicals have not replaced the natural counterparts one-for-one. On the contrary, drug abuse and drug-related deaths have boomed as these chemicals have gotten cheaper and more potent. According to the Centers for Disease Control, total overdose deaths from hard drugs in the US were at an all-time high in 2020, topping more than 100,000 for the first time and rising nearly 30% from the previous year. The amount of fentanyl seized by the US Drug Enforcement Agency in 2022 was enough to give a lethal dose to every American. Deaths from fentanyl increased tenfold from 2014 to 2020, from about five thousand per year to over fifty thousand, including the musician Prince, who died from a fentanyl overdose at his home in 2016. More adults between 18 and 45 in the US died of fentanyl overdoses in 2020 than COVID-19, motor vehicle accidents, cancer, or suicide. The increase in deaths, particularly from COVID-19 and opioids, have driven the US life expectancy to its lowest level in a quarter century.

As of late 2022, drugmakers, distributors, and other organizations have reached about $30 billion in settlements, including pharmacies CVS and Walgreens who have each agreed to pay nearly $5 billion each and global consulting firm McKinsey nearly $600 million, for their contributions to promoting the opioid crisis.

But now fentanyl itself might get disrupted. On the one hand, scientists at the University of Houston have identified a vaccine that, according to a university press release, could block the ability of fentanyl "to enter the brain, thus eliminating the drug's 'high.'" (Fickman, 2022) As well, a study by researchers at the University of Pennsylvania published in the *Journal of the American Medical Association* found that opiate-related deaths decreased by approximately one-third in states where medical marijuana was legalized in the years after legalization, suggesting that legal marijuana might compete with harder drugs. (Bachhuber, 2014) On the other hand, a new class of synthetic opioids, nitazenes, which can be ten times stronger than fentanyl, has "been found in street drugs across the Midwest and Northeast since 2019", according to NBC News. (Edwards, 2022) Another opioid, carfentanil, is ten times stronger still. Only time will tell which of these will dominate the future market.

Sports Disruptions

> *Disruptions have affected the world of sports as much as any other facet of life. The cheap, oxygen-enriched air that enabled the post-War revolution in steelmaking also made scuba diving much more popular. Petroleum-based materials quickly replaced solid wood in surfing, opening up the sport to new participants (and helping to get the Apollo mission to the Moon). The man who invented nylon also invented the neoprene that wetsuits are made from. Even things as fundamental as jumping and eating are open to radical reinvention.*

Many sports in the past century have been revolutionized by better equipment designs and better materials, particularly light-weight and durable materials derived from petroleum. Some have also been disrupted purely by new ideas about how they should be performed. Let's look at a few examples: surfing, scuba diving, swimming, the high jump, and competitive eating.

Surfing

The popularity of surfing took off in the 1960s due to a convergence of factors, especially demographic ones: peacetime prosperity, the post-war generation entering their teenage years, and rapidly rising populations in naturally good surfing areas like Hawaii and California. But also playing a large role were new, petroleum-derived products like polyurethane foam, which quickly replaced wood as the primary material from which boards were made, and wetsuits made from neoprene, an artificial rubber brought to market by DuPont chemist Wallace Carothers (who also made nylon a household name).

American surfer Dave Sweet is credited with first experimenting with polyurethane foam boards starting in 1956, when he was 27 years old. At the time, surfboards were mass-produced. According to Peter Neushul and Peter Westwick in 'Blowing Foam and Blowing Minds: Better Surfing Through Chemistry', a chapter in the 2016 book *Groovy Science: Knowledge, Innovation and American Counterculture*, "a handful of large labels … exerted a virtual monopoly" on board production. Wooden boards took a lot of work to make, and at scale that required specialized equipment, but Sweet's polyurethane foam cores could be shaped in a garage or shed. "Where a balsa blank required laborious laminating and chipping," Neushul and Westwick say, "a power planer cut through foam like butter".

It took people far outside the dominant California surfboard manufacturing hubs, mostly Australians, to ride this new material (quite literally) to a revolution, what *Surfer* editor Drew Kampion called the "greatest conceptual shift in surfing history". Working in foam opened new possibilities in board shape, style, and length. "The shortboard revolution occurred between 1967 and 1970", Neushul and Westwick wrote, a little more than a decade after Sweet began his tinkering. "Within a couple years, the average board length shrank by a third, or three feet: from a ten-foot board... to six or seven feet".

This was not just a revolution in board design, shortboards also had severe implications for the incumbent business of the surfboard production industry. According to Neushul and Westwick, shortboards "introduced radically new surfboard designs and shifted the prevailing mass-production model in the surfboard industry to a backyard-craftsman ideal. Instead of surfboards knocked out by the thousands in a few standard models, surfboard shapers now worked alone with a surfer to develop a custom design tailored to the individual's style."

According to *GearPatrol.com*, polyurethane boards "put a lot of companies out of business." Ironically enough, one of those affected was Dave Sweet himself. Neushul and Westwick recount that "Sweet opened a store on Fourteenth and Olympic in Los Angeles and served a steady stream of customers until the late 1960s, when the shortboard revolution cut sales drastically and forced him out of business."

But having a pool of people who had experience with polyurethane foam proved critical to the Apollo missions to the Moon. When NASA engineers had trouble with the insulation that coated the extremely cold tanks in the Saturn V rocket, they hired surfers to apply it.

Foam was far from the only revolution to happen in surfing, and not the only one from Down Under either. In the early 1980s, Australian surfer Simon Anderson was dissatisfied with the predominant designs, which had one or two stabilizing fins mounted to the underside of the board. In a 2013 interview available on YouTube called '3 Fins 3 Decades', he explained the origin of his invention: "At the time, back in 1980-81, there was two schools of thought: there was a single-fin camp and the twin-fin camp and pretty much everyone was in the single-fin camp." Anderson developed the three-fin (or "Thruster") surfboard, which according to skateboarder David Hill, a contestant on Shark Tank Australia, "took 90% of the market after two years". According to 'adventure and action sports' website Rad Season, "since 1982 the winner of every ASP [Association of Surfing Professionals] world surfing title has ridden some variation of a three-fin configuration."

Scuba Diving

The rise of surfing coincided with the rise of other aquatic activities like scuba diving, largely for the same reasons, like general economic prosperity and a demographic boom in young people.

But scuba also benefited from the convergence of several other technologies such as new and cheaper materials like wetsuit neoprene (the first synthetic rubber, invented in 1931 by Wallace Carothers), cheaper steel for equipment due to the post-War rise of the oxygen blast furnace, and affordable oxygen-enriched air. As we learned earlier in Chapter 3, the Russian physicist Pyotr Kapitsa won the Nobel Prize in Physics for his work in 1939 inventing the turboexpander, a device that produces oxygen-enriched air in industrial quantities. These inventions, plus the Aqua-Lung, invented by Jacques Cousteau, whose books and documentaries made him a household name worldwide in the 1950s, turned scuba diving from an activity done only by Navy 'frogmen' into something nearly any healthy adult could do.

Swimming

The 100-meter backstroke in swimming underwent a revolution in the 1970s and 80s. In that event, competitors start in the water, holding the edge of the pool with their hands and pressing their feet against the pool wall. In 1976, Jesse Vassallo began each race by swimming the first "5 to 7 meters", or four to five kicks, underwater. By the 1984 Olympics in Los Angeles, Vassallo says, "nobody was using" the 'dolphin kick' technique except Vassallo himself. But swimming underwater, using the legs to kick in unison, gives a significant speed advantage over simply doing the backstroke at the surface.

According to Igor Polyansky, who won the gold medal in the men's 200-meter backstroke at the 1988 Summer Olympics in Seoul, South Korea, by 1987, "many, many swimmers" had adopted the dolphin kick technique for the simple reason that it was "much faster than the backstroke swim on the top level of the water." The 100-meter event involves swimming the length of a 50-meter pool, turning, and then swimming back again. During the 100-meter men's backstroke final event at that same Olympics, at twenty meters into the race, only three of the swimmers were above the water and five were still under. The winner of the gold in that race, Daichi Suzuki of Japan, used what he called 'the Vassallo technique' for the first thirty meters, more than half of the length of the pool.

The Olympic Committee changed the rules afterwards so that each swimmer's head must break the surface of the water in the first ten meters of the backstroke event (a rule they extended in 1991 to fifteen meters). In fifteen years, the underwater 'dolphin kick' went from being a technique used by essentially a single swimmer, to one used by 'many, many swimmers', including the majority of the 1988 men's Olympic backstroke final field, to being curtailed by the Olympic Committee.

High Jump

Other sports have seen similar, or even more dramatic, changes in a short time. High jumping was transformed irrevocably on a single day, October 20, 1968. At the Summer Olympics in Mexico City, to the crowd's surprise, the 21-year-old engineering student Dick Fosbury threw himself over the bar backwards, a stark contrast to the ways high jumpers typically performed. Fosbury's jump did not set a world record, but it was the best at that Games, winning him the gold medal.

Until that point, competitive high jumpers had used various techniques with names like the 'Western roll', the 'scissors', or the 'straddle', preferring to go forwards over the bar. But when he was a 16-year-old high school student who was not particularly good at the high jump, Fosbury first began experimenting with an alternate approach.

Even though his technique won gold, it did not immediately win over the hearts of all observers. In a blog post written shortly after Fosbury's death in March of 2023, Ian Leslie, author of *Conflicted: How Productive Disagreements Lead to Better Outcomes*, notes that Fosbury said of his win, "everyone laughed at me," and that newspapers of the time compared Fosbury to "a guy being pushed out of a 30-storey window" and said he looked like "a fish flopping into a boat".

But over the course of the ten years after the Olympics virtually all high jumpers adopted the 'Fosbury Flop'. West German Ulrike Meyfarth was one of the first few people to adopt Fosbury's technique by the time of the 1972 Olympic Games in Munich, just four years after Fosbury's Olympic win. There, she won gold and set a new women's high jump record, becoming at the age of sixteen the *youngest* Olympic champion ever in athletics at an individual event.

In 1982 and 1983, Meyfarth set the women's high jump record two more times at European and World championship events and, in 1984, won gold again at the Olympic Games in Los Angeles, becoming at age 28 the *oldest* woman ever to win the Olympic high jump title.

In 1973, Dwight Stones was the first man to flop his way to a world record, reaching 2.30 meters (7 ft, 6.5 in) and then beating his own record twice more in 1976. The last Olympic gold medal in the high jump events that used the straddle technique was won by East Germany's Rosemarie Ackermann in 1976. The last straddle jumper to hold the World Record was 19-year-old Ukrainian Vladimir Yashchenko in 1978, a record that was beaten less than two years later by Polish jumper Jacek Wszoła. All record-setters since Yashchenko have used the Flop technique.

It was just fifteen years from Fosbury's invention of the flop in 1963 to him winning Olympic gold in 1968 to the last of the non-flop jumpers holding the World Record in 1978. Many high jumpers claim that Fosbury's flop is easier to learn than other techniques, requires less flexibility in the legs than the 'straddle', and is less likely to result in an injury if a jump does not go well. So, the Fosbury technique was probably not a one-for-one replacement: it likely opened the sport to new participants, resulting in an increase in the total number of young people attempting to become Olympic high jumpers and an increase in the average length of any individual's high jump career.

The high jump had been an Olympic sport for over seventy years when Fosbury introduced his new technique. So why was it Fosbury who invented the Flop in the late 1960s, and not someone else ten years before or ten years after? In his blog post Ian Leslie notes several reasons, including a recent change in technology.

He writes, "even a small change in technology, for instance, opens up new possibilities which aren't immediately apparent. High jumpers had always landed in a pit of sand or sawdust. Anyone trying to do a flop jump into a sawdust pit would have risked breaking their neck. It wasn't a possibility worth considering. In the 1960s, foam rubber replaced the pits. Fosbury was the first to seize the new adjacent possible."

Competitive Eating

The spectators at the 2001 Nathan's Hot Dog Eating Contest, which has been a fixture of Coney Island (part of New York City) since 1972, knew they were witnessing one of the greatest performances in competitive history when the contest organizers ran out of pre-printed numbers and had to resort to handwritten signs to continue counting how many hot dogs one competitor was eating.

In the year 2000, the contest was won by 101-pound Kazutoyo Arai of Saitama, Japan, who had set a new record by eating twenty-five hotdogs with buns in twelve minutes, or more than two dogs per minute. (In 1981, Thomas DeBerry, ate eleven dogs in five minutes, which was a faster rate than Arai, but a smaller total number of dogs. Most of the events in the first decade of the contest were short – just 3½, 5, or 6½ minutes – allowing winners who only put away eight or ten dogs to have relatively high rates of consumption. Since 1982, the competition has been set at either ten or twelve minutes.)

In 2001, Arai broke his own record to eat thirty-one hotdogs with buns, which would have been a new world record except for one issue: also in the competition was Arai's fellow countryman 128-pound Takeru "Kobi" Kobayashi.

Kobi ate differently from any other competitor on the stage. While some crushed the bun around the dog to make the bread denser, Kobayashi realized that nowhere in the rules did it say that the bun and sausage need to be eaten together. So he separated the bun from the sausage entirely. He broke the sausage in half (which does some of the work of chewing) and then ate both halves at the same time, side by side, essentially turning each of his bites into two bites. Rather than eat the dry bread and then drink water to rehydrate, he dipped the bun in warm water and then squeezed out the excess. And he did different tasks with each hand, dividing the labor to ensure a continuous supply of food. One competitor said of his performance that "he was like a conveyor belt".

When the time ran out, Kobayashi had eaten an astonishing fifty hot dogs with buns, doubling the previous record. (Arai came in second. American Eric Booker was third with twenty-two.) To call Kobi's performance stunning is an understatement. As Ryan Norz wrote in *Eat This Book: A Year of Gorging and Glory on the Competitive Eating Circuit*, "what had long been considered a publicity-driven sideshow now looked more like a sport."

Economists Stephen Dubner and Steven Levitt interviewed Kobayashi for their book *Think Like a Freak: How to Think Smarter About Almost Everything,* the follow-up to their best-selling Freakonomics. According to them, Kobi turned hotdog eating from a 'sideshow' into a sport by treating it like a sport. They explain that "Kobayashi experimented with different eating techniques leading up to this event. He recorded in spreadsheets how much sleep the night before mattered, how much time eating the buns separately could save, how to eat the buns the fastest, etc. He doubled the previous record because he massively out-thought his competition and thought in creative new ways to eat competitively... His experimentation was endless. He videotaped his training sessions and recorded all his data in a spreadsheet."

Advancements in competitive eating by Kobayashi were possible not only due to a convergence of new ideas, but also because newer consumer technologies like video cameras that, like airfare to New York City, might not have been affordable to a college student just a few years prior, allowed him to test and sift through his eating ideas in a scientific way.

Before Kobayashi, most competitors' strategies boiled down to this: eat hotdogs like when you eat a meal, just eat them as quickly as possible. Reporter Amy Stone, associated with a CBS affiliate, asked other competitors at the event to describe their strategies. One said, "my strategy is to take the hot dog, just squish all the air out of it and just eat it", and another said, "eat as quick and as many hot dogs as I can". But in a 2014 episode of Dubner and Levitt's popular podcast (also called *Freakonomics*) Kobayashi said through a translator, "the key to me was that I had to change the mentality that it was a sport, it wasn't having a meal... So maybe I gave them a reason for everyone at the same moment to rethink the problem again."

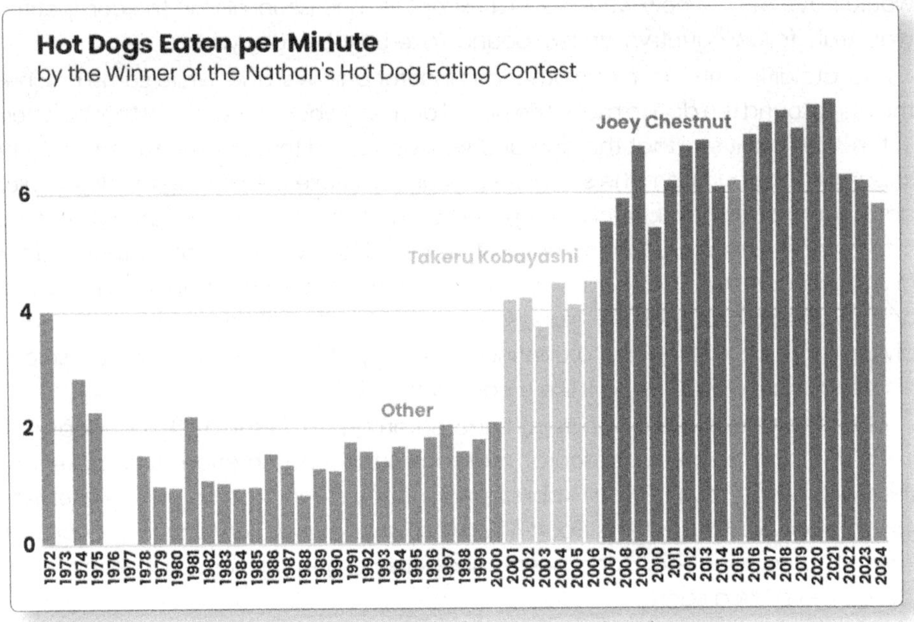

Figure 129 – Hot Dogs Eaten per Minute by the Winner of the Nathan's Hot Dog Eating Contest

Kobayashi won not only in 2001, but every year until 2006, a total of six times in row, breaking his own record in 2002, 2004, and 2006, when he ate 53.75 hotdogs with buns, his best ever. But Kobayashi's systematic experiments to find the fastest way to eat hotdogs was just the beginning, not only for competitive eaters in general, but also for Kobayashi himself. He now holds world records in eating hamburgers (ninety-three in eight minutes), rice balls (twenty pounds in thirty minutes), lobster rolls (forty-one in ten minutes), grilled cheese sandwiches (thirteen in one minute), pizza (sixty-two slices in twelve minutes) and cow brains (17.7 pounds in fifteen minutes), among others.

Other people adopted his techniques to win records for eating pickles and other hotdog-like foods. But no one has improved upon Kobayashi's techniques more than American Joey Chestnut, who won the 2007 Nathan's Hot Dog Eating Contest by eating sixty-six, topping Kobayashi's record by more than twelve dogs. In 2005, Chestnut ate thirty-two to Kobayashi's forty-nine. In 2006, he ate fifty-two, winning second place and falling less than two dogs behind Kobayashi. Since 2007, he has dominated the contest. Of the seventeen Nathan's competitions from 2007 through 2023 (the last year he competed), Chestnut won sixteen.

In 2008, the competition was cut from twelve minutes down to ten, but Chestnut's performance kept going up. In 2016 he downed seventy dogs. In 2017, he ate seventy-two. In 2018, he ate seventy-four. In 2020, despite a COVID-altered format, he ate seventy-five and in 2021, he set another record with 76, which meant eating at a rate more than three times faster than record-holder Arai's 2000 performance and more than five times the average of the 1972 through 2000 winners.

ESPN Sport Science put Joey Chestnut up against TV show host John Brenkus in a five-minute hotdog eating contest. "In the time it takes me to eat one hot dog," Brenkus said, "Joey scarfs down ten."

However, in 2024, Chestnut was banned from the competition for endorsing vegetarian company Impossible Foods, makers of burgers, nuggets and, yes, hot dogs. The competition on the Fourth of July, 2024, was won by Patrick Bertoletti, who downed fifty-eight hot dogs with buns in ten minutes, a rate still well above Kobayashi's revolutionary performance two decades before or the performance of anyone who came before him.

In September 2024, Chestnut and Kobayashi faced off against each other in a live event streamed on Netflix, using rules similar (but not identical) to the rules of the Nathan's competition. In a ten-minute contest Kobayashi bested his personal record set in 2009 of 64.5 hots dogs with buns to eat sixty-six. But Chestnut set a new world record by eating eighty-three hot dogs with buns. It seems that even after twenty years of competing, neither man has reached the limits of their performance.

If something as fundamental as eating can be re-invented over just a few years, what might not be?

The Counterculture Revolution Launched by Foam

Prior to the 1970s, few Americans said that they had ever tried marijuana. In 1969, according to the Gallup survey company, only 4% said that they had. But over about the next fifteen years, rising in an S-shaped curve, the portion of Americans who reported having ever tried marijuana reached one-third of the adult population, a level it would stay at for the next thirty years until another step up above 40% began when US states individually started decriminalizing marijuana possession starting with Washington state and Colorado in late 2012. Today, nearly one-half of Americans report having ever tried the drug.

What might have prompted this sudden rise in popularity from a niche drug to one which a significant portion of adults had tried? One possibility is surfing. As we have already seen, surfing became much more popular in the US after World War II, when petroleum-based rigid polyurethane foams revolutionized surfboard manufacturing away from expensive, solid wood, factory-made boards, to cheaper, foam-based, handmade ones.

One of the unintended consequences of this change was that young surfers quickly realized that hollowed-out polyurethane foam and frequent trips to

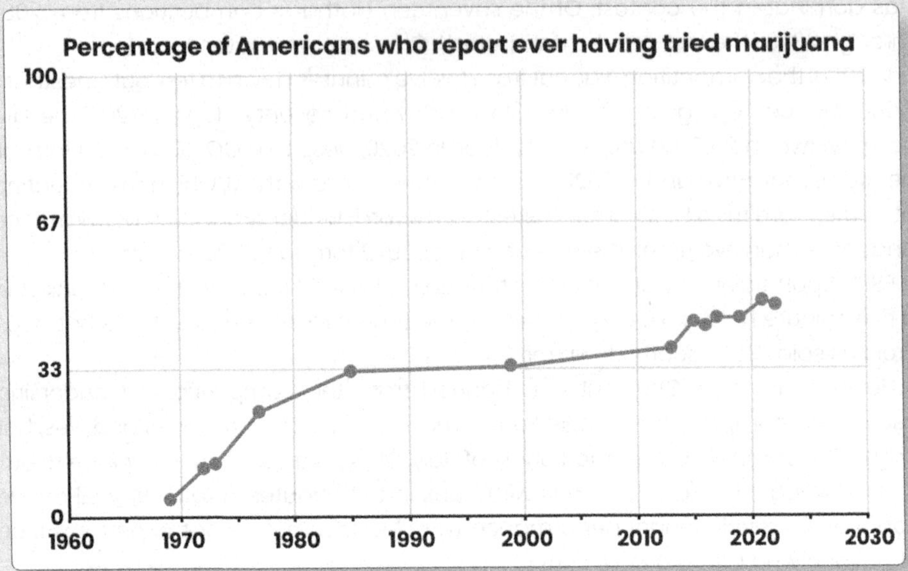

Figure 130 – Percentage of Americans who report ever having tried marijuana.
Data from (Saad, 2022)

exotic drug-filled destinations provided a great way for them to make a lot of money by smuggling.

In *Groovy Science: Knowledge, Innovation and American Counterculture*, Peter Neushul and Peter Westwick wrote, "a group of Laguna Beach surfers raised this technique to the peak of sophistication in an enterprise that mushroomed, so to speak, into a global drug-smuggling ring worth $200 million, responsible for half the LSD and hash in the United States. The group was known as the Brotherhood of Eternal Love."

"The brotherhood's network", they write, "brought in hash and pot measured not by the ounce or the pound but by the ton. The feds refused to believe that a bunch of hippie surfers were capable of running such a complicated operation. The US government belatedly recognized it in the 1970s as 'one of the largest and most complex drug systems in the history of this country's narcotic law enforcement efforts' and added that the drug ring had made Laguna Beach 'the psychedelic drug capital of the world.'"

So, the rise of petroleum-foam-based surfboards in the late 1960s led to a surfing-based drug culture that spread from a handful of young men in southern California to much of the adult population in just fifteen or so years, and which is still unfolding today. The recent trend toward decriminalizing marijuana possession and use has helped make smoking weed more popular than smoking tobacco. That Gallup survey found that in 2022, for the first time, more Americans (16%) said they smoke marijuana than had smoked a tobacco cigarette (11%) in the past week.

Chapter 5:
The Tangible Future

◆ ◆ ◆

"One thing a person cannot do, no matter how rigorous his analysis or heroic his imagination, is to draw up a list of things that would never occur to him."

— Nobel Prize-winning economist Thomas Schelling

◆ ◆ ◆

In his 1971 lecture when he accepted the Sveriges Riksbank Prize in Economic Sciences (the "Nobel Prize in Economics"), Simon Kuznets described 'epochal innovations' as those concepts "that constituted dominant sources of sustained growth over long periods and spread to a substantial part of the world". We are now undergoing technological and social disruptions as dramatic as any we have seen before in world history, ones that we are likely to describe as 'epochal innovations' when we look back on them decades from now. The revolutions underway in energy, transportation, agriculture, materials, communications, artificial intelligence, robotics, and other areas are having profound effects on their own, are influencing and accelerating each other, and are promising enormous implications for health, well-being, the environment, climate change, the protection of rights, and even how we structure our economies and societies.

For millions of years, we relied largely on burning things as a source of energy. We burnt wood, peat, or dried dung for direct heating, and later coal, oil, and fossil gas for heat or to make steam to generate electricity. Literally only in the past few years, for the first time ever, solar photovoltaic (PV) panels and wind turbines (combined with batteries) have reached an economic tipping point where they are now the cheapest sources of new electric power and constitute virtually all new capacity additions. Calculations by RethinkX show that already, in most parts of the world, the lowest-cost solar, wind, and battery power (SWB) system is cheaper than one based on combusting fossil fuels. This system, which must be designed to produce power at the worst times of the year, would therefore produce an abundance of cheap power (called "super power") at all other times of the year.

What might we do with that cheap power? Cheaper batteries have enabled electric cars, which in areas of SWB power will be essentially free to charge during sunny and windy times. Combined with cheap cameras and powerful artificial intelligence software, vehicles that drive themselves would mean road transportation so cheap and convenient that most people would be financially much better off renting a robo-taxi by the trip than owning a car. The effect on the car industry, the oil industry, international shipping and, thus, on global geopolitics would be swift and devastating. But urban areas now devoted to the car would be freed up for new purposes and cheap personal and good transport would mean more money in most people's pockets. The implications of a revolution in motor transport in societies built around motor transport are so profound that they are difficult to comprehend.

Those implications extend beyond economics to include human health, wildlife, and the climate. Cheap power means cheap recycling of metals and cheap filtering and purification of water. When combined with genetically modified microbes, cheap energy also means cheap production of new materials, including pharmaceuticals and food. Thread by thread, the worst problems humanity is facing could quickly unravel. Growing meat without animals would mean much less need for natural gas to make fertilizer as well as less demand for irrigation water. Ending combustion of fuels would have profoundly beneficial health and environmental implications. And ending the use of humans for labor could unleash an age where everyone's basic needs are met.

Energy: Solar, Wind, and Batteries

> *Cheap solar panels, wind turbines, and batteries are revolutionizing our electric system. Over the course of the next couple of decades, not only will they reduce greenhouse gas emissions and the cost of electric power production, but the even the smallest system that will generate sufficient power for any region during the few worst weeks of the year will necessarily create enormous amounts of extra power, what we call 'super power' for all of the rest of the year, giving rise to a new problem – what to do with all of this excess energy*

The story of economic growth and prosperity isn't entirely the story of producing more-abundant energy, but it is close. The OECD, the Organisation for Economic Co-operation and Development provides estimates of the world's total production of primary energy (the energy consumed, including waste heat and inefficiencies that do not result in actual output). The World Bank compiles data on the global gross domestic product (GDP), the sum of the value of all products and services produced on Earth.

Plotting the two data sets against each other reveals a remarkable relationship. In 1971, we consumed the energy of about 5.5 billion tons of oil and produced about $3.3 trillion in economic output, or roughly $4000 (in 2020 dollars) for each ton of oil worth of energy. In 2020, we consumed nearly fourteen billion tons of oil worth of energy, or two-and-a-half times as much as 1971, but produced more than four times as much economic activity, or more than $6000 for each ton of oil worth of energy. All the years between fall near a gradually rising line that passes through the data. (Only two major deviations are apparent: one starting in 1979 that lasted about seven or eight years, and another starting in 1995, also lasting about seven or eight years.)

All years prior to about 1970 fit somewhere in that box in the lower-left, where energy consumption and GDP were low, with most of human history falling close to zero for both compared to today. For good or for bad, the answer to every question as to why every event in human history happened (or failed to happen) has essentially the same answer: energy. The cheaper and more abundant it is, the easier and more plentiful everything else is.

And now we are finally unlocking the ability to collect solar power at scale. The sun is an incredibly powerful and abundant source of energy. As much solar power reaches the Earth in an hour as humans use in a year. Harnessing even a small fraction of its power could potentially meet all our current energy usage, or radically increase the energy available to us.

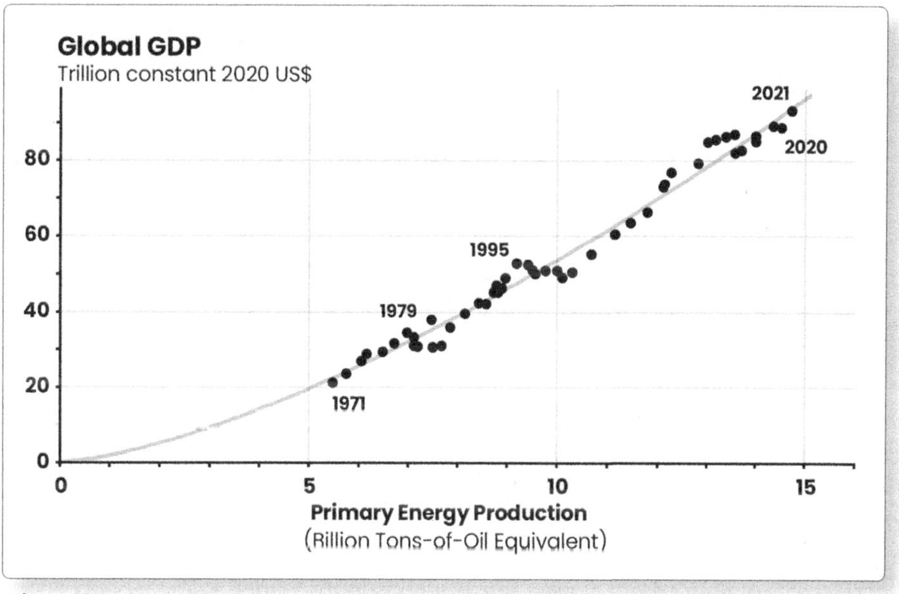

Figure 131 – Global GDP, in trillion constant 2020 US dollars. *Economic data from The World Bank. Energy data from the Organisation for Economic Co-operation and Development (OECD).*

Right now, we are still in the Stone Age of energy. We harvest sunshine that landed on Earth millions of years ago by burning fossil molecules, turning them into pollution after a single use. But we are also in the process of reforming our energy system from one that works as a one-way flow to one based on the number (or 'stockpile') of solar panels, wind turbines, and battery storage systems we have.

The percentage of new electrical capacity installations worldwide went from 20% solar panels or wind turbines to 70% in the thirteen years from 2008 to 2021. Also in 2021, for the first time ever, solar and wind combined accounted for more than 10% of global electric power production. We have passed the 'take-off' point where these technologies – solar panels, wind turbines, and battery-based storage systems – will revolutionize how we harvest, store, and use electric power.

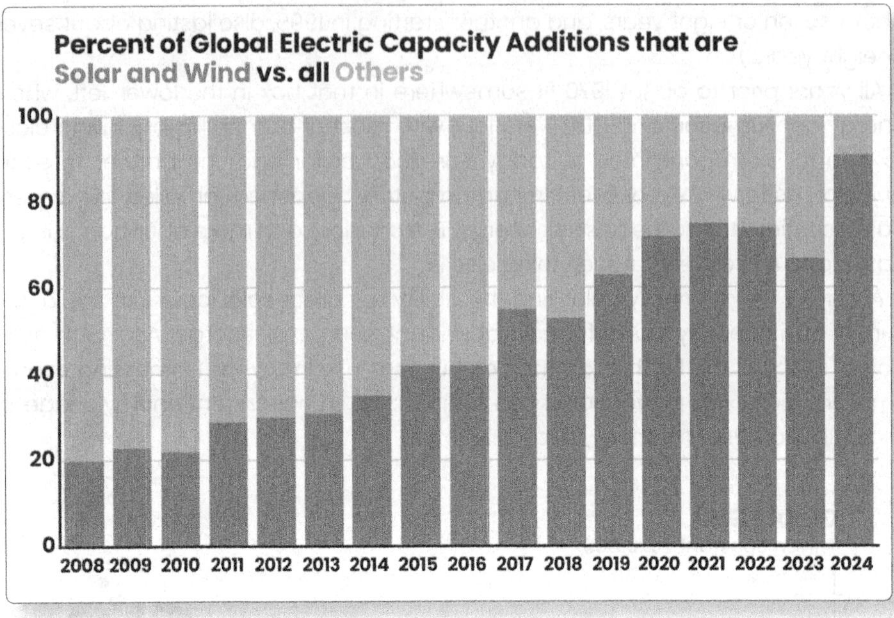

Figure 132 – Percent of Global Electricity Capacity Additions that are Solar and Wind vs. all Other sources. *Data from (Demôro, 2022), BNEF, and Wood Mackenzie*

Solar Photovoltaic (PV) Panels are Already the Cheapest Source of Electricity and Continue to Get Better

In Stavanger, Norway, on Professor Olav Hanssens Street, there is a building that is remarkable for two reasons: the first is that its exterior is completely covered in solar panels in a country whose four seasons consist of a short summer and what amounts to three different kinds of winter. The second is that it is the headquarters of the Norwegian Petroleum Directorate, the national governmental authority charged with overseeing the nation's largest and most-lucrative industry. The building's existence testifies that Norwegians understand both the cost-effectiveness of solar panels these days and the direction the world's energy sources are developing.

Even more remarkable is that there are similar buildings even at even higher latitude in Norway. The Petroleum Directorate's headquarters is north of all the mainland of Great Britain and further north than Juneau, Alaska. But there is also an office building in the Norwegian city of Trondheim, three hundred miles north of Stavanger, that its architects say is 'energy positive' – the building's solar panels create "more power over its lifetime than it consumes", including production of the construction materials. "Powerhouse Brattørkaia is expected to generate approximately 485,000 kWh of energy per year," according to a report on the website Life In Norway, "more than half of that is excess energy, i.e. more than the building itself needs to operate." (Nikel, 2019) What are they doing with the excess power? According to the report, it will be used to help charge "a new fleet of electric and hybrid buses" serving the waterfront area. Even further north, there is a plan to build Svart, an 'energy positive off-grid hotel' above the Arctic Circle. And Norwegian authorities are considering converting the entire 2600-person island archipelago of Svalbard, at 78 degrees north and the last land before the North Pole, to 100% clean energy (including solar, wind, geothermal and other renewable energy sources) by 2030.

Since the late 1800s, the world has produced electric power primarily by burning fuels (coal, natural gas and, to a lesser extent, oil). Globally, these three sources have accounted for more than 60% of electric power production for decades. Nuclear energy and hydropower, until recently, have made up most of the remainder. Twenty years ago, these energy sources combined accounted for about 99% of global electric power production. But in the past couple of decades, that has begun to change quickly, as even Norway is showing us.

The price of electricity from solar panels declined almost 90% between 2009 and 2019. According to the International Energy Agency's *World Energy Outlook 2020*, "for projects with low-cost financing that tap high-quality resources," solar photovoltaic (PV) panels are "now the cheapest source of electricity in history." The cost changes have paralleled the computer chip industry, in part because the technologies are similar. The first silicon photovoltaic (PV) cell was developed by Daryl Chapin, Calvin Fuller, and Gerald Pearson at Bell Labs in 1954, the same year and place as the first silicon transistor, using essentially the same technology.

As with computer technology, as solar power has gotten cheaper, it has become much more popular. It took from its invention until 2021 for the world's first terawatt of solar panels to be installed. The second terawatt was installed by the end of 2024. If present trends continue, we will soon be installing a terawatt every year.

And panels have not just gotten cheaper, they have gotten better. Early panels were just not very efficient. They "started at 2% efficiency," wrote Dr. Nafeez Ahmed in a report for The Club of Rome, but by 2020 "had increased to an average of around 22%." Canadian Solar, one of the world's largest solar photovoltaic panel manufacturers, reported in December 2022 that they will introduce a cell with 25% efficiency and Meyer Burger, a solar module manufacturer based in Switzerland announced that they will produce cells with an efficiency of nearly 30% and some laboratories have reported efficiencies over 30%.

Higher efficiency, of course, directly makes solar panels more cost effective, but also means that they require less land area or roof area to generate the same amount of power. And they now have longer functional lifespans than they have in the past as well. A five-year-long study by the U.S. Department of Energy found that most solar modules tested "have effective lifetimes exceeding 30 years". (Kennedy, 2022)

The improvements in solar panel performance are continuing beyond cost and efficiency, opening new usage possibilities. Saule Technologies in Warsaw, Poland, has developed a printing process for making silicon-free solar cells encased in flexible plastic. A story by MIT's Technology Review reports that "a panel containing Saule's cells is about a tenth as heavy as a silicon panel of the same size." (Crownhart, 2021) Other companies are producing panels as transparent as glass or ones in the form of fabrics. At MIT in Cambridge, Massachusetts, engineers have produced durable and flexible solar panels that "are one-hundredth the weight" and which generate nearly twenty times as much power-per-weight as conventional solar panels. (Zewe, 2022) According to a university press release, these panels 'can turn any surface into a power source'. In the coming decades, not only might every building look like those in Norway, but every other human-made surface might too.

Wind Power

Head south from Norway across Skagerrak, the water that separates Norway from Denmark, and you can reach the former site of the world's first offshore wind farm, Vindeby, which started operations in 1991. It was only ten years prior to that, in late 1980, when the world's first land-based wind farm was installed in southern New Hampshire with a total capacity of six hundred kilowatts, produced by twenty wind turbines rated at thirty kilowatts each. A decade later, each turbine at Vindeby had about 75% of the power production capacity of the entire New Hampshire wind farm, for a total capacity of about five megawatts. Today, individual turbines are commonly rated at five megawatts each, and some are double that size or more. (Engineers in China have built a prototype turbine that is 18 MW.) That is, each one has ten to twenty times the power output potential as that entire New Hampshire wind farm put together.

As wind turbines have gotten larger, the price of the power they produce has become cheaper. According to *Rethinking Energy 2020-2030*, a report by Tony Seba and Adam Dorr of RethinkX, "for onshore wind power, the decline in capital costs has not been as spectacular as that of solar PV, but in the United States these costs have nevertheless fallen at an average rate of 6.2% each year over the last decade" That is, while the cost of solar power fell 82% from 2010 to 2020, for wind power the decline was still a remarkable 46%.

Wind turbines have become more efficient as their support towers have gotten taller. Taller towers allow the turbine blades to reach higher into the atmosphere, where the wind is both stronger and more consistent. This can increase the amount of energy that the turbine is able to generate, making it more efficient.

Though wind power is not as cheap as solar, wind turbines make an excellent complement to solar panels. Wind speeds tend to be highest at times when solar power is not available, such as in the mornings, evenings, and on stormy days. This is because the movement of air is driven by temperature differences in the atmosphere. During the day, the sun heats the earth's surface, causing the air near the ground to warm and rise. As the warm air rises, it creates low pressure at the surface, which draws in cooler air from surrounding areas.

If only the sun and wind perfectly matched the level of local demand for electricity at every moment, our energy problems would be solved. Fortunately, battery costs have decreased more than the costs of solar panels or wind turbines. From 2010 to 2020, "solar PV capacity costs have fallen over 80%, onshore wind capacity costs have fallen more than 45%, and lithium-ion battery capacity costs have fallen almost 90%," according to RethinkX, and they estimate that "by 2030 their costs will have decreased a further 70%, 40%, and 80% respectively." So, by building solar panels, wind turbines, and battery systems, perhaps our energy problems will be solved after all.

◆ ◆ ◆

"It has been found that within a 100-mile radius a wind is always blowing. Windmills installed around the world converting their direct current into alternating current and feeding the electric energy into the world network can harvest the planet Earth's prime daily energy income source – the wind – and adequately supply all the world's energy needs."

— Buckminster Fuller, *Critical Path*

◆ ◆ ◆

Batteries

For the first time ever, we are within arm's length of being able to build enough solar panels, wind turbines, and energy storage systems that can meet our energy needs and do so at a lower cost than using fuel-based systems.

Battery packs are not much to look at. They do not really do anything. There aren't even a lot of lights that blink. But the numbers they put up can be very impressive indeed.

When the state of South Australia offered a tender in early 2017 for a battery to accompany a wind farm, they did not expect the ninety proposals they received. But one stood out among the others. Tesla said they would build the largest lithium-ion battery system in the world and do so in under one hundred days after the contract was signed. The Hornsdale Power Reserve was online just sixty-three days after the contract was signed. With 129 megawatt-hours (MWh) of energy storage capacity, if demand spiked or if the wind farm's supply suddenly dropped the battery could provide short bursts of power while other generators were started, and it could also store energy when prices were low and then sell power when prices were high.

Batteries do not create power – they only store it. So, a battery system that consistently generates profits is taking money from customers while literally creating nothing. There was a genuine concern that the Hornsdale Power Reserve might be making money simply by gouging customers.

However, an assessment by an independent third-party, Aurecon, found that the opposite was the case – because of the presence of the battery, "costs from South Australian generators fell from $470/MWh to less than $40/MWh (a 91% reduction), where they remain today... This has resulted in considerable savings in South Australian energy costs." At a cost of a little over $170 million Australian (about $135 million US at the time), it paid for itself in under two years.

Numbers like those were irresistible. The Hornsdale Power Reserve was enlarged to

194 MWh, but it still soon lost its title as the world's largest lithium-ion battery system. Taking the title was the Gateway Energy Storage, outside San Diego, California, which charged with solar power and with a capacity of 250 MWh, and which came online in the summer of 2020. The following year, the Victorian Big Battery, near Geelong, Australia, was inaugurated with 450 MWh of storage capacity. As of 2025, there are at least eleven grid-scale battery systems in the world with at least 1000 MWh of storage capacity, and nine more under construction, six of those in Australia alone.

According to the Rocky Mountain Institute, global annual battery sales have increased exponentially, from essentially zero in 2011 to four hundred Gigawatt-hours in 2021, with an annual growth rate of almost 70%, representing a doubling every three years. (Butler-Sloss, 2022) Since a Gigawatt-hour is one thousand times larger than a Megawatt-hour, the 450 MWh Victorian Big Battery, though the world's largest at the time, represented only a small fraction of all batteries sold that year.

It is clear that a massive transformation of the world's electric system, in particular, and energy system in general is already well underway. It is being driven by the rapidly declining costs of key technologies including solar panels, wind power, and batteries (SWB), resulting in an exponential increase in their rate of adoption. But the replacement of conventional energy sources with these new ones is just the beginning. As with any product or service, a declining cost of power means much more will be consumed. But that is OK because any collection of solar panels, wind turbines, and batteries that can meet the electric power demand during the darkest, calmest few weeks of the winter will produce an abundance of extra power in the remaining time of the year. And that will require a rethink of not only our energy production, but also of everything that uses energy... which is everything.

The Clean Energy U-Curve: 100% Clean Energy is Just the Beginning

Imagine you have a stream beside your house that supplies *exactly* as much water as your household uses in a year, but this stream flows more at certain times of the year and at certain times of the day than at other times. If you wanted this water to supply *all* your household needs, you would need a lot of storage capacity, as you could not afford to lose a single drop. All that meltwater in the springtime or rainfall during the heaviest storms would need to be stored for later use, as the stream supplies only exactly as much water as you need in a year.

But if the flowrate through the stream could somehow be increased greatly, then you would need much less storage capacity. The stream might supply much more water than you use in a year, so you would only need to store water to get you through the driest times.

Likewise, if you wanted solar power and wind power to supply all your electrical needs, you would need a lot of electrical storage capacity (rechargeable batteries) if those solar panels and wind turbines could only supply exactly the amount of electric power you consume in a year. But if you simply built more panels and turbines, you would need far fewer batteries. We already have streams of energy far in excess of what we use in a year – streams of sunshine and of wind – we simply need to harvest them.

This is not intuitive to understand, but it is true. There is a fundamental trade-off between how much electrical generation ability (solar panels and wind turbines) you need, and how much electrical storage ability (rechargeable batteries) you need to supply all your power needs. And in a world where solar panels, wind turbines, and rechargeable batteries have all gotten dramatically cheaper in the past years, that system is increasingly financially beneficial to build.

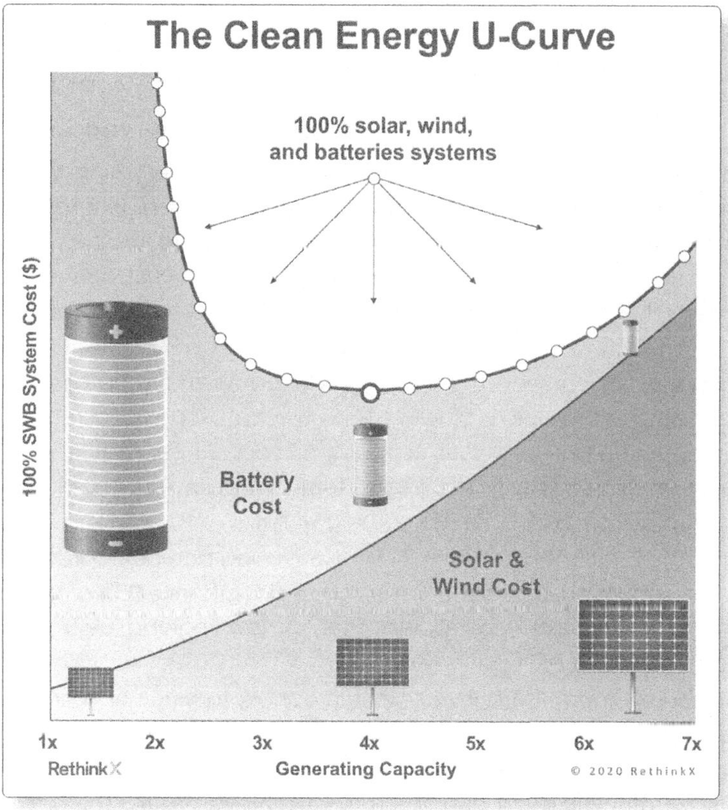

Figure 133 – The Clean Energy U-Curve. *From RethinkX*

RethinkX has done analyses of the electrical systems in places as widespread as California, Germany, Alaska, Texas, India, New England, and others, that asked a very simple question, "Based on the actual measured historical data from the production of electric power solar and wind installations in each of these places, how many solar panels, wind turbines, and batteries would it take for these places to each provide their current hourly demand for electric power?" The results were surprising.

RethinkX's work has shown that the cheapest system is one that produces *four or five times* as much power as an area's average level of demand. That cost is less than even buying fuel for existing coal- or natural gas-fired power plants. That is, if someone gave you a coal-fired power plant for free, you would still be better off using the fuel money to make monthly payments on solar panels, wind turbines, and batteries instead. And you would have money left over.

And it is not just RethinkX that has discovered this. Dr. Nafeez Ahmed wrote in his report for The Club of Rome that "other researchers are now corroborating these findings. For instance, Marc Perez at Columbia University found that building overcapacity of solar and wind by a factor of three times peak load not only reduces the need for battery storage, but lowers the cost of electricity by as much as 75%, while eliminating intermittency challenges... Global energy firm Wartsila similarly found that overbuilding solar and wind by four times peak load requires no seasonal storage, and needs only 4–10 days of multi-day storage capacity, making it the lowest-cost system."

Unlocking 'Super Power': A Hot Tub for Everybody!

An electric system based on solar and wind power turns everything we know about energy on its head. The reasoning is simple: we do not choose when the sun shines or the wind blows, so any combination of solar panels, wind turbines, and batteries that can meet the electric power demand during the darkest, calmest few weeks of the winter will produce an abundance of extra power in the remaining time of the year.

We call this 'super power' – the power produced when the batteries are full but the sun is shining or the wind is blowing. This power must be used immediately, but since its production burns no fuel, its marginal cost is free. That is, when you get your power from the sun and wind, it is wasteful to conserve energy. It is wasteful to let power be produced and then do nothing with it.

Dr. Dorr's study *Rethinking Energy 2020-2030*, using actual records of solar and wind power production, found that for Texas and California, 93% of all days would have at least one hour of super power. Even in New England, 89% of days would have at least one hour of super power.

But wait, there's more! Extra investment in panels, turbines and batteries would lead to a disproportionately large increase in this extra power. A 20% additional incremental investment could mean 200 to 300% more super power. Tony Seba asks, "what investment can you make that if you make a 20% incremental investment, it's going to return 200%+ more super power? No form of extraction resource energy can compete with this. None."

What would we do with all that clean, cheap power? In short, anything we want that can be started or stopped in a short time could be worth doing. Charging electric vehicles and running electrical machines would be free in the middle of the day. We could build machines that sort and recycle garbage dumps from decades gone by to reclaim the waste glass, metal, and plastic. We could desalinate saltwater or treat wastewater to make it fresh again. We could take carbon dioxide from the atmosphere to make lab-grown diamonds and then use them as landscaping gravel.

In the 1890 book *Mizora: A Prophesy*, Mary Bradley Lane writes of a feminist utopia where clean energy is abundant. "The extreme cheapness of artificial heat made a conservatory a necessary luxury of every dwelling. The same pipes that supplied the dwelling rooms with warmth, supplied the hot-house also". In these indoor greenhouses, the women of Mizora grew all their vegetables and fruit.

What might we consider a 'necessary luxury' in the future that we do not have today? What 'necessary luxuries' do we have today that we did not have just a few decades ago? High-speed internet access? Flat-screen TVs? Refrigerators with an ice dispenser built into the door? A telephone per person? In regions powered by the sun and wind, any place that has sunshine would have abundant clean water and any place that has clean water could heat water so that every household could have a hot tub. What today is considered a luxury could in the future become a 'necessary luxury'. Perhaps in the future we will consider it standard to have a hot tub in every household, heated during the mid-day hours with super power.

More importantly, super power might help unlock new trillion dollar business ideas. To give just one example, just as excess power from an office building in Norway is used to charge electric buses, robotaxis could be operated under a new business model, Transport-as-a-Service, and charged with power that is essentially free.

◆ ◆ ◆

"The future is already here – it's just not very evenly distributed."

— William Gibson

◆ ◆ ◆

Transportation: Cheap, Electric, Autonomous, and Operated as Fleets

We are in the midst of the greatest revolution in transportation for people and goods since the first Model T's rolled off Ford's production line, maybe since the first people rode horses. This current revolution is being propelled by one thing more than anything else: electric cars enabled by powerful lithium-ion batteries. The decade of the 2010s saw lithium-ion batteries fall in price, according to Dr. Adam Dorr, by 87%, and from 2020 to 2030, he contends they might fall an additional 80%, making them 45x cheaper in 2030 than they were in 2010.

Originally designed for consumer electronics, the use of batteries for automobiles surpassed their use in consumer electronics in about the year 2018, according to Bloomberg New Energy Finance. Combined with powerful and efficient electric motors, electric cars are already replacing combustion-driven cars quickly, especially in wealthier areas. But autonomous driving ability, unlocked by cheap sensors like cameras and lasers and by powerful artificial intelligence software, could really blow the doors off the entire business model on which car companies have operated for the past century, ushering in a new era of 'Transportation-as-a-Service' (TaaS), where consumers rent robot taxis by the mile, rather than own expensive devices they must operate and maintain themselves.

Electric Propulsion is Already Disrupting the Automotive Industry

The automotive business is undergoing a once-in-a-century revolution. In late 2010, Nissan began selling the all-electric Leaf in the US. Tesla had already brought their battery-powered Roadster to market in 2008 and then introduced the full-sized Model S in 2012. These were far from the first electric cars ever sold in the US – the first battery-powered cars were made in the 1800s and, by the year 1900, one-third of cars sold in the US were electric.

But what makes the current generation of electric vehicles different from previous iterations are the batteries. The lithium-ion batteries developed in the 1990s and used in consumer electronics like laptops, digital cameras, and mobile phones have gotten dramatically cheaper and better-performing, unlocking the possibility of moving cars and even heavier vehicles with them.

The inherent advantages of electric vehicles have made them a tantalizing possibility from the start. Even the earliest electric cars were advertised as being quieter than combustion cars, with no noxious fumes to smell. Fundamentally, EVs are simpler in design too, having no transmission, no gearbox, no timing belt, no alternator, no spark plugs, no engine air filters, and no oxygen sensor, many of which are the most common parts to fail in an internal combustion engine car. In fact, electric cars typically have only about twenty key moving parts instead of about two thousand for combustion vehicles.

The contained explosions that power a combustion engine also cause the wear and tear that degrades the engine over time. Electric drivetrains, in comparison, do not suffer from this design issue and therefore are likely to last longer. This might not be a huge issue for many purchasers of personal cars, but for commercial fleet operators, a significantly longer vehicle lifetime translates into the question of whether to buy a single electric vehicle that will last for many years or to purchase several combustion vehicles over the same period.

Electric drivetrains also make electric vehicles typically cheaper to maintain than combustion ones. In addition to being simpler in design and cheaper to maintain, electric vehicles are cheaper to operate per mile than combustion vehicles. Electric motors are fundamentally more efficient than combustion ones. While combustion engines convert less than half the energy contained in their fuel to the power that turns the wheels, for electric cars this value is more like 90%. Even if the power to charge an electric car's batteries came solely from the combustion of coal, the electric vehicle would still use less energy and emit less carbon dioxide per distance driven than a gasoline or diesel vehicle.

This means that electric vehicles cost less to run and can therefore be driven further for any given amount of money. In 2021 British consumer price comparison firm Uswitch calculated the distance that the electric Nissan Leaf and the combustion-engine Volkswagen Golf could move for £50 in more than thirty countries. In all the countries examined, the Leaf could be driven further than the Golf. Even in Australia, Slovenia, the US, and Japan, the Leaf could move nearly three times as far as the Golf for the same amount of money. In Sweden, Estonia, Lithuania, and Iceland, the electric vehicle could move more than eight times the distance of the combustion-engine car. And in Norway, the country with the highest ratio of electric distance to combustion distance, the Leaf could move more than ten times as far. Understandably, the countries with the highest ratio have been the ones with the fastest adoption of electric vehicles, while the places with the lowest ratios have adopted the new technology slower.

Cost of travel is not the only measure of superior performance. Owners of electric vehicles also report liking the quieter interior, odor-free operation, and convenience of charging at home. High-torque motors mean a short time to accelerate to highway speed. And with their batteries under the floor of the vehicle, electric cars tend to be less likely to roll over in a crash.

The stumbling point for most consumers up to now for purchasing an electric vehicle has been the initial purchase price. A vehicle can be cheaper to operate and less costly to maintain, but consumers look first at the price tag. In 2010, lithium-ion batteries still cost more than $1000 per kilowatt-hour. This was good enough if you wanted to make an iPhone, which debuted in 2007, or a sports car like the original Tesla Roadster, which was produced from 2008 until 2012. With a 53 kWh battery pack, the batteries alone for these vehicles cost more than $53,000. No wonder Tesla only sold a few thousand of them.

By the time he was writing his 2014 book *Clean Disruption of Energy and Transportation: How Silicon Valley Will Make Oil, Nuclear, Natural Gas, Coal, Electric Utilities and Conventional Cars Obsolete by 2030*, Tony Seba could see where lithium-ion battery prices were heading and what that meant for the future of energy and transportation. "When batteries fall to the $100/kWh level," Seba wrote, "the batteries for electric vehicles with a range of 200 miles could cost as low as $5000." As of late 2024, BloombergNEF, a news service that is part of the Bloomberg organization, reported that global average lithium-ion battery prices for EVs had reached that level. By 2030, Seba estimated, batteries could be about one-third of that price, meaning that a 200-mile-range vehicle would cost $5000 to manufacture in total.

It does not take too much imagination to realize that Seba's forecast of cheap electric cars is well within reach. In some sense, it has already been realized.

Your $5000 Car Has Arrived

When it first rolled off the production line in July 2020, the Wuling Hongguang Mini EV became an instant hit. A collaboration between Chinese manufacturers Wuling and SAIC as well as American manufacturer General Motors, the two-door Wuling Mini is no sports car. Small and boxy, it weighs less than 1500 pounds and its 27-horsepower motor with 13.8kWh battery gives it a range of 110 miles and stated top speed that just barely cracks sixty miles per hour. But one other number is what gets the most attention: a price tag of about $5600.

Actually, the Wuling Mini has put up some more numbers that have raised eyebrows: it sold more than 125,000 units in its first six months and one million more in the year after that, making it the best-selling electric car in China, the world's largest automobile market. The 2025 edition sports a beefier 40-horsepower motor and a longer 128-mile range, but still the same price tag, around $5500 US.

It has already spawned several clones from other manufacturers in China and is already being sold in Europe (as the Freze Nikrob, manufactured in Lithuania, one of the countries we saw earlier with one of the best cost-of-travel ratios compared to combustion cars). In Japan, Honda introduced a 125-mile range electric van called the 'N-van e' in 2024 for a price of $7400, which is right in line in terms of the performance-to-price ratio with Seba's 2014 forecast.

If the price of electric car batteries, which were one-seventh the price in 2021 that they were in 2011, falls again by half, then Seba's prediction of a $5000 electric vehicle with over two hundred miles of range is possible. And that could upend the automobile industry beyond recognition. In just the five years from 2020 to 2025, China went from the smallest of the five major car exporters (the others being Germany, Japan, South Korea and the US), to the largest by a wide margin, on the basis of good-quality, affordable cars that are posing a real threat to incumbents around the world.

Manufacturers design a vehicle around the motor. They do not put a cheap powertrain into an expensive body, nor an expensive powertrain into a cheap body. As motors and batteries get cheaper, they are putting them into cheaper bodies. Already, electric bicycles and electric two- and three-wheeled scooters are outselling (and starting to eat into the demand for) electric cars. But one thing could impact the future of transportation even more than the cost of batteries – autonomy software to operate those vehicles.

Robotaxis will make Transport Cheap and Individual Car Ownership Obsolete

A convergence of technologies is working to overturn the way that the automobile industry has been operating for the past century and, with it, the dependence on oil as a source of locomotion and even on the private, individual ownership of vehicles that spend most of their time unused.

We have already seen that vehicles powered by lithium-ion batteries and electric motors are quickly taking over the passenger car market as consumers replace their old internal combustion engine (ICE) vehicles with electric ones. But cheap sensors (particularly high-resolution cameras and, to a lesser extent, lidar laser-based range detectors) combined with cheaper, more powerful, and more energy-efficient computer processors and the artificial intelligence software they can run, are promising to reshape the world of transport and energy over the course of just a few years.

Companies like Gatik (based in Palo Alto, California), Waymo (founded by Google), Cruise (founded by General Motors), Tesla, Einride, Baidu, and others have already demonstrated vehicles that can drive themselves. A 2017 freely distributed report, *Rethinking Transportation 2020-2030: The Disruption of Transportation and the Collapse of the Internal-Combustion Vehicle and Oil Industries* by Tony Seba and James Arbib, lays out in meticulous detail how these vehicles could be adopted and what the implications are.

Seba and Arbib's thesis at its core is simple: a car that can drive itself will be many times more useful than one a person must drive. We build these complicated, beautiful machines and then do not use them very much. Allowing each vehicle to be used much more will mean we will need to build and maintain far fewer vehicles.

An electric, autonomous car would have a low cost of fuel and low cost of maintenance, and the cost of the vehicle itself would be spread over many more miles of travel in its lifetime, making the per-mile cost of operating a vehicle very cheap. Since it could drive itself, it could operate as a taxi, available on-demand to anyone with a smartphone.

This would do more than make road travel become very cheap, it would make it financially unwise for most people to even own a car. Why would anyone own a vehicle themselves that can travel 100,000 miles per year? Anyone who did own that vehicle would still be better off renting it out, on a pay-per-mile basis, most of the time. The transition from individually owned, combustion-based vehicles to electric, autonomous cars operated as taxi fleets in a business model Seba and Arbib call 'Transportation-as-a-Service' (TaaS) could happen as quickly as Model T's disrupted horses more than a century ago. And that was fast.

Even in the most extreme case, where someone owned an autonomous car and simply refused to lend it out, these vehicles would still at least spell the end of the multi-car family, since there would be little need for most families to own more than one. Surprisingly, in Seba and Arbib's calculations, about half of the vehicles on the road in the 2030s would still be human-driven personal cars. But the vast majority of passenger-miles would be served by autonomous cars. So, even if you say that you would not, or could not, give up your car for an on-demand taxi service, if only half the cars on the road are taxis, then taxis will be doing about 90% of the traveling.

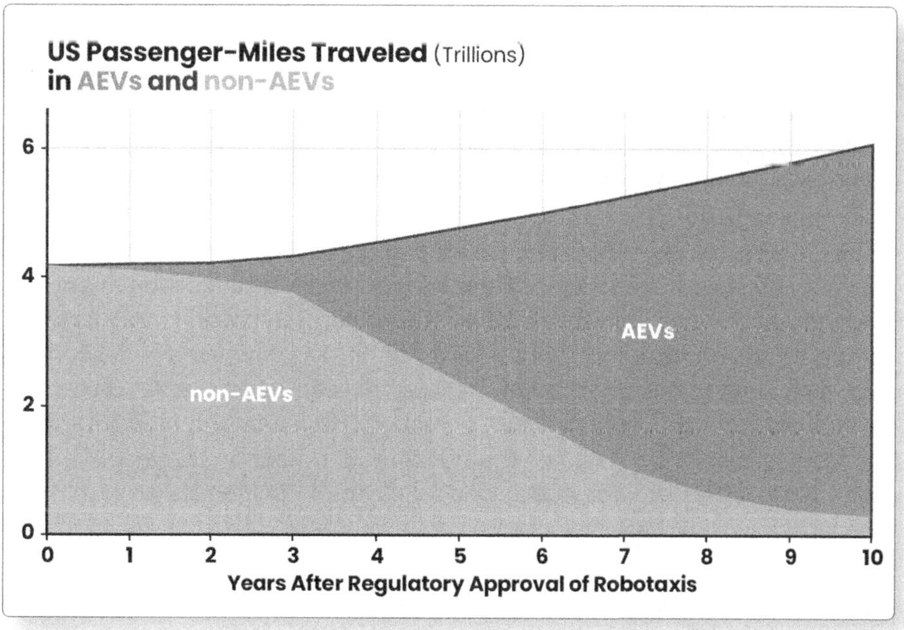

Figure 134 – Passenger-Miles Traveled (Trillions) in Autonomous, Electric Vehicles (AEVs) and non-AEVs. *From (Seba, 2017)*

Imagine that there are two cars, one autonomous and the other human-driven. If the autonomous car travels ten times as much per year as the human-driven car, then the autonomous car would do about 90% of the combined traveling of the two cars. Put another way, when autonomous cars make up only about 10% of the cars on the road, they will be doing the majority of the total traveling.

This would not be a one-for-one replacement. The total market (in terms of passenger-miles traveled per year) of course will increase as transportation becomes cheaper and more convenient, from the roughly four trillion miles Americans currently travel per year to what Seba and Arbib estimate to be more than six trillion passenger-miles per year within a decade after the start of the disruption. Even at times of current peak traffic, only about one in five cars are being used. So we could still handle morning commutes, holidays and other busy times with fewer cars than we have now.

One implication of Seba and Arbib's analysis is that we are likely already at, or beyond, the peak in ownership of internal combustion cars and the peak in the sales of those vehicles. Powerful self-reinforcing feedback loops will hasten the collapse of internal combustion car sales and car ownership, as vehicles get harder to insure, harder to maintain, and eventually harder to even refuel. Just as usage and ownership of typewriters and film cameras quickly unraveled, owners of internal combustion cars could see the value of their vehicles "plunge to zero", according to Seba and Arbib, or even become negative as owners pay to dispose of cars that are non-electric and non-autonomous.

The implications of this rapid shift to electric, TaaS transport for the major industrial economies that are essentially built around car production, car ownership, and car use would be staggering. To begin with, there would be far less oil consumption and also less consumption of fuel additives like corn- or sugar-derived ethanol. Plus, there would be a greatly reduced demand for the materials to make new cars: including steel, glass, plastics, and leather. Jobs would be lost among everyone from professional drivers to car assemblers. All of this would fracture the foundations of some of the largest industries in the largest national economies on Earth.

But it would also mean thousands of extra dollars a year in the pockets of most families. As families are unshackled from the burden of monthly car payments, and paying for vehicle maintenance, parking, and fueling, Seba and Arbib write, "the average American family will save more than $5,600 per year in transportation costs, equivalent to a wage raise of 10%. This will keep an additional $1 trillion per year in Americans' pockets by 2030, potentially generating the largest infusion of consumer spending in history." The adoption of autonomous, electric vehicles would mean better urban air quality and far fewer motor crashes. And Transport-as-a-Service would present incredible new business opportunities that could reshape entire industries and allow us to rethink much of urban land use.

Before we get into those things, let us remember that the disruption of transportation would not be limited to just passenger cars. Every form of transportation would be affected, on land, in the air, and on the water. Let us look at some of those first.

Delivery Vehicles: from Trucks to Pods

Electric vehicles that do not need a driver could change the business of moving goods as dramatically as they could change moving people. Trucking is an $800 billion a year industry in the US alone and according to a 2015 article by National Public Radio, 'truck, delivery, and tractor driver' was the most common job in twenty-nine states, far ahead of primary school teacher (five states), software developer (four states), and farmer (two states). (Bui Q. , 2015)

The cost advantage of electric trucks makes them appealing. Replacing the driver with autonomous driving ability would make electric trucks unbeatable. Vehicles like this are already on the open roads. The Swedish company Einride makes an electric-powered delivery 'pod' that not only does not have a driver, it does not even have a cab. They have already been operating their trucks in their native country and are now also expanding to the US and Germany. Though the technology is not perfect yet, the company employs remote drivers who can use high-speed communications to take control of a vehicle whenever it encounters an unfamiliar situation and help get it back to where the software can take over again. In this way, one driver can operate many vehicles a day.

Einride is far from the only company pursuing this lucrative opportunity. There are startup companies like TuSimple, who are hauling loads in Arizona for Union Pacific, Aurora, who are moving freight in Texas for FedEx, and Gatik, who have been hauling freight for grocery giant Loblaw in Canada and Walmart in the US. There are also better-known names in the business, like Google's Waymo division, Tesla, and Daimler Truck, all of whom see the opportunities that electric and autonomous vehicles offer.

Electric propulsion not only lowers the cost per mile compared to diesel vehicles, it also reduces downtime. Autonomy enables vehicles to be used more hours per day and opens long routes that a human is not legally allowed to do in a day. These are not just matters of degree, shaving a few cents off of the cost of logistics – they have the possibility to completely reshape how retail spaces, restaurants, and other businesses work.

"They're cute," said a student at Fairfield University in Connecticut about the cooler-sized delivery robots that are now rolling around campus. (Hopwood, 2022) Their maker, Starship Technologies, has partnered with GrubHub to deliver food orders from local businesses to students using autonomous, electric vehicles. The company operates not only in the US, but also the UK and Estonia. (Like Einride, they rely on remote drivers for the trickiest situations.) In Ann Arbor, Michigan, another university town, RethinkX's Director of Research Dr. Adam Dorr had the chance to test a similar delivery service by a startup company at the nearby University of Michigan. "It was definitely one of those, 'oh yeah, this is how the future will be' moments," he said of the experience.

The COVID epidemic saw a boom in demand for 'ghost kitchens' (delivery-only restaurants) and 'ghost stores' (delivery-only retailers). The combination of delivery-only order fulfilment centers and autonomous delivery robots could usher in a near-total collapse in travel to run errands and with that, the destruction of value of many retail spaces and restaurant locations. If food and shopping orders can be delivered autonomously, then the locations of the warehouses and kitchen will not matter from the consumers' perspective. (All of McDonald's $40 Billion in real estate value could be at risk, which is perhaps why McDonald's is experimenting with more-automated restaurants that take orders by phone and deliver meals by a conveyor belt.) There could be much less food waste, as consumers could order frequently, in small quantities, rather than purchasing food in large packages where a sizable portion will inevitably go to waste. Cheap and automated deliveries would then effectively end the existence of 'food deserts', populated places with few supermarkets. We shall escape the absurdity of building and operating an entire multi-passenger car just to bring home broccoli.

It is not just food that is being delivered by robots. Zipline, based in South San Francisco, builds electric drone aircraft and uses them to operate delivery services in the US, Rwanda, Ghana, and Japan. They began commercial deliveries of blood, platelets, plasma, and other blood products to medical centers in Rwanda in 2016. As of late 2021, barely five years later, more than 75 percent of blood deliveries outside of Kigali used Zipline drones. Engineer and YouTube video creator Mark Rober reports that Zipline's drones "have reduced in-hospital maternal mortality by 88%" in that country. In 2024, Zipline celebrated their millionth delivery.

Recreational Vehicles (RVs) Will Disrupt Airlines and Hotels

The air travel industry has the potential to not only be revolutionized by small, electric aircraft, but also to be undercut by autonomous and electric ground-based transport. An autonomous recreational vehicle (RV) would effectively be a 'rolling hotel room'. Let us say we are planning to go visit friends or relatives several hundred miles away. If we drive in a car, it might take eight or ten hours of driving, a full day of unpleasant travel on each end of our trip. But imagine instead that we could reserve a self-driving, electric RV and have it arrive directly at our home. We could board at our leisure and leave whenever we felt ready. The vehicle would drive itself through the night, going slowly enough that we might not even notice it was moving. Maybe we would make ourselves dinner, watch TV and relax. When we woke up in the morning, we would be at our destination. Then it could drive off to get cleaned and head to another customer. Perhaps Walmart stores, which have large parking lots and ample supplies of goods available, could act as RV refurbishment centers and we could see Walmart competing with airlines due to the existence of autonomous vehicles.

Without a driver or front passenger seat, the RV would be more spacious inside than current RVs. Even if the vehicle costs several hundred thousand dollars to make and lasts several hundred thousand miles over its lifetime, that would still only amount to about one dollar per mile for the hardware. Depending on the number of people in the group, this could be a very affordable way to travel.

This would not only affect the air travel industry, but the hotel industry as well. It is entirely possible that some people would trade staying in a fixed apartment for a life of travel, especially as rooftop solar panels and high-speed mobile internet service make it easier to work from anywhere.

Electric Propulsion Will Mean a Private Plane for Everyone

It is not that difficult to imagine: You and your family take an autonomous taxi to a small airport not far from your home and wait for the next aircraft to be available. Because the planes only seat a few passengers and are electric, they do not need a long runway and they are quiet, so there are several airports like this in the area.

There would be no stress in getting to the airport, since departure times would not be scheduled. Because the planes are small, there would be little need for airport security, which means no need to arrive at the airport two hours early. With no pilot and a low cost of fuel, the cost-per-mile to operate the plane would also be low. The toughest part would be deciding on a destination.

Wright Electric, based in Los Angeles, is developing a 100-passenger plane that can reach San Francisco, while Archer Aviation, Wisk Aero, and eHang (in Guangzhou, China) are developing small aircraft that take off and land vertically, eliminating even the need for a runway. And Merlin Labs, based in Boston, is building equipment to make existing planes autonomous.

Whether all, or even any, of these companies will succeed remains to be seen. Aviation is a notoriously difficult industry to start a new company and the rate of failure is high. But if even one company succeeds, flying would be radically transformed forever.

Space Travel Can Disrupt Transoceanic Flights and Beyond

While electric aircraft might disrupt combustion-powered planes in the shortest half of routes, longer flights are facing competition from the possibility of rocketing around the globe in literally only an hour or two.

We are in a new golden era of space travel. In 2021, more orbital rocket launch attempts were made worldwide than in any year in history, 144, of which 133 were successful. This surpasses the previous record for most attempts made, set in 1967, and the record for most successful launches, set in 1976. The year 2022 then topped 2021's records by wide margins, 2023 then topped 2022, and 2024 topped 2023.

In new reports about the revolution that is underway in space launches, one company's name comes up more than any other: SpaceX. As of early 2023, more than half of all active satellites in orbit were put there by that one company.

One factor explaining why SpaceX can launch so many satellites is that electronics and solar panels have simply gotten smaller, cheaper, and more powerful, allowing satellites to be smaller and more numerous. While a Juno rocket launched the single Explorer 1 satellite into orbit in 1958, today SpaceX regularly launches fifty to sixty of its Starlink communications satellites at a time and some launches have carried more than one hundred 'microsatellites'. Of the 7,560 active satellites in orbit as of May 2023, according to a database compiled by The Union of Concerned Scientists, more than half of them had been launched by SpaceX and more than 90% of the active satellites SpaceX launched had been launched since the start of 2020.

But SpaceX are not the only ones shaking up the industry. According to Eric Berger, senior space editor at the website Ars Technica and author of the book Liftoff about the rise of SpaceX, "the total number of global launch attempts has

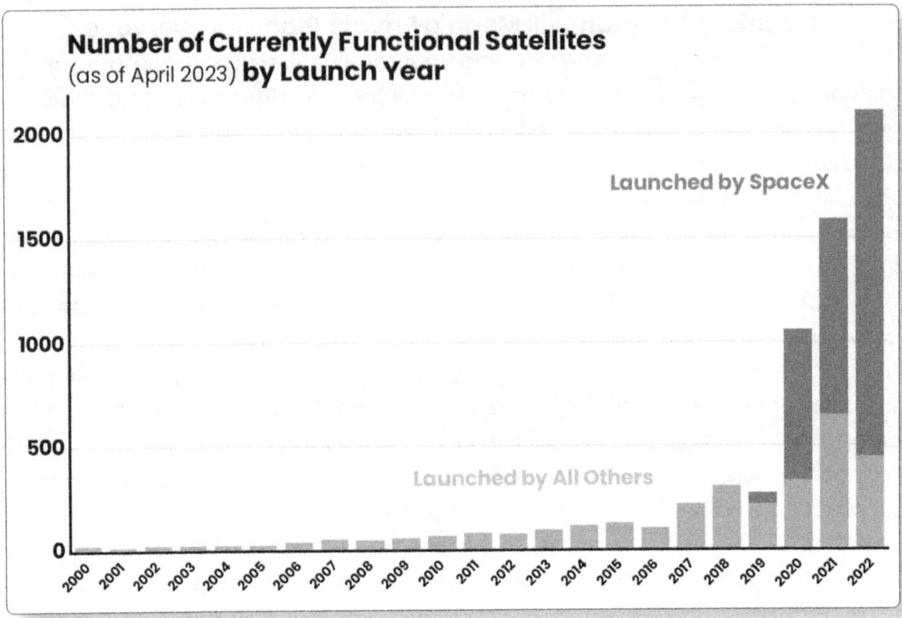

Figure 135 – Number of Currently Functional Satellites (as of April 2023) by Launch Year.
Data from (UCS, 2022)

doubled during the last decade. Before 2010, China averaged fewer than 10 rocket launches a year. ... The Chinese government and a handful of private operators launched 56 rockets in 2021." Up by a factor of five in just a decade.

Another factor explaining the rising number of launches is that they too are getting cheaper. Harry Jones of the NASA Ames Research Center in Palo Alto, California, presented a surprising compilation of launch cost data at a conference in 2018.

The Vanguard rocket, the successor to Juno, cost more than $400,000 per pound of payload mass launched, according to Jones. Barely ten years later the Saturn V rocket that carried the astronauts to the Moon was nearly three hundred times more powerful and cost less than 1% the cost per pound of the Vanguard. But then, launch costs stalled for the next four decades.

The 1970s were a time of rising per-pound costs, as NASA developed the Titan-Centaur (used for launching the Viking and Voyager interplanetary missions) and China built the first Long March rockets. The Space Shuttle, a beautiful distraction, was intended to make near space accessible again by being re-usable. But from its first launch in 1981 to its last launch thirty years later in 2011, its problematic design instead made it the most expensive way to reach orbit.

Until SpaceX's Falcon 9 first launched in 2010, we had largely resigned ourselves to the assumption that space would always be difficult to access. The economics of getting to low Earth orbit had never been below $2,000 per pound, levels that had been reached by the Soviet Proton in the mid-1960s and by the Ukrainian Zenit 2 in the mid-1980s.

While through the 2010s, NASA developed the Space Launch System (SLS), the largest rocket ever built, SpaceX pursued a strategy of making their vehicles rapidly

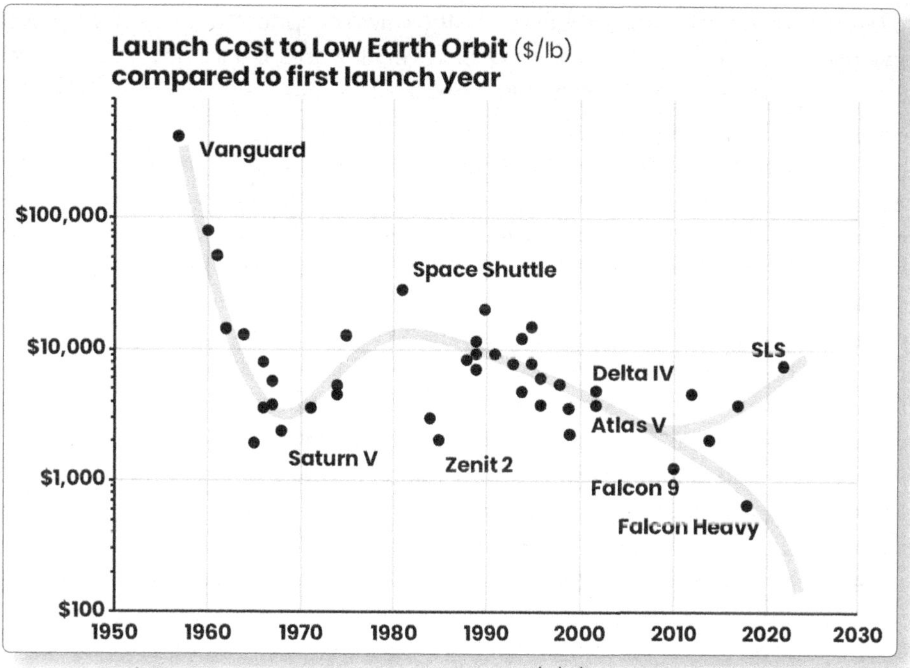

Figure 136 – Launch Cost to Low Earth Orbit ($/lb). *Data from (Jones, 2018)*

re-usable and flights both cheap and reliable. This was done in part by having the first stage of their rockets (that is, the lower part, which carries the rocket through the lowest layer of the atmosphere) land standing upright, just as it had launched several minutes prior. SpaceX's first ground landing of a rocket's first stage succeeded in December 2015, and the first landing at sea on a drone ship in April 2016. In 2024, SpaceX celebrated the twenty-fourth flight of a single first-stage rocket. (In comparison, the space shuttle Challenger flew ten times in its career, including its fatal last flight, which only lasted 73 seconds.)

SpaceX's Falcon Heavy is roughly one thousand times cheaper to fly per pound than the Vanguard, and SpaceX says its cost could go lower with a high cadence, or rate of launches. According to an October 2020 investor presentation by Momentus, a California-based company that helps move satellites between orbits, SpaceX's Starship, designed for orbital and interplanetary flight and which became the largest rocket ever to fly in 2023, might cost $110 per pound, fifteen times cheaper than SpaceX's own Falcon-9. SpaceX has stated that its cost might even reach $10 per kilogram, less than $5 per pound of payload.

The combination of low launch cost and high reliability of landings opens new possibilities for moving cargo and even passengers, not only to orbit but also from point-to-point on Earth in only an hour or two. In a 2015 interview on *The Late Show with Stephen Colbert*, the CEO of SpaceX said: "If you throw away the rockets every time, it's crazy expensive to go to space. But if you can re-fly the rockets, it could be comparable to air flight in its costs."

Dr. Greg Spanjers, a U.S. Air Force scientist and the Rocket Cargo Program Manager, is quoted by Eric Berger in *Ars Technica* as saying, "This idea [using rockets to deliver cargo] has been around since the dawn of spaceflight... It's always been an intriguing idea. We've looked at it about every 10 years, but it's never really made sense. The reason we're doing it now is because it looks like technology may have caught up with a good idea."

Sending goods to or from Earth orbit (which takes about ten minutes to reach) or the Moon (which takes about three days) will effectively become easier than sailing a container ship to the other side of the Earth (which takes about three weeks). A Moon colony would become economically feasible. So might space manufacturing. In June 2022, Rocket Lab, founded in New Zealand but now based in the US, launched a small 'cubesat' to the Moon to test the future possible orbit that could be used by NASA's planned Gateway space station. In February 2024, the IM-1 mission from Intuitive Machines became the first commercial lunar lander to successfully soft-land on the Moon, the first US mission to land on the Moon in fifty years, after being launched on a SpaceX Falcon 9 rocket.

Varda Space Industries, founded by a former SpaceX engineer in late 2020, has announced contracts with both Rocket Lab and SpaceX for manufacturing high-value products, like pharmaceuticals, in weightless conditions. With lower launch costs, many long-standing dreams can become realities. For example, in early 2024, Varda returned to Earth crystals of an HIV drug that were manufactured in orbit, in a craft launched on a SpaceX Falcon 9 rocket, demonstrating the possibilities for research and manufacturing.

Implications of Clean, Cheap Transport

In November 2022, hundreds of climate activists at Amsterdam's Schiphol airport caused a disturbance at a runway for private jets. Being the Netherlands, they did this in the most Dutch way possible – by riding around the tarmac on bicycles. The protest was short-lived, but the message they sent was clear, 'flying in small aircraft is disproportionately damaging to the environment and therefore this kind of travel must stop.'

There was only one problem with protests like these: flying is good. Traveling is good. Transport is good. It is the pollution that is bad. Massive amounts of land devoted to vehicles is bad. Crashes that result in the loss of life or that cause permanent injury are bad. But being able to move ourselves and our things from where they are to where we want them to be, fundamentally, is good.

Fortunately, solutions are not only on the horizon, they are already rolling up beside us. Waymo, a spin-off from Google, is operating fully driverless robo-taxis in San Francisco, Phoenix, Los Angeles, and Austin, and Chinese search engine giant Baidu operating similar vehicles in Beijing. Various companies are testing electric 'vertical take-off and landing' (VTOL) helicopters and autonomous electric ferries.

The implications of the rapidly unfolding revolution in electric, autonomous, fleet-owned cars and trucks, electric aircraft, and accessible rocket travel are staggering, both individually and in combination.

Combustion-based motor vehicles and their petroleum-derived fuels are the backbone of the modern global economy. A 2018 investigation by the World

Economic Forum found that the top two most-traded goods internationally in the world in 2016 (by value) were cars and refined petroleum, adding up to about 8% of all trade. Crude petroleum and vehicle parts were another $1.2 trillion, or 4.5%, of trade. That is, just these four products – crude petroleum, refined petroleum, cars, and vehicle parts – together constituted one-eighth of all global trade by value.

Coal is a relatively low value product, but it is used in large quantities. As a result, 40% of all ships, by weight, are used for moving coal, crude oil, oil products, natural gas, and other fossil fuels. That is, if you took all the world's large ocean-going ships, emptied them out and put them on a scale, the ones used for moving fossil fuels would be nearly half the weight of all the ships. Maybe you have seen a car bumper sticker that says "Save the Whales". Well, now solar panels and electric vehicles offer a way to significantly reduce our impact on the seas, as we reduce our use of coal and oil, and the maritime traffic needed to move those products.

Roughly seventy percent of every barrel of oil produced is refined into gasoline or diesel fuel. Electric road transportation could eliminate most of the demand for the oil industry's main products. It does not take much destruction of demand to throw the oil market into chaos. In late 2014, an estimated 2% supply-demand discrepancy caused the global oil price to collapse from over $100 per barrel to under $50 over the course of just a few months and they did not recover to their former level until Russia's expansion of its invasion of Ukraine in early 2022. Oil prices, which were over $60 a barrel in early 2020, before the COVID epidemic became widespread, briefly turned negative on the expectation of a strong drop in demand for fuel for cars and airplanes.

All indications suggest the global sales of internal combustion engine (ICE) vehicles had already peaked by 2018 and are unlikely to ever recover. The decline in use of crude oil to make liquid fuels will affect all aspects of the oil and gas industry. The components of crude oil are separated and refined into products in a similar way, conceptually, to how beef cattle are butchered. A portion of each animal is turned into ground beef, a portion into steaks, a portion becomes leather, and much of the rest is the low-value 'drop' – the blood and bones and bowels. There is no way to raise a head of cattle that yields only filet mignon steaks. Getting any one product means producing (and selling) the others, too.

Likewise, a portion of each barrel of crude oil can be refined into car, truck, and airplane fuels, a portion into asphalt or the heavy sludge-like fuels used by large ships, a portion into waxes and lubricants, a portion into plastic. There is no way to take away the demand for the primary product of crude oil, liquid fuels, without significantly changing the supply, price, and profitability of all other oil products. Tony Seba calls this a 'composition disruption'. Petroleum is not only used in a bewildering array of products, its exploration, production, transport, and usage are intimately intertwined with the products of the fossil gas industry.

As electric vehicles take over the road, incumbents in car- and oil-related industries are being left scrambling. Electric vehicles, being simpler in design than combustion cars, simply require fewer assembly line workers. The entire value chain of the current transportation industry, from petroleum geologists to oil industry workers, to automobile assemblers, to car repair and maintenance technicians, professional drivers, insurance and car loan originators, and emergency room doctors and rehabilitation therapists could see the demand for their work, or the nature of their work, change dramatically in the coming years. Even the locations where jobs are located could change as the centers of car design and production move from the US Midwest, Japan, South Korea, and Germany to places like China, California, India, Brazil, and Vietnam.

Electric vehicles will have an enormous impact on the urban environment and human health. Poor urban air quality contributes to millions of deaths per year, not only from transport, but also from burning fuels to make electricity and from agricultural operations, both of which are also poised to change dramatically in the coming years. Robotaxis and autonomous delivery vehicles, which could stop faster and more safely than human-operated vehicles, could return the roads to bicyclists, pedestrians, children, and pets. Car crashes worldwide currently claim more than one million lives per year, and many more people are seriously injured. Plus, there are an estimated one to two million collisions with wildlife per year in the US alone. Most of this could quickly end as autonomous vehicles and vehicles with advanced safety systems become more common.

Transport-as-a-Service will require us to rethink much of urban land use and rural land use as well. Roughly one-third of the land area of modern cities is taken up by roads and parking spaces. America has an average of one thousand square feet of parking for each car – eight parking spaces per car – compared to only eight hundred square feet of housing per person. (Evans D. , 2021) We literally devote more space to our cars than to ourselves.

A radical drop in the number of cars in our cities would free up parking and street space, which could be used for anything we want – more bicycle lanes, more parks and athletic spaces, or more new housing, even in small lots. The world's skinniest skyscraper, Steinway Tower, opened in New York City in 2022. At fifty-seven feet wide, it is only about the width of six or seven parking spaces, but it is also more than a quarter-mile tall. And it is not just parking and residential space that will be affected by autonomous vehicles, all existing retail space would be affected too. Retailers might eliminate showrooms and switch to delivery-only. How will our downtowns look if they are no longer primarily for shopping?

In rural areas, the change could be just as dramatic. The number one crop in the US, corn, is primarily made into ethanol as a fuel additive, and into animal feed. Only 10% is eaten by people. A sharp reduction in the demand for crops like these would liberate an area worldwide equal to the size of China, the US, and Australia combined. How will our countryside look if it is no longer primarily for agriculture?

And the impacts on our society could be profound as well. Think of Jayland Walker. Patrick Lyoya. Duante Wright. Ronald Greene. Philando Castile. Mumia Abu-Jamal. Keenan Anderson. George Floyd. Tyre Nichols.

All were African-American men involved in deadly police altercations that began with vehicle stops. Social rights advocates have sarcastically labelled the supposed crime some of these men committed to be 'DWB' – Driving While Black.

But in a world of fully autonomous vehicles, there would simply be no need for these interactions to have happened and therefore, hopefully, no more cases like these.

Autonomous food delivery vehicles would mean there would be no more 'food deserts', urban regions where it is difficult to find a well-stocked grocery store. No more elderly shut-ins, lonely and unable to travel or socialize if they cannot drive. Much of what we think of now as social issues will be revealed to be nothing but technological problems as our ability to supply cheap and clean transport, energy, and food increases.

◆ ◆ ◆

"It's almost impossible to predict the future. But it's also unnecessary, because most people are living in the past. All you have to do is see the present before everyone else does."

— Jason Crawford, founder of The Roots of Progress foundation

◆ ◆ ◆

Food: From Extraction to Creation

If the world of transportation is undergoing its greatest transformation in the past one hundred years, then plant and animal agriculture is undergoing its greatest transformation in the past ten thousand. Archaeological evidence suggests that people have been harvesting wild grains for baking and brewing since before we domesticated grain, perhaps somewhere in central Türkiye, ten millennia ago. Since then, we have relied on only about twelve plants and about five animals, all of which were domesticated thousands of years ago, to produce most of our calories.

Fundamentally, this has been a system based on extraction. We grow corn or wheat simply for their kernels and discard the rest of the plant. We raise cattle, chickens, and pigs for their muscles, skins, fat, milk, and eggs. As a result, we have developed breeds of plants that are as short as possible and animals that produce as much meat (or milk or eggs) as possible, as quickly as possible. We raise plants and animals largely for one product, in some cases, like tobacco or sugar cane, for a single kind of molecule. The rest is used as by-products or simply discarded.

The 2019 report *Rethinking Food and Agriculture 2020-2030* said: "Currently, farmers essentially grow an entire cow before breaking it down to into specific products, such as steak, leather or collagen, and the process is nearing its limits in terms of resource efficiency, with little potential to improve costs of production."

But in recent years, a new system of production has started emerging based on creation rather than extraction, on building up, rather than tearing apart, and the implications for agriculture, the environment, and human health could be astonishing.

Part of this system is based on new uses of organisms found in nature. In Bath, England, a company called Clean Food Group is commercializing a strain of yeast they developed from one "commonly found on the surface of grapes used in organic winemaking." Just as ranchers for thousands of years have bred their best animals to produce even better offspring, the scientists from Clean Food Group have bred this strain of yeast to produce an oil that can substitute for palm oil. In Finland, Solar Foods grows a microbe they found in the forest that is high in protein and can be used in making pancakes, breads, pasta, smoothies, and other foods. And Nature's Fynd, based in Chicago, discovered a fungus in the geothermal springs of Yellowstone National Park that they grow to make dairy-free cream cheese and meatless breakfast patties.

Another part of the new system uses techniques for growing cells in laboratories to make a new field called 'cellular agriculture'. Scientists are now raising plant and animal cells to make meat or other products without having to kill any further plants or animals. In December 2020, the California-based company Eat Just became the first in the world to serve cellular agriculture meat – nuggets of real chicken meat grown outside a chicken – to regular customers at a restaurant in Singapore. Other companies have presented beef and pork, seafoods like salmon, tuna, lobster, and crab, as well as coffee, foie gras (duck liver), chocolate cocoa, chicken fat, leather, sweeteners, flavorings, coloring agents, and a variety of other products. Whatever nature grows inside of animals and plants, people are now trying to grow outside.

But what could really blow the doors open to a new system of creation-based foods and materials would be the ability to 'brew' specific molecules as easily as we now brew alcohol using only yeast and sugar. Combining precision biology with the age-old technology of fermentation to produce specific molecules, or 'precision fermentation' (PF) as Seba and Tubb called it, has already been used to make cow's milk proteins like casein, whey, and lactoferrin, the proteins found in chicken egg whites, the 'heme' protein found in blood that makes red meat taste like red meat, and extremely sweet proteins found in rare fruits. Some of these products are already on the market. Colorifix, based in Norwich, England, partnered with clothing manufacturer Pangaia to produce a track suit dyed with a pink dye produced from the DNA of a microbe found in a hot spring in Yellowstone National Park. California-based Geltor's PF collagen is available in some South Korean skin care products. And the Japanese company Spiber has produced silk used in a limited-edition parka.

The simple fact is, though fermentation is now being used in new ways, it is an old technology and many products are already on the market that use this approach. If you are diabetic, your insulin comes from a genetically modified microbe, not the pancreas of a cow or pig, and has since the 1980s. Even if you are not diabetic, precision fermentation is used to make enzymes that keep bread soft, ones to clarify fruit juices, and ones used in deli meats and imitation crabmeat. Since the 1990s, 80% of the milk-clotting enzyme used to make cheese has come from precision fermentation. Unilever has used 'ice structuring protein', found naturally in fish, carrots, and grass, and produced by the company using genetically modified baker's yeast, in their ice creams since 2010. According to the company's website,

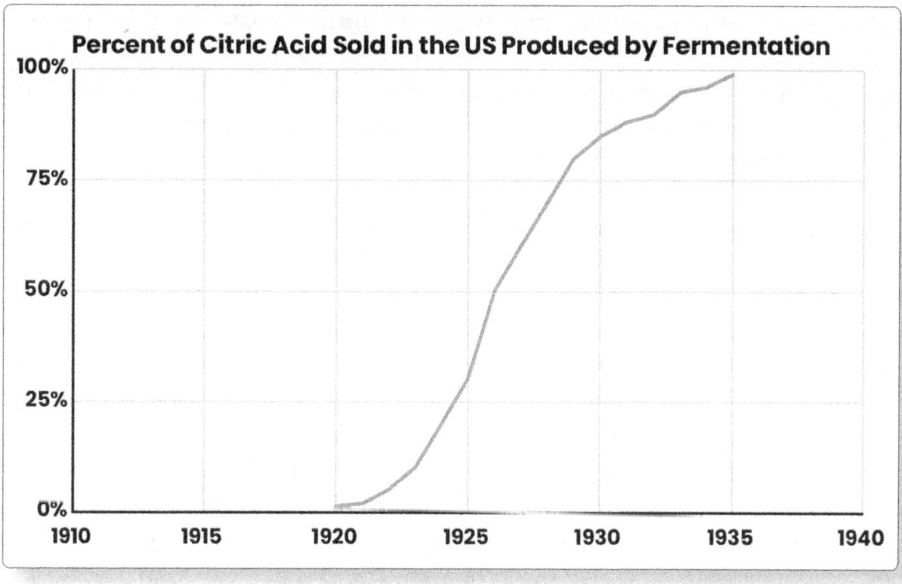

Figure 137 – Percent of Citric Acid Sold in the US Produced by Fermentation

the protein, used in 'tiny amounts', allows them to make frozen treats "which are lower in fat, sugar and calories and at the same time include more fruit".

Precision fermentation promises not only the ability to produce proteins and other chemicals at low cost and in essentially unlimited quantities but also products not found in nature, such as those from extinct species, as well as products that nature has never produced. Ginkgo Bioworks, a Boston-based company, re-created fragrances from extinct flower species by extracting DNA from dried dead plants and inserting the DNA into yeast. Since then, other companies have produced mastodon collagen and mammoth heme. We have even produced products that no living thing has ever made before, like the 'analog' insulin that is similar, but not identical to, human insulin, and that is now the majority of insulin on the market.

With these technologies, we are on the cusp of not just another agricultural revolution, but a true 'Second Domestication' of plants and animals. Rather than marginally improving existing breeds by producing hornless cattle or chickens that mature a few days sooner, or that are slightly more muscular, or that yield slightly more milk per day, we are now achieving step-change improvements in the performance, cost, and environmental impacts of the products we desire.

In an interview in the September 1894 edition of *McClure's Magazine*, chemistry professor Marcellin Berthelot said, "when we are able to make starch direct, what will hinder us from making a potato? And what is to prevent us, once we have gained the mastery, from making better milk, better meat, and better potatoes, at any season of the year, than those which nature gives us...?"

Professor Berthelot continued, "carbon from carbonic acid [carbon dioxide], oxygen and hydrogen from water, and nitrogen from the air will be a source of food for all the world. What the animals and vegetables have produced through the energy of nature, we shall produce as well, if not better, by our study of nature's laws." After more than a century-and-a-quarter, we are now seeing these fantasies become a reality. "Do you mean to predict that all our milk, eggs, meat, and flour will in the future be made in factories?", the interviewer asked. "Why not, if it proves cheaper and better to make the same materials than to grow them?"

Berthelot's words were staggering, but he was not the first to express ideas like them. In 1880 to 1881, Mary Bradley Lane published a serial novel in the *Cincinnati Commercial* newspaper, later re-published in 1890 in the form of a book called *Mizora: A Prophesy*, about a feminist utopia where "they took certain chemicals and converted them into milk, and cream, and cheese, and butter, and every variety of meat, in a vessel that admitted neither air nor light... The cream manufactured in Mizora was a perfect imitation of the finest dairy product. It was the same with meats; they combined the elements, and the article produced possessed no detrimental flavor. It was a more economical way of obtaining meat than by fattening animals." Synthetic food was not the only wonder found in this paradise. According to Lane, the women of Mizora used a "small apparatus" that provided a picture "exactly similar to the image of a mirror" to communicate at great distances, their "carriages were propelled by compressed air or electricity" and they had "air ships and railroads" whose heat was provided by electricity. Lane wrote that "boats are propelled in Mizora either by electricity or compressed air, and glide through the water with soundless swiftness." It has taken almost a century and a half, but Lane's fantastical ideas about food, transportation, communications, and energy are only just now being realized.

The Protein Disruption, Like Infancy Itself, Starts with Milk

What to Professor Berthelot was informed speculation is now becoming reality. This is no longer science fiction. Dairy products made without livestock are already on the market. Not soy milk. Not oat milk. Real cow's milk, made without cows.

California-based company Perfect Day has taken the genetic material from cattle that produces dairy whey protein and inserted it into a fungus. (Actually, they did not even take the genes directly. They simply synthesized a new fragment of whey-encoding DNA without even involving an animal in that step.) As a result, the company makes whey protein that is identical to what is made by cows, but without the cows. And even though the fungus is genetically modified, the finished product is pure protein, so does not contain genetically modified organisms (GMOs). The company is already selling cow-free ice cream, cow-free liquid milk, and even cow-free dried whey powder under its own labels. Starting in 2024, Breyer's began selling ice cream containing Perfect Day's proteins marked 'Lactose-free Plus'. Boston-based Brown Foods plans to sell the first cow's milk made from mammal cells, UnReal Milk, in the US starting in 2025.

Monica Massey, an executive vice president and chief of staff at the Dairy Farmers of America, the largest U.S. dairy cooperative by membership, was among the first to taste Perfect Day's ice cream when it debuted in 2019. She said of the experience, "We sat down in a dairy cooperative headquarters and ate it and I said 'Oh, we're screwed' because it tasted just like ice cream".

Milk proteins are found in more than just dairy products. They are added to a surprising number of other foods that you might not expect. Check the labels on the food packages in your house. Milk proteins are often found in potato chips and frozen chicken, they are sometimes used as a binder in deli meats and sausages, and you might even find them in canned tuna fish and chewing gum.

Other companies are taking a similar approach to what Perfect Day is doing. Israeli company Remilk, South Africa's De Novo Foodlabs, the UK's Better Dairy and in Australia's All G Foods, are all using a similar method to 'brew' milk proteins. New York City-based Helaina uses microbes to produce human breast milk while Israeli company Wilk makes both human breast milk and dairy-identical cow milk. Wilk has already produced "the world's first yogurt developed with cell-cultured milk fat", according to a press release. San Francisco Bay Area company New Culture Foods makes the milk casein protein that is key to making cheeses that stretch and melt – properties that plant-based cheese products have typically had trouble replicating. Others, like Earth First Food Ventures, are building facilities to make lactoferrin, a nutritious milk protein that can be used as a food ingredient or supplement. "Over the next seven years we aim to make it [All G's milk] cheaper to buy than cow-based dairy," Jan Pacas, founder of All G Foods, said in a 2022 interview.

Whole cow's milk contains a little over 3% protein and up to 5% milkfat. Make those cheaper than what cows can do and, in the words of Tony Seba, "you disrupt the whole bottle." The shift away from animal agriculture as a source of milk proteins could be swift. According to *Rethinking Food and Agriculture 2020-2030*, in the US about 35% of milk is used as an ingredient in other products. Consumers buy it as cheese on pizza or as an ingredient in a snack bar, without specifically choosing that particular dairy product. Thus, restaurants and food manufacturers could switch to the newer, cheaper ingredient source without anyone needing to convince consumers that the new product is worth trying.

Dairy products made by precision fermentation can not only be cheaper than what cattle produce, they can also be better. Manufacturers could make dairy products with any milkfat content, protein content, or lactose sugar content. And lactose-free PF milk would be cheaper and easier to make than lactose-containing milk, which would benefit the majority of people on Earth who cannot digest lactose.

Manufacturers could choose not only the ratios of ingredients, but the specific chemical forms of those ingredients too. For example, while Asian and African cattle produce a form of casein protein called 'A2', some cattle who are the ancestors of modern European cattle began producing a different form of casein protein. As a result, Europeans (and thus, Americans, Canadians, Australians, and New Zealanders) typically drink this mutated milk. With precision fermentation, manufacturers of dairy products would be able to use whichever ingredients they thought best, not simply the ones that modern cattle supply.

It's not just cow's milk that can be produced this way, goat milk could be too, as well as donkey milk, yak milk, buffalo milk, even camel milk. Camels only produce 1/10th of the milk in their lifetime as cows. Plus, the young need to be kept with their mother for her to keep giving milk, so you must feed and care for two animals to get milk from one. The result is that the market for camel milk is much smaller than cow's milk, and the prices are higher. But it need not be that way in the future.

In years gone by, barns were often painted red because paint containing iron oxide (rust) was cheap and easy to produce. Today, any paint store can mix ingredients to produce any one of millions of colors. Likewise, milk and dairy products could be produced having whatever profile food producers and end consumers want, freed from the constraints of what is currently available.

With a system based on microbes rather than large animals, the production of any desired milk could be rapidly scaled up and down in accordance with consumer demand, far faster than herd sizes can be increased and decreased. Gone would be the days when dairy farmers need to dump excess milk because their herds are producing too much to sell.

And it could all be done indoors, by well-paid people working in conditions similar to beer breweries. A roughly 120 liters (about thirty-two gallons) fermentation space can replace one cow. So, replacing the nearly ten million cows in the US would take about three hundred million gallons of milk fermenters, or roughly one gallon of fermenter space per person. Total US beer production amounts to about six billion gallons (about twenty gallons of beer per person), which means that there are roughly six hundred million gallons of beer fermenters in the US, or two gallons of beer fermenters per person. Though the total replacement of the entire US dairy industry sounds daunting, it would only be the equivalent of building *half* of the current beer brewing industry.

How much land would this take? Israeli company Remilk announced (and then paused) plans to build a factory in Denmark that would have been 750,000 square feet and that they claim would have made the dairy output equal to 50,000 cows. That is about fifty-five square inches per person served. An adult's hand from the base of the palm (where it meets the wrist) to the tip of the longest finger is about seven inches and the width (including the thumb) is about four inches. Put both of your hands together, side by side, palms up. If your hands are about the same size as mine, then that is about the area of land needed to hold one person's worth of a precision fermentation dairy facility.

If the cost of space could be built the same as Remilk's facility, then the facilities to replace all ten million dairy cows in the US would cost about $24 Billion. This seems like a lot of money, but actually is about the same as the amount spent to build the Space Launch System (SLS), a rocket that so far has only launched once.

Not only will precision fermentation milk require less land, but assessments from companies like Perfect Day have shown that it will also take less water, less energy, and less labor. Today, dairy farmers act as food service personnel for cows, custodial engineers for cows, and nursing caregivers for cows. Taking cattle out of the milk-production process, but delivering the same high-quality and nutritious product, would mean that the effort (and therefore, money) that currently goes to caring for cattle could instead be used to provide higher salaries and better working conditions for the people involved.

Protein Molecules and Fat Molecules Will be Grown, Not Raised

It is straightforward to understand that being able to make dairy proteins using genetically modified microbes unlocks an entire universe of protein production, with the chicken eggs being an excellent example of another industry devoted to producing proteins that are amenable to being produced by precision fermentation.

The weight of an average egg is about 12% protein, with much of the rest being water. So, it only takes a few grams of egg protein made by precision fermentation to displace from the market a chicken egg that weighs about ten times as much. And, as with milk proteins, this replacement is likely to start as a business-to-business decision to switch ingredients in other products. "Just like the milk market," Seba and Tubb wrote in *Rethinking Food and Agriculture 2020-2030*, "the modern food industry just needs to disrupt the egg protein ingredient market to push the primary egg production industry into a financial tailspin."

Some PF egg products are already on the market. In 2021 the EVERY Company, based in San Francisco, signed a deal with Anheuser-Busch InBev, the world's largest brewing company, to begin brewing egg proteins at commercial scale. The company now offers two products to food manufacturers: a soluble egg protein that can be used in liquid foods (like smoothies, kombucha, and carbonated beverages) and replacer for egg whites that can be used, for example, in baked goods, protein bars, noodles, or as a binder in meat products.

Unlike chicken egg producers, PF protein manufacturers can make single proteins. Egg white is a rich mixture of proteins, the principal one of which is ovalbumin. But there are others. One has anti-microbial properties while another is an excellent foaming agent. Of course, it is not economic to produce large numbers of eggs just to extract a single kind of protein. But with precision fermentation, it would be possible to produce the constituent parts in essentially unlimited quantities.

With PF, it would also be possible to produce egg proteins from other species, or to make new varieties of proteins like those found in eggs, but which are not found in nature at all. Seagull eggs are offered in some high-end restaurants, but because gulls only lay once per year, harvesting is strictly regulated, and the eggs are therefore expensive and only available at certain times of the year. In various parts of the world, the eggs from ostrich, pheasant, quail, turkey, and many other birds are considered delicacies. Some of these are reported to be delicious while others are an acquired taste. The 'white' part of the eggs of African penguins, instead of ovalbumin, contains penalbumin protein, which has anti-freeze properties and is translucent when cooked. Who knows what dishes chefs might prepare when any egg protein is cheap and readily available?

It is not just food where abundant egg proteins might find uses. In 2022, researchers from Princeton University used egg whites to create a lightweight and porous material that can be used in "water filtration, energy storage, and sound and thermal insulation." When freeze-dried and heated, the proteins created a network of carbon fibers and sheets of graphene. The authors showed that the resulting material can remove salt and microplastics from seawater with high efficiency. (Ozden, 2022)

Just as electricity can be produced in abundance as its sources get cheaper, proteins, fats, and other molecules could also be produced in abundance as well. In addition to 'super power', we will have 'super food' – new products, available in abundance, at low cost and with little environmental impact.

But one kind of chemical could change the culinary world more than almost any other: hormones. With cheap, plentiful hormones we could, in principle, grow meat in abundance without using animals at all.

◆ ◆ ◆

"We shall escape the absurdity of growing a whole chicken in order to eat the breast or wing, by growing these parts separately under a suitable medium."

— Winston Churchill, 1931

◆ ◆ ◆

We Shall Escape the Absurdity of Growing a Whole Chicken

In August 2013, Dutch doctor and professor Mark Post presented beef hamburger patties at a press conference in London, prepared by a professional chef and then eaten by two food critics. What made this burger remarkable was that it was grown in a laboratory from beef cells taken from a cow without killing the animal. One of the tasters, Chicago-based food writer Josh Schonwald, said afterward, "It had a very sort of bland, neutral flavor. What made it most similar to real beef was the texture. When I bit into it, I was impressed with the bite and how it did have a kind of density that was familiar."

The project had been funded by Google's billionaire co-founder Sergey Brin, which is fortunate because Post estimates that the meat for that first demonstration cost more than $325,000 to produce. But while the cost to make that meat was enormous, so is the potential business opportunity in disrupting the industrial animal-based meat industry.

Meat has been a staple of the human diet for at least two million years. As we saw in Chapter 1, a dense source of calories and nutrients might have been instrumental to the self-reinforcing feedback loop of brain development and tool/weapon development. In short, the better we got at making tools and weapons, the better we got at harvesting meat, which contributed to further brain development.

For many thousands of years, we have raised animals for meat. In the last couple of hundred years we have used scientific principles of breeding to make breeds with specific characteristics. Only in the last few decades have we relied on industrialized animal production and processing, heavily dependent on fossil fuels, antibiotics, land, water, and cheap labor.

But now we seem to be well beyond the limits of what the Earth can sustain, and the demand for meat only keeps growing as the human population and economic prosperity do. According to *Our World in Data*, more than one-third of all land on

Earth is used for plant and animal agriculture. Our livestock are about 60% of the mass of all the mammals on Earth while people are another 36%. Only about 4% of the mass of all the mammals on Earth are wild mammals.

However, new technologies promise to deliver more meat and better meat, at a lower financial cost of production, and with far less impact on the environment. Perhaps one day, eating rare Japanese delicacy wagyu beef will be as common as formerly-exotic foods, like oranges or bananas, are to us.

After the burger demonstration, Dr. Post told an interviewer that "the aim of course is to make a consumer product out of it [his cultivated burger]. It may take ten years, maybe even earlier, but ten years is probably a good guess." In fact, it was an excellent guess. In December 2020, San Francisco-based company Eat Just became the first company in the world to sell and serve cultivated meat, chicken nuggets, at a restaurant in Singapore. But many other companies are trying to do similar things. Israeli Prime Minister Benjamin Netanyahu became the first sitting head of state to eat cultivated meat, a slice of minute steak made by Israeli company Aleph Meats.

Dr. Post founded Mosa Meat, focusing not only on ground beef but also on beef fat, a key part of giving meat more than a 'bland, neutral' flavor. In Australia, Vow Foods, has started operating a facility that they claim can produce over 50,000 pounds of meat per year, while Upside Foods in San Francisco has built a facility of similar size. In early 2024, Vow introduced 'Forged Parfait', a meat based on quail but which they describe as 'not an imitation of chicken or beef, but a new flavor and texture combination altogether.' Upside Foods executive Amy Chen described her first experience eating the company's cultured chicken: "I had that first bite, and realized that it was simultaneously one of the most unremarkable things and one of the most remarkable things I've ever eaten. It's just meat." (Watson, 2021) Maybe people will be more impressed when they try offerings from London-based Primeval Foods, which is focusing on cultivating exotic meats like lion, tiger, and zebra.

Not all companies are working on lean meat. Some, like Swiss company Cultivated Biosciences, UK-based Hoxton Farms, Israel's Steakholder Foods and Yali Bio, based in San Francisco, are focusing solely on animal fats. A Paris-based startup called Gourmey is using cell cultivation, much like how Eat Just produces chicken meat, to instead make foie gras, a French delicacy made from duck or goose liver from animals that have been force-fed. Others are growing leather and fur, all without killing animals.

These approaches will help take the cruelty out of meat production, and much else besides. Cultivated meat production requires less water and less energy, and less time. Eat Just's process to grow chicken meat takes about 14 days, rather than the forty or more it takes to raise a whole chicken. Cattle and pigs take much longer, giving cultivated meat and even greater advantage.

And the process requires no antibiotics. Today, about 70% of the antibiotics used in the world are given to animals, not people, primarily in the production of meat, which is leading to the development of strains of bacteria that are resistant to antibiotics. But growing meat from cells, or vegetarian meat substitutes that use no animal products at all, would essentially eliminate the need for antibiotics in food production.

It's not magic that we can now make meat without killing an animal, it's just science. Techniques developed in laboratories to raise cells for research are being repurposed and rethought as these techniques get cheap enough to conceivably be used to produce meat for consumption. "I can say with confidence," cultivated meat startup SuperMeat CEO Ido Savir said, "we can reach cost parity with chicken." (Watson, 2022) He added: "in the long run, the basal [growth] media will be the main cost driver, which relates to our partnership with Ajinomoto [the company that first brought MSG to commercial production], which is the largest global amino acid producer."

The idea of growing meat is not new. In the September 1894 issue of *McClure's Magazine*, chemistry professor Marcellin Berthelot is quoted as imagining a world where people could do this, rather than raise animals. "Herds of cattle, flocks of sheep, and droves of swine will cease to be bred, because beef and mutton and pork will be manufactured direct from their elements," he said. "One can foresee the disappearance of the beasts from our fields, because horses no longer be used for traction or cattle for food."

Nearly four decades later, in 1931, Winston Churchill, before he first became Prime Minister, wrote an essay called 'Fifty Years Hence' where he said: "Microbes, which at present convert the nitrogen of the air into the proteins by which animals live, will be fostered and made to work under controlled conditions, just as yeast is now. New strains of microbes will be developed and made to do a great deal of our chemistry for us. With a greater knowledge of what are called hormones, i.e. the chemical messengers in our blood, it will be possible to control growth. We shall escape the absurdity of growing a whole chicken in order to eat the breast or wing, by growing these parts separately under a suitable medium."

Churchill got the timing wrong, but the mechanism he describes – greater knowledge of hormones – is much closer to what Dr. Post used to make his burger than Professor Berthelot's suggestion of manufacturing meat 'from its elements'. We can give cells the chemicals that tell them when to divide and grow, essentially tricking them into thinking they are inside a baby animal and should grow. Unfortunately, these "growth factors" are expensive, which is why Dr. Post's burger samples cost over $300,000 to make.

So, while some companies are trying to figure out how to make delicious animal products from cells, others are racing to figure out how to make growth factors cheaply. In Israel, start-up company BioBetter is using genetically modified tobacco plants to grow these chemicals. Others are using genetically modified microbes to try to make them by precision fermentation. But if we can make any protein or fat molecule we want from precision fermentation, we might not need to grow animal cells at all.

Enter the Impossible Burger

During a 2009 sabbatical year, Stanford University professor Dr. Patrick Brown began unravelling the chemistry behind what makes up the sensory experience of eating meat. One key ingredient that gives meat its flavor and aroma, he found, was the presence of heme, a molecule in blood that transports oxygen, but that is also found in other animals, like cattle and pigs and in some plants, like soy. When just a couple of percent heme, by weight, is mixed with plant proteins and plant-derived fats, the product superficially resembles beef in appearance. The heme is naturally red in color, but turns brown during cooking, giving the plant-based patty the appearance, aroma and even flavor that resembles animal meat.

Dr. Brown's company, Impossible Foods, claims that their burger product uses 95% less land and 74% less water, and its production emits about 87% less greenhouse gas than cow-derived beef. As well, the plant-based product has more protein than a beef burger, with less total fat and no cholesterol. It is now both kosher- and halal-certified.

To obtain heme in large quantities, Dr. Brown tried to isolate the chemical from soy plants. But this 'extractive' approach could not yield enough of the desired molecule. Genetically modifying yeast to contain the heme gene meant that he could obtain the molecule he wanted by precision fermentation, 'brewing' a batch of heme in much the same way that beer brewers make beer.

Impossible Foods began selling their first product, Impossible Burger, in limited quantities at high-end restaurants in 2016 and the popularity has grown exponentially as more people have tried it and seen how closely it resembles animal beef in flavor and texture. It is now available for sale through their website and in Burger King restaurants as the Impossible Whopper.

"Our mission is to completely replace the use of animals as a food technology by 2035," said Brown. "We're dead serious about it and we believe it's doable." He might be right. Precision fermentation and cultivated food products are getting better over time, but cows and chickens are not. It is not just burger that the company is targeting, but pork and chicken nuggets, too. Brown even claims that Impossible Foods has "unlocked the flavor of fish". The entire spectrum of meat products is up for grabs, and perhaps some new niches will be opened in the process.

As Bruce Friedrich of the Good Food Institute said: "We went very, very quickly from landlines to cell phones. We went from film to digital cameras pretty quickly. The transition from horse and buggy to automobiles happened very, very quickly... So I think both the optimistic but also realistic trajectory is in 30 years we're just not making meat from animals anymore." (Klein, 2019)

We Will Turn Loaves into Fishes

Judging solely by appearances, San Francisco-based Wildtype's cultivated salmon meat looks remarkably like salmon meat from a fish. It even has the 'woodgrain' pattern seen in natural salmon meat, with alternating layers of pink tissue separated by whiter layers. The company claims that the two materials are "fundamentally two different structures, as they are in the conventional salmon we eat today." But food is more than appearance. In a tasting demonstration in June 2021 for students from MIT and Stanford, one of the tasters of Wildtype's salmon remarked, "it tastes like fish!" At the tasting demonstration, CEO Justin Kolbeck said, "I think in ten or fifteen years things get really interesting because then there really is no need for large-scale fish farming."

Around the world, entrepreneurs are salivating at the possibility of producing seafood products through cellular agriculture and precision fermentation, just like for land-based animals. In Berlin, Germany, Bluu Seafood is growing cells from Atlantic salmon, rainbow trout, and carp. BlueNalu, based in San Diego, says its first product will be bluefin tuna while Israel's Forsea Foods is focusing on eel meat, both of which are commonly used in sushi. Singapore's Shiok Meats and Umami Bioworks (which merged in 2024) and the US's Cultured Decadence (now a part of Upside Foods) are focusing on shellfish.

With farm-raised and wild-caught seafood, we are restricted to obtaining fish products in the proportions that living animals produce. But cellular agriculture and precision fermentation allow us to make ingredients and components in any proportions desired. We can make products with more lean meat and less fat, or vice versa. In an interview in late 2022, BlueNalu's chief technology officer Lauran Madden said, "We can actually create cells, if we want, that have exactly the same profile [as wild-caught fish], or we can create cells that, for example, might have 90 percent omega-3 [fatty acids]." (Torrella, 2022) Many people claim that they do not like certain kinds of fish because of the distinctive smell and flavor. But precision-fermented or cell-cultivated products could potentially be made to not produce those aromas and flavors in the first place, greatly increasing the market for fish products.

Soon You Might Smell Like a Dinosaur, Eat One, and Look Like One Too

It was an art exhibit unlike any other when *'Resurrecting the Sublime'* opened at the Centre Pompidou in Paris in February 2019. Starting a few years before, a team from Ginkgo Bioworks, a Boston-based biotechnology company, began a project to re-create the aromas from flowering plants that are now extinct.

Searching through samples from the more than five million specimens in the Harvard Herbarium, the scientists took tiny bits of leaf from *Hibiscadelphus wilderianus*, a flowering plant that is known only from a single specimen that was found on Maui, Hawaii, in the early 1900s, *Orbexilum stipulatum*, whose habitat near Louisville, Kentucky, was intentionally flooded in the 1920s, and *Leucadendron grandiflorum*, whose habitat near Table Mountain, Cape Town, South Africa, was lost to colonial vineyards.

The Ginkgo scientists worked with UC Santa Cruz to identify sequences of DNA involved in fragrance production and then used those sequences to make strains of yeast containing the extinct DNA. Ginkgo's efforts culminated in the art exhibit, in which visitors could smell the fragrances that are chemically identical to those produced by the original flowers while standing in re-creations of the lost landscapes.

In 2023, fragrance brand Future Society began selling perfumes using these three scents and three others: *Macrostylis villosa*, a small aromatic shrub once found in South Africa, *Wendlandia angustifolia*, a small flowering tree lost forever to drought in 1917 in India, and *Shorea cuspidata*, a towering tree last recorded in 1998 in Borneo. All of these scents are available individually or as a sampler pack at the company's website.

The implications of technology like this go well beyond unique perfumes or an art exhibit. In *Wired* magazine, Maria Perez Ortiz describes how researchers at the University of Lund, in Sweden, found a method to mass-produce pheromones to control insects. Until now, the article reported, "creating artificial pheromones has been pretty expensive, so it's only made economic sense to use them to protect high-value crops, such as fruits." But now, yeast could be used to make sex pheromones at scale, radically changing pest control in the process.

What other aromas might be created this way? What extinct aromas, or ones that no flower has ever produced? What flavors? What meats? What milks? What fats? What leathers? What silks? What woods? What shells? If we can produce, in any quantity we want, fragrant molecules from nature (or psychotropic molecules, or cancer-fighting molecules, etc), and if we can produce fragrant molecules from plants that are now extinct, can we not make rare molecules plentiful? Can we not create new molecules that have never before existed?

Proteins are sequences of amino acids and there are twenty amino acids in the genetic code, which means that there are 20 x 20, or four hundred, possible combinations of two amino acids. For strings of three amino acids, there are twenty times that number, or eight thousand possible sequences. A typical protein is about four hundred amino acids long, so the number of possible amino acid sequences of that length is twenty multiplied by itself four hundred times. That is a number far in excess of the number of grains of sand in all the beaches of the world. Greater than the number of electrons in the Universe. In their report *Rethinking Food and Agriculture 2020-2030*, authors Tony Seba and Dr. Catherine Tubb write simply, "the number of possible proteins is effectively infinite."

If we know the sequence of amino acids we desire, we know the string of DNA that will create that sequence. In 2015, Alex Lorestani and Nick Ouzounov, co-founders of Geltor, produced mastodon collagen, a protein found in skin, hair, and bone, using genetically modified microbes. The sequence of mastodon collagen had only first been reported in scientific literature thirteen years prior and the ability to make collagen protein using microbes was brand new. Using a candy mould, Lorestani and Ouzounov did what any reasonable people would do in that situation – they made gummy candies from a protein that humans had not eaten in thousands of years.

They did not stop there. Geltor now offers a product called PrimaColl, which the company claims is "the world's first vegan collagen for food and beverage markets... a nature-identical replica of poultry collagen derived using precision fermentation technology", Elastapure, which is human elastin, and Collume, a marine collagen. (Wolf, 2021) They also produce HumaColl21, a type of human collagen which Geltor claims not only directly improves skin appearance, but which also prompts the body itself to produce more collagen, resulting in a greater effect than regular collagen.

If we can make molecules from extinct flowers and extinct mammals, why not from more ancient creatures? We have discovered dinosaur collagen in nature, so why not make dinosaur candy? How about aromatic and flavorful molecules from species that are under pressure, so that they do not become extinct at all?

Eat and Drink to Be Merry

Maybe we are loving our treats to death. Both coffee and chocolate are delicious, perhaps a little too delicious, as the expanding areas used for growing coffee beans and cacao beans is one of the primary drivers of tropical rainforest deforestation. Also, both kinds of bean production are known for using child labor. According to *Our World in Data,* coffee and cocoa produce higher greenhouse gas emissions per weight of product than palm oil, pig meat, poultry meat, eggs, milk, cane sugar and a variety of other common foods. Of the products *Our World in Data* considered, only beef, lamb, and cheese rated higher. (Ritchie, 2020)

Changes in global temperature and precipitation patterns from greenhouse gas emissions are affecting coffee-growing areas. Considering that one-third of the world's coffee beans come from Brazil and nearly half of all cacao is grown in the Ivory Coast (Côte d'Ivoire), this geographic concentration alone poses a significant risk to the sustainability of future production of these products.

In addition, not just for coffee and chocolate, but for an enormous portion of our food supply we are becoming dangerously reliant on just a few varieties of species. There are more than one hundred species in the *Coffea* genus, but over 90% of the coffee sold in the world is from only two kinds: Robusta (which is used in instant coffees) and Arabica (which is used for everything else). Most bananas, corn (maize), milk, beef, chicken meat, and eggs come not just from a single species, but from single breeds within each species, raised at massive scale. Eventually, something has to give.

Enter companies like California Cultured, based in Davis, California, who are using cellular agriculture to raise cacao and coffee plant cells. According to their website, they select "rare varieties of coffee and cacao" and grow them in batches over the course of just three or four days. Compared to raising animal meat or fat cells, CEO and co-founder Alan Perlstein claims his company's process is simpler and cheaper. "We only use plant hormones that are found in agricultural waste," he said in an interview. As a result, "the basic input costs for media are two orders of magnitude cheaper on bench scale than meat cell culture." It is not just the ingredients that are cheaper, Perlstein claims that the equipment is cheaper too. "The complexity of the bioreactors needed to grow plant cells is far less than what is needed to grow meat cells," he explained, so the capital expenditure "costs are anywhere from a fifth to a tenth of the cost" of meat bioreactors.

Because his company produces real chocolate which "will be price competitive with the chocolate and coffee found in the grocery store", CEO Perlstein claims that his company's products are "not chocolate or coffee substitutes. We are the future of coffee and chocolate." If he is right, the countries that are currently the largest customers of the world's coffee and chocolate producers could quickly become their largest competitors.

This would not be a one-for-one substitution, but instead could greatly expand the market for chocolate and coffee products. Studies have found that nutrients in cocoa have "many health benefits" including showing a "reduction in the total number of major cardiovascular events including strokes, heart attacks and deaths". (McDermott, 2020) (Sesso, 2022) Thus, California Cultured's first planned product, according to the company, will be a cocoa powder used as a nutritional supplement.

With precision fermentation, other companies are producing individual compounds with nutritional, health or, um, let's call it 'recreational' benefits. Cannabinoids are a family of more than one hundred different chemical compounds found in cannabis plants, the most famous of which is probably THC (tetrahydrocannabinol, the primary psychoactive chemical in marijuana). But numerous other chemicals have known or suspected physiological effects, but without the psychoactive properties. CBD (cannabidiol), for example, is used by some people to deal with anxiety or pain issues and has been approved by US and European Union authorities for treating some seizures. Cannabigerol (CBG) is used to treat pain and nausea and cannabichromene (CBC) is an anti-inflammatory agent.

Genetically modified microbes could be a cheap way to produce these chemicals, without having to grow entire plants to get them. "Making artificial cannabinoids is a lot like making fake meat," according to an article from Bloomberg News. "The process revolves around genetically engineered yeast, the same technology used to produce heme, the ingredient that gives Impossible Foods' faux burgers their meaty qualities." It is not only "cheaper to engineer cannabinoids than to extract them from plants", the article claims, it also yields a purer, more-consistent product. According to the article, the target production cost for CBG is under $1 per gram, "far below CBG's current wholesale price of $20 a gram." (Kary, 2020) These products are already on the market. Cronos Group, based in Toronto, Canada, sells gummy candies infused with THC, CBG, and CBC that they manufacture by precision fermentation.

Precision fermentation came from the convergence of precision biology and brewing. Now it is changing the business of brewing beer itself. Danish company EvodiaBio uses precision fermentation to produce a proprietary aroma it calls Yops, which it says can improve the taste of non-alcoholic beer. Genetically modified yeasts are already on the market from companies like Berkeley Yeast that produce the chemicals otherwise found in hops, the plant that gives beers their distinctive aromas and bitterness, that prevent unwanted off-flavors, or that allow brewers to combine multi-step processes, like fermenting and souring, into single steps.

The Future of Sweeteners and Flavorings

The revolutions in food and transport underway right now are also affecting the world of sweeteners and flavorings. Soon we could have a lot more natural sweeteners available as the electric vehicle revolution means that we will feed less sugar to our cars. One of the primary uses of cane sugar in Brazil is to make ethanol (alcohol) as a fuel additive. In the US, we ferment corn (maize) instead, but the end use is the same. In fact, feeding cars instead of humans is the number one use of corn, not making cornmeal or high-fructose corn syrup for human consumption. Electric cars, of course, do not burn ethanol, so as they become more popular, the reduction in demand for cane sugar and corn sugar that we now make into fuel additives could be available for people instead.

Additionally, precision fermentation is unlocking a greater supply of sweeteners like honey and sweet proteins. California-based MeliBio debuted bee-free honey at a restaurant in San Francisco in 2022. Fooditive, a company with offices in the Netherlands and Sweden, has a similar goal, making honey without bees.

But sweeteners do not need to be high-calorie. For example, brazzein is a small protein hundreds of times sweeter than cane sugar. But it is found in a rare fruit, the oubli, which grows in west Africa. Starting up an entire new industry to grow this fruit and extract a single sweet protein from it would be daunting. This is why Sweegen, a company based in California, has teamed up with Conagen, based in Bedford, Massachusetts, to use precision fermentation to produce just that single protein. Since it is so much sweeter than sugar, it can also be used sparingly, and therefore adds essentially no calories to food. In 2024, Abu Dhabi-based company Novel Foods Group announced an agreement to build a $500 million biotech production hub in the United Arab Emirates, the first phase of which will involve building a facility to produce brazzein using precision fermentation.

The startup company Oobli, based in Davis, California, is already offering chocolate bars made with sweet protein, the first-ever food product brought to market containing sweet proteins. They claim their bars contain 70% less sugar than most chocolate bars and that their sweet proteins "deliver a sugar-like sweetness that is diabetic-, gut-, and climate-friendly."

Israeli company Amai Proteins think they might be able to do even better. Their scientists have developed a sweet protein called 'sweelin' that is up to three thousand times sweeter than sugar, using artificial intelligence to make a yeast strain that expresses the molecule with precision fermentation. Similar to monellin, a protein found in serendipity berries, sweelin is slightly different in structure, making it more stable to heat and pH level. CEO Dr Ilan Samish says "it's inspired by monellin, but it is not identical to any protein that is found in the wild... Sweelin is the sweetest and most stable sweet protein in the world."

This is not the only application of precision fermentation that these companies have in mind. Amai is working with customers making meat, milk, and plant-based foods, according to their CEO, "but in each case we take a protein which is found in the wild, and we design a novel protein that will work in the mass food market."

Conagen has also taken the genetic material from hot chili peppers that produces capsaicin, the chief chemical responsible for pepper's spiciness and burning sensation, and transferred that DNA into microbes. This has uses far beyond hot sauces. Infused into electrical insulation, capsaicin could help prevent rodents and other small mammals from chewing through electrical wiring.

Capsaicin is chemically similar to vanillin, the main flavoring chemical in vanilla which is useful in making chocolate and for masking unpleasant notes in pea protein and in soy for plant-based meats. Back in 1894, professor Marcellin Berthelot saw what the future held for distinctive aromas and flavorings when he said, "the tea plants, coffee shrubs, and cocoa trees must some day follow the lead of madder and indigo." Now, after more than a century, those things are starting to come true.

◆ ◆ ◆

"Political leaders are at the tail end of affairs. To ask change of political leaders is like asking the cow's tail to redesign the cow."

— **Buckminster Fuller,** from Lloyd Sieden's

A Fuller View, Buckminster Fuller's Vision of Hope and Abundance for All

◆ ◆ ◆

Materials: From Linear to Circular

Cheap energy means more than cheap transport and cheap food – it will redefine our relationship with the materials we currently use and open new possibilities for ones we do not yet use. Single-use materials, like thermal coal or heating oil, will become worthless to us. We will move from extraction – breaking down plants or animals to get their parts – to creation, literally designing materials on computers and then building them up to get the properties we want. Instead of disposal, we will recycle.

Some materials we use now will be both abundant (and therefore cheap) but also produced with far less impact on the environment, like palm oil, rare earth elements, and dyes, as well as leather and fur. Fur coats might become fashionable again when no animals need to be harmed to make them.

And new material possibilities will be discovered and developed. There is already a company making spider silk as a replacement for plastic wrap for fruits, vegetables, and meat. Soon you might be eating flavorless, invisible coatings made of spiderweb and not even know or care about it. Scientists from the University of Kent (in the United Kingdom) have developed a gel made from a human skin protein that can stop bullets. Talin Shock Absorbing Material, or 'TSAM', in lab experiments was able to stop projectiles fired at nearly a mile per second, twice as fast as bullets travel from firearms and about the speed that particles in space can impact satellites. The researchers suggest that their material could be used to make body armor for police or soldiers and shielding for airplanes and space satellites. These things sound strange to us today, but to people from just a century ago, nylon and transistors would have been just as difficult to envision or explain.

Let us look at a few examples of some of the materials whose production and uses are being rethought due to clean energy and precision fermentation – plant oils, metals, fabric, fibers, and dyes.

Palm Oil

Palm oil is estimated to be in half of the products on supermarket shelves, from chocolate to potato chips to shampoo and toothpaste. Its global production has grown dramatically, rising more than three-fold in the past two decades. Today around the world an area about the size of New England in the US (almost as large as England and Scotland combined) are used to cultivate oil palm trees, mostly in tropical areas near the Equator.

But of course, all of this has environmental consequences. Deforestation to make way for palm plantations causes CO_2 emissions and destroys the habitat of orangutans. Shipping palm oil to the places where it will be consumed results in more CO_2 emissions, and in many cases, palm oil itself is burned, causing further emissions. In fact, in the EU-28 countries slightly more than half of all palm oil used is as a 'biofuel' additive to diesel, where it is burned to move trucks and personal automobiles.

Electric cars and trucks, which do not burn fuel, can cut some of that demand, but precision fermentation could disrupt the supply, making palm oil and oils like palm oil, without most of the environmental destruction.

In San Mateo, California, a company called Zero Acre Farms already offers 'Cultured Oil', a cooking oil made by feeding sugar "to a community of microorganisms... that eat the sugar and convert it into healthy fats – primarily monounsaturated fat", according to their website. In the UK, Clean Food Group, the US's C16 Biosciences, and Dutch company NoPalm are all using a similar approach – subjecting wild yeast to Directed Evolution to increase the yeast's oil production. C16 has already debuted their first commercial product, Palmless. And Wisconsin-based Xylome makes two versions of Yoil, a refined, bleached, deodorized (RBD) palm oil, one for food uses and one for cosmetics.

Making oils with microbes changes the inputs required. NoPalm says that they can use "potato peels, rejected vegetables, or any other biomass that contains either sugar, organic acids, or alcohol", turning waste into a valuable product. But it also expands the range of products that are possible. Natural palm oil is a mixture of similar molecules. By choosing which ones to produce, palm oil brewers can choose the properties of the final product. NoPalm CEO Lars Langhout says that they "can make a fat that is solid at room temperature, an oil that's liquid at room temperature" and, unlike oil from palm trees, they "can make different variations on the fatty acid profile." (Southey, 2022)

Using microbes greatly accelerates the process of learning and experimentation in producing new products. C16 co-founder David Heller says that they "can design an experiment and start it and get a learning about whether that helped us produce better and more oil within about seven days." In comparison, experimenting with a new seed at a palm oil plantation takes years. (Clifford, 2022) As the price of production comes down, tree-based producers of palm oil will find it increasingly difficult to compete. Orangutans will thank us for that.

Steel

The coming years will be as disruptive to the steelmaking industry as the oxygen blast furnace was in the last century. The increasing affordability (and therefore, popularity) of lithium-ion batteries, combined with the rapidly improving performance of autonomous driving systems will make passenger travel predominantly by autonomous, electric vehicles ('AEVs'). This alone will dramatically decrease the rate of car production, decreasing demand for steel. The accompanying decrease in demand for liquid fuels will also decrease demand for steel for ships (half of which, by tonnage displacement, are used for transporting fossil fuels), demand for steel for oil rigs and oil refineries, and demand for steel for oil pipelines.

The disruptions of the automobile and oil industries represent a two-pronged attack on steelmaking: a dramatic decline in the demand for steel for new vehicles and oil infrastructure coupled with the simultaneous increased availability of high-quality scrap steel from stranded individually owned vehicles and stranded oil assets. Additionally, the coming decentralization of electricity production (which is also driven in part by cheaper batteries) will make most electric transmission and distribution infrastructure and central power generation infrastructure obsolete, adding even more scrap steel (and copper) to the market.

For steelmakers, there are however some positive aspects to the coming disruptions. The increased availability of urban land will spark a boom in urban construction and renovation. New buildings will require steel. Photovoltaic solar panels and precision fermentation systems will also require a large amount of steel.

Those steelmakers who recognize the shift in customers away from the vehicle, oil and centralized electric industries and toward construction, those who work with scrap steel rather than virgin materials, and those who take advantage of electric powered furnace technologies will weather the coming disruption better than those who do not.

The steelmaking industry underwent an unprecedented disruption in the 1960s driven from within by new technology. It will soon see another, driven from rapidly changing demand for its product.

Leather and Fur

We have long had not-very-good-quality, petroleum-based alternatives to leather. Manufacturers of genuine leather products, especially high-quality ones, have not had to worry too much about competing materials.

But now new sources of leather are emerging just as the demand for, and supply of, genuine animal leather is being threatened. Companies like Modern Meadow and Bolt Threads have developed high-quality, leather-like products made from mycelium, the root structures of mushrooms. But Bolt and others are also taking genuine animal leathers to the next level by culturing them in labs the way that meats and fats are being lab cultured.

VitroLabs, based in San Jose, California, and Qorium, a Dutch biotech company co-founded by Dr. Mark Post, who made the first cell-cultured hamburger, are making cow leather in continuous sheets from tiny samples of cow cells. Qorium claims that its cell-cultured leather "takes 99% less water and 66% less energy to produce than conventional leather." Bolt Threads claims to be able to make "leathers of any size, shape, or thickness derived from any organism (including those that are rare, extinct or not typically farmed)." And Geltor claims to be able to make leather from "dinosaurs, giant moas, or Atlantic grey whales – for lower cost than cow leather today."

At the same time, the disruption of beef and dairy production threatens the supply of cattle hides. Leather production is largely a by-product of beef production.

Farmers do not raise cattle to make leather – they raise cattle for beef and then sell the resulting hides as an extra revenue source. But low-cost beef and dairy products made from precision fermentation and cellular agriculture will cause the unravelling of animal agriculture and, with it, the supply of cattle hides. As well, automated driving will radically change the demand for leather. Robotaxis are unlikely to have leather interiors and about 15% of all leather is used in car interiors.

Silk and Nylon

In 2016, Benedetto Marelli, then a researcher at Tufts University, returned to the lab to find that strawberries he had dipped in silk protein a week prior still looked fresh, while uncoated ones nearby had spoiled. He and co-workers described their observation in the academic journal Nature Scientific Reports.

Since then, Mori, the company they founded, has been commercializing the use of silk proteins as an edible, flavorless, odorless coating that slows dehydration, inhibits oxidation, and reduces microbial growth on vegetables and meats. The company claims that their coating could reduce the use of plastic packaging for foods and double shelf life, reducing spoilage. By extension, less food spoilage would mean less food would need to be grown, and therefore less use of land, water, energy, fungicides, herbicides, and pesticides. Worldwide, an estimated 250,000 deaths occur each year due to pesticide poisoning alone. And coatings made from cheap silk proteins could help reduce all of that. This is an example of what Tony Seba calls a 'disruption from the edge'. Who knew that silk could remove some of the need for chemicals and even some of the impetus for genetically modified, spoilage-resistant plants?

Silk is so versatile that a couple dozen species make their own. We have harvested silk from silkworms and some kinds of moths to make luxurious fabrics. Polish priest Casimir Zeglen made the first bulletproof vest from silk. Some ants make silk, so do some bees, and some caterpillars too. Some mussels make silk to attach themselves to rocks. Carp do the same to attach their eggs to rocks. Even goats have been genetically modified to produce milk that contains silk proteins.

On a weight basis, spider silk is stronger than steel and has the advantage of also being biodegradable. It is also notoriously difficult to produce. Starting in 2004, Englishman Simon Peers and American Nicholas Godley, employing a team of eighty people, began producing a stunning robe from naturally gold-colored silk. The piece took eight years to create, harvesting silk from 1.2 million wild female Golden Orb Weaver spiders who were collected in the highlands of Madagascar. It is understandable why such articles are one-of-a-kinds.

It is also understandable, then, why companies like Spiber in Japan and Germany's AMSilk are trying to synthesize silk at low cost and in high volume. In 2016, Adidas used AMSilk's Biosteel fiber in one of their shoes and, in 2017, precision fermented Microsilk by California-based Bolt Threads was used in a gold dress by Stella McCartney. Since then, Bolt Threads has even produced eyeglass frames from silk.

On his personal website, Tsung Xu, an Australian engineer who lives in Austin, Texas, wrote in 2022 that "spider silk protein startups have managed to almost double production on average every six months since 2008. These startups are closing in on about eight tons of production this year, over eight orders of magnitude [that is, one hundred million times] more than in 2008." But some companies, like Spiber, have their sights set on much larger numbers than that.

According to a life cycle analysis conducted by the company, Higashi says that the company's "biodegradable textiles are predicted to generate just one-fifth of the carbon emissions of animal-based fibers once they are in full-scale production". And the company plans to convert discarded garments made from cotton into the raw materials needed for fermentation.

But companies that are making silks are not limiting themselves to what is found in nature. Bolt Threads, who made silk eyeglass frames and the gold fabric for Stella McCartney's dress, also make a silk protein powder, b-SILK™, similar but not identical to what is found in nature. You can already buy vegan b-SILK in Lady Gaga's B Structural Volumizing & Lengthening Mascara.

Dyes and Coloring Agents

It is not only fibers and fabrics that could be made by biological techniques, but also the dyes and colors for those fabrics, and even colors for foods. For thousands of years, dyes and pigments were extracted from plant matter. With the rise of fossil fuels, dyes were then synthesized from the chemicals found in coal and petroleum, eventually giving rise to the pharmaceutical industry. Now, companies are looking to biology to produce these dyes and colors. California-based Huue, French biotech company Pili Bio, and UK-based Colorifix are all using precision fermentation microbes to produce fabric dyes. In an interview with Fast Company magazine, Colorifix says that "the conventional dye process takes between five and eight hours at high temperatures, and often requires as many as five washes; Colorifix's process takes three hours and a single wash." (Peters, 2021) Considering that about 20% of global water pollution stems from the textile industry, any improvement to how fabrics are dyed could have enormous benefits for the environment and human health.

Other companies, like Denmark's Chromalogics and Israeli startup Phytolon are attempting similar things, but for food colorings. Both companies are focusing on red colors, as those colors make up about half of demand for food colorings.

With these new technologies for producing and recycling materials, it is reasonable to think that we are only just beginning to unlock the possibilities of getting the materials we want with a fraction of the energy, cost, and environmental impact of ones we use now.

◆ ◆ ◆

"Electricity is the thing. There are no whirring and grinding gears with their numerous levers to confuse. There is not that almost terrifying uncertain throb and whirr of the powerful combustion engine. There is no water-circulating system to get out of order – no dangerous and evil-smelling gasoline and no noise."

— Thomas Edison, on electric cars

◆ ◆ ◆

Information: Communication and Intelligence

In March 2016 in Seoul, South Korea, the crowd burst into applause when Aja Huang resigned to thirty-three-year-old professional Go player Lee Sedol in the fourth match of The DeepMind Challenge, a five-game competition. But Huang himself was not playing against Sedol. He was merely making moves that were chosen by AlphaGo, an artificial intelligence (AI) computer program developed by London-based Google subsidiary DeepMind.

Sedol had already lost the first three matches of the competition to AlphaGo's brilliant and unorthodox playing style, ensuring that the computer would win the tournament, and he would later lose the last match as well. But when AlphaGo made a series of poor plays late in Game Four, and then resigned to Sedol, we were reminded that people can still – sometimes – beat machines in tasks of sheer brain power.

In an interview recorded in a 2017 documentary called simply AlphaGo, Sedol said after the fourth game, "I heard people shouting in joy when it was clear that AlphaGo had lost the game. I think it is clear why. People felt helplessness and fear. It seemed like we humans are so weak and fragile. And this victory meant we could still hold our own. As time goes on, it will probably be very difficult to beat AI. But winning this one time, it felt like it was enough."

The Challenge represented the first time a computer Go program had beaten a top-ranked ('9 dan') professional without handicap.

The competition echoed the six-game chess tournament in May 1997, nearly twenty years before, in New York City between Garry Kasparov and Deep Blue, created by IBM, the first defeat of a reigning world chess champion by a computer under tournament conditions. Chess can still be exciting, as Norway's Magnus Carlsen showed in a 2014 match against Bill Gates when he checkmated the billionaire using up only twelve seconds off his own clock. But since the Kasparov competition, we have recognized that even the best human players cannot regularly beat computers.

Go, on the other hand, is much more complex than chess. While a chessboard contains sixty-four squares and has twenty possible first moves by the White player, a full-sized Go board has 361 squares and therefore 361 opening moves. Since the opposing Black player in chess also has twenty possible opening moves, there are four hundred ways for the two players to make their first moves. In Go, there are nearly 130,000 ways. After three moves in chess (White's second move), there are more than five thousand possible positions. In Go, more than forty-six million. "The rules are simpler than those of chess," a BBC News article about DeepMind reported, "but a player typically has a choice of 200 moves at most points in the game, compared with about 20 in chess." (Cellan-Jones, 2017)

As a result, prior to 2015, the best Go programs in the world only managed to reach amateur level, according to an article written by Huang and some of his collaborators from DeepMind. (Silver, 2016)

But then AI computer programs got radically better. In October 2015, AlphaGo beat Fan Hui, the Go champion of Europe, five times out of five in tournament conditions, which according to Huang was "the first time that a computer program has defeated a human professional player in the full-sized game". It was only a few months later that it beat Sedol four times.

The Sedol tournament was chosen by the journal Science at the end of 2016 as one of the top stories of the year in their 'Breakthrough of the Year' compilation. And for good reason. For many people, the story of the computer technology revolution has been one of more powerful processors, clearer screens, larger hard drives, and faster internet connections. But the real story might be the quality of the underlying algorithms which have enabled chatbots, computational fluid dynamics, computer generated imagery, machine learning, speech recognition, automatic language translation, and many other applications.

Sedol retired from playing Go professionally in 2019, saying that AI programs were 'invincible', but AlphaGo was retired first. (Vincent, 2019) In 2017, the DeepMind team unveiled a new program, AlphaGo Zero, which started with just the rules of the game and did not learn from a repository of games humans had played, as AlphaGo did. Instead, the new program played nearly five million games with itself over the course of three days. Then, it beat the AlphaGo version that defeated Sedol by 100 games to zero.

"It shows it's the novel algorithms that count, not the compute power or the data," said David Silver, leader of DeepMind's Go efforts.

Go might be just a game but DeepMind's ambitions, and their algorithms, go far beyond games. In 2020 according to an article in Nature, DeepMind "made a gargantuan leap in solving one of biology's grandest challenges – determining a protein's 3D shape from its amino-acid sequence". (Callaway, 2020) Their AlphaFold program, according to the article, performed "head-and-shoulders above other teams" on a standard set of protein structures and even well above their own results from the previous year. As with AlphaGo, DeepMind's AlphaFold got dramatically better in a short period of time. In 2022, DeepMind released a database of two hundred million protein structures determined with their program, the structure of nearly every protein known to science. (Wetsman, 2022)

But the number of possible unique proteins is effectively infinite. By examining proteins found in nature, we have only just scratched the surface of what is possible. It is not only AlphaFold, but a variety of rapidly improving AI algorithms are making the idea of rational protein design a real possibility. The diffusion model, the same technology that is being used to generate complicated pictures from text prompts, is being used "to conjure up designs for new types of protein never seen in nature", according to an article in MIT Technology Review. (Heaven, 2022) Because plastics do not occur in nature, there are few proteins known that break them down for recycling. But companies like Epoch BioDesign, based in London, England, uses AI to engineer plastic-degrading enzymes.

All that we lump together under the title of 'information technology' are affecting and accelerating each other – the ability to gather information with cheap (and therefore numerous) sensors, to process and store information with better hardware, to make sense of it all with more powerful algorithms, and to transmit information across long distances quickly. The technologies that are conquering ancient games and novel protein design are being applied to many other complex tasks, like making music, writing novels, and even programming computers. Companies like Waymo, Tesla, Cruise and others are applying them to driving cars.

SpaceX is using them to make rockets that land standing upright, making it cheaper to access space and therefore to build an orbiting network of thousands of powerful communications satellites. It is estimated that 90% of internet cable breaks worldwide each year are from commercial fishing boat nets or anchors. (Wakefield, 2022) Information technology is helping to solve this problem from both sides, replacing undersea cables with orbiting satellites while also helping us design proteins that could eliminate the need for the fishing boats in the first place.

These technologies are revolutionizing how we are living now, what we will be able to do in the future, and our understanding of our own past. Jeremy DeSilva, the author of First Steps: How Upright Walking Made Us Human, says that satellite imagery is driving an acceleration in the rate at which ancient hominin fossil sites are being found, as the entrances to caves and clusters of trees associated with fossil find sites. And these technologies are revolutionizing other fields of science, and thus our understanding of many other aspects of life.

The New Scientific Revolutions

> *Cheaper and better sensors, computers, and data storage are allowing amateur astronomers to make images that rival the quality of what professionals were able to do just a couple of decades ago, making the work of professional scientists easier, and reshaping our understanding of many different fields.*

Astronomers claim SpaceX's numerous Starlink satellites are interfering with their observations just as amateur astronomy is undergoing a revolution thanks to techniques like 'lucky imaging'. Now, rather than make long-duration exposures, amateurs are simply taking hundreds if not thousands of short-duration shots, and then keeping the 'lucky' best images. Professional scopes use large mirrors and expensive systems to correct for atmospheric distortion. But with lucky imaging, amateurs are getting results that rival the big scopes, using equipment that is literally thousands of times cheaper, simply by collecting data for much more time.

This is real science they are doing. In a 2021 interview amateur astronomer Damian Peach said, "a lot of the storms that have been discovered on Neptune over, I'd say, probably the last five, six years have actually been discovered from amateur imaging of the planet."

This new ability for amateur astronomers to be able to do what was formerly only available to professionals is being driven by a convergence of new technologies, particularly so-called CMOS (Complementary Metal-Oxide Semiconductor) image sensors, which took over the market from Charge-Coupled Device (CCD) sensors in under ten years, and cheap hard drives for data storage.

Peach says, in "a really productive night out there imaging the planets, you can easily come away with a couple of hundred gigabytes of video data." Just a few years ago it was not practical for an amateur to afford the data storage ability or computing power necessary to collect, sort, and blend the images taken, but now digital image processing software internet communications (which allows amateurs to learn from each other and to collaborate) means that thousands of amateur astronomers around the world are banding together into a distributed network.

Galileo's telescopes themselves, made in the early 1600s, were disruptive technologies. The website of the Museo Galileo in Florence, Italy, reports that two of Galileo's telescopes had magnification factors of 14x and 21x.

Now, revolutions in professional astronomy are reshaping our understanding of everything from local asteroids to distant supernovas. The first-known asteroid, Ceres, named after the goddess of the harvest and of brewing (who gave her name to Saccharomyces cerevisiae, the key yeast used in brewing beer) was discovered in 1801. It took nearly fifty years, until 1849, for us to discover nine more asteroids. By 1989, we knew of ten thousand and, by 2020, we knew of one million. That is, 99% of all known asteroids have been discovered in the past thirty years or, put more simply, the number of known asteroids is increasing by 10X every fifteen years.

But that number could soon rapidly increase. In 2022, the New York Times reported on an algorithm developed by Dr. Ed Lu, a former NASA astronaut, which searches through old astronomy photographs and identifies new asteroids by their appearances in those old photos. As of 2022, they have identified over one hundred asteroids using just a small fraction of the available data. "It's one of the coolest applications I've seen," said Scott Penberthy, director of applied artificial intelligence at Google. And, the algorithm could also be used to help choose future observations "that could benefit other research, including the search for supernovas."

Astronomy is not the only science that has benefited from the convergence of powerful sensors and cheap computer processing. In a cave in Alabama, scientists have used a technique a lot like 'lucky image' astronomy – taking many digital photos and then combining them with software – to produce images of artwork scratched into the cave ceiling somewhere between 1600 and 1900 years ago. The convergence of cheap microphones, the ubiquitous smartphone, and AI apps are allowing amateur bird watchers to identify species in their area, even at night, recording useful scientific data measuring local bird populations. (A similar convergence enabled the defense forces of Ukraine for just a few million dollars to assemble a connected network of thousands of smartphones to detect incoming Russian drones. General James B. Hecker, commander of the US Air Forces in Europe reported in early 2024 that the system was able to detect 100% of incoming drones and then anti-aircraft guns were able to shoot down more than 95% of those.) And in the rainforests of the Yucatan Peninsula, scientists have used lidar – the same technology that engineers are using to make self-driving cars – to discover that the density of human settlement during the Mayan era was far higher than we had previously suspected. From deep underground to far above our heads, cheaper and better sensors, computers, and data storage are revolutionizing many fields of science, including ones that have a direct impact on developing the disruptive new technologies that are shaping the future of clean transportation, clean energy, and beyond.

"Improving batteries has always been hampered by slow experimentation and discovery processes," Daniel Oberhaus wrote in an article in Wired magazine. "Machine learning is speeding it up by orders of magnitude." In the article, Oberhaus describes the work of William Chueh, a materials scientist at Stanford University, who used AI to "optimize a fast-charging protocol for a lithium-ion battery in less than a month; to achieve those same results without the aid of AI would usually take around two years." A more than 20X acceleration.

Improvements in artificial intelligence (including speech recognition and translation, and text creation), batteries, autonomous systems, high-speed computation and communications, clean power, and other fields, reinforce each other and have brought us to the edge of a world where many tasks now undertaken by people, including both ones that are mundane and repetitive, but also ones that require creativity and inventiveness, will soon be within the reach of human-engineered devices.

"A machine that can learn to drive in Los Angeles or Shanghai", Adam Dorr writes in his 2023 book Brighter: Optimism, Progress, and the Future of Environmentalism, "can also learn to operate in a warehouse, a factory floor, a supermarket, a kitchen, and thousands of other workplaces, and so self-driving vehicles are the pebble that will trigger the automation avalanche." The implications for the use of human labor will be profound.

◆ ◆ ◆

"There are a thousand hacking at the branches of evil to one who is striking at the root."

— Henry David Thoreau

◆ ◆ ◆

Clean Labor: This Time, We Are the Horses

We like to think of slavery as something confined to the distant past, but unfortunately it is not. Quite the contrary. A September 2022 report by the International Labor Organization (ILO) estimated that fifty million people around the world today live in a state of bondage or near-bondage, an increase of ten million people from 2016. That is, there are five times as many people on Earth who are enslaved right now as the number brought across the Atlantic over more than three hundred and fifty years, and that number has recently been growing.

While enslaved people centuries ago worked in areas like agriculture (sugar, tobacco), textile production (cotton) and domestic labor, people in bondage today tend to work in areas like agriculture (sugar, cocoa), textile production (cotton, silk) and domestic labor. Fortunately, these are all areas that could be disrupted by new technologies in the coming decades, both ones that reduce the demand for human labor and ones that increase the supply of labor from intelligent machines.

In *Mizora: A Prophesy*, a science-fiction story published in 1880-1881, Mary Bradley Lane describes a feminist utopia where "in every vocation of life machinery was called upon to perform the arduous physical labor. The whole domestic department was a marvel of ingenious mechanical contrivances. Dishwashing, scouring and cleaning of every description were done by machinery."

Advances in computer object recognition, artificial intelligence, and other aspects of general-purpose robotics are promising to bring about this end to the use of humans in 'arduous physical labor'. Perhaps you have already seen demonstrations of the eerily life-like robots by Boston Dynamics, like Spot, a dog-like robot, BigDog, a mule-like robot with oddly human-looking legs, and Atlas, a humanoid robot that can run, jump, tumble, and dance. Automaker Tesla has unveiled a plan to make a humanoid robot they call Optimus.

These new machines go way beyond the robotic arms already in widespread use in automotive assembly and intelligent robots would do far more than Lane's envisioned 'dishwashing, scouring and cleaning'. Factories, residential and commercial construction zones, warehouses, farms, forests, mines, shipyards, dockyards, fighting fires, battlefields, outer space and anywhere else where people currently do dangerous and difficult work would all be prime locations for robots.

Humanoid robots would not only reduce the danger of hard labor to humans, but as Adam Dorr explains in his 2023 book *Brighter. Optimism, Progress, and the Future of Environmentalism*, intelligent automated labor would reduce the cost of nearly everything. "Energy is only expensive because someone has to build the solar panels and batteries to capture it," he writes, "and the raw materials with which to do so are only expensive because someone has to mine and refine them, and mines and refineries with which to do so are only expensive because someone has to build and operate them, and so on. It is therefore labor which is ultimately the limiting factor of production. As Adam Smith wrote more than two centuries ago, 'The real price of everything ... is the toil and trouble of acquiring it.' It follows that if we had an unlimited supply of dirt-cheap labor, almost everything else would be dirt cheap too."

The connection between cheap, mechanized labor and living lives of abundance was clear to Mary Lane in Mizora: "I lived long enough in Mizora to comprehend that the absence of pauperism, genteel and otherwise," she wrote, "was largely due to the ingenious application of machinery to all kinds of physical labor. When the cost of producing luxuries decreases, the value of the luxuries produced must decrease with it. The result is they are within reach of the narrowest incomes."

Cheap energy, cheap labor, cheap communications, cheap transport, cheap materials, and cheap food would mean that the basic needs of every person on Earth could be met. What would we do with our time if we did not need to work to supply our basic needs? Buckminster Fuller, who among other things designed the iconic geodesic building at EPCOT Center, once said "we should do away with the absolutely specious notion that everybody has to earn a living. It is a fact today that one in ten thousand of us can make a technological breakthrough capable of supporting all the rest. The youth of today are absolutely right in recognizing this nonsense of earning a living. We keep inventing jobs because of this false idea that everybody has to be employed at some kind of drudgery because, according to Malthusian Darwinian theory he must justify his right to exist. So we have inspectors of inspectors and people making instruments for inspectors to inspect inspectors. The true business of people should be to go back to school and think about whatever it was they were thinking about before somebody came along and told them they had to earn a living."

◆ ◆ ◆

"You never change things by fighting the existing reality. To change something, build a new model that makes the existing model obsolete."

— **Buckminster Fuller**

◆ ◆ ◆

A New Era for Health and the Environment

Clean energy, clean transportation, and clean agriculture offer the best chance we have ever had to continue human progress and live lives of prosperity while also respecting the atmosphere, land, water, environment, and all living things, ourselves included. We could not only reduce the rate of deforestation but stop and reverse it to allow the chance for reforestation. We could stop the dangerous experiment we are performing in our atmosphere. We could allow marine populations to rebound. We could do what we choose to, if we have the courage.

Healthy Air

According to a 2022 assessment by the World Health Organization, 99% of the world's population lives in areas where air quality is below recommended limits. (Keaten, 2022) The major pollutants causing the problem include soot and small particles from combustion that go deep into the lungs, as well as chemicals and sand-like dust. Causes differ around the world, and by the time of the year, but the primary contributors to poor air quality include combustion by vehicles in cities and electrical power plants, burning crop residues after harvests, dust from construction as well as dust from agricultural soils, ammonia from fertilizer, cooking smoke from burning sticks or dung, and natural sources, like desert dust storms.

Even in the Middle East, according to a study by researchers who used ship-based sensors moving around the Arabian Peninsula, more than 90% of the most-damaging pollution particles come from human activities and not natural desert dust. (Osipov, 2022)

The known effects of air pollution on human health are staggering. According to a study at University of Birmingham in the UK, about eight million deaths per year worldwide – one thousand people per hour, or about one in every five deaths – are attributable to the burning of fossil fuels. (Vohra, 2021) This makes air pollution the world's single largest environmental risk factor for disease and premature death.

An MIT study found that there are approximately 200,000 excess deaths per year in the United States due to fine particulate matter in the atmosphere. (Chu, 2013) Eliminating air pollution just from the production of energy could prevent the deaths of 50,000 Americans per year and $600 billion in annual economic damages (or about $1750 per person per year in the US), according to a 2022 study by Nicholas

Mailloux and co-workers from the University of Wisconsin in Madison.

It is notable that all these sources will be dramatically impacted by the coming disruptions that we are concerned with. Even low levels of air pollution can have enormous effects on human health. Some of the known effects, like lung cancer and asthma are obvious, but there are also many that are not as well known, including possible causal connections to dementia, Alzheimer's disease, heart attacks, chronic obstructive pulmonary disease (COPD), stroke, and depression, among others.

The effects of exposure to air pollution in childhood can be large and life-long. A study by Diane Alexander of the University of Pennsylvania and Hannes Schwandt of Northwestern University examined the health effects of the Volkswagen 'clean diesel' scandal, where vehicles sold by the European automaker were rigged to cheat emissions tests from 2008 to 2015, when engineers from the University of West Virginia discovered the malfeasance. Alexander and Schwandt found that one extra cheating vehicle per every one thousand vehicles on the road resulted in an almost 2% increase in low birth weight among babies and also an almost 2% increase in infant mortality. Another study by Janet Currie and Reed Walker of Columbia University found that when electronic tolling ("E-ZPass") was introduced in New York, congestion at toll plazas dropped, lowering local air pollution. The rate of premature births to mothers within about a mile-and-a-half of the plaza dropped by up to 7-9% and the incidence of low birth rate babies by 8-11%.

The effects of air pollution are not evenly distributed, even within a single country or city. A 2012 study by the National Association for the Advancement of Colored People (NAACP) found that 78% of African-Americans live within thirty miles of a coal-fired power plant. Air pollution is cited as the primary reason why the east side of most major cities in the Northern hemisphere often have lower property values than the west side. Because winds blow predominantly from west to east, pollution emitted within the city tends to carry toward the east side of the city, while the west side receives fresh air from the surrounding countryside. Significantly reducing emissions from transport and energy production could for the first time shift land values within cities based on this one factor alone.

Air pollution does not just affect human health, it also reduces crop yields. A 2019 study led by the Indian Institute of Technology-Madras found that more than 20% of India's wheat crop is destroyed each year by air pollution from power plants, vehicles, industries, and biomass burning. Similarly, a study conducted by Stanford University found that reductions in pollution in the US between 1999 and 2019 contributed about twenty percent of the increase in corn and soybean yield gains during that time. (Fernandes, 2019) (Lobell, 2021)

Reducing air pollution can also boost solar panel efficiency, increasing the output from the existing stock of panels in a self-reinforcing feedback loop: as air gets cleaner, solar panels will produce more power, making them more cost effective and incentivizing more installations of solar panels, which will make the air cleaner. Again, a study from the Indian Institute of Technology found that "India lost almost a third of its solar power potential between 2001 and 2018 due to air pollution". (Barber H. , 2022)

Healthy Land

The new technologies we have talked about could mean the end of deforestation and the start of reforestation. According to Our World in Data, approximately ten thousand years ago, before the widespread adoption of agriculture, about 57% of the Earth's habitable land was forested. By 1800, that number had fallen to 50% and today is under 40%. (Not all land is habitable, so of the Earth's total land area, only about 26% is forest.)

Today only about 1% of the Earth's land is the built-up area of cities. More land is for livestock, an area equal to North America and South America combined, than is forest. Producing proteins and fats from precision fermentation and cellular agriculture would bring about the end of net deforestation. Florence Pendrill and co-workers at Chalmers University of Technology in Gothenburg, Sweden, found that agriculture is currently responsible for 90% of deforestation, but only about half of land that gets deforested gets converted into productive agricultural use. Precision fermentation and cellular agriculture could put an end to that and free up an area of land the size of the United States, China, and Australia combined from their current use in supporting livestock production.

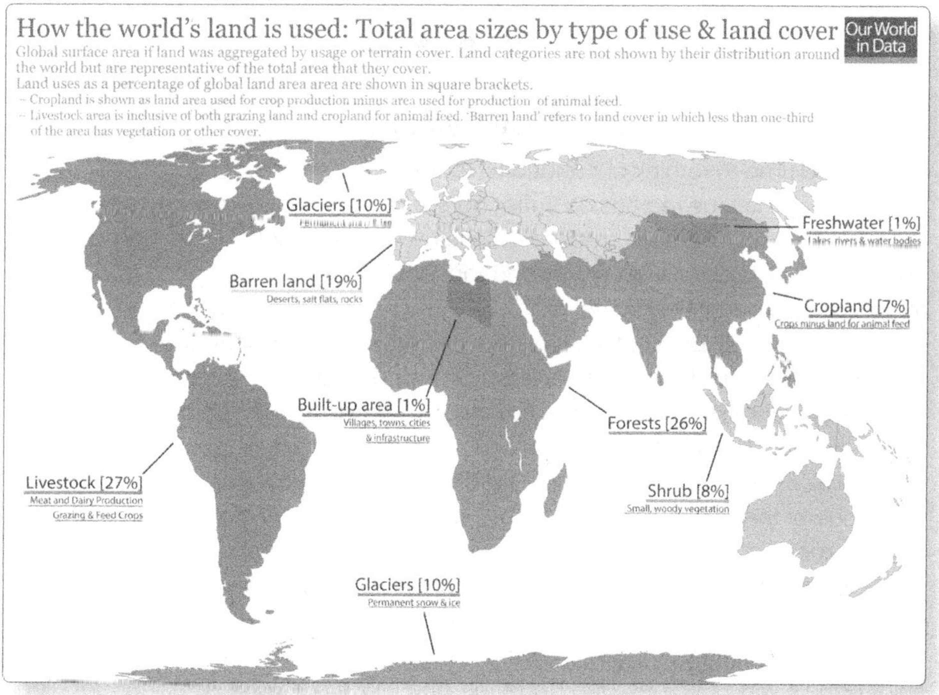

Figure 138 – Global Land Use. *Figure from Our World in Data*

Precision fermentation and cellular agriculture mean that we could be the first people since the invention of agriculture to see an increase in the worldwide area of forested land.

Professor Marcellin Berthelot foresaw this in his interview by Henry Dam in McClure's Magazine, published in September 1894. "At some more or less distant period in the future," he said, "synthetic chemistry will destroy all the great agricultural industries, and put to new uses the grain fields and cattle ranges of to-day."

What might we do with the land that is freed up from growing so many crops? In short, anything we want. Anyone who wanted to live off-grid with abundant energy and mechanized labor, as they do in Mizora, could do that. Those who wanted to live in large urban areas, with clean air and little space devoted to retail could do that instead. We could return large areas of land to the native peoples of the Americas. Or we could simply leave land alone. Studies have shown that tropical forests can get back to roughly 80% of their old-growth status in just twenty years without human interference. (Quaglia, 2021) We could just let that happen.

◆ ◆ ◆

"With every drop of water you drink, every breath you take, you're connected to the sea, no matter where on Earth you live. Most of the oxygen in the atmosphere is generated by the sea."

— Sylvia Earle

◆ ◆ ◆

Healthy Waters

It is beyond time to rethink our relationship with the world's waters. For too long we have treated the blue regions as buffet bars, as floating highways, and as garbage dumps with little long-term regard for their well-being. But the coming decades can be ones of restoration for the oceans, benefiting not only the life that lives there, but also benefiting ourselves.

To anyone who pays attention, the negative news about the oceans can be overwhelming: rising sea levels, collapsing animal populations, declining ecosystem richness, coral reefs in distress, warmer and more-acidic waters, 'dead zones' caused by a lack of oxygen, oil spills, microplastics, persistent pollutant chemicals, a 'Great Pacific Garbage Patch' of swirling detritus, and much more.

The bulk of these problems result from a few core issues: how we get our energy, how we move ourselves and our goods, and how we get our food.

Economists estimate that 90% of all goods that are traded internationally go aboard a ship at some point in their travel. 40% of all large ocean-going trading ships (by dry weight tonnage) are for moving fossil fuels. If we get our power from solar panels, wind turbines, and batteries, and move our vehicles using that clean electricity, then much of the movement of large commercial ships would cease. The most heavily trafficked lanes would die down, as would the ocean noise from shipping and from seismic surveys from hunting for more fossil fuels.

In 2024, researchers at Nanyang Technological University in Singapore reported that they had modified bacteria to produce a barnacle protein that forms a protective layer on steel, inhibiting corrosion. Right now, ships use paints that contain toxic chemicals to prevent 'fouling' on their hulls. But these paints flake off into the water over time and they are terrible for the environment. But anti-fouling paints based on barnacle proteins could be a huge step forward for the oceans. (Bui M. , 2024)

The seas are impacted not only by what we put into them, but also what we take out. Industrial fishing has pushed global fish stocks as far as they can go. According to *Our World in Data*, wild catch of seafood has been constant at around ninety million tons per year since the late 1980s. The UN Food and Agriculture Organization estimates that 60% of global fisheries are fished at their maximum sustainable yield (MSY) and an additional 34% are overfished. The oceans simply cannot give us more per year than they do now and should reasonably be asked to give less. Much less.

The problem is not *overfishing* per se – the problem is *fishing*. The maximum sustainable yield of a fishery happens when the population being exploited is kept at about half of its natural, unfished level. At that level, the population collectively has its highest rate of annual increase, for the same reason that an S-shaped curve is steepest at the middle. So, a 'sustainably' exploited population is held to an artificially low level which, as a result, also keeps down the population of predators that rely on that species. (Ritchie, Biodiversity, 2021)

This is not the only negative effect that fishing has. Far from it. Laurent Lebreton and co-workers from The Ocean Cleanup, an organization that has designed and built devices to tackle the Great Pacific Garbage Patch (GPGP), a plastic-filled region of the northern Pacific Ocean, estimates that more than 75% of the GPGP plastic "originates from fishing activities at sea".

Fishing affects even species we do not deliberately hunt. Scientists have determined the top three causes of whale deaths to be malnutrition, ship strikes, and entanglement in fishing equipment. In fact, five out of every six identified right whales have been entangled at least once in fishing gear, and more than half have been snagged twice. (Ebersole, 2020) None of the world's commercial fisheries targets sea turtles, but six out of seven species of sea turtle are threatened or endangered, mostly from being caught in trawler nets. Because an estimated one hundred million sharks each year are being killed by people, mostly through fishing for other species, more than one-third of all known shark and ray species are currently threatened with extinction, with some populations having been reduced by more than 96 per cent in the last fifty years alone.

And fishing harms many of the people who do it. A 2022 study funded by The Pew Charitable Trusts and the Walmart Foundation found that 100,000 people die each year in the global commercial fishing industry. (Holliday, 2022) This is a business that is not only cruel to the fish it catches, but also to the people who give their limbs and lives to put food on our plates.

As a result of maxing out what the oceans can provide, all growth in the production of seafood worldwide since the late 1980s has come from aquaculture. But this too creates significant problems. Feeding captive fish relies on ocean bycatch, the leftover parts from wild-caught fish that people do not eat, as well as additional protein and fats, often in the form of soy and canola grown on deforested land using fertilizers derived from natural gas and using pesticides. Fish farms create fecal waste that devastates the local environment and house maladjusted breeds of fish bred only to grow quickly. They pit fish farmers against wildlife like sea lions, who are often killed to protect the fish.

The rise of precision fermentation and cellular agriculture would spell the end of commercial fishing and aquaculture and will therefore initiate an unprecedented recovery of marine ecosystems. Studies have shown that when fishing pressure is removed, wild fish populations can reach 80% of their ultimate recovery level in under ten years. (Pew, 2021)

When we can grow meat from cells, rather than raise entire animals and kill them to obtain their meat, we will suddenly realize that the land, oceans, and skies do not belong to us and the creatures that live in them do not exist to serve us.

We could live in a future where all people live lives of respect and fulfilment. There is no time to waste. The future has already started.

◆ ◆ ◆

"All our dreams can come true,
if we have the courage to pursue them."
— Walt Disney

◆ ◆ ◆

Works cited

AAMC. (2016). American Association of Medical Colleges. *https://www.aamcdiversityfactsandfigures2016.org/report-section/applicants-enrollment/*

Abbott, J. (1849). *Alexander the Great.*

Adams, W. H. (2002). Machine cut nails and wire nails: American production and use for dating 19th-century and early-20th-century sites. *Historical Archaeology*, 36, 66–88.

Ahktar, R. (2022, August 5). 'Hertz says Tesla fleet of rental EVs are 50 pct cheaper to maintain than fossil fuel cars'. The Driven: *https://thedriven.io/2022/08/05/hertz-says-tesla-fleet-of-rental-evs-are-50-pct-cheaper-to-maintain-than-fossil-fuel-cars/*

Ahmed, N. (2022, August). 'The Clean Energy Transformation: A New Paradigm for Social Progress Within Planetary Boundaries'. Club of Rome: *https://www.clubofrome.org/wp-content/uploads/2022/08/Earth4All_Deep_Dive_Ahmed.pdf*

Alexander, D. (2022). 'The Impact of Car Pollution on Infant and Child Health: Evidence from Emissions Cheating'. *The Review of Economic Studies*, doi.org/10.1093/restud/rdac007.

Anaya, A. (2021). 'Evolutionary trends of the lateral foot in catarrhine primates: Contextualizing the fourth metatarsal of Australopithecus afarensis'. *Journal of Human Evolution*, doi.org/10.1016/j.jhevol.2021.103078

Andreas, P. (2020). *Killer High: A History of War in Six Drugs.* Oxford: Oxford University Press.

Anthony, D. (2007). *The Horse, the Wheel, and Language: How Bronze-Age Riders from the Eurasian Steppes Shaped the Modern World.* Princeton University Press.

Anthony, S. (2016, July 15). 'Kodak's Downfall Wasn't About Technology'. *Harvard Business Review.*

Armstrong, G. (1999, January 6). 'Trends in Infectious Disease Mortality in the United States During the 20th Century'. *Journal of the American Medical Association, pp. 61-66.*

Atwood, A. (2022). 'The Long-Term Effects of Measles Vaccination on Earnings and Employment'. *American Economic Journal: Economic Policy*, doi:10.1257/pol.20190509

Aurecon. (2020). 'Hornsdale Power Reserve: Year 2 Technical and Market Impact Case Study'. *https://www.aurecongroup.com/-/media/files/downloads-library/thought-leadership/aurecon-hornsdale-power-reserve-impact-study-2020.pdf*

Ayers, R. U. (1989). 'Technological Transformations and Long Waves'. Laxenburg, Austria: International Institute for Applied Systems Analysis (IIASA).

Bachhuber, M. (2014). 'Medical Cannabis Laws and Opioid Analgesic Overdose Mortality in the United States, 1999-2010'. *JAMA Intern Med.*, doi:10.1001/jamainternmed.2014.4005

Baker, I. (2018). *Fifty Materials That Make the World.* Springer International Publishing.

Bar-On, Y.M. (2018). The biomass distribution on Earth. Proceedings of the National Academy of Sciences (PNAS), 115 (25) 6506-6511.

Barak, O. (2020). *Powering Empire: How Coal Made the Middle East and Sparked Global Carbonization.* University of California Press.

Barber, E. (1991). *Prehistoric Textiles: The Development of Cloth in the Neolithic and Bronze Ages with Special Reference to the Aegean.* Princeton University Press.

Barber, H. (2022, March 21). 'Air pollution cut India's solar energy output by a third'. The Telegraph: *https://www.telegraph.co.uk/global-health/climate-and-people/air-pollution-cut-indias-solar-energy-output-third/*

Basker, E. (2012). 'Raising the Barcode Scanner: Technology and Productivity in the Retail Sector'. National Bureau of Economic Research (NBER), *http://www.nber.org/papers/w17825*

BBC World Service. (2020, July 29). 'Ingenious: The milkshake and the cyclops gene'. *https://podcasts.apple.com/no/podcast/the-documentary-podcast/id73802620?i=1000486483181*

BBC. (2021, July 12). 'Florida breaks manatee death record in first six months of 2021'. *https://www.bbc.com/news/world-us-canada-57802165*

Beaton, K. (1955). 'Dr. Gesner's Kerosene: The Start of American Oil Refining'. *Business History Review*, doi:10.2307/3111597

Bebber, M. (2019, April 8). 'The Exceptional Abandonment of Metal Tools by North American Hunter-Gatherers, 3000 B.P.', *Nature*, doi.org/10.1038/s41598-019-42185-y

Bennett, C. (2018). 'The Broiler Chicken as a Signal of a Human Reconfigured Biosphere'. *Royal Society Open Science*, doi.org/10.1098/rsos.180325

Berger, D. (2019). 'Isotope Systematics and Chemical Composition of Tin Ingots from Mochlos (Crete) and Other Late Bronze Age Sites in the Eastern Mediterranean Sea: An Ultimate Key to Tin Provenance?', *PLOS ONE*, doi.org/10.1371/journal.pone.0218326

Berger, E. (2021, June 4). 'Yes, the military is serious about rocketing supplies around the planet'. Ars Technica: *https://arstechnica.com/science/2021/06/the-military-wants-to-use-rockets-for-cargo-delivery-anywhere-on-earth/*

Berger, E. (2022, January 3). 'The world just set a record for sending the most rockets into orbit'. Ars Technica: *https://arstechnica.com/science/2022/01/thanks-to-china-and-spacex-the-world-set-an-orbital-launch-record-in-2021/*

Berk, G. (2009). *Louis D. Brandeis and the Making of Regulated Competition, 1900-1932.* Cambridge University Press.

Bernstein, W. (2008). *A Splendid Exchange: How Trade Shaped the World.* Grove Press.

Bertrand, D. A. (2019). 'Building the Medieval Trebuchet'. Utah State University: doi.org/10.26076/0da1-8257

Bius, J. (2018). *Smoke 'Em If You Got 'Em.* Naval Institute Press.

Blanchard, M. (1999). *History of the Mass Media in the United States: An Encyclopedia.* Routledge.

Blasi, D. (2019, March 15). 'Human sound systems are shaped by post-Neolithic changes in bite configuration'. *Science,* doi:10.1126/science.aav3218.

Bleasdale, M. (2021). 'Ancient proteins provide evidence of dairy consumption in eastern Africa'. *Nature Communications,* pp. 12, 632.

Blum, D. (2010). *The Poisoner's Handbook: Murder and the Birth of Forensic Medicine in Jazz Age New York.*

Bogaard, A. (2019). 'The Farming-Inequality Nexus: New Insights from Ancient Western Eurasia'. *Antiquity,* doi.org/10.15184/aqy.2019.105

Booth, R. (2022, November 29). 'England and Wales now minority Christian countries, census reveals'. The Guardian: *https://www.theguardian.com/uk-news/2022/nov/29/leicester-and-birmingham-are-uk-first-minority-majority-cities-census-reveals*

Bourke, A. (1964). 'Emergence of Potato Blight'. *Nature,* doi:10.1038/203805a0

Bown, S. (2005). *A Most Damnable Invention: Dynamite, Nitrates, and the Making of the Modern World.* New York: Thomas Dunne Books.

Braudel, F. (1973). *The Structures of Everyday Life.* New York: Harper & Row.

Braun, E. (1978). *Revolution in Miniature: The History and Impact of Semiconductor Electronics.* Cambridge University Press.

Brock, G. (2003). *The Second Information Revolution.* Harvard University Press.

Brown, W. L. (1983). 'H.A. Wallace and the Development of Hybrid Corn'. *The Annals of Iowa, State Historical Society of Iowa,* Vol 47 (2), 167-179.

Bryson, B. (2010). *At Home: A Short History of Private Life.*

Bui, Q. (2015, February 5). Planet Money. Map: The Most Common Job In Every State: *https://www.npr.org/sections/money/2015/02/05/382664837/map-the-most-common-job-in-every-state*

Burger, J. (2020, September 03). 'Low Prevalence of Lactase Persistence in Bronze Age Europe Indicates Ongoing Strong Selection over the Last 3000 Years'. *Current Biology,* doi:10.1016/j.cub.2020.08.033

Buringh, E. (2009). 'Charting the "Rise of the West": Manuscripts and Printed Books in Europe, A Long-Term Perspective from the Sixth through Eighteenth Centuries'. *The Journal of Economic History,* doi.org/10.1017/S0022050709000837

Burmeister, S. (2019). 'Some Notes on Pictograms Interpreted as Sledges and Vehicles Wheeled in the Archaic Texts from Uruk'. In P. *Raulwing, Equids and Wheeled Vehicles in the Ancient World. Essays in Memory of Mary A. Littauer* (pp. 49-70). Oxford.

Burnett, C. (2005). 'The Edges of Empire and the Limits of Sovereignty: American Guano Islands'. *American Quarterly, https://www.jstor.org/stable/40068316*

Butler-Sloss, S. (2022). The Energy Transition and the Global South: A More Attractive Energy Future for All. Rocky Mountain Institute.

Cairns, R. (2022, June 1). 'Inspired by spiders, one Japanese startup is inventing a new generation of textiles'. CNN: *https://edition.cnn.com/2022/05/22/business/spiber-spider-silk-fabric-hnk-spc-intl/index.html*

Callaway, E. (2020, November 30). "It will change everything': DeepMind's AI makes gigantic leap in solving protein structures'. Nature, *https://www.nature.com/articles/d41586-020-03348-4*

Carter, R. (2005). 'The History and Prehistory of Pearling in the Persian Gulf'. *Journal of the Economic and Social History of the Orient,* p. 139-209.

Carvajal, D. (1999, April 28). 'The New York Times Bans Cigarette Ads'. *The New York Times.*

Cassidy, C. (2020, May 6). 'Who Invented the Wheel? And How Did They Do It?' Wired: *https://www.wired.com/story/who-invented-wheel-how-did-they-do-it/*

CDC. (1999, September 30). 'Achievements in Public Health, 1900-1999: Healthier Mothers and Babies'. *https://www.cdc.gov/mmwr/preview/mmwrhtml/mm4838a2.htm*

Cellan-Jones, R. (2017, October 18). 'Google DeepMind: AI becomes more alien'. *BBC News, https://www.bbc.com/news/technology-41668701*

Charlton, S. (2019). 'New insights into Neolithic milk consumption through proteomic analysis of dental calculus'. *Archaeological and Anthropological Sciences,* doi:10.1007/s12520-019-00911-7

Chattopadhyaya, D. (1999). *History of Science, Philosophy and Culture in Indian Civilization.*

Cheng, S. (2021, July). 'What's Another Year?: The Lengthening Training and Career Paths of Scientists'. *https://conference.nber.org/conf_papers/f159298.pdf*

Works cited

Cherrico, J. (2016, March 11). 'Former Pottery World Record: 159 Pots In One Hour On A Kick-Wheel'. YouTube: *https://www.youtube.com/watch?v=BvciVXijupc*

Christensen, C. (1997). *The Innovator's Dilemma: When New Technologies Cause Great Firms to Fail.* Boston, Massachusetts: Harvard Business School Press.

Christiansen, T. (2020). 'New Insights into the Composition of Ancient Egyptian Red and Black Inks on Papyri Achieved by Synchrotron-based Micro-analyses'. *Proceedings of the National Academy of Sciences*, doi.org/10.1073/pnas.2004534117

Chu, J. (2013, August 29). 'Study: Air pollution causes 200,000 early deaths each year in the U.S.' MIT: *https://news.mit.edu/2013/study-air-pollution-causes-200000-early-deaths-each-year-in-the-us-0829*

CIPA. (2021). Camera and Imaging Products Association. *https://www.cipa.jp/e/stats/dc.html*

Clark, R. (1993). 'Indigo, woad, and Tyrian Purple: important vat dyes from antiquity to the present'. *Endeavour*, doi.org/10.1016/0160-9327(93)90062-8

Clausen, J. (2017). 'Sticky Transformation: How path dependencies in socio-technical regimes are impeding the transformation to a Green Economy'. *Journal of Innovation Management*, 111-138.

Clifford, C. (2022, November 3). 'Palm oil is in almost everything, and it's devastating rainforests. This Bill Gates-backed company used microbes to create an alternative'. CNBC: *https://www.cnbc.com/2022/11/03/gates-backed-c16-biosciences-uses-yeast-to-create-palm-oil-substitute.html*

Cline, E. (2014). *1177BC: The Year Civilization Collapsed.*

Colquhoun, J. (1997). 'Why I Changed my Mind About Water Fluoridation'. *Perspectives in Biology and Medicine*, doi.org/10.1353/pbm.1997.0017

Comin, D. (2009). 'The CHAT Dataset'. National Bureau of Economic Research (NBER). *http://www.nber.org/papers/w15319*

Compton, T. (2021). 'The morphology of the Late Pleistocene hominin remains from the site of La Cotte de St Brelade, Jersey (Channel Islands)'. *Journal of Human Evolution*, 152: 102939; doi: 10.1016/j.jhevol.2020.102939

Condo, R. (1985). *The Latin American Economies: Growth and the Export Sector, 1880-1930.*

Considine, G. (2008). *Van Nostrand's Scientific Encyclopedia.* Hoboken NJ: Wiley.

Cooper, G. (1968). *The Invention of the Sewing Machine.*

Cornish, K. (2001). 'Biochemistry of natural rubber, a vital raw material, emphasizing biosynthetic rate, molecular weight and compartmentalization, in evolutionarily divergent plant species'. *Nat. Prod. Rep.*, doi:10.1039/a902191d

Cowie, C. (August 2018). *Diabetes in America. 3rd Edition.* National Institute for Health: *https://www.niddk.nih.gov/about-niddk/strategic-plans-reports/diabetes-in-america-3rd-edition*

Crider, J. (2020, June 20). 'How Tesla Became The World's Most Valuable Automaker In Just 16 Years'. Clean Technica: *https://cleantechnica.com/2020/06/20/how-tesla-became-the-worlds-most-valuable-automaker-in-just-16-years/amp/*

Crownhart, C. (2021, June 29). 'Can the Most Exciting New Solar Material Live Up to its Hype?', *MIT Technology Review, https://www.technologyreview.com/2021/06/29/1027451/perovskite-solar-panels-hype-commercial-debut/*

Cruz-Fernández, P. d. (2013). *Atlantic Threads: Singer in Spain and Mexico, 1800 1940.* FIU Electronic Theses and Dissertations.

Currie, J. 'Traffic Congestion and Infant Health: Evidence from E-ZPass'. *https://economics.yale.edu/sites/default/files/files/Workshops-Seminars/Labor-Public/currie-101008.pdf*

Curtin, P. (1990). *The Rise and Fall of the Plantation Complex: Essays in Atlantic History.* New York: Cambridge University Press.

Cutlip, K. (2015, May 11). 'How Nylon Stockings Changed the World'. *Smithsonian, https://www.smithsonianmag.com/smithsonian-institution/how-nylon-stockings-changed-world-180955219/*

Dam, H. (September 1894). 'Foods in the Year 2000'. *McClure's Magazine*, 311.

Davies, J. (2006). 'Where Have all the Antibiotics Gone?', *Canadian Journal of Infectious Diseases and Medical Microbiology*, 287-290.

De Liso, N. (2021). 'The "Sailing-Ship Effect" as a Technological Principle', *Industrial and Corporate Change*, doi:10.1093/icc/dtab021

Desjardins, J. (2018, February 23). 'These are the world's most traded goods'. World Economic Forum: *https://www.weforum.org/agenda/2018/02/the-top-importers-and-exporters-of-the-world-s-18-most-traded-goods*

Devine, W. (1982). *An Historical Perspective on the Value of Electricity in American Manufacturing.* Oak Ridge, Tennessee: Institute for Energy Analysis.

Devine, W. (1983, June). 'From Shafts to Wires: Historical Perspective on Electrification'. *The Journal of Economic History*, pp. 347-372.

Devitt, N. (1977). 'The Transition from Home to Hospital Birth in the United States: 1930–1960'. *Birth and the Family Journal, doi.org/10.1111/j.1523-536X.1977.tb01207.*

Donkin, R. (1998). *Beyond Price: Pearls and Pearl-fishing: Origins to the Age of Discoveries.*

Dorr, A. (2023). *Brighter. Optimism, Progress, and the Future of Environmentalism.* RethinkX.

Dorr, A. (2016). 'Common errors in reasoning about the future: Three informal fallacies'. *Technological Forecasting & Social Change, doi.org/10.1016/j.techfore.2016.06.018*

Dove, A. (2005). 'Maurice Hilleman'. *Nature Medicine,* S2.

Draznin, Y. C. (2001). *Victorian London's Middle-Class Housewife: What She Did All Day.*

Drum, K. (2013, Jan/Feb). 'Lead: America's Real Criminal Element'. Mother Jones: *https://www.motherjones.com/environment/2016/02/lead-exposure-gasoline-crime-increase-children-health/*

DuBoff, R. (1967, December). 'The Introduction of Electric Power in American Manufacturing'. *The Economic History Review, https://ideas.repec.org/a/bla/ehsrev/v20y1967i3p509-518.html*

Dunbar, R. (1992). 'Neocortex size as a constraint on group size in primates'. *Journal of Human Evolution,* doi:10.1016/0047-2484(92)90081-J

Durant, W. (1935). *The Story of Civilization, VII: The Age of Reason Begins.* New York: Simon & Schuster.

Dvorsky, G. (2014, October 01). 'Scientists Watch How New Technologies Spread Among Wild Chimps'. Gizmodo: *https://io9.gizmodo.com/scientists-watch-how-new-technologies-spread-among-wild-1641172343*

Ebersole, R. (2020, November 17). 'At Sea and in Court, the Fight to Save Right Whales Intensifies'. YaleEnvironment360: *https://e360.yale.edu/features/at-sea-and-in-court-the-fight-to-save-right-whales-intensifies*

Economist, (2012, January 14). 'The last Kodak moment? Kodak is at death's door; Fujifilm, its old rival, is thriving. Why?', *https://www.economist.com/node/21542796*

Edwards, E. (2022, September 15). 'Two powerful drugs are making their way into the illicit drug supply'. NBC News: *https://www.nbcnews.com/health/health-news/two-powerful-drugs-are-making-way-illicit-drug-supply-rcna47674*

Ennos, R. (2020). *The Age of Wood: Our Most Useful Material and the Construction of Civilization.*

Evans, B. (2020, June 15). 'News by the ton: 75 years of US advertising'. *https://www.ben-evans.com/benedictevans/2020/6/14/75-years-of-us-advertising*

Evans, D. (2021, August 31). 'Free Parking Is Killing Cities'. Bloomberg: *https://www.bloomberg.com/news/features/2021-08-31/why-free-parking-is-bad-according-to-one-ucla-professor*

Fahey, M. (2015, December 11). 'How Much is Moby Dick Worth Today?' CNBC: *https://www.cnbc.com/2015/12/11/how-much-is-moby-dick-worth-today.html*

"Fall of Civilizations". (2020, August 16). 'The Sumerians - Fall of the First Cities'. YouTube: *https://www.youtube.com/watch?v=d2lJUOv0hLA*

Farcau, B. (2000). *The Ten Cents War: Chile, Peru, and Bolivia in the War of the Pacific, 1879-1884.* Westport: Greenwood Publishing Group.

Fazeli, H. (2002). 'Stone Tool Production, Distribution and Use during the Late Neolithic and Chalcolithic on the Tehran Plain, Iran'. *Journal of Persian Studies, https://www.jstor.org/stable/4300616*

Feder, B. (1997, June 22). 'Defining Fast Food: Between 55 Seconds and 30 Minutes', *The New York Times.*

Fernandes, S. (2019, March 11). 'Pollution destroys 21% wheat, 6% rice crop every year: IIT-M study'. Hindustan Times: *https://www.hindustantimes.com/mumbai-news/pollution-destroys-21-wheat-6-rice-crop-every-year-iit-m-study/story-FfsjiUstkx62FL7ALe5uxl.html*

Fickman, L. (2022, November 14). 'Fentanyl Vaccine Potential 'Game Changer' for Opioid Epidemic Study Suggests New Vaccine Could Prevent Deadly Opioid from Entering the Brain'. University of Houston: *https://www.uh.edu/news-events/stories/2022-news-articles/november-2022/11142022-fentanyl-vaccine-haile-kosten.php*

Fisher, J. (1971-1972). 'A Simple Substitution Model of Technological Change'. *Technological Forecasting and Social Change, doi:10.1016/s0040-1625(71)80005-7*

Follows, S. (2016, January 11). 'Film vs digital – What is Hollywood shooting on?', *https://stephenfollows.com/film-vs-digital/*

Fosse, P. (2020, September 28). Tesla Tire Costs 10X Battery Cost, & My Plan To Deflate Tire Costs 90%. CleanTechnica: *https://cleantechnica.com/2020/09/28/tesla-tire-costs-10x-battery-cost-my-plan-to-deflate-tire-costs-90/*

Foster, G. (1959). 'The Potter's Wheel: An Analysis of Idea and Artifact in Invention'. *Southwestern Journal of Anthropology,* 15(2), 99-119.

Fouquet, R. (2008). *Heat Power and Light: Revolutions in Energy Services.* Northampton, MA.

Francis, J. A. (2021). *King Cotton, the Munificent Slavery and (Under) development in the United States, 1789-1865. https://joefrancis.info/pdfs/Francis_US_slavery.pdf*

Fuller, R. (2022). 'Pollution and health: a progress update'. *The Lancet,* doi.org/10.1016/S2542-5196(22)00090-0

Galchen, R. (2021, October 4). 'Can Nuclear Fusion Put the Brakes on Climate Change?', *The New Yorker, https://www.newyorker.com/magazine/2021/10/11/can-nuclear-fusion-put-the-brakes-on-climate-change*

Galloway, J. H. (1977, April). 'The Mediterranean Sugar Industry'. *Geographical Review,* pp. 177-194.

Works cited

Goldin, C. A. (2002). 'The Power of the Pill: Oral Contraceptives and Women's Career and Marriage Decisions'. *Journal of Political Economy*, 110, 4, 730–770.

Gordon, R. J. (2016). *The Rise and Fall of American Growth.*

Gosho, M. (1984). 'The Sperm Whale: Physeter macrocephalus'. *Marine Fisheries Review.*

Graeber, D. (2021). *The Dawn of Everything: A New History of Humanity.*

Green, N. (1997). *Ready-to-Wear and Ready-to-Work: A Century of Industry and Immigrants in Paris and New York.*

Griffiths, J. (1992). *The Third Man: The Life and Times of William Murdoch, 1754–1839, The Inventor of Gas Lighting.*

Griliches, Z. (1957). *Hybrid Corn: An Exploration of the Economics of Technological Change.* New York.

Grubler, A. (1997). 'Time for a Change: On the Patterns of Diffusion of Innovation'. *Technological Trajectories and the Human Environment*, 14–32.

Grubler, A. (2010). 'The Costs of the French Nuclear Scale-up: A Case of Negative Learning by Doing'. *Energy Policy*, doi.org/10.1016/j.enpol.2010.05.003

Handley, S. (2000). *Nylon: The Story of a Fashion Revolution.*

Hanson, S. (1951). *Economic Development in Latin America: An Introduction to the Economic Problems of Latin America.* Washington: Inter-American Affairs Press.

Harari, Y. N. (2021, February 21). 'Lessons from a year of Covid'. Financial Times, *https://www.ft.com/content/f1b30f2c-84aa-4595-84f2-7816796d6841*

Hardy, B. (2020). 'Direct evidence of Neanderthal fibre technology and its cognitive and behavioral implications'. *Scientific Reports*, doi.org/10.1038/s41598-020-61839-w

Harford, T. (2017, August 28). 'Why did we use leaded petrol for so long?'. BBC World Service: *https://www.bbc.com/news/business-40593353*

Harford, T. (2019, September 11). 'Can solar power shake up the energy market?'. BBC News: *https://www.bbc.co.uk/news/business-49344595.amp*

Harford, T. (2020, February 5). 'How McDonald's revolutionised business'. BBC: *https://www.bbc.com/news/business-51208592*

Harley, K. (2015). 'The shift from sailing ships to steamships, 1850–1890: A study in technological change and its diffusion'. In *Essays on a Mature Economy: Britain After 1840* (pp. 215–237). Princeton University Press.

Hasell, J. (2017, December 7). 'Famines'. Our World in Data: *https://ourworldindata.org/famines*

Hayami, Y. (1971). *Agricultural Development: An International Perspective.* Johns Hopkins Press.

Heaven, W. (2022, December 1). 'Biotech labs are using AI inspired by DALL-E to invent new drugs'. *MIT Technology Review, https://www.technologyreview.com/2022/12/01/1064023/biotech-labs-are-using-ai-inspired-by-dall-e-to-invent-new-drugs/*

Hersh, J. (2009). *Sweet Diversity: Colonial Goods and the Rise of European Living Standards after 1492.* doi.org/10.2139/ssrn.1402322

Hobaiter, C. (2014, September 14). 'Social Network Analysis Shows Direct Evidence for Social Transmission of Tool Use in Wild Chimpanzees'. Public Library of Science: *https://journals.plos.org/plosbiology/article?id=10.1371/journal.pbio.1001960*

Holappa, L. (2017). 'Recent Achievements in Iron and Steel Technology'. *Journal of Chemical Technology and Metallurgy*, v2., 159–167.

Holliday, E. (2022). *Triggering Death: Quantifying the True Human Cost of Global Fishing. https://www.the-human-cost-of-fishing.org*

Hope, A. (1979, February 1). 'The Battles of Armstrong – Radio's Forgotten Man'. *New Scientist*, pp. 306–309.

Hopwood, O. (2022, September 21). 'Futuristic Food Deliverers Touch Down on Fairfield's Campus'. Fairfield Mirror, *https://fairfieldmirror.com/news/futuristic-food-deliverers-touch-fairfields-campus/*

Hornborg, A. (2012). *Ecology and Power: Struggles over Land and Material Resources in the Past, Present and Future.* Routledge.

IEA. (2021, November 3). 'Global Lighting Sales, Historical and in the Net-Zero Scenario, 2010–2030'. *https://www.iea.org/data-and-statistics/charts/global-lighting-sales-historical-and-in-the-net-zero-scenario-2010-2030*

IKEA Museum. (2020). *https://ikeamuseum.com/sv/startsida/*

Ivar Kreuger. Wikipedia: *https://en.wikipedia.org/wiki/Ivar_Kreuger*

Jones, H. (2018). 'The Recent Large Reduction in Space Launch Cost'. *48th International Conference on Environmental Systems.* NASA Ames Research Center.

Jönköpings Tändsticksfabrik. Wikipedia: *https://sv.m.wikipedia.org/wiki/Jönköpings_Tändsticksfabrik*

Joosse, T. (2021, December 9). 'Meet the Scientist Studying How Humans Started Eating Meat'. Smithsonian: *https://www.smithsonianmag.com/blogs/national-museum-of-natural-history/2021/12/09/meet-the-scientist-studying-how-humans-started-eating-meat/*

Judt, T. (2011). *Postwar: A History of Europe Since 1945.*

Kary, T. (2020, October 4). 'Race for Artificial Cannabinoids Heats Up'. *Bloomberg News.*

Kaszynski, W. (2012). *The American Highway. The History and Culture of Roads in the United States.*

Keaten, J. (2022, April 4). 'WHO says 99% of world's population breathes poor-quality air'. AP News: *https://apnews.com/article/science-health-asia-united-nations-environment-dafbedfa74d46297e013a5f47ced010d*

Kelly, B. (2018). 'Measuring Technological Innovation Over the Long Run. Working Paper 25266', National Bureau of Economic Research: *http://www.nber.org/papers/w25266*

Kennedy, R. (2022, November 8). 'Most New Solar Panels Retain 80% Production After 30 Years'. *PV Magazine, https://www.pv-magazine.com/2022/11/08/most-new-solar-panels-retain-80-production-after-30-years/*

Kennett, D. J. (2020). 'Early isotopic evidence for maize as a staple grain in the Americas'. *Science Advances.*

Kinealy, C. (1994). *This Great Calamity.* Gill & Macmillan.

Klein Goldewijk, K. (2010). 'Long-term dynamic modeling of global population and built-up area in a spatially explicit way: HYDE 3.1'. *The Holocene*, doi.org/10.1177/0959683609356587

Klein, E. (2019, January 29). 'Ending the age of animal cruelty'. Vox: *https://www.vox.com/future-perfect/2019/1/29/18197907/clean-meat-cell-plant-impossible-beyond-animal-cruelty*

Klemm, F. (1959). *A History of Technology.*

Klimscha, F. (2012-2022). Digital Atlas of Innovations. *https://atlas-innovations.de/en/*

Kline, R. (2000). *Consumers in the Country. Technology and Social Change in Rural America.* Johns Hopkins University Press.

Kniesly, C. A. (2019, November). 'How Corn Conquered the World'. TED.com: *https://www.ted.com/talks/chris_a_kniesly_how_corn_conquered_the_world/*

Kominski, R. (1988). *Computer Use in the United States: 1984.* U.S. Department of Commerce.

Kovarik, W. (2005). 'Ethyl-leaded Gasoline: How a Classic Occupational Disease Became an International Public Health Disaster'. *International Journal of Occupational and Environmental Health*, 11:384–397.

Kristensen, H. (2018). The Bulletin of the Atomic Scientists' Nuclear Notebook. *https://thebulletin.org/nuclear-notebook/*

Landes, D. (1969). *The Unbound Prometheus.* Cambridge: Cambridge University Press.

Lane, B. (1975). 'Electret Microphones'. *DB Magazine*, p 24-26.

Lebreton, L. (2022). 'Industrialised fishing nations largely contribute to floating plastic pollution in the North Pacific subtropical gyre'. *Nature*, doi.org/10.1038/s41598-022-16529-0

Lee, C. (1998). *This Sceptred Isle: 55 BC – 1901.* London: Penguin Books.

Lewtas, I. (2015). 'The ballistic performance of the bombard Mons Meg'. *Defence Technology*, doi:10.1016/j.dt.2015.12.001

Librado, P. (2021). 'The origins and spread of domestic horses from the Western Eurasian steppes'. *Nature*, doi.org/10.1038/s41586-021-04018-9

Linstone, H. (1976). *Technological Substitution: Forecasting Techniques and Applications.* American Elsevier Publishing.

Lobell, D. (2021). 'Cleaner air has contributed one-fifth of US maize and soybean yield gains since 1999'. *Environmental Research Letters*, doi.org/10.1088/1748-9326/ac0fa4

Lopez, J. (1989). *The transition from natural madder to synthetic alizarine in the American textile industry, 1870-1890.* Ph.D. Thesis, Iowa State University.

Loveday, A. (1983). *The Rise and Decline of the American Cut Nail Industry.* Greenwood Press.

Ma, D. (2004). 'Why Japan, Not China, Was the First to Develop in East Asia: Lessons from Sericulture, 1850–1937'. *Economic Development and Cultural Change.*

Ma, D. (2005). 'Between Cottage and Factory: the Evolution of Chinese and Japanese: Silk-Reeling Industries in the Latter Half of 19th Century'. *Journal of the Asia Pacific Economy*, doi.org/10.1080/13547860500071451

Magnier, A. (July 2010). Product Life Cycles and Innovation in the US Seed Corn Industry. *https://www.ifama.org/resources/Documents/v13i3/Magnier-Kalaitzandonakes-Miller.pdf*

Maia, S. (2022, September 21). Power Transition Trends. BloombergNEF.

Mailloux, N. (2022). 'Nationwide and Regional PM2.5-Related Air Quality Health Benefits From the Removal of Energy-Related Emissions in the United States'. *GeoHealth*, doi.org/10.1029/2022GH000603

Make it Primitive. (2020, March 17). 'Potter's wheel made from limestone slabs'. YouTube: *https://youtube.com/watch?v=B8ltcRPNau4*

Marelli, B. (2016). 'Silk Fibroin as Edible Coating for Perishable Food Preservation'. *Nature Scientific Reports*, doi:10.1038/srep25263

Marr, P. (2007). 'Ghosts of the Atacama: The Abandonment of Nitrate Mining in the Tarapaca Region of Chile'. *Middle States Geographer*, 40:22-31.

Works cited

Martin, K. (2018, March 26). 'The Big Inch: Fueling America's WWII War Effort'. National Institute of Standards and Technology (NIST): *https://www.nist.gov/blogs/taking-measure/big-inch-fueling-americas-wwii-war-effort*

Mathew, W. (1970). 'Peru and the British Guano Market'. *The Economic History Review, http://www.jstor.org/stable/2594566*

McCullough, D. (2015). *The Wright Brothers*. Simon & Schuster.

McDermott, M. (2020). 'Cocoa to Improve Walking Performance in Older People With Peripheral Artery Disease'. *Circulation Research,* doi.org/10.1161/CIRCRESAHA.119.315600

McDowell, M. (2020, January 28). 'A top Silicon Valley futurist on how AI, AR and VR will shape fashion's future'. Vogue Business, *https://www.voguebusiness.com/technology/ai-ar-and-vr-shaping-fashions-future-peter-diamandis*

McNeill, J. (2000). *Something New Under the Sun: An Environmental History of the Twentieth-Century World.* New York: W. W. Norton & Company.

Mehring, A. (1957). *Statistics on Fertilizers and Liming Materials in the United States.* University of California.

Metz, L. (2023, February 22). 'Bow-and-arrow, technology of the first modern humans in Europe 54,000 years ago at Mandrin, France'. *Science Advances,* doi:10.1126/sciadv.add4675

Milne, G. (2000). *Trade and Traders in Mid-Victorian Liverpool: Mercantile Business and the Making of a World Port.*

MIT. (2020). 'Wallace Carothers, Polymeric Materials: Nylon and Neoprene'. *https://lemelson.mit.edu/resources/wallace-carothers*

Molla, R. (2017, June 26). 'How Apple's iPhone changed the world: 10 years in 10 charts'. Vox: *https://www.vox.com/2017/6/26/15821652/iphone-apple-10-year-anniversary-launch-mobile-stats-smart-phone-steve-jobs*

Molla, R. (2018, March 26). Advertisers will spend $40 billion more on internet ads than on TV ads this year. Retrieved from Vox: *https://www.vox.com/2018/3/26/17163852/online-internet-advertisers-outspend-tv-ads-advertisers-social-video-mobile-40-billion-2018*

Molyneaux, I. (2021, July 19). 'Why cattle troughs are scattered across London and the places where you'll find them'. MyLondon: *https://www.mylondon.news/news/nostalgia/cattle-troughs-scattered-across-london-19951408*

Moore, J. (1988). *The Emergence of the Cotton Kingdom in the Old Southwest: Mississippi, 1770-1860.* Baton Rouge, LA: Louisiana State University Press.

Morris, P. (1992). 'A history of the international dyestuff industry'. *American Dyestuff Reporter*, 81(11): 59–100, 192–195.

Morrison, R. (2021, February 3). 'Air pollution can damage your brain'. Mail Online: *https://www.dailymail.co.uk/sciencetech/article-9218617/Exposure-smog-childhood-affect-cognitive-skills-60-years-later-study-warns.html*

Mowery, D. (1998). *Paths of Innovation: Technological Change in 20th-Century America.*

Mowery, D. (1999). *Sources of Industrial Leadership*. Cambridge University Press.

Mui, C. (2012, January 18). 'How Kodak Failed'. *Forbes, https://www.forbes.com/sites/chunkamui/2012/01/18/how-kodak-failed/?sh=7393ea36f27a*

Murmann, J. (2001). 'Comparing Evolutionary Dynamics Across Different National Settings: The Case of the Synthetic Dye Industry, 1857–1914'. *Journal of Evolutionary Economics.*

NAACP. (2012). Coal Blooded: Putting Profits Before People. *https://naacp.org/resources/coal-blooded-putting-profits-people*

Naam, R. (2019, February 19). 'How to decarbonize America — and the world'. TechCrunch: *https://techcrunch.com/2019/02/15/how-to-decarbonize-america-and-the-world/*

Nakicenovic, N. (1987). 'The Automobile Road to Technological Change: Diffusion of the Automobile as a Process of Technological Substitution'. Laxenburg, Austria: International Institute for Applied Systems Analysis (IIASA).

National Cancer Institute. (October 2001). 'Risks Associated with Smoking Cigarettes with Low Machine-Measured Yields of Tar and Nicotine'. Tobacco Control Monograph No. 13. NIH Pub. No. 02-5047: Bethesda, MD: U.S. Department of Health and Human Services, National Institutes of Health, National Cancer Institute.

National Research Council. (1998). *Fostering Research on the Economic and Social Impacts of Information Technology.* National Academy Press.

Nelson, R. (1982). *Government and Technical Progress: A Cross-Industry Analysis.* Pergamon Press.

Nerz, R. (2006). *Eat This Book: A Year of Gorging and Glory on the Competitive Eating Circuit.* St. Martin's Griffin.

Neushul, P. (2016). 'Blowing Foam and Blowing Minds: Better Surfing Through Chemistry'. In D. Kaiser, *Groovy Science: Knowledge, Innovation and American Counterculture.*

New York Times. (1975, April 17). 'Transmission Problems in Cars Linked to Ban on Whale Killing'. *https://www.nytimes.com/1975/04/17/archives/transmission-problems-in-cars-linked-to-ban-on-whale-killing.html*

Newitz, A. (2021). *Four Lost Cities: A Secret History of the Urban Age.* W. W. Norton & Company.

Newport, F. (2013, September 25). 'Desire for Children Still Norm in U.S.'. Gallup: *https://news.gallup.com/poll/164618/desire-children-norm.aspx*

NHTSA. (2008). 'How States Achieve High Seat Belt Use Rates'. National Highway Traffic Safety Administration. *https://crashstats.nhtsa.dot.gov/Api/Public/ViewPublication/810962*

Nicholas, T. (2019). *VC: An American History*. Harvard University Press.

Nikel, D. (2019, September 1). 'Trondheim's New 'Energy Positive' Landmark'. Life in Norway: *https://www.lifeinnorway.net/trondheims-new-energy-positive-landmark/*

Nippon.com. (2021, May 18). 'Life of Crime: Yakuza Membership Hits New Record Low'. *https://www.nippon.com/en/japan-data/h01007/*

Oberhaus, D. (2020, October 12). 'AI Is Throwing Battery Development Into Overdrive'. Wired: *https://www.wired.com/story/ai-is-throwing-battery-development-into-overdrive/*

OECD. (2021). *https://stats.oecd.org/Index.aspx?DataSetCode=PDB_LV*

Offit, P. (2007) *Vaccinated: One Man's Quest to Defeat the World's Deadliest Diseases*. Smithsonian Books.

Olmstead, A., & Rhode, P. (2000). 'The Diffusion of the Tractor in American Agriculture: 1910-60'. Cambridge, MA: National Bureau of Economic Research.

Olszynko-Gryn, J. (2020). 'Predictor: The First Home Pregnancy Test'. *Journal of British Studies*, doi:10.1017/jbr.2020.70

Ortiz, M. P. (2022, November 23). 'To Ditch Pesticides, Scientists Are Hacking Insects' Sex Signals'. Wired: *https://www.wired.com/story/insect-pheromones-pest-control/*

Osgood, B. (2021, October 22). 'Dying for debt relief: Why are NYC taxi drivers on hunger strike?'. Al Jazeera: *https://www.aljazeera.com/economy/2021/10/22/dying-for-debt-relief-why-are-nyc-taxi-drivers-on-hunger-strike*

Osipov, S. (2022). 'Severe atmospheric pollution in the Middle East is attributable to anthropogenic sources'. *Communications Earth & Environment*. doi.org/10.1038/s43247-022-00514-6

Oudijk, G. (2010). 'The Rise and Fall of Organometallic Additives in Automotive Gasoline'. *Environmental Forensics*. doi:10.1080/15275920903346794

Ozden, S. (2022). 'Egg protein derived ultralightweight hybrid monolithic aerogel for water purification'. *Materials Today*. doi:10.1016/j.mattod.2022.08.001

Pakenham, T. (1992). *The Scramble for Africa: White Man's Conquest of the Dark Continent from 1876 to 1912*. Avon Books.

Pargeter, J. (2019). Going Big Versus Going Small: Lithic Miniaturization in Hominin Lithic Technology. *Evolutionary Anthropology*, pp. 1-14.

Parker, S. (1988). *Acoustics Source Book*. New York: McGraw-Hill.

Peck, G. (1854). *Melbourne, and the Chincha Islands; with Sketches of Lima, and a Voyage Round the World*. New York: Scribner.

Pendrill, F. (2022, September 9). 'Disentangling the numbers behind agriculture-driven tropical deforestation'. *Science*, doi:10.1126/science.abm9267

Penzel, F. (1978). *Theatre Lighting Before Electricity*.

Perez, C. (2002). *Technological Revolutions and Financial Capital: The Dynamics of Bubbles and Golden Ages*.

Perkin, F. M. (1900). 'The Present Condition of the Indigo Industry'. *Nature*, v. 63, 7-9.

Perry, J. (1979). *Technological Innovation in American Local Governments: The Case of Computing*. Pergamon Press.

Perry, R. (1997). *Perry's Chemical Engineers' Handbook, Volume 2*.

Peters, A. (2020, July 15). 'This Melinda Gates-backed biotech startup is growing bacteria that make sustainable dye for denim'. *Fast Company*.

Peters, A. (2021, November 16). 'The Dye that Makes this Tracksuit Pink was Grown in a Lab'. *Fast Company, https://www.fastcompany.com/90697169/the-dye-that-makes-this-tracksuit-pink-was-grown-in-a-lab*

Pew Research Center. (2010). 'Tolerance and Tension: Islam and Christianity in Sub-Saharan Africa'. *https://www.pewresearch.org/religion/2010/04/15/executive-summary-islam-and-christianity-in-sub-saharan-africa/*

Pew. (2021, April 7). Pew Research Center. Mobile Fact Sheet: *https://www.pewresearch.org/internet/fact-sheet/mobile/*

Pew. (2021, November 9). 'How the WTO Can Save Fish for the Future'. The Pew Charitable Trusts: *https://www.pewtrusts.org/en/research-and-analysis/fact-sheets/2021/11/how-the-wto-can-save-fish-for-the-future*

Pollard, A. (2021, November 22). 'Whole Foods' Animal Welfare Certifier to Cut 90% of U.S. Chicken From Approved Supplies'. *Bloomberg. https://www.bloomberg.com/news/articles/2021-11-22/whole-foods-chicken-certifier-upends-market-with-more-humane-breed-rules*

Poole, S. (2016). *Rethink: The Surprising History of New Ideas*. New York: Scribner.

Porter, P. (1969). 'Origins of the American Tobacco Company'. *Business History Review*, p. 59-76.

PortNews. (2007, May 8). 'Korean shipbuilders hold off China threat on high-value orders'. *https://portnews.ru/digest/print/7445/?backurl=/digest/*

Postrel, V. (2020). *The Fabric of Civilization: How Textiles Made the World*. Basic Books.

Quaglia, S. (2021, December 9). 'Tropical forests can partially regenerate in just 20 years without human interference'. The Guardian. *https://www.theguardian.com/environment/2021/dec/09/tropical-forests-can-regenerate-in-just-20-years-without-human-interference*

Works cited

Rajan, R. (2000). 'The Eclipse of the U.S. Tire Industry'. In S. N. Kaplan, *Mergers and Productivity* (pp. 51 - 92). Chicago: University of Chicago Press.

Rapier, G. (2019, March 15). 'Elon Musk says building a factory is '100 times' as hard as building a car'. Business Insider: *https://www.businessinsider.com/elon-musk-says-building-factory-100-times-harder-than-making-car-2019-3*

Ratcliffe, S. (2016). *Oxford Essential Quotations*. Oxford University Press.

Raworth, K. (2017). *Doughnut Economics: Seven Ways to Think Like a 21st-Century Economist.*

Reid-Green, K. (1989, February). 'The History of Census Tabulation'. *Scientific American*, pp. 98-103.

Revelin, D. (2022, August 8). 'India at the global forefront in digital payments'. Banque de France: *https://publications.banque-france.fr/en/india-global-forefront-digital-payments*

Reynolds, S. (2018, September 17). 'How Auto-Tune Revolutionized the Sound of Popular Music'. Pitchfork: *https://pitchfork.com/features/article/how-auto-tune-revolutionized-the-sound-of-popular-music/*

RIAA. *https://www.riaa.com/u-s-sales-database/*

Riordan, M. (1998). *Crystal Fire: The Invention of the Transistor and the Birth of the Information Age.* W. W. Norton & Company.

Ritchie, H. (2014). 'Energy'. Our World in Data: *https://ourworldindata.org/energy*

Ritchie, H. (2017). 'Meat and Dairy Production'. Our World in Data: *https://ourworldindata.org/meat-production*

Ritchie, H. (2017). 'Technology Adoption'. Our World in Data: *https://ourworldindata.org/technology-adoption*

Ritchie, H. (2020, February). 'Food: Greenhouse Gas Emissions Across the Supply'. Our World in Data: *https://ourworldindata.org/uploads/2020/02/Environmental-impact-of-food-by-life-cycle-stage-1536x1380.png*

Ritchie, H. (2021, April 20). 'Wild mammals have declined by 85% since the rise of humans, but there is a possible future where they flourish'. Our World in Data: *https://ourworldindata.org/wild-mammal-decline*

Ritchie, H. (2021). 'Biodiversity'. Our World in Data: *https://ourworldindata.org/biodiversity*

Ritchie, H. (2022, January 11). 'How the world eliminated lead from gasoline'. Our World in Data: *https://ourworldindata.org/leaded-gasoline-phase-out*

Roberts, M. (2013, October 15). 'Estimate doubled for vCJD carriers in UK:'. BBC News: *https://www.bbc.com/news/health-24525584*

Rodrigues, P. (2020). 'The Importance of Guano Birds to the Inca Empire and the First Conservation Measures Implemented by Humans'. *International Journal of Avian Science*, doi.org/10.1111/ibi.12867

Rogers, C. (1993). 'The Military Revolutions of the Hundred Years' War'. *The Journal of Military History*, 57 (2) 241-78.

Rogers, E. M. (1962). *Diffusion of Innovations*. New York: Free Press.

Rosenfeld, M. (2019). 'Disintermediating your friends: How online dating in the United States displaces other ways of meeting'. *PNAS*. doi.org/10.1073/pnas.1908630116

Roser, M. (2018). 'Light at Night'. Our World in Data: *https://ourworldindata.org/light-at-night*

Rothman, L. (2015, October 29). 'Why the Invention of the Ballpoint Pen Was Such a Big Deal'. *Time* magazine: *https://time.com/4083274/ballpoint-pen/*

Roy, R. (2018). 'When dinosaurs fly: The role of firm capabilities in the 'avianization' of incumbents during disruptive technological change'. *Strategic Entrepreneurship Journal*, doi.org/10.1002/sej.1278

Rozado, D. (2022). 'Longitudinal analysis of sentiment and emotion in news media headlines using automated labelling with Transformer language models'. *PLOS ONE*, doi.org/10.1371/journal.pone.0276367

Rua, G. (2014). 'Diffusion of Containerization'. doi.org/10.2139/ssrn.2514849

Saad, L. (2022, August 16). 'Americans Not Convinced Marijuana Benefits Society'. Gallup: *https://news.gallup.com/poll/396893/americans-not-convinced-marijuana-benefits-society.aspx*

Sano, C. (2009). 'History of Glutamate Production'. *The American Journal of Clinical Nutrition*, doi:10.3945/ajcn.2009.27462F.

Schmidt-Nielsen, K. (December 1962). 'Comparative Physiology of Desert Mammals'. University of Missouri Agricultural Experiment Station, Special Report 21. *https://core.ac.uk/download/pdf/75906319.pdf*

Schnaars, S. P. (2002). *Managing Imitation Strategies: How Later Entrants Seize Markets from Pioneers*. New York: Free Press.

Schneider, T. (2022). 'Taxi and Ridehailing Usage in New York City'. *https://toddwschneider.com/dashboards/nyc-taxi-ridehailing-uber-lyft-data/*

Schwab, R. (1983). 'Cyclotron Analysis of the Ink in the 42-Line Bible'. *The Papers of the Bibliographical Society of America*, doi:10.1086/pbsa.77.3.24302918

Schwab, R. (1985). 'New Evidence on the Printing of the Gutenberg Bible: The Inks in the Doheny Copy'. *The Papers of the Bibliographical Society of America, https://www.jstor.org/stable/24303664*

Schwartz, S. (1998). *Atomic Audit: The Costs and Consequences of U.S. Nuclear Weapons.* Washington, D.C: Brookings Institution Press.

Schwartz, S. (2008, October 1). 'The Costs of U.S. Nuclear Weapons'. Nuclear Threat Initiative: *https://www.nti.org/analysis/articles/costs-us-nuclear-weapons/*

Seba, T. (2017). 'Rethinking Transportation 2020 - 2030: The Disruption of Transportation and the Collapse of the Internal-Combustion Vehicle and Oil Industries'. RethinkX: *https://www.rethinkx.com/transportation-report*

Segall, G. (2001). *John D. Rockefeller. Anointed With Oil.* Oxford University Press.

Seiboth, B. (2011). 'Chapter 13: Trichoderma reesei: A Fungal Enzyme Producer for Cellulosic Biofuels'. In *Biofuel Production-Recent Developments and Prospects* (ISBN 978-953-307-478-8).

Selden, M. (2009). 'A Forgotten Holocaust: U.S. Bombing Strategy, the Destruction of Japanese Cities, and the American Way of War from the Pacific War to Iraq'. In Y. Tanaka, *Bombing Civilians: A Twentieth-Century History* (pp. 77–96). New York: New Press.

Sephton, J. (1880). *Eirik the Red's Saga: A Translation.* Liverpool: D. Marples & Co.

Sesso, H. (2022). 'Effect of cocoa flavanol supplementation for the prevention of cardiovascular disease events: the COcoa Supplement and Multivitamin Outcomes Study (COSMOS) randomized clinical trial'. *The American Journal of Clinical Nutrition*, doi.org/10.1093/ajcn/nqac055

Shadbolt, P. (2015, April 2). 'How Stone Age Blades are Still Cutting it in Modern Surgery'. CNN: *https://edition.cnn.com/2015/04/02/health/surgery-scalpels-obsidian/index.html*

Sichel, D. (2022). 'The Price of Nails Since 1695: A Window into Economic Change'. *Journal of Economic Perspectives*, 125-150.

Silver, D. (2016, January 27). 'Mastering the game of Go with deep neural networks and tree search'. *Nature*, doi:10.1038/nature16961

Skippon, W. (2013). The animal health and welfare consequences of foie gras production'. *Canadian Veterinary Journal*, 403–404.

Smil, V. (2006). *Transforming the 20th Century. Technical Innovations and Their Consequences.* Oxford University Press.

Smith, D. (1964). 'Wood Pulp and Newspapers, 1867-1900'. *Business History Review*, 328-345.

Smith, H. (1936). 'The Origin of the Horsepower Unit'. *American Journal of Physics*, doi: 10.1119/1.1999088

Smythe, T. (2002). 'The Diffusion of the Urban Daily, 1850–1900'. *Journalism History*, 73-84.

Sockol, M. (2007). 'Chimpanzee locomotor energetics and the origin of human bipedalism'. *Proc Natl Acad Sci USA*, 104:12265-12269.

Southey, F. (2022, August 9). 'NoPalm Ingredients: Meet the 'circular by nature' start-up disrupting tropical oils with microbes'. FoodNavigator.com: *https://www.foodnavigator.com/Article/2022/08/09/nopalm-ingredients-meet-the-circular-by-nature-start-up-disrupting-tropical-oils-with-microbes*

Standage, T. (2021). *A Brief History of Motion: From the Wheel, to the Car, to What Comes Next.* Bloomsbury.

Stemmer, J. (1989). 'Freight Rates in the Trade between Europe and South America, 1840–1914'. *Journal of Latin American Studies*, doi:10.1017/S0022216X00014413

Strochlic, N. (2020, March 6). ''Wash your hands' was once controversial medical advice'. National Geographic: *https://www.nationalgeographic.com/history/article/handwashing-once-controversial-medical-advice*

Symons, J. (2003). *The Mining and Smelting of Copper in England and Wales, 1760-1820.* Coventry University.

Szreter, S. (2020). 'The Pox in Boswell's London: An Estimate of the Extent of Syphilis Infection in the Metropolis in the 1770s'. *Economic History Review*, 1-28.

Tabuchi, H. (2022, January 23). 'Old-Fashioned, Inefficient Light Bulbs Live On at the Nation's Dollar Stores'. *The New York Times.*

Tedlow, R. S. (2010). 'Why Business Leaders Fail to Look Facts in the Face - and What to Do About It'. *Portfolio.*

Tenold, S. (June 2021). 'International Transfer of Tacit Knowledge: The Transmission of Shipbuilding Skills from Scotland to South Korea in the Early 1970s'. *Enterprise & Society*, pp. 335 – 367, doi.org/10.1017/eso.2019.68

Thompson, C. (2016, January). 'How the Phonograph Changed Music Forever'. *Smithsonian.*

Thompson, C. (July/August 2022). 'When Coal First Arrived, Americans Said 'No Thanks''. *Smithsonian, https://www.smithsonianmag.com/innovation/americans-hated-coal-180980342/*

Thompson, D. (2012, February 22). 'The Spectacular Rise and Fall of U.S. Whaling: An Innovation Story'. *The Atlantic, https://www.theatlantic.com/business/archive/2012/02/the-spectacular-rise-and-fall-of-us-whaling-an-innovation-story/253355/*

Thun, H. C. (2002). 'Tobacco use and cancer: an epidemiologic perspective for geneticists'. *Oncogene*, 21, 7307–7325.

Tiffany, K. (2021). 'The Rise and Fall of an American Tech Giant'. *The Atlantic.*

Timmerman, A. (2020). 'Quantifying the potential causes of Neanderthal extinction: Abrupt climate change versus competition and interbreeding'. *Quaternary Science Reviews*, v. 238.

Torrella, K. (2022, December 14). 'The First Lab-grown Seafood will be Fancy'. Vox: *https://www.vox.com/future-perfect/23507372/lab-grown-seafood-fish-bluenalu-wildtype-cultivated-cultured-meat/*

Toske, S. (2019). 'Recent methamphetamine profiling trends: Tracking the nitrostyrene method used for P2P production'. *Forensic Chemistry,* doi.org/10.1016/j.forc.2018.12.003

Twist, J. (2005, July 14). 'Eco-designs on future cities'. BBC News: *http://news.bbc.co.uk/2/hi/science/nature/4682011.stm*

UCS. (2022, May 1). Union of Concerned Scientists Satellite Database. *https://www.ucsusa.org/resources/satellite-database*

Ulaby, N. (2011, February 23). 'Cassette Tapes Get A West Coast Rewind'. NPR: *https://www.npr.org/sections/therecord/2011/02/24/133995073/cassette-tapes-get-a-west-coast-rewind*

U.S. Army. (June 2004). 'Special Forces Use of Pack Animals'. FM 3-05.213 (FM 31-27). *https://fas.org/irp/doddir/army/fm3-05-213.pdf*

U.S. Bureau of Economic Analysis. (2020, November 1). 'Sales of imported new autos [B149RC1Q027SBEA]' and 'Domestic output of new autos [B148RC1Q027SBEA]'. Retrieved from FRED, Federal Reserve Bank of St. Louis: *https://fred.stlouisfed.org/series/B149RC1Q027SBEA and https://fred.stlouisfed.org/series/B148RC1Q027SBEA*

U.S. Department of Commerce. (1933). Fifteenth Census of the United States, Census of Agriculture, The Farm Horse. *https://www2.census.gov/library/publications/decennial/1930/agriculture-farm-horse/1930sr-farm-horse.pdf*

U.S. Department of Health and Human Services. (2014). The Health Consequences of Smoking - 50 Years of Progress: A Report of the Surgeon General. *https://www.ncbi.nlm.nih.gov/books/NBK179276/pdf/Bookshelf_NBK179276.pdf*

US Census. (1975). *Historical Statistics of the United States, Colonial Times to 1970. https://www.census.gov/library/publications/1975/compendia/hist_stats_colonial-1970.html.*

USDA. (1962). National Agricultural Statistics Service. 'Agricultural Statistics': *https://usda.library.cornell.edu/concern/publications/j3860694x*

USGAO. (June 2018). 'F-35 Joint Strike Fighter'. United States Government Accountability Office.

Utterback, J. (1994). *Mastering the Dynamics of Innovation.*

Vanderslott, S. (2013). 'Vaccination'. Our World in Data: *https://ourworldindata.org/vaccination*

Vardanyan, R. (2014). 'Fentanyl-related compounds and derivatives: current status and future prospects for pharmaceutical applications'. *Future Med Chem,* doi:10.4155/fmc.13.215

Vélez, D. (2022, May 19). 'Brazilians are adopting digital payments faster than anyone else — what lessons can we learn?'. World Economic Forum: *https://www.weforum.org/agenda/2022/05/brazilians-are-adopting-digital-payments-faster-than-anyone-else-what-lessons-can-we-learn/*

Vergano, D. (2017, June 5). 'This Drug Is Cheaper And More Powerful Than Heroin — And May Be Killing Way More People'. BuzzFeed News: *https://www.buzzfeednews.com/article/danvergano/fentanyl-by-the-numbers*

Vincent, J. (2019, November 27). 'Former Go champion beaten by DeepMind retires after declaring AI invincible'. The Verge: *https://www.theverge.com/2019/11/27/20985260/ai-go-alphago-lee-se-dol-retired-deepmind-defeat*

Vohra, K. (2021). 'Global mortality from outdoor fine particle pollution generated by fossil fuel combustion: Results from GEOS-Chem'. *Environmental Research,* doi.org/10.1016/j.envres.2021.110754

Wakefield, J. (2022, January 24). 'How will Tonga's broken internet cable be mended?'. BBC News: *https://www.bbc.com/news/technology-60069066*

Watkins, J. (1908). *King Cotton: A Historical and Statistical Review 1790 to 1908.* New York.

Watson, E. (2021, June 24). 'Upside Foods brings in PepsiCo SVP as it gears up for commercial launch of cell-cultured chicken'. *https://www.foodnavigator-usa.com/Article/2021/06/24/UPSIDE-Foods-brings-in-PepsiCo-SVP-as-it-gears-up-for-commercial-launch-of-cell-cultured-chicken*

Watson, E. (2022, December 1). 'SuperMeat CEO on cultivated meat metrics: 'We have a trajectory to cost parity, and I'm not just talking about beef, I'm talking about chicken...''. Food Navigator: *https://www.foodnavigator-usa.com/Article/2022/12/01/watch-supermeat-ceo-on-cultivated-meat-metrics-we-have-a-trajectory-to-cost-parity-and-i-m-not-just-talking-about-beef-i-m-talking-about-chick*

Wendt, P. (1947, January). 'The Control of Rubber in World War II'. *Southern Economic Journal,* pp. pp. 203-227.

Wertz, R. (1977). *Lying-in: A History of Childbirth in America.* New Haven and London: Yale University Press

Wetsman, N. (2022, July 28). 'DeepMind found the structure of nearly every protein known to science'. The Verge: *https://www.theverge.com/2022/7/28/23280743/deepmind-alphafold-protein-database-alphabet*

Whitbeck, R. (1931). 'Chilean Nitrate and the Nitrogen Revolution'. Economic Geography, *http://www.jstor.org/stable/140893*

Whitehead, H. (2021). 'Adaptation of sperm whales to open-boat whalers: rapid social learning on a large scale?'. *Biology Letters,* doi.org/10.1098/rsbl.2021.0030

Wilburn, D. (2008). 'Material use in the United States—Selected case studies for cadmium, cobalt, lithium, and nickel in rechargeable batteries'. U.S. Geological Survey Scientific Investigations Report 2008-5141: *https://pubs.usgs.gov/sir/2008/5141/*

Williams, A. (2003). *The Knight and the Blast Furnace: A History of the Metallurgy of Armour in the Middle Ages &*

the Early Modern Period.

Williams, S. (2021). 'New fossils of Australopithecus sediba reveal a nearly complete lower back'. *Evolutionary Biology*, doi:10.7554/eLife.70447

Wilson, E. (2021). *The Social Conquest of Earth.* Liveright.

Winchester, S. (2018). Exactly: *How Precision Engineers Created the Modern World.* London: William Collins.

Wingfield, C. (1873). 'The China coolie traffic from Macao to Peru and Cuba: speech of Sir Charles Wingfield, M.P., in the House of Commons, on May 23rd, 1873'.

Wolf, M. (2021, June 8). 'Geltor Debuts Animal-Free Collagen For Food and Beverage Markets'. The Spoon: *https://thespoon.tech/geltor-debuts-animal-free-collagen-for-food-and-beverage-markets/*

World Bank. (2020, October). Commodity Markets Data: *https://www.worldbank.org/en/research/commodity-markets*

Yong, E. (2017, May 4). 'How a Frog Became the First Mainstream Pregnancy Test'. *The Atlantic.*

Zewe, A. (2022, December 9). 'Paper-thin solar cell can turn any surface into a power source'. Retrieved from MIT News: *https://news.mit.edu/2022/ultrathin-solar-cells-1209*

Zuidhof, M. J. (2014). 'Growth, Efficiency, and Yield of Commercial Broilers from 1957, 1978, and 2005'. *Poultry Science*, 2970-2982.

Made in the USA
Monee, IL
28 May 2026

51172281R00216